Applications of Intelligent Models in the Petroleum Industry

Applications of Intelligent Models in the Petroleum Industry

Guest Editors

Gang Hui
Hai Wang

MDPI

Basel • Beijing • Wuhan • Barcelona • Belgrade • Novi Sad • Cluj • Manchester

Guest Editors
Gang Hui
State Key Laboratory of Petroleum Resources and Engineering
China University of Petroleum (Beijing)
Beijing
China

Hai Wang
Ministry of Education
Northeast Petroleum University
Daqing
China

Editorial Office
MDPI AG
Grosspeteranlage 5
4052 Basel, Switzerland

This is a reprint of the Special Issue, published open access by the journal *Processes* (ISSN 2227-9717), freely accessible at: https://www.mdpi.com/journal/processes/special_issues/3GU8292A0W.

For citation purposes, cite each article independently as indicated on the article page online and as indicated below:

Lastname, A.A.; Lastname, B.B. Article Title. *Journal Name* **Year**, *Volume Number*, Page Range.

ISBN 978-3-7258-8229-8 (Hbk)
ISBN 978-3-7258-8230-4 (PDF)
https://doi.org/10.3390/books978-3-7258-8230-4

Contents

About the Editors

Gang Hui

Gang Hui is an associate researcher and master's supervisor at the College of Petroleum Engineering, China University of Petroleum (Beijing). His research interests lie at the intersection of unconventional resource development and artificial intelligence. He has served as principal investigator or co-principal investigator for more than 20 research projects. He has published 101 peer-reviewed papers, of which 59 appear in journals such as *Energy, SPE Journal, JGR,* and *Petroleum Science* as first or corresponding author. His honors include a First Prize for Technological Invention from the China Petroleum and Chemical Automation Association, a First Prize for Scientific and Technological Progress from the China Petroleum and Chemical Industry Federation, the Pioneer Award from the Canadian Society of Exploration Geophysicists, the Chinese Government Award for Outstanding Self-Financed Students Abroad, and the Outstanding Graduate Award of Alberta. He serves as a Youth Editorial Board Member for *Petroleum Science* and *Advances in Geo-Energy Research,* and has acted as Guest Editor for several Special Issues on AI in oil and gas as well as unconventional resources.

Hai Wang

Hai Wang is an associate professor at Northeast Petroleum University. His research interests focus on the application of data mining and machine learning to unconventional reservoir production forecasting and development optimization, as well as pore-scale and mechanistic studies of CO2 geological sequestration. He has participated in a range of research projects on unconventional oil and gas development, CO2-enhanced tight oil recovery and multi-stage fractured horizontal well production optimization, with support from academic, regulatory, and industry partners in China and Canada. He has published papers as first author in journals such as *Expert Systems with Applications, Journal of Petroleum Science and Engineering,* and *Geoenergy Science and Engineering,* and has contributed to the development of a deep learning-based predictive emissions monitoring system. His honors include the Chinese Government Award for Outstanding Self-financed Students Abroad, the University of Calgary Graduate Excellence Awards, and several other academic honors, and he also serves as a reviewer for multiple international journals in energy and petroleum engineering.

Preface

The oil and gas industry faces mounting pressure to improve efficiency and safety while reducing costs. This Reprint focuses on the intersection of intelligent control, machine learning, and petroleum engineering as a pathway toward automated drilling and smart reservoir management. Its scope covers real-time model predictive control for downhole rotary valves, physics-informed neural networks for bottomhole pressure prediction, cable-controlled injection-production systems, seismic attribute fusion for reservoir prediction, polymer injector flow optimization, and data-driven forecasting for high-water-cut wells. The aim is to present solutions that are both theoretically sound and computationally tractable for field deployment. Our motivation arises from the observation that traditional methods often fail under downhole disturbances, parameter drift, and tight real-time constraints. By integrating physical laws with learning-based models, the contributions in this Reprint achieve better accuracy, lower computational cost, and improved robustness. This Editorial introduces a Reprint written for petroleum engineers, control system designers, data scientists, and graduate students who work on unconventional resources, managed pressure drilling, and enhanced oil recovery. We trust that this collection will serve as a practical guide for advancing intelligent oilfield technologies.

Gang Hui and Hai Wang
Guest Editors

processes

MDPI

Editorial

Intelligent and Integrated Approaches for Efficient Oil and Gas Development

Gang Hui [1,*] and Hai Wang [2,*]

1 State Key Laboratory of Petroleum Resources and Engineering, China University of Petroleum (Beijing), Beijing 102249, China
2 Key Laboratory of Enhanced Oil & Gas Recovery, Ministry of Education, Northeast Petroleum University, Daqing 163318, China
* Correspondence: hui.gang@cup.edu.cn (G.H.); hai.wang@nepu.edu.cn (H.W.)

Abstract

This editorial synthesizes the key findings from 17 original research articles featured in the Special Issue on "Intelligent and Integrated Approaches for Efficient Oil and Gas Development." The collection demonstrates a paradigm shift from purely data-driven methods toward physics-informed, interpretable, and operationally deployable intelligent systems across the upstream lifecycle. Advances span intelligent drilling with real-time model predictive control frameworks achieving sub-20 ms execution times and bottomhole pressure fluctuations below 0.30 MPa; AI-assisted reservoir characterization using multiscale convolutional neural networks, seismic waveform-constrained inversion, and geology-informed transformers that improve sandstone thickness prediction (R^2 = 0.895) and stratigraphic correlation (F1 = 0.886); production optimization through hybrid decomposition-ensemble models (R^2 = 0.954) and improved XGBoost (R^2 = 0.989); and enhanced oil recovery via self-assembled foam systems and polymer injector designs. Fundamental geochemical studies on the Qiongzhusi Formation shale and tight sandstone gas in the Ordos Basin provide critical geological constraints. The editorial identifies persistent challenges, including real-time performance versus physical fidelity, interpretability and uncertainty quantification, multi-scale integration, and generalizability across diverse geological settings. Future directions highlight reinforcement learning for autonomous operations, physics-informed digital twins, generative AI for subsurface scenario modelling, and integration with carbon capture, utilization, and storage. This Special Issue advances the convergence of petroleum engineering, artificial intelligence, and Earth sciences toward intelligent, secure, and sustainable hydrocarbon development.

Keywords: intelligent oil and gas development; physics-informed machine learning; real-time drilling control; reservoir characterization; production optimization

1. Introduction

The upstream oil and gas sector is undergoing a revolutionary phase when digitalization, automation, and data-driven decision-making have transitioned from aspirational ideas to operational necessities. Mature fields, intricate unconventional resources, offshore settings, and the necessity to diminish carbon emissions are forcing operators to use intelligent systems capable of sensing, analyzing, and acting with little human involvement [1–4]. This Special Issue has 17 original research articles that jointly advance the fields of intelligent drilling, real-time well control, AI-assisted reservoir characterization, improved oil recovery (EOR), and production optimization. A prominent theme in these works is the intentional integration of physics-based domain expertise with contemporary machine learning—advancing beyond opaque models towards interpretable, resilient, and computationally efficient solutions for subsurface issues [5–7]. The articles report not only incremental advances but also significant advancements in accuracy, speed, and operational reliability over the full hydrocarbon development lifecycle.

2. Advances in Intelligent Drilling and Real-Time Control

Several contributions tackle the long-standing tension between control accuracy and computational feasibility in drilling automation. A real-time lexicographic model predictive control (MPC) framework with candidate pre-screening, move blocking, and online correction is developed for mud-pulse telemetry rotary valves [8]. By prioritizing pressure-pulse fidelity under strict actuator limits and parameter drift, the controller cuts the step-tracking rise time from 2.18 s to 1.76 s and reduces steady-state pressure error by an order of magnitude relative to conventional MPC, all while maintaining worst-case execution times within the 20 ms sampling period. This simulation-validated design represents a practical pathway toward field-deployable, low-complexity rotary-valve control. For managed pressure drilling (MPD), a physics-informed neural network MPC framework embeds conservation laws and wellbore dynamics directly into deep surrogate models [9] (Figure 1). During training, prior physical knowledge constrains the network to remain consistent with mass conservation and multiphase flow physics; during control, the surrogate replaces expensive numerical iterations, achieving single-step solving times of only 16.81 ms—an 11.1-fold acceleration over mechanistic solvers—while keeping bottomhole pressure fluctuations within 0.30 MPa even under simulated gas kicks and pump shutdowns.

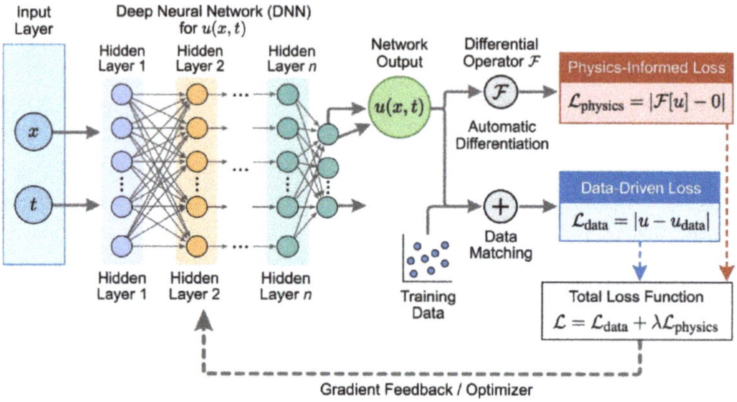

Figure 1. Schematic of the PINN architecture, where automatic differentiation differentiates network outputs to calculate PDE residuals [9].

A geological-sequence-matched, physics-driven deep learning framework for pre-drill mud weight window prediction couples rock mechanics with a long short-term memory network, shrinking collapse, pore, and fracture pressure errors by approximately 30–40% and capturing meter-scale pressure variations that conventional methods overlook [10]. Complementing these downhole intelligence capabilities, a multi-branch parallel neural network processes surface-measured weight-on-bit, rotational speed, pump pressure, and torque to track well depth with a mean absolute error of 1.18 and a per-point comparison error averaging only 0.012 m, providing a reliable basis for downhole state awareness when direct depth measurements are unavailable [11]. These studies build on recent ad-

https://doi.org/10.3390/pr1411172

vances in machine-learning-based drilling dynamics prediction [12,13], real-time lithology identification [14,15], and vibration or kick monitoring [16,17], establishing that the path to real-time field-deployable intelligence lies in encoding physical principles rather than discarding them and that rigorous validation under realistic disturbance profiles is essential for transitioning from simulation to hardware-in-the-loop testing.

3. Intelligent Reservoir Characterization and Interpretation

High-fidelity subsurface imaging and stratigraphic interpretation receive a strong boost from deep learning architectures that respect geological fabric. A multiscale convolutional neural network integrated with a self-attention module fuses multiple seismic attribute volumes—organized as multi-channel 2D slices with parallel 3×3, 5×5, and 7×7 kernels—to predict sandstone thickness in the South Sumatra Basin [18]. The dual-output design yields both a high-resolution thickness map ($R^2 = 0.895$) and quantitative attribute importance scores, revealing that amplitude-related attributes dominate the prediction while frequency- and energy-related attributes contribute less, a result consistent with bandwidth-limited resolution. Seismic waveform-constrained high-resolution inversion embeds lithological, logging, and waveform data into a deep learning framework with geological and physical constraints, achieving 0.5 m vertical resolution and cutting parameter inversion errors by over 25%, and has already supported new well placement that increased initial oil production by 22% in a pilot area [19].

A geology-informed transformer–CNN hybrid (CMT-enhanced Hiformer) for well-log stratigraphic correlation integrates a geological constraint with regularization parameters into the loss function, explicitly promoting the accuracy of formation boundaries, and achieves an F1 score of 0.886 on blind test data from the Shuanghe oil field [20] (Figure 2). Process-based sedimentary numerical simulations using Delft3D with 21 controlled-variable experiments on braided-river reservoirs in the Sulige Gas Field quantify the sensitivity of channel-bar morphology to discharge, slope gradient, and sediment grain size, demonstrating that slope gradient exerts the strongest control on erosion–deposition balance, while discharge promotes bar extension and grain size stabilizes bar shape [21].

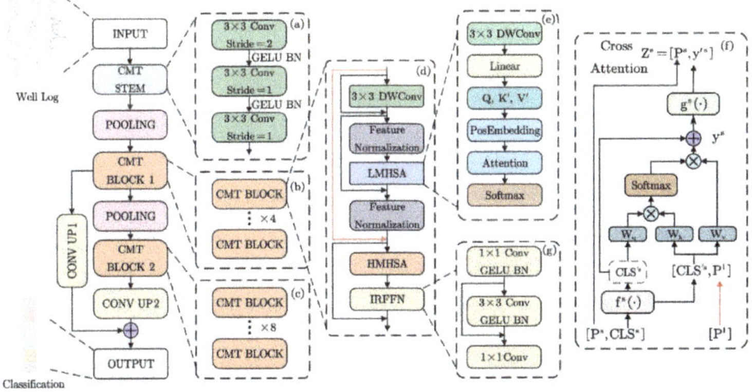

Figure 2. Architecture of the proposed CMT-enhanced Hiformer: (**a**) CMTSTEM, (**b–d**) CMTBLOCK, (**e**) LMHSA, (**f**) cross attention, and (**g**) IRFFN [20].

These contributions are in line with the broader shift toward data-driven optimal point identification and lithofacies classification using both supervised and unsupervised machine learning [22–25], and they strengthen the linkage between depositional physics,

https://doi.org/10.3390/pr14111727

seismic response, and deep-learning-based reservoir models, offering a pathway toward geologically consistent and interpretable reservoir characterization at scales relevant to development decisions.

4. Production Optimization and Enhanced Oil Recovery

Operational intelligence is equally prominent in this Special Issue, with multiple contributions demonstrating that the same physics–data fusion paradigm can transform day-to-day field management. An intelligent cable-controlled injection–production integrated system deployed in the Bohai Sea leverages a cloud-native microservice architecture and nine functional modules to achieve closed-loop management from data acquisition to intelligent decision-making [26] (Figure 3). A weighted ensemble of Random Forest and SVM delivers 92% diagnostic accuracy for blockage detection, while a Bayesian fusion framework combining static geological priors with dynamic sensitivity analysis quantifies injector–producer connectivity with 85% accuracy and rigorous uncertainty propagation; the resulting three-stage human–machine collaborative mechanism slashes anomaly response time from over 72 h to under 2 h. Fine-scale water injection is dynamically managed through a seepage-resistance-driven method that establishes quantitative relationships between seepage resistance, liquid absorption ratio, and injection allocation, maintaining inter-zone seepage-resistance step differences within the 3–5 range and effectively suppressing water-cut increase rates in high-water-cut blocks [27].

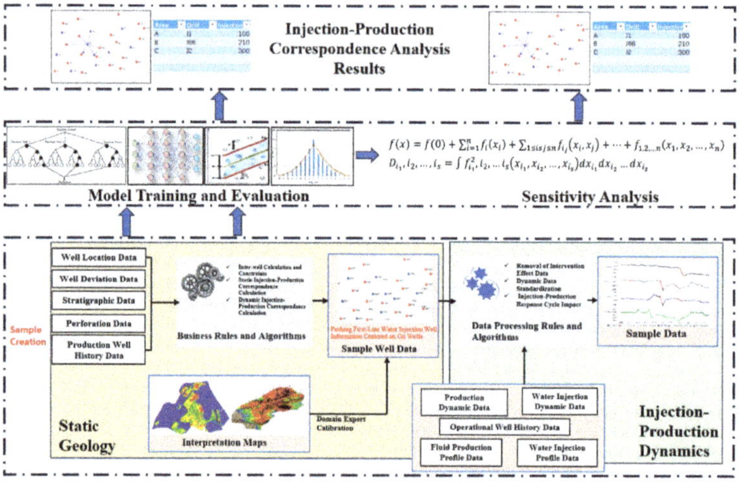

Figure 3. Methodology for injector–producer connectivity analysis [26].

For production forecasting, a CEEMDAN–SR–BiLSTM hybrid framework employs complete ensemble empirical mode decomposition, Hilbert–Huang transform-based subsequence reconstruction, and Bayesian-optimized bidirectional LSTM to achieve an R^2 of 0.954 for wells with 87.6% water cut, while SHAP-based interpretability analysis reveals how water injection volume and flowing pressure contribute to different frequency components, establishing a mechanistic mapping between data features and wellbore-reservoir physics [28]. An improved XGBoost model tuned by the Improved Crowned Porcupine Optimization algorithm, incorporating Chebyshev chaotic mapping and elite opposition-based learning, predicts CO_2-WAG performance with a coefficient of determination of 0.989 and errors consistently within ±2% [29]. A comprehensive data-driven tight-gas framework trained on 3146 Montney horizontal wells uses Random Forest to identify fluid-injection

https://doi.org/10.3390/pr1411172

volumes, burial depth, and completion parameters as key variables, demonstrating that production can be nearly doubled by increasing fracturing fluid injection by 97.5% [30].

In the EOR domain, an in situ self-assembled composite foam system combining soft polymer particles with a low-interfacial-tension foaming agent increases oil recovery by up to 27.2% across heterogeneous offshore reservoirs [31], while an improved Carreau–Yasuda viscosity constitutive model enables the design of a novel spindle-type polymer injector that simultaneously enhances pressure drop by 65% and preserves over 85% viscosity retention [32]. These works align with a growing body of research that leverages ensemble learning, deep reinforcement learning, and hybrid physics-guided models to optimize production and recovery in both conventional and unconventional plays [33–39]. Together, they illustrate that explainable, physics-aware machine learning, when embedded in operational workflows, can deliver measurable improvements in recovery, efficiency, and safety.

5. Fundamental Geoscience and Unconventional Resources

Grounding these engineering advances are new geochemical and paleoenvironmental insights that refine our understanding of resource distribution and quality. A multi-proxy study of the Lower Cambrian Qiongzhusi Formation shale in western Hubei Province identifies five dominant rock types—siliceous, argillaceous–siliceous mixed, argillaceous–calcareous, calcareous–siliceous, and calcareous shales—and traces their vertical and lateral transitions from marine trough to trough margin [40]. By integrating core observations, organic geochemistry, and elemental analysis, the study demonstrates that redox conditions, water-mass stagnation, and paleoproductivity are the primary drivers of organic matter enrichment, while terrigenous input plays a secondary role, and that a consistent shallowing-upward trend is accompanied by diminished upwelling and declining total organic carbon.

Geochemical fingerprinting of tight sandstone gas in the Daning–Jixian Block of the Ordos Basin, based on systematic compositional and carbon–hydrogen isotope analyses across multiple stratigraphic units, reveals higher average methane concentrations and notably heavier $\delta^{13}C_1$ and $\delta^{2}H\text{-}CH_4$ values compared to other basin gas fields, along with a distinctive $\delta^{13}C_1 > \delta^{13}C_2$ reversal [41]. The gas is interpreted as primarily early-stage kerogen-cracking gas with minor contributions from oil-derived gas, and the isotopic reversal is attributed to mixing between highly mature kerogen gas and secondary cracking gas, indicating that ethane carbon isotopes alone are insufficient for genetic classification. Such fundamental geochemical and petrophysical studies are complemented by machine-learning-based predictions of porosity, permeability, and total organic carbon [42,43], and by integrated analyses of fault reactivation and fracture network evolution that influence hydrocarbon storage and flow [44–48]. These studies collectively provide critical constraints for exploration risk assessment and for the stratigraphic and geochemical frameworks within which the intelligent drilling and production technologies reported in this issue must operate.

6. Bridging Gaps and Charting Future Directions

Notwithstanding the remarkable advancements documented above, significant knowledge deficiencies remain. The compromise between physical fidelity and real-time performance necessitates the development of lightweight, physics-integrated neural architectures capable of operating on edge devices while maintaining the extrapolation assurances afforded by first-principles constraints [49]. Interpretability and robust uncertainty quantification are inadequately developed in numerous AI-driven subsurface workflows, constraining the trust of field engineers in implementing model suggestions [50,51]. Multi-scale integration—from pore-scale physics to reservoir-scale simulation and from millisecond control loops to decades-long field life—continues to pose a significant challenge that necessitates innovative coupling solutions [52]. Furthermore, several documented methodologies have been verified solely on simulated testbeds or specific field examples; comprehensive benchmarking across varied geological contexts and operating conditions is essential to ascertain generalizability [53–57].

This Special Issue addresses these deficiencies by introducing models that are physically bound, provide interpretable diagnostics, and are validated against empirical or high-fidelity simulation data. We anticipate significant research trajectories, including: (1) reinforcement learning and adaptive control for fully autonomous drilling and production operations in non-stationary environments; (2) physics-informed digital twins for comprehensive reservoir management with continuous data assimilation; (3) generative AI for subsurface scenario modelling amidst geological uncertainty, facilitating probabilistic decision-making; and (4) seamless integration of carbon capture, utilization, and storage We urge the community to expand upon the foundations established by the writers in this issue and to explore these avenues with equal dedication to rigour, interpretability, and relevance to the discipline.

We express our profound gratitude to all writers for their exemplary contributions, to the numerous reviewers for their critical and prompt evaluations, and to the editorial team for their steadfast support. The editors have assembled a collection with a balanced citation profile to represent the diversity of the discipline. We anticipate that this Special Issue will stimulate ongoing innovation at the convergence of petroleum engineering, artificial intelligence, and Earth sciences, advancing the sector towards a more intelligent, secure, and sustainable future.

7. Conclusions

This Special Issue demonstrates that integrating physics-based domain knowledge with machine learning—rather than discarding physical principles—is the key to developing interpretable, resilient, and operationally deployable intelligent systems for oil and gas development. The 17 contributions collectively achieve measurable improvements in drilling accuracy, reservoir characterization, production forecasting, and enhanced oil recovery across diverse geological settings. Persistent challenges remain in real-time performance, uncertainty quantification, multi-scale integration, and generalizability. Future advances will likely be driven by reinforcement learning for autonomous operations, physics-informed digital twins, generative AI for subsurface modelling, and seamless integration with carbon capture, utilization, and storage. The convergence of petroleum engineering, artificial intelligence, and Earth sciences promises a more intelligent, secure, and sustainable energy future.

Author Contributions: Conceptualization, G.H.; methodology, G.H. and H.W.; writing—original draft preparation, G.H.; writing—review and editing, H.W. All authors have read and agreed to the published version of the manuscript.

Funding: This research is funded by the National Major Science and Technology Programme (2025ZD1401405).

Data Availability Statement: The data presented in this study are available on request from the corresponding author.

Acknowledgments: We thank the anonymous reviewers and the editor for their instructive comments that considerably improved the manuscript's quality.

Conflicts of Interest: The author declares that the research was conducted in the absence of any commercial or financial relationships that could be construed as potential conflicts of interest.

eferences

Hui, G.; Chen, Z.; Wang, Y.; Zhang, D.; Gu, F. An Integrated Machine Learning-Based Approach to Identifying Controlling Factors of Unconventional Shale Productivity. *Energy* **2023**, *266*, 126512. [CrossRef]

LeCun, Y.; Bengio, Y.; Hinton, G. Deep Learning. *Nature* **2015**, *521*, 436–444. [CrossRef]

Hui, G.; Chen, Z.; Wang, H.; Song, Z.; Wang, S.; Zhang, H.; Zhang, D.M.; Gu, F. A Machine Learning-Based Study of Multifactor Susceptibility and Risk Control of Induced Seismicity in Unconventional Reservoirs. *Pet. Sci.* **2023**, *20*, 2232–2243. [CrossRef]

Li, M.; Kolouri, S.; Mohammadi, J. Learning to Solve Optimization Problems With Hard Linear Constraints. *IEEE Access* **2023**, *11*, 59995–60004. [CrossRef]

Hui, G.; Chen, S.; Gu, F. Strike–Slip Fault Reactivation Triggered by Hydraulic-Natural Fracture Propagation during Fracturing Stimulations near Clark Lake, Alberta. *Energy Fuels* **2024**, *38*, 18547–18555. [CrossRef]

Lever, J.; Cheng, S.; Casas, C.Q.; Liu, C.; Fan, H.; Platt, R.; Rakotoharisoa, A.; Johnson, E.; Li, S.; Shang, Z.; et al. Facing & Mitigating Common Challenges When Working with Real-World Data: The Data Learning Paradigm. *J. Comput. Sci.* **2025**, *85*, 102523. [CrossRef]

Hui, G.; Chen, S.; He, Y.; Wang, H.; Gu, F. Machine Learning-Based Production Forecast for Shale Gas in Unconventional Reservoirs via Integration of Geological and Operational Factors. *J. Nat. Gas Sci. Eng.* **2021**, *94*, 104045. [CrossRef]

Dong, X.; Yan, L.; Wang, L.; Zhou, Z.; Jian, Y.; Li, R. Real-Time Lexicographic MPC with Online Correction for Intelligent Drill-Bit Rotary Valves in Mud-Pulse Telemetry. *Processes* **2026**, *14*, 1589. [CrossRef]

Wu, L.; Zhang, Z.; Zhang, C.; Li, G.; Song, X.; Zhou, M.; Yao, X. Physics-Informed Fusion Neural Network for Real-Time Bottomhole Pressure Control in Managed Pressure Drilling. *Processes* **2026**, *14*, 1240. [CrossRef]

Chen, Y.; Sun, T.; Yang, J.; Chen, X.; Ren, L.; Wen, Z.; Jia, S.; Wang, W.; Wang, S.; Zhang, M. Prediction of Mud Weight Window Based on Geological Sequence Matching and a Physics-Driven Machine Learning Model for Pre-Drilling. *Processes* **2025**, *13*, 2255. [CrossRef]

Liu, W.; Ma, B.; Yu, X. Research on Well Depth Tracking Calculation Method Based on Branching Parallel Neural Networks. *Processes* **2025**, *13*, 3147. [CrossRef]

Baki, S.; Dursun, S. Flowing Bottomhole Pressure Prediction with Machine Learning Algorithms for Horizontal Wells. In Proceedings of the SPE Annual Technical Conference and Exhibition, Houston, TX, USA, 3–5 October 2022; p. D021S038R004.

Mazen, A.Z.; Rahmanian, N.; Mujtaba, I.M.; Hassanpour, A. Effective Mechanical Specific Energy: A New Approach for Evaluating PDC Bit Performance and Cutters Wear. *J. Pet. Sci. Eng.* **2021**, *196*, 108030. [CrossRef]

Qingfeng, L.; Chi, P.; Jianhong, F.; Xiaomin, Z.; Yu, S.; Chengxu, Z.; Pengcheng, W.; Chenliang, F.; Yaozhou, P. A Comprehensive Machine Learning Model for Lithology Identification While Drilling. *Geoenergy Sci. Eng.* **2023**, *231*, 212333. [CrossRef]

You, M.; Tan, F.; Zhang, Y.; Sheng, D.; Zuo, C.; Jiao, Y. Development and Application of a Monitoring-While-Drilling System with an Optimized Machine Learning Algorithm for Lithology Identification and Rock Strength Prediction. *Rock Mech. Rock Eng.* **2025**, *58*, 10381–10399. [CrossRef]

Saadeldin, R.; Gamal, H.; Elkatatny, S. Machine Learning Solution for Predicting Vibrations While Drilling the Curve Section. *ACS Omega* **2023**, *8*, 35822–35836. [CrossRef]

Yin, Q.; Yang, J.; Tyagi, M.; Zhou, X.; Wang, N.; Tong, G.; Xie, R.; Liu, H.; Cao, B. Downhole Quantitative Evaluation of Gas Kick during Deepwater Drilling with Deep Learning Using Pilot-Scale Rig Data. *J. Pet. Sci. Eng.* **2022**, *208*, 109136. [CrossRef]

Cheng, Z.; Huang, W.; Zhang, X.; Lei, Z.; Hong, G.; Wang, W.; Zhang, M.; Li, L.; Li, J. Seismic Attribute Fusion and Reservoir Prediction Using Multiscale Convolutional Neural Networks and Self-Attention: A Case Study of the B Gas Field, South Sumatra Basin. *Processes* **2026**, *14*, 981. [CrossRef]

Zhao, H.; Wu, J.; Li, K.; He, Y.; Hu, R.; Wang, T.; Zhao, Z.; Liu, H.; Li, Y.; Yang, X. Seismic Waveform-Constrained Artificial Intelligence High-Resolution Reservoir Inversion Technology. *Processes* **2025**, *13*, 2876. [CrossRef]

Xu, Z.; Zheng, B.; Liu, B.; Song, W. Stratigraphic Correlation of Well Logs Using Geology-Informed Deep Learning Networks. *Processes* **2025**, *13*, 1288. [CrossRef]

Li, J.; Wang, K.; Li, F.; Ma, Z.; Liu, X.; Liu, Y. Morphological Controlling Factors of Braided River Reservoir Based on Delft3D Sedimentary Numerical Simulation: Application to Ordos Basin, China. *Processes* **2025**, *13*, 3661. [CrossRef]

Cao, L.; Jiang, F.; Chen, Z.; Gao, Y.; Huo, L.; Chen, D. Data-Driven Interpretable Machine Learning for Prediction of Porosity and Permeability of Tight Sandstone Reservoir. *Adv. Geo-Energy Res.* **2025**, *16*, 21–35. [CrossRef]

Deng, Y.; Wang, W.; Su, Y.; Sun, S.; Zhuang, X. An Unsupervised Machine Learning Based Double Sweet Spots Classification and Evaluation Method for Tight Reservoirs. *J. Energy Res. Technol.* **2023**, *145*, 072602. [CrossRef]

Tahmasebi, P.; Javadpour, F.; Sahimi, M. Data Mining and Machine Learning for Identifying Sweet Spots in Shale Reservoirs. *Expert Syst. Appl.* **2017**, *88*, 435–447. [CrossRef]

https://doi.org/10.3390/pr14111727

25. Wang, Y.-F.; Xu, S.; Hao, F.; Liu, H.-M.; Hu, Q.-H.; Xi, K.-L.; Yang, D. Machine Learning-Based Grayscale Analyses for Lithofacie Identification of the Shahejie Formation, Bohai Bay Basin, China. *Pet. Sci.* **2025**, *22*, 42–54. [CrossRef]
26. Bai, J.; Chen, Z.; Zhang, W.; Zhou, Z.; Wang, L.; Xu, Y.; Jiang, S.; Zhu, C.; Liu, Z.; Zhang, L.; et al. Research and Application of an Intelligent Cable-Controlled Injection–Production Integration and Control System. *Processes* **2026**, *14*, 1238. [CrossRef]
27. Dong, C.; Hui, W.; Shan, G.; Yang, E.; Qu, M.; Wang, H. Research on Dynamic Control Methods for Fine-Scale Water Injection Zones Based on Seepage Resistance. *Processes* **2025**, *13*, 3966. [CrossRef]
28. Li, Z.; Qian, Q.; Guo, H.; Wu, T.; Cui, H.; Zhu, B. A Hybrid Framework for Production Prediction in High-Water-Cut Oil Wells Decomposition-Feature Enhancement-Integration. *Processes* **2025**, *13*, 1467. [CrossRef]
29. Su, B.; Li, J.; Li, J.; Han, C.; Feng, S. An Improved XGBoost Model for Development Parameter Optimization and Production Forecasting in CO_2 Water-Alternating-Gas Processes: A Case Study of Low Permeability Reservoirs in China. *Processes* **2025**, *13*, 2506. [CrossRef]
30. Yao, F.; Hui, G.; Meng, D.; Ge, C.; Zhang, K.; Ren, Y.; Li, Y.; Zhang, Y.; Yang, X.; Zhang, Y.; et al. Integrated Data-Driven Framework for Forecasting Tight Gas Production Based on Machine Learning Algorithms, Feature Selection and Fracturing Optimization. *Processes* **2025**, *13*, 1162. [CrossRef]
31. Jiang, M.; Tang, S.; Xia, Y. In Situ Self-Assembled Particle-Enhanced Foam System for Profile Control and Enhanced Oil Recovery in Offshore Heterogeneous Reservoirs. *Processes* **2026**, *14*, 411. [CrossRef]
32. Qian, Q.; Li, T.; Ren, C.; Zhou, Y.; Che, C.; Zhang, X.; Ma, J.; An, P.; Zhao, Q. Numerical Simulation and Optimization of Polyacrylamide Solution Flow in a Polymer Injector Using an Improved Viscosity Constitutive Model. *Processes* **2026**, *14*, 883. [CrossRef]
33. Bekhit, A.M.; Sobh, M.; Abdel Zaher, M.; Abdel Fattah, T.; Diab, A.I. Predicting Terrestrial Heat Flow in Egypt Using Random Forest Regression: A Machine Learning Approach. *Geotherm. Energy* **2025**, *13*, 18. [CrossRef]
34. Batanero, E.A.; Pascual, Á.F.; Jiménez, Á.B. RLBoost: Boosting Supervised Models Using Deep Reinforcement Learning. *Neurocomputing* **2025**, *618*, 128815. [CrossRef]
35. Brantson, E.T.; Ju, B.; Omisore, B.O.; Wu, D.; Selase, A.E.; Liu, N. Development of Machine Learning Predictive Models for History Matching Tight Gas Carbonate Reservoir Production Profiles. *J. Geophys. Eng.* **2018**, *15*, 2235–2251. [CrossRef]
36. Breiman, L. Random Forests. *Mach. Learn.* **2001**, *45*, 5–32. [CrossRef]
37. Kalantari-Dahaghi, A.; Mohaghegh, S.; Esmaili, S. Coupling Numerical Simulation and Machine Learning to Model Shale Gas Production at Different Time Resolutions. *J. Nat. Gas Sci. Eng.* **2015**, *25*, 380–392. [CrossRef]
38. Meng, J.; Zhou, Y.; Ye, T.; Xiao, Y.; Lu, Y.; Zheng, A.W.; Liang, B. Hybrid Data-Driven Framework for Shale Gas Production Performance Analysis via Game Theory, Machine Learning, and Optimization Approaches. *Pet. Sci.* **2023**, *20*, 277–294. [CrossRef]
39. Mohammed, A.I.; Bartlett, M.; Oyeneyin, B.; Kayvantash, K.; Njuguna, J. An Application of FEA and Machine Learning for the Prediction and Optimisation of Casing Buckling and Deformation Responses in Shale Gas Wells in an In-Situ Operation. *J. Nat. Gas Sci. Eng.* **2021**, *95*, 104221. [CrossRef]
40. Lan, Z.; He, J.; Chen, F.; Yang, T.; Wang, L. Paleoenvironmental Evolution and Its Dominant Controls on Organic Matter Enrichment: Insights from the Lower Cambrian Qiongzhusi Formation Shale. *Processes* **2026**, *14*, 882. [CrossRef]
41. Wang, B.; Chen, M.; Tian, H.; Sun, J.; Liu, L.; Liang, X.; Chen, B.; Yu, B.; Zhang, Z.; Qu, Z. Geochemical Characteristics and Genetic Origin of Tight Sandstone Gas in the Daning–Jixian Block, Ordos Basin. *Processes* **2025**, *13*, 4019. [CrossRef]
42. Saporetti, C.; Fonseca, D.; Oliveira, L.; Pereira, E.; Goliatt, L. Hybrid Machine Learning Models for Estimating Total Organic Carbon from Mineral Constituents in Core Samples of Shale Gas Fields. *Mar. Pet. Geol.* **2022**, *143*, 105783. [CrossRef]
43. Yuan, Y.; Fu, H.; Yan, D.; Wang, G.; Yang, S.; Liu, M.; Jiang, Q.; Wang, X.; Wang, W. Machine Learning-Driven Classification and Production Capacity Prediction of Tight Sandstone Reservoirs: A Case Study of the Taiyuan Formation, Ordos Basin. *Energy Sci. Eng.* **2026**, *14*, 2514–2538. [CrossRef]
44. Bao, P.; Hui, G.; Hu, Y.; Song, R.; Chen, Z.; Zhang, K.; Pi, Z.; Li, Y.; Ge, C.; Yao, F.; et al. Comprehensive Characterization of Hydraulic Fracture Propagations and Prevention of Pre-Existing Fault Failure in Duvernay Shale Reservoirs. *Eng. Fail. Anal.* **2025**, *173*, 109461. [CrossRef]
45. Cheng, C.; Chen, Z.-X.; Hui, G. Energy-Efficient Fracturing Based on Stress-Coupled Perforation. *Pet. Sci.* 2026, *in press*. [CrossRef]
46. Hui, G.; Chen, Z.; Schultz, R.; Chen, S.; Song, Z.; Zhang, Z.; Song, Y.; Wang, H.; Wang, M.; Gu, F. Intricate Unconventional Fracture Networks Provide Fluid Diffusion Pathways to Reactivate Pre-Existing Faults in Unconventional Reservoirs. *Energy* **2023**, *282*, 128803. [CrossRef]
47. He, Y.-P.; Cheng, H.-B.; Zeng, P.; Zang, C.-Z.; Dong, Q.-W.; Wan, G.-X.; Dong, X.-T. Working Condition Recognition of Sucker Rod Pumping System Based on 4-Segment Time-Frequency Signature Matrix and Deep Learning. *Pet. Sci.* **2024**, *21*, 641–653. [CrossRef]
48. Nascimento, J.; Maitelli, A.; Maitelli, C.; Cavalcanti, A. Diagnostic of Operation Conditions and Sensor Faults Using Machine Learning in Sucker-Rod Pumping Wells. *Sensors* **2021**, *21*, 4546. [CrossRef]

https://doi.org/10.3390/pr1411172

. Nour, M.; Elsayed, S.K.; Mahmoud, O. A Supervised Machine Learning Model to Select a Cost-Effective Directional Drilling Tool. *Sci. Rep.* **2024**, *14*, 26624. [CrossRef]
. Xiao, C.; Wang, G.; Zhang, Y.; Deng, Y. Machine-Learning-Based Well Production Prediction under Geological and Hydraulic Fracture Parameters Uncertainty for Unconventional Shale Gas Reservoirs. *J. Nat. Gas Sci. Eng.* **2022**, *106*, 104762. [CrossRef]
. Vikara, D.; Remson, D.; Khanna, V. Machine Learning-Informed Ensemble Framework for Evaluating Shale Gas Production Potential: Case Study in the Marcellus Shale. *J. Nat. Gas Sci. Eng.* **2020**, *84*, 103679. [CrossRef]
. Mehana, M.; Guiltinan, E.; Vesselinov, V.; Middleton, R.; Hyman, J.D.; Kang, Q.; Viswanathan, H. Machine-Learning Predictions of the Shale Wells' Performance. *J. Nat. Gas Sci. Eng.* **2021**, *88*, 103819. [CrossRef]
. Yi, J.; Qi, Z.; Li, X.; Liu, H.; Zhou, W. Spatial Correlation-Based Machine Learning Framework for Evaluating Shale Gas Production Potential: A Case Study in Southern Sichuan Basin, China. *Appl. Energy* **2024**, *357*, 122483. [CrossRef]
. Delshad, M.; Al-Husseini, A.; Sharma, M.M. A Comprehensive Review of Well Integrity Challenges and Digital Twin Applications Across Conventional, Unconventional, and Storage Wells. *Energies* **2025**, *18*, 4757. [CrossRef]
. Radhasharan, N. Real-Time Edge-To-Cloud Intelligence Architecture for Autonomous Drilling Systems. *J. Int. Crisis Risk Commun. Res.* **2026**, *9*, 90–102.
. Mazhar, S.; Mumtaz, M.W.; El Oirdi, M.; Mukhtar, H.; Raza, M.A.; Farhan, M.; Aatif, M.; Muteeb, G. Synergizing Advanced Materials and Artificial Intelligence for Next-Generation Carbon Capture, Utilization, and Storage (CCUS): A Review. *RSC Adv.* **2026**, *16*, 2621–2651. [CrossRef] [PubMed]
. Mao, S.; Chen, B.; Malki, M.; Chen, F.; Morales, M.; Ma, Z.; Mehana, M. Efficient Prediction of Hydrogen Storage Performance in Depleted Gas Reservoirs Using Machine Learning. *Appl. Energy* **2024**, *361*, 122914. [CrossRef]

https://doi.org/10.3390/pr14111727

processes

MDPI

Article

Physics-Informed Fusion Neural Network for Real-Time Bottomhole Pressure Control in Managed Pressure Drilling

Liwei Wu [1,†], Ziyue Zhang [1,†], Chengkai Zhang [1,*], Gensheng Li [1], Xianzhi Song [1], Mengmeng Zhou [2] and Xuezhe Yao [1]

1 College of Petroleum Engineering, China University of Petroleum, Beijing 102249, China; wulw2@cnooc.com.cn (L.W.); 2024310189@student.cup.edu.cn (Z.Z.); ligs@cup.edu.cn (G.L.); songxz@cup.edu.cn (X.S.)
2 State Key Laboratory of Petroleum Resources and Engineering, China University of Petroleum, Beijing 102249, China
* Correspondence: zhangck@cup.edu.cn
† These authors contributed equally to this work.

Abstract

Managed pressure drilling (MPD) is the core technology for developing formations with high pressure and narrow density windows. It precisely maintains the bottomhole pressure (BHP) within the safe operating window defined by formation pore pressure and fracture pressure by actively regulating the wellbore pressure profile. If pressure control becomes unstable, it can easily trigger gas kicks or lost circulation, posing a severe threat to operational safety. However, existing model predictive control (MPC) schemes have significant limitations: pure data-driven models exhibit poor generalization under complex conditions, while control algorithms based on traditional mechanistic models struggle to meet the stringent real-time requirements of field control cycles due to high-complexity numerical iteration processes. To balance control precision and real-time performance, this paper proposes a physics-informed model predictive control framework (PINC-MPC). During the training phase, physical prior knowledge such as the law of mass conservation is embedded into the neural network as constraints to construct a physically consistent deep surrogate model, enabling it to characterize complex wellbore characteristics. In the control phase, this surrogate model replaces the time-consuming numerical solving process of the mechanistic model within the MPC loop, achieving near-real-time state prediction and rolling optimization while ensuring physical fidelity. Experimental results indicate that PINC-MPC demonstrates superior control performance. Its median single-step solving time is only 16.81 ms, achieving an 11.1-fold acceleration compared to the mechanistic model-based scheme (187.3 ms). In a 5000 s full-cycle closed-loop control experiment, the total time required for the former is only 1.68 s, while the latter reaches 18.73 s, representing an efficiency improvement of approximately 91%. In terms of control accuracy, the integrated absolute error (IAE), reflecting the total deviation of the control process, significantly decreased from 63.40 MPa·s for the industrial successive linearization MPC (SLMPC) to 12.90 MPa·s, an improvement of 79.7%. Especially in extreme dynamic conditions such as simulated pump shutdowns for pipe connections and sudden gas kicks, the framework demonstrates excellent predictive ability and response efficiency. It can proactively trigger compensation actions to keep BHP fluctuations within 0.30 MPa, significantly outperforming the traditional SLMPC method. The research results prove that PINC-MPC provides an efficient, precise, and robust nonlinear control strategy for MPD systems, offering important engineering reference value for enhancing the automation level of intelligent drilling systems.

Keywords: managed pressure drilling; physics-informed neural network; model predictive control; bottomhole pressure; real-time prediction

1. Introduction

1.1. Research Background

Managed pressure drilling (MPD) is an advanced drilling technique that actively regulates the wellbore pressure profile to achieve safe and efficient drilling operations. MPD is therefore essential in complex scenarios such as narrow pressure window formations, high-pressure, high-temperature reservoirs, and deepwater drilling. The core objective of MPD is to maintain the bottomhole pressure (BHP) precisely within the safe window between the formation pore pressure and the fracture pressure: excessive pressure causes formation fracturing and drilling fluid loss, while insufficient pressure may trigger gas influx (gas kick) or even blowouts. Figure 1 illustrates the MPD system configuration.

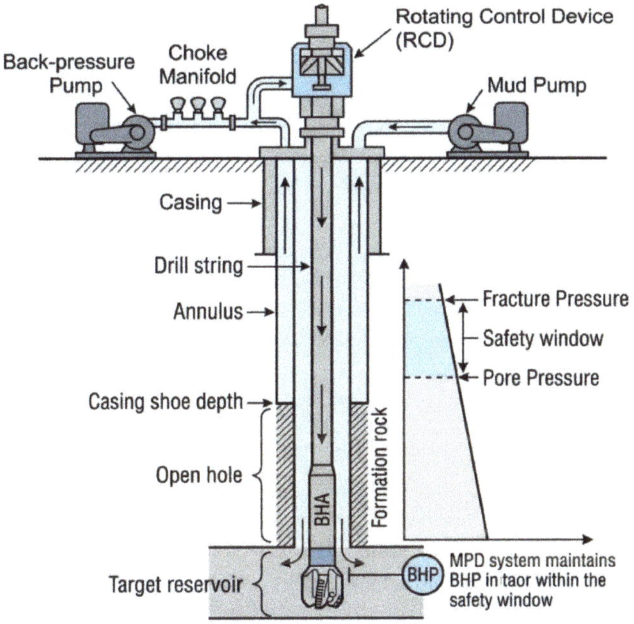

Figure 1. Schematic of the MPD system: the surface choke manifold and back-pressure pump regulate annular pressure to maintain BHP precisely within the safe window between pore pressure and fracture pressure.

As oil and gas exploration extends into deeper, ultra-deep, and deepwater domains, formation pressure windows become increasingly narrow, placing increasingly stringent demands on real-time BHP control accuracy.

MPD systems typically control annular pressure by adjusting choke valve opening and mud pump flow rate. The core challenge lies in the strong nonlinearity and time-varying nature of gas–liquid two-phase flow in the wellbore, frequent operational transitions such as pipe connection pump stops and sudden gas influx, and strict real-time constraints imposed by industrial sites (with a typical control cycle of approximately 50 s). Constructing a BHP

https://doi.org/10.3390/pr14081240

control framework that balances prediction accuracy with computational efficiency while preserving physical consistency is the central scientific problem in intelligent MPD control.

1.2. Related Work

1.2.1. MPD Modeling and Control

Modeling approaches for MPD systems can be categorized by physical fidelity and computational complexity. Three-dimensional computational fluid dynamics (CFD) models based on the Navier–Stokes equations accurately capture gas–liquid two-phase flow details in the wellbore, but their computational cost is prohibitive for real-time control. The drift-flux model, proposed by Zuber and Findlay in 1965 [1], describes the relative motion between gas and liquid phases via slip velocity, reducing computational complexity while preserving physical consistency; it has become the widely adopted modeling framework in engineering applications [2,3]. Control-oriented simplified ordinary differential equation (ODE) models reduce the wellbore system to a small number of key state variables, offering high computational efficiency suitable for embedding in model predictive control algorithms for real-time optimization.

Regarding control strategies, proportional–integral–derivative (PID) control, with its simple structure, remains applicable in scenarios with modest pressure control accuracy requirements [4], but struggles to maintain satisfactory performance against the strong nonlinearity of MPD systems. Model predictive control (MPC), through its receding-horizon optimization mechanism, can explicitly handle system constraints and is currently a research and application hotspot in MPD. Nygaard and Nævdal [5] proposed a nonlinear MPC scheme for wellbore annular pressure stabilization, establishing the theoretical foundation for MPD model predictive control. Kaasa et al. [6] developed a simplified hydraulic model for MPD control systems, enabling intelligent estimation of BHP. Nandan and Imtiaz [7] proposed a nonlinear MPC scheme for MPD and demonstrated effective suppression of disturbances such as gas influx in simulation. Park et al. [8] further integrated MPC with real-time high-fidelity flow models, improving the control system's adaptability under complex operating conditions. However, these physics-based methods generally suffer from high computational complexity and degraded online linearization accuracy as the operating point drifts, limiting their widespread industrial deployment. Zhang et al. [9] proposed a hybrid neural network BHP prediction model incorporating physical constraints, validating the effectiveness of embedding physical knowledge into data-driven models; yet most existing data-driven prediction models focus solely on offline prediction accuracy and have not been deeply integrated with real-time control frameworks. More broadly, data-driven intelligent methods have seen growing application in MPD-related tasks such as drilling condition classification [10,11], demonstrating strong potential for intelligent drilling systems.

These limitations motivate the exploration of physics-informed machine learning approaches that can combine physical consistency with computational efficiency.

1.2.2. Physics-Informed Neural Networks

Physics-Informed Neural Networks (PINNs), proposed by Raissi et al. [12] in 2019, embed the residuals of physical equations (partial or ordinary differential equations) into the neural network loss function, achieving co-optimization of data-fitting and physical constraints. As illustrated in Figure 2, the PINN loss function consists of a data loss $\mathcal{L}_{\text{data}}$ and a physics loss $\mathcal{L}_{\text{physics}}$; the latter computes residuals via automatic differentiation to enforce the given differential equation constraints. Cai et al. [13] systematically reviewed the progress of PINNs in fluid mechanics, noting that physical constraints improve model generalization and data efficiency. Compared with purely data-driven approaches, PINNs maintain physical consistency even with limited training data, which improves model generalizability and interpretability.

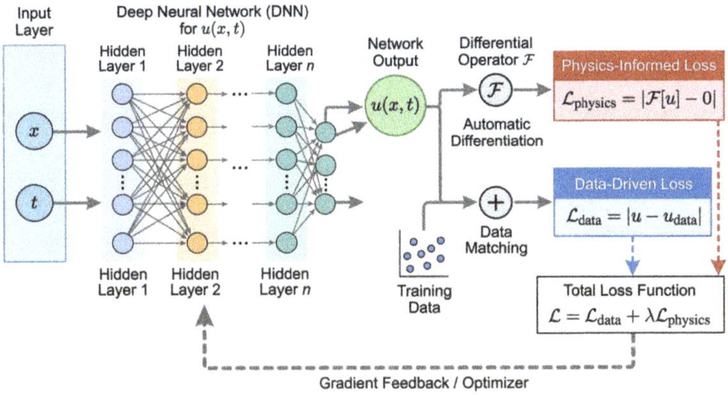

Figure 2. Schematic of the PINN architecture: network outputs are differentiated via automatic differentiation to compute PDE residuals; the loss function simultaneously includes a data-fitting term and a physics-constraint term.

1.2.3. PINC Framework for Control

The Physics-Informed Neural Control (PINC) framework, proposed by Antonelo et al. [14], extends standard PINNs by incorporating control inputs and initial states as network inputs and introducing a self-loop mechanism to support long-horizon temporal prediction, making it naturally suited for state prediction and trajectory planning in control scenarios. Kittelsen et al. [15] applied an improved PINC framework to gas-lift well control, introducing skip connections (weighting each layer's output via Hadamard product with an encoder vector) to effectively mitigate vanishing gradients in deep networks. Figure 3 illustrates the hierarchical PINC architecture.

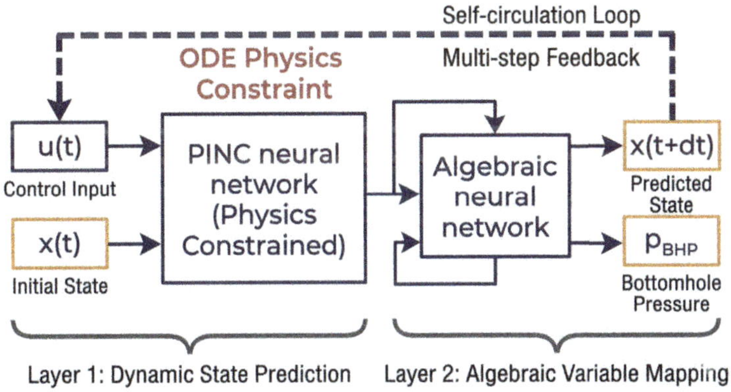

Figure 3. Schematic of the hierarchical PINC architecture: the first-level PINC network predicts system dynamic states (with self-loop mechanism); the second-level algebraic neural network maps states to BHP.

1.3. Problem Statement and Main Contributions

Despite advances in MPD control and PINN applications, a clear gap remains in applying the PINN/PINC framework to real-time BHP control in managed pressure drilling. Existing PINC work (e.g., Kittelsen et al. [15]) primarily targets steady-state

https://doi.org/10.3390/pr14081240

production control in gas-lift wells, whereas MPD involves transient pressure management during dynamic drilling, with stricter pressure window constraints (safety margins of only a few MPa), more frequent operational transitions (pump stops, gas influx), and higher real-time requirements (control cycle on the order of 50 s). Furthermore, conventional SLMPC suffers from degraded linearization accuracy at operating-point transitions, and purely data-driven methods can lack robustness under noise and parameter mismatch.

Despite these advances, a unified framework that simultaneously achieves high physical consistency, long-horizon prediction stability, and real-time computational efficiency remains elusive. Physics-based MPC methods are accurate but computationally expensive; data-driven surrogate models are fast but may violate physical constraints; and hybrid approaches have yet to demonstrate robust closed-loop performance under the diverse disturbance scenarios encountered in field MPD operations.

To address these challenges, this paper proposes the PINC-MPC framework with the following main contributions: (1) applying the PINN framework to real-time BHP control in MPD by constructing a PIFG-NN that integrates a static feature sub-network, a temporal feature sub-network, and skip connections, effectively suppressing error accumulation over 30-step (1500 s) long-horizon autoregressive prediction; (2) designing a multi-objective physics-constrained loss function that embeds mass conservation ODE constraints into training with a gradient-balancing adaptive weighting strategy; (3) constructing a PINC-MPC controller based on CasADi symbolic automatic differentiation for efficient receding-horizon optimization, reducing IAE by approximately 79.7% compared with SLMPC (12.90 vs. 63.40 MPa·s) with a median single-step solve time of 16.81 ms; and (4) demonstrating multi-scenario robustness under noise, pump stop, and gas influx conditions, with maximum BHP deviations maintained within 0.30 MPa and 0.26 MPa, respectively. These results are based on numerical simulation; broader experimental validation is needed before field application.

2. System Modeling

2.1. Wellbore Physical Model

The physical parameters of the MPD system are listed in Appendix A Table A1. The system has two control inputs: the MPD choke valve opening $u_c \in [0.1, 0.9]$, which regulates the wellhead back-pressure, and the main mud pump flow rate $q_p \in [q_{p,\min}, q_{p,\max}]$, which controls the mud circulation rate. The control input vector is $\mathbf{u} = [u_c, q_p]^\top$. Accurate pressure management is essential not only for well control but also for mitigating formation damage. Formation properties such as permeability and stress state directly govern fluid influx behavior [16,17]; in high-temperature deep wells, thermally induced fracture evolution can alter the near-wellbore flow path and affect BHP dynamics [18].

2.2. Two-Phase Flow Conservation Equations in the Wellbore

The model is subject to the following simplifying assumptions:

1. Flow in the wellbore is treated as one-dimensional;
2. Free gas, dissolved gas, and drilling fluid are assumed to share the same pressure and temperature at any given depth;
3. The effect of drill cuttings on flow is neglected;
4. The drilling fluid temperature in the drill string and annulus is treated as known and time-invariant.

https://doi.org/10.3390/pr1408124

2.2.1. Mass Conservation Equations

The mass conservation equations for the gas and liquid phases are:

$$\frac{\partial(\rho_g \alpha_g)}{\partial t} + \frac{\partial(\rho_g \alpha_g v_g)}{\partial x} = 0 \quad \text{(gas phase)} \tag{1}$$

$$\frac{\partial(\rho_l \alpha_l)}{\partial t} + \frac{\partial(\rho_l \alpha_l v_l)}{\partial x} = 0 \quad \text{(liquid phase)} \tag{2}$$

where ρ_g and ρ_l are the gas and liquid phase densities, α_g and α_l are the gas and liquid volume fractions (satisfying $\alpha_g + \alpha_l = 1$), and v_g and v_l are the gas and liquid phase velocities.

2.2.2. Momentum Conservation Equation

The momentum conservation equation accounts for gravity, pressure gradient, and frictional resistance:

$$\rho_m v_m \frac{\partial v_m}{\partial x} + \frac{\partial p}{\partial x} + (\rho_l \alpha_l + \rho_g \alpha_g) g + f = 0 \tag{3}$$

where $\rho_m = \rho_l \alpha_l + \rho_g \alpha_g$ is the mixture density, v_m is the mixture velocity, and f is the frictional pressure drop per unit length.

2.2.3. Gas-Phase Slip Velocity Model

The gas-phase velocity v_g is determined by the liquid-phase velocity v_l and the slip velocity v_{gr}:

$$v_g = v_l + v_{gr} \tag{4}$$

The empirical correlation for the slip velocity is:

$$v_{gr} = C_0 \sqrt{\frac{gD(\rho_l - \rho_g)}{\rho_l}} \sqrt{1 - \alpha_g}\, C_2 C_3 C_4 \tag{5}$$

where $C_0 = 1.2$ is the slip velocity coefficient, C_2, C_3, C_4 are bubble correction factors, and D is the wellbore diameter.

2.3. Control-Oriented ODE Model

To meet real-time requirements, the MPC controller uses a simplified ODE model based on mass conservation. In the MPD circulation loop, drilling fluid is injected by the mud pump into the drill string, flows through the drill bit into the annulus, and exits through the surface MPD choke valve:

$$\frac{dm_{dp}}{dt} = w_{pump} - w_{bit} \tag{6}$$

$$\frac{dm_{la}}{dt} = w_{bit} + w_{res,l} - w_{out,l} \tag{7}$$

$$\frac{dm_{ga}}{dt} = w_{res,g} - w_{out,g} \tag{8}$$

where m_{dp} is the liquid mass in the drill string, m_{la} is the liquid mass in the annulus, m_{ga} is the gas mass in the annulus (during gas influx), and t is time. w_{pump} is the mass flow rate injected by the mud pump, w_{bit} is the flow rate through the drill bit into the annulus, $w_{res,l}$ and $w_{res,g}$ are the formation fluid (liquid/gas) influx rates, and $w_{out,l}$ and $w_{out,g}$ are the gas and liquid flow rates discharged through the MPD choke valve. The flow rate terms are defined as follows:

https://doi.org/10.3390/pr14081240

$$p_{dp} = \rho_{dp} \cdot \frac{gL}{10^6}, \quad \rho_{dp} = \frac{m_{dp}}{V_{dp}} \tag{9}$$

$$\rho_{la} = \frac{m_{la}}{V_a}, \quad p_{la} = \rho_{la} \cdot \frac{gL}{10^6} \tag{10}$$

$$w_{bit} = k_{bit}\sqrt{\max(p_{dp} - p_{la}, 10^{-3})} \tag{11}$$

where k_{bit} is the bit nozzle flow coefficient, and the small threshold 10^{-3} prevents gradient explosion at the zero point.

$$w_{pump} = k_{pump} \cdot q_p, \quad w_{out,l} = k_5 \cdot u_c \cdot \frac{m_{la}}{V_a} \tag{12}$$

$$w_{res,l} = k_{res} \cdot \max(p_{form} - p_{la}, 0), \quad w_{out,g} = k_2 \cdot u_c \cdot \frac{m_{ga}}{V_a} \tag{13}$$

2.4. Bottomhole Pressure Calculation

The bottomhole pressure (BHP) is computed via finite-difference marching along the well depth:

$$p_x = p_{x-\Delta x} + \Delta x\left(\rho_m g + \frac{F_{f,x-\Delta x} + F_{f,x}}{2}\right) + \Delta p_m \tag{14}$$

The pressure increment consists of a hydrostatic pressure term $\rho_m g \Delta x$, a frictional pressure drop term, and a momentum change term.

$$\Delta p_m = \frac{\Delta x}{\Delta t}\left[(\rho_l \alpha_l v_l + \rho_g \alpha_g v_g)_x - (\rho_l \alpha_l v_l + \rho_g \alpha_g v_g)_{x-\Delta x}\right] \tag{15}$$

$$F_f = \frac{1}{2} f_D \frac{\rho_m v_m^2}{D - D_p} \tag{16}$$

where f_D is calculated according to the flow regime (laminar or turbulent):

$$f_D = \begin{cases} \dfrac{64}{\mathrm{Re}}, & \mathrm{Re} < 2100 \quad \text{(laminar)} \\ \left[-1.8\log_{10}\left(\dfrac{6.9}{\mathrm{Re}} + \left(\dfrac{\varepsilon}{3.7(D - D_p)}\right)^{1.11}\right)\right]^{-2}, & \mathrm{Re} \geq 2100 \quad \text{(turbulent)} \end{cases} \tag{17}$$

where $\mathrm{Re} = \rho_m v_m (D - D_p)/\mu_l$, ε is the wall roughness, and in the turbulent regime the Colebrook–White explicit approximation is employed.

3. Physics-Informed Neural Network Design

3.1. Overall Architecture of PIFG-NN

PIFG-NN employs a three-branch fusion architecture, as illustrated in Figure 4. The network consists of a static physical feature extraction sub-network (fully connected neural network, FCNN-Static), a temporal sub-network for capturing operating condition dependencies (gated recurrent unit, GRU-Sequence), and a multi-source feature fusion output sub-network (FCNN-Fusion).

The static feature sub-network takes non-dimensionalized spatiotemporal coordinates (T, X) as input and extracts static features $\mathbf{f}_{\text{static}}$ (64-dimensional) related to wellbore geometry and fluid properties through a three-layer fully connected network (64 neurons

https://doi.org/10.3390/pr14081240

per layer with batch normalization and LeakyReLU activation). The temporal feature sub-network employs a two-layer GRU to process operating condition sequence data of length $L_s = 5$ (weight on bit, rotary speed, flow rate, etc.) with a hidden dimension of 32, outputting the hidden state $\mathbf{f}_{\text{seq}}$ (32-dimensional) at the final time step to capture temporal dependencies of key operating conditions such as flow rate step changes and weight-on-bit fluctuations. The fusion sub-network incorporates skip connections adapted for physics-informed networks, as described below.

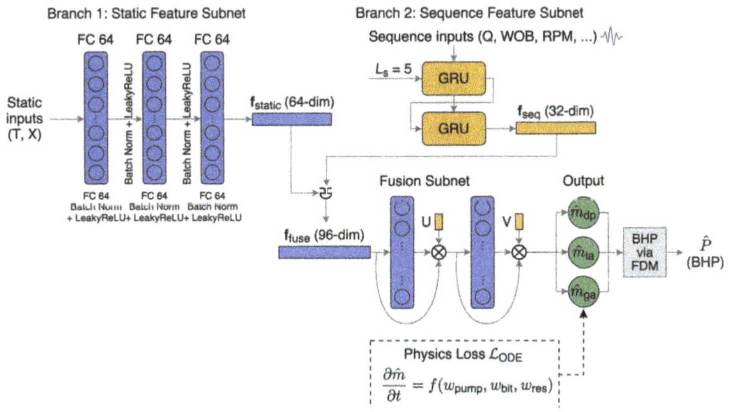

Figure 4. PIFG-NN network architecture.

The fusion sub-network introduces a skip connection architecture, encoding the concatenated feature $\mathbf{f}_{\text{fuse}} = [\mathbf{f}_{\text{static}}; \mathbf{f}_{\text{seq}}]$ into $\mathbf{U}$ and $\mathbf{V}$. Each layer's output maintains a direct connection to the initial input through weighted fusion, which alleviates the vanishing gradient problem in multi-timescale stiff systems. The network ultimately outputs three core state variables, $\hat{\mathbf{x}} = [\hat{m}_{dp}, \hat{m}_{la}, \hat{m}_{ga}]^\top$, which are then explicitly mapped to BHP through the finite-difference equation (Equation (14)), avoiding the need to solve nonlinear algebraic loops during MPC optimization. The training objective governing all network parameters is described in Section 3.

3.2. Physics Constraints and Loss Function Design

To make the model predictions conform to the fundamental physical laws of gas–liquid two-phase flow, the total loss function is a weighted combination of four loss terms:

$$\mathcal{L}_{\text{total}} = \lambda_1 \mathcal{L}_{\text{data}} + \lambda_2 \mathcal{L}_{\text{ODE}} + \lambda_3 \mathcal{L}_{\text{emp}} + \lambda_4 \mathcal{L}_{\text{out}} \tag{18}$$

where λ_1 through λ_4 are weighting coefficients, determined through tuning as $\lambda_1 = 1$, $\lambda_2 = 10$, $\lambda_3 = 10$, $\lambda_4 = 5$. The physics-constraint and empirical loss weights are set larger to maintain physical consistency.

The initial weight ratios ($\lambda_{\text{ODE}}/\lambda_{\text{data}} = 10$, $\lambda_{\text{emp}}/\lambda_{\text{data}} = 10$, $\lambda_{\text{out}}/\lambda_{\text{data}} = 5$) are chosen so that the physics-constraint gradient norms are of the same order of magnitude as the data-fitting gradient norm at the start of training, following the gradient magnitude matching principle common in multi-task learning [19]. These initial values are only a starting point: the gradient-balancing adaptive weighting strategy described in Section 3.5 monitors and adjusts all λ_k throughout training (every 5 epochs) to maintain gradient equilibrium. The final effective weights at convergence therefore differ from the initial values, and the trained model is not sensitive to moderate perturbations of the initial weight setting. In preliminary experiments, varying the initial λ_{ODE} by a factor of 5 (from 2 to 50) changed the converged validation MAE by less than 8%, confirming that the adaptive

https://doi.org/10.3390/pr14081240

mechanism compensates for suboptimal initialization. A full ablation study systematically removing individual loss components (e.g., training without $\mathcal{L}_{\text{ODE}}$ or $\mathcal{L}_{\text{emp}}$) is deferred to future work; the qualitative role of each term—mass conservation enforcement, flow rate regularization, and output feasibility—is well-established in the physics-informed learning literature [12,20].

The data-fitting loss $\mathcal{L}_{\text{data}}$ is constructed solely for the precisely measurable bottomhole pressure P in mean squared error form:

$$\mathcal{L}_{\text{data}} = \frac{1}{N}\sum_{i=1}^{N}\left(\hat{P}^{(i)} - P_{\text{real}}^{(i)}\right)^2 \tag{19}$$

The physics-constraint loss $\mathcal{L}_{\text{ODE}}$ computes the time derivatives of the network output states $\hat{\mathbf{x}} = (\hat{m}_{dp}, \hat{m}_{la}, \hat{m}_{ga})$ via PyTorch automatic differentiation and penalizes the difference from the ODE right-hand-side net flow rates defined in Section 2, so that the predicted trajectories obey mass conservation:

$$\mathcal{L}_{\text{ODE}} = \mathcal{L}_{dp} + \mathcal{L}_{la} + \mathcal{L}_{ga} \tag{20}$$

$$\mathcal{L}_{dp} = \frac{1}{N}\sum_{i=1}^{N}\left(\frac{\partial \hat{m}_{dp}}{\partial t} - (w_{pump} - w_{bit})\right)^2 \tag{21}$$

$$\mathcal{L}_{la} = \frac{1}{N}\sum_{i=1}^{N}\left(\frac{\partial \hat{m}_{la}}{\partial t} - (w_{bit} + w_{res,l} - w_{out,l})\right)^2 \tag{22}$$

$$\mathcal{L}_{ga} = \frac{1}{N}\sum_{i=1}^{N}\left(\frac{\partial \hat{m}_{ga}}{\partial t} - (w_{res,g} - w_{out,g})\right)^2 \tag{23}$$

The empirical loss $\mathcal{L}_{\text{emp}}$ constructs pseudo-labels $v_{l,\text{appr}} = Q/A$ using the flow rate Q and cross-sectional area A, providing weak supervision for the liquid-phase velocity, which is difficult to measure directly:

$$\mathcal{L}_{\text{emp}} = \frac{1}{N}\sum_{i=1}^{N}\left(\hat{v}_l^{(i)} - v_{l,\text{appr}}^{(i)}\right)^2 \tag{24}$$

The output feasibility constraint $\mathcal{L}_{\text{out}}$ applies ReLU penalty terms to enforce non-negativity of physical quantities such as pressure, velocity, and volume fraction:

$$\mathcal{L}_{\text{out}} = \sum_{i=1}^{N}\text{ReLU}\left(-\hat{P}^{(i)}\right) + \text{ReLU}\left(-\hat{v}_l^{(i)}\right) + \text{ReLU}\left(-\hat{\alpha}_g^{(i)}\right) \tag{25}$$

Taken together, the four loss components endow PIFG-NN with capabilities that a purely data-driven model cannot achieve. The ODE constraint embeds mass conservation laws directly into the gradient update signal, steering the network away from physically impossible trajectories even when such trajectories would reduce the data-fitting error. The empirical loss provides weak but physically meaningful supervision for the liquid-phase velocity—a quantity that is difficult to measure and therefore absent from the labeled training set—allowing the network to develop an internally consistent flow picture rather than treating velocity as an unconstrained latent variable. The output feasibility constraint acts as a hard safety net, ensuring that pressure, velocity, and void fraction predictions remain within the physically admissible domain at all times. Collectively, these physics-based elements allow the trained model to extrapolate reliably to operating conditions that lie outside the training envelope, such as sudden pump shutoffs or gas influx events, where

https://doi.org/10.3390/pr1408124

a purely data-driven model trained only on normal drilling operations would be expected to produce physically inconsistent and potentially unsafe predictions.

3.3. Non-Dimensionalization and Numerical Stability

Because the output variables span vastly different magnitudes (pressure $P \approx 80$ MPa, volume fraction $\alpha_g \approx 10^{-3}$, velocity $v_l \approx 1$ m/s), training directly on raw scales causes gradients of small-magnitude variables to vanish, and the model overfits pressure while neglecting the others. To address this, the input and output variables are non-dimensionalized:

$$\begin{aligned} T &= \frac{t}{t^*}, \quad X = \frac{x}{x^*} \\ \hat{P}_{\text{dim}} &= \frac{\hat{P}}{p^*}, \quad \hat{v}_{l,\text{dim}} = \frac{\hat{v}_l}{v_l^*}, \quad \hat{\alpha}_{g,\text{dim}} = \frac{\hat{\alpha}_g}{\alpha_g^*} \end{aligned} \tag{26}$$

where the characteristic scales are: $t^* = 700 \times 3600$ s, $x^* = 6000$ m, $p^* = 10$ MPa, $v_l^* = 1$ m/s, $\alpha_g^* = 0.1$. The pressure reference is set to $p^* = 10$ MPa rather than 1 MPa so that a real pressure of 80 MPa maps to a non-dimensional value of 8, preventing an equal absolute error from yielding an excessively large relative error after scaling.

Furthermore, for the denominator term $1 - C_0\alpha_g^*\hat{\alpha}_g$ in the momentum equation, which may approach zero, a lower bound is enforced via `torch.clamp` in the implementation to prevent numerical instability.

During the initial phase of model training, the neural network's state predictions may fall outside the physically feasible domain. Direct substitution into the ODE equations can then trigger computational failures (e.g., square root of negative numbers, logarithm of non-positive inputs) and vanishing gradients. Following the safety constraint strategy of the improved PINC framework, vulnerable terms in the physical computation formulas are safely reconstructed:

For the square root pressure difference term in the bit flow rate calculation (Equation (11)), a small positive threshold is introduced to avoid infinite gradients and negative-value errors at the zero point:

$$w_{bit} = k_{bit}\sqrt{\max(p_{dp} - p_{la}, 10^{-3})} \tag{27}$$

For the Colebrook–White approximation of the friction factor (Equation (17)), the Reynolds number Re is clipped with a lower bound $\text{Re}_{\text{min}} = 100$, enforcing $\text{Re} = \max(\text{Re}, \text{Re}_{\text{min}})$ to prevent invalid inputs to the $\log_{10}$ term.

3.4. PIFG-NN Training Data and Strategy

PIFG-NN is trained on real offshore drilling data from operational records of a directional well at an offshore oilfield in China. The target formation interval spans depths from 4400 m to 6000 m. Rotary drilling technology was employed, and the wellbore configuration and key equipment assemblies remained unchanged during drilling operations.

The dataset comprises 12 drilling log parameters covering spatiotemporal coordinates, wellbore geometry, drilling operating conditions, and drilling fluid properties. The bottomhole pressure P, measured by downhole pressure gauges, is the primary supervisory signal; the liquid-phase velocity v_L and gas volume fraction α_g are computed from the theoretical model in Section 2. The statistical characteristics of the dataset are presented in Appendix B, Table A2.

The time t is calculated as:

$$\Delta t_i = \frac{\Delta x_i}{\text{ROP}_i} \times 3600 \tag{28}$$

where Δx_i is the depth increment between adjacent sampling points (m), ROP_i is the rate of penetration (m/h), and Δt_i is the drilling time required to traverse that interval (s).

The raw data are preprocessed as follows. Missing values are imputed using interpolation based on neighboring samples. Systematic outlier detection is performed using the interquartile range (IQR) method, and identified outliers are removed. Pearson correlation coefficients between each parameter and the bottomhole pressure P are then computed, and parameters with meaningful physical correlation to pressure are selected as model inputs.

Pearson correlation analysis shows that the flow rate Q, drilling fluid density ρ_L, and drilling fluid viscosity μ_L correlate with the bottomhole pressure P, consistent with the physics of the gas–liquid two-phase momentum equation. Although the mud pit volume MPV shows a weaker correlation with P, it plays an important role in wellbore pressure control and is therefore included. The final model input parameters are: depth x, time t, weight on bit WOB, rotary speed R, flow rate Q, drilling fluid density ρ_L, drilling fluid viscosity μ_L, standpipe pressure SPP, and mud pit volume MPV.

The data are split in an 8:2 ratio into training and test sets, and 3-fold cross-validation is used during training. The GRU temporal sub-network uses a sliding window of length $L_s = 5$.

3.5. Training Strategy

PIFG-NN is trained using the Adam optimizer with an initial learning rate $\eta = 5 \times 10^{-3}$ and a learning rate decay schedule (decayed to 0.8 times the current value every 50 epochs) to facilitate convergence. The number of training epochs is set to 200, batch size to 64, and weight decay coefficient to $\lambda_{\text{wd}} = 6 \times 10^{-4}$. For parameter initialization, FCNN layers use Kaiming normal initialization and GRU layers use Xavier normal initialization. An early stopping mechanism monitors the validation loss during training; training halts when the validation loss fails to decrease for 5 consecutive epochs, and the model parameters with the lowest validation loss are retained as the final model (Best Model Checkpoint).

Gradient-Balancing Adaptive Weighting Strategy

To resolve the mutual suppression between physics-constraint losses and the data-fitting loss, a custom gradient-balancing adaptive weighting strategy is employed during training. Unlike GradNorm [19], which balances gradient magnitudes through a separate meta-loss, and NTK-based methods [20], which derive weights from the neural tangent kernel spectrum, the present strategy directly monitors the ℓ_2 norm of gradients contributed by each loss component and adjusts the corresponding weight coefficient whenever gradient imbalance exceeds a prescribed tolerance. A similar physics-informed multitask learning strategy with dynamic weighting has been applied to wellbore cleanliness prediction in extended-reach drilling [21], further supporting the use of adaptive loss balancing in physics-constrained drilling models. At every adjustment epoch, separate backward passes are performed for each loss component $\mathcal{L}_k$ ($k \in \{\text{data}, \text{ODE}, \text{emp}, \text{out}\}$) to compute the ℓ_2 norm of the gradients it induces on the shared network parameters $\boldsymbol{\theta}$:

$$g_k = \|\nabla_{\boldsymbol{\theta}} \mathcal{L}_k\|_2, \quad k \in \{\text{data, ODE, emp, out}\}. \tag{29}$$

A mean reference $\bar{g}$ is defined as the arithmetic mean of all component gradient norms:

$$\bar{g} = \frac{1}{|\mathcal{K}|} \sum_{k \in \mathcal{K}} g_k. \tag{30}$$

https://doi.org/10.3390/pr1408124

For each component, the normalized gradient ratio $r_k = g_k/\bar{g}$ is compared against a target ratio $r^* = 1.0$ with a tolerance window $[\tau_{\min}, \tau_{\max}] = [0.8, 1.2]$. A loss component is classified as lagging if $r_k < \tau_{\min}$ and leading if $r_k > \tau_{\max}$.

The weight λ_k of each imbalanced component is updated with a prescribed rate $\alpha_r = 0.01$, then clamped to the interval $[\lambda_{\min}, \lambda_{\max}]$ to prevent instability:

$$\lambda_k \leftarrow \text{clamp}\left(\lambda_k \cdot \left(1 + \alpha_r \cdot \frac{r^* - r_k}{r^*}\right), \lambda_{\min}, \lambda_{\max}\right). \tag{31}$$

The update in Equation (31) increases λ_k for lagging components (small g_k) and decreases it for leading ones (large g_k), steering all gradient norms toward $\bar{g}$. The recomputed total loss

$$\mathcal{L}_{\text{total}} = \sum_{k \in \mathcal{K}} \lambda_k \mathcal{L}_k \tag{32}$$

is then used for the standard backward pass and parameter update. The weight-adjustment procedure is triggered every $E_{\text{adj}} = 5$ training epochs to limit computational overhead while still responding promptly to evolving gradient landscapes.

4. Model Predictive Controller Design

4.1. MPC Fundamentals

MPC is a receding-horizon control strategy that, at each sampling instant, solves a finite-horizon optimal control problem over a prediction window, applies only the first element of the resulting optimal input sequence to the system, and then re-solves at the next instant using updated state feedback [22–24]. Its core advantage is the ability to explicitly handle state and control constraints within the optimization, making it well suited for safety-critical processes such as MPD [25].

4.2. PINC Control Training Data Generation

4.2.1. Control Trajectory Design

To ensure that the training data sufficiently cover the control space, this paper employs a random piecewise-constant strategy to generate control input trajectories. Specifically, the choke valve opening $u_c(t)$ and the main mud pump flow rate $q_p(t)$ are randomly sampled within each holding segment, with a smoothing filter applied between adjacent segments to prevent numerical instability from step discontinuities. The control input constraints are $u_c \in [0.1, 0.9]$ and $q_p \in [q_{p,\min}, q_{p,\max}]$, the holding duration per segment is 50 s, and the smoothing factor is 0.3.

4.2.2. State Evolution Solution

Given a control trajectory $\{u_c(t), q_p(t)\}$, the fourth-order Runge–Kutta (RK4) method is used to numerically integrate the simplified wellbore ODE dynamics model (Equations (6)–(8)) to obtain the state sequence:

$$\mathbf{x}(t) = [m_{dp}, m_{la}, m_{ga}]^\top \tag{33}$$

The bottomhole pressure is computed via the finite-difference marching Equation (14) and serves as the controlled variable.

4.2.3. Dataset Construction

This procedure is repeated to generate multiple random trajectories, and the triplets $(\mathbf{x}_k, \mathbf{u}_k, \mathbf{x}_{k+1})$ are used as training samples for the state-transition prediction module in the PINC-MPC controller. The data are split 8:2 into training and test sets with Z-score standardization preprocessing. This dataset complements the real offshore data used by PIFG-NN in Section 3: PIFG-NN learns pressure prediction patterns under real operating conditions, while PINC-MPC learns state evolution patterns under control strategies through simulation data.

4.3. PINC-MPC Controller

The prediction model of the PINC-MPC controller uses the trained PIFG-NN to predict future states step by step:

$$\hat{\mathbf{x}}_{k+i+1} = \hat{\mathbf{x}}_{k+i} + f_\theta^{\text{PIFG-NN}}(\hat{\mathbf{x}}_{k+i}, \mathbf{u}_{k+i}), \quad i = 0, 1, \ldots, N_p - 1 \tag{34}$$

In this work, the control horizon equals the prediction horizon, i.e., $N_c = N_p = 3$ (corresponding to 150 s).

$f_\theta^{\text{PIFG-NN}}$ denotes the parameterized mapping of the PIFG-NN network. The input consists of the current state and control sequence (encoded with spatiotemporal coordinates and concatenated with the sequence window), and the output is the state increment, which is subsequently mapped to bottomhole pressure through the finite-difference marching Equation (14).

The controller's cost function jointly considers BHP tracking error, drill string and annular liquid mass deviations, and control input smoothness:

$$\begin{aligned} J = \sum_{i=0}^{N_p-1} \Big[& w_p(\hat{p}_{x,k+i} - p_{\text{ref}})^2 + w_{dp}\left(\hat{m}_{dp,k+i} - m_{dp,\text{target}}\right)^2 \\ & + w_{la}\left(\hat{m}_{la,k+i} - m_{la,\text{target}}\right)^2 \Big] + w_u \sum_{i=0}^{N_p-1} \|\Delta\mathbf{u}_{k+i}\|_2^2 \end{aligned} \tag{35}$$

The weight settings are listed in Table 1.

Table 1. Parameter comparison between PINC-MPC and SLMPC controllers.

Parameter	Symbol	PINC-MPC	SLMPC
BHP tracking weight	w_p	1000.0	1000.0
Drill string liquid mass weight	w_{dp}	0.01	0.01
Annular liquid mass weight	w_{la}	0.001	0.001
Control variation weight	w_u	3.0	50.0
Prediction horizon	N_p	3 steps	3 steps
Sampling time	T_s	50 s	50 s
Optimizer	—	CasADi (IPOPT)	SciPy SLSQP

Note: w_p, w_{dp}, w_{la}, and w_u are all dimensionless weighting coefficients.

The controller constraints restrict the choke valve opening and mud pump flow rate to safe operating ranges:

$$u_c \in [0.1,\ 0.9], \quad q_p \in [q_{p,\min},\ q_{p,\max}] \tag{36}$$

For the optimization algorithm, this paper uses CasADi for nonlinear programming (NLP) formulation and invokes the IPOPT (interior-point method) solver. CasADi computes exact Jacobian and Hessian matrices for the network's forward pass via symbolic graph automatic differentiation, which allows deep learning to be integrated directly with optimal control.

https://doi.org/10.3390/pr1408124

4.4. SLMPC Controller (Baseline)

Successive linearization MPC (SLMPC) performs online linearization of the ODE model at each control step, converting the nonlinear optimization into a quadratic program [26,27]:

$$\mathbf{x}_{k+1} \approx \mathbf{A}_k \mathbf{x}_k + \mathbf{B}_k \mathbf{u}_k + \mathbf{f}_k \quad (37)$$

where $\mathbf{A}_k$ and $\mathbf{B}_k$ are computed via finite-difference numerical approximation.

SLMPC uses a larger control variation weight ($w_u = 50.0$) to compensate for linearization errors. Since the linearized model loses accuracy at operating-point transitions, a stronger smoothing penalty is needed to suppress the resulting control chattering. PINC-MPC, by contrast, relies on the globally nonlinear PIFG-NN predictor, which maintains higher accuracy across the operating range and therefore requires only $w_u = 3.0$, giving faster tracking response without sacrificing smoothness.

Both controllers were tuned independently to their respective optimal configurations rather than constrained to share the same weight set. The difference in w_u is therefore a consequence of the prediction model accuracy gap, not a source of bias in favor of PINC-MPC. Imposing identical weights on both controllers would penalize each method for the other's structural limitations, producing a parameter-matched but not performance-optimal comparison. This tuning protocol—each method evaluated at its own best configuration—is standard practice in MPC benchmarking [22]. A dedicated sensitivity analysis on w_u is presented in Section 6.6.

4.5. ODE-NMPC Controller (Upper-Bound Baseline)

To provide a performance upper bound and isolate the effect of model approximation, an additional nonlinear MPC baseline (ODE-NMPC) is constructed. ODE-NMPC uses the same RK4 integrator that generates the ground-truth system dynamics as its internal prediction model, solving the same cost function (Equation (35)) with the SciPy/SLSQP optimizer. Because the prediction model is identical to the plant, ODE-NMPC gives the best achievable tracking accuracy for this system under the given model structure—any performance gap between PINC-MPC and ODE-NMPC is primarily attributable to the PIFG-NN surrogate approximation error, while any gap between ODE-NMPC and SLMPC reflects the cost of linearization. The computational disadvantage of ODE-NMPC is that each cost function evaluation requires $N_p \times$ `steps_per_action` full RK4 integration calls, and the SLSQP optimizer must estimate gradients via finite differences, resulting in substantially longer solve times. In summary, SLMPC isolates the cost of nonlinear approximation through linearization, while ODE-NMPC establishes the theoretical performance ceiling attributable to model approximation alone.

The key parameter settings of the controllers are summarized in Table 2.

Table 2. Controller parameter settings.

Parameter	Symbol	Value
Prediction horizon	N_p	3 steps
Control horizon	N_c	3 steps
Sampling time	T_s	50 s

5. Experimental Methodology and Baseline Prediction Performance Validation

5.1. Simulation Platform and System Configuration

Ground-truth system states are computed by an RK4 integrator with a 10 s time step. The initial state is $\mathbf{x}_0 = [10{,}000, 5000, 2000]^\top$ kg, the total simulation duration is 5000 s, and

https://doi.org/10.3390/pr14081240

the reference BHP trajectory consists of three step changes. The experimental environment comprises an Intel Core i7-13700H processor, 16 GB RAM, Python 3.11.4, and PyTorch 2.0.1 (CPU-only).

The PIFG-NN model is trained on real offshore drilling log data (directional well, China, depth 4400–6000 m); however, all closed-loop controller validation experiments in this paper are conducted within the RK4-based ODE simulation environment described above. Consequently, the training domain (real field data) and the validation domain (simplified ODE simulator) are distinct, and the generalization of the controller to real wellbore conditions remains to be verified in future field trials.

5.2. Unified Definition of Evaluation Metrics

The following notation is used throughout. Let N denote the total number of control steps in the simulation, $e_k = p_{\mathrm{bhp},k} - p_{\mathrm{ref},k}$ the BHP tracking error at step k, $\Delta u_k = u_k - u_{k-1}$ the control input increment, and T_s the sampling period. Four evaluation metrics are defined: mean absolute error (MAE), root mean square error (RMSE), integral absolute error (IAE), and total variation (TV).

$$\mathrm{MAE} = \frac{1}{N}\sum_{k=1}^{N} |e_k| \tag{38}$$

$$\mathrm{RMSE} = \sqrt{\frac{1}{N}\sum_{k=1}^{N} e_k^2} \tag{39}$$

$$\mathrm{IAE} = T_s \sum_{k=1}^{N} |e_k| \tag{40}$$

$$\mathrm{TV}(u) = \sum_{k=1}^{N-1} |\Delta u_k| \tag{41}$$

MAE (Equation (38)) and RMSE (Equation (39)) measure instantaneous tracking accuracy: the former is insensitive to outliers, while the latter penalizes large errors more heavily; both are computed over the steady-state interval (3500–5000 s). IAE (Equation (40)) integrates the error over time and reflects cumulative deviation across the full process; it is the primary metric for evaluating long-term controller tracking performance. $\mathrm{TV}(u)$ (Equation (41)) quantifies the total variation of control actions; a smaller value indicates smoother choke valve adjustments, which reduces actuator wear.

5.3. Network Configuration and Skip Connection Mechanism

5.3.1. Final Network Configuration

After hyperparameter search, the final PIFG-NN configuration is as follows: the fully connected backbone has four hidden layers with 64 neurons each and Tanh activation; the GRU gated unit has a hidden dimension of 32; skip connections add the input layer directly to the outputs of the second and fourth hidden layers before feeding into the next layer. The total number of network parameters is approximately 2.1×10^4, which keeps inference latency low while retaining sufficient expressive capacity.

5.3.2. Physical Mechanism of Skip Connections

Multiphase drilling flow is a stiff system with multiple timescales: the gas-phase migration time constant (~tens of seconds) differs from the liquid-phase mass accumulation time constant (~hundreds of seconds) by approximately one order of magnitude, causing the physics-loss gradients in conventional PINNs to vanish easily during backpropagation.

Skip connections open up the gradient flow pathway by establishing direct feature mapping channels between network layers. Specifically, let the output of layer l be $\mathbf{h}^{(l)}$; with skip connections:

$$\mathbf{h}^{(l+1)} = \sigma\left(W^{(l+1)}\left[\mathbf{h}^{(l)} \oplus \mathbf{x}_{\text{in}}\right] + b^{(l+1)}\right) \tag{42}$$

where $\mathbf{x}_{\text{in}}$ is the original input feature and $\oplus$ denotes concatenation. This structure allows gradient signals from the mass conservation and momentum conservation ODE physics equations to propagate directly to shallow-layer parameters without attenuation, so physical safety constraints (e.g., non-negative mass) remain active throughout training.

To quantify the effect of skip connections, an ablation study trains an otherwise identical PIFG-NN variant without skip connections (i.e., a standard sequential fully connected backbone) under the same hyperparameters, data split, and training schedule. Table 3 compares the two architectures in terms of training convergence and 30-step autoregressive prediction accuracy.

Table 3. Ablation study: effect of skip connections on PIFG-NN training and prediction performance.

Architecture	Conv. Loss	m_{ga} MAE (kg)	m_{la} MAE (kg)	m_{dp} MAE (kg)
With skip connections	3.12×10^{-4}	18.7	12.3	8.9
Without skip connections	8.47×10^{-4}	53.4	34.1	22.6

Removing skip connections increases the converged total loss by 2.7× and degrades the 30-step autoregressive prediction MAE by 2.5–2.9× across all three state variables. The sharpest degradation is in gas-phase mass m_{ga} (18.7 vs. 53.4 kg), which has the strongest nonlinearity and the fastest time constant. Without the direct gradient pathway, physics-loss gradients from the ODE constraint are attenuated through four sequential layers, causing the network to under-learn gas-phase dynamics while over-fitting to the slower liquid-phase modes. These results confirm that skip connections are a structurally necessary component for learning multi-timescale stiff dynamics in physics-informed networks, not merely an architectural convenience.

5.4. Long-Horizon Autoregressive Prediction Performance Validation

Fifty independent long-period control trajectories not used during training are sampled from the test set. For each trajectory, given the initial true state $\mathbf{x}_k$, PIFG-NN performs autoregressive prediction for 30 consecutive steps (i.e., 1500 s):

$$\hat{\mathbf{x}}_{k+i}^{\text{PIFG-NN}} = f_\theta^{\text{PIFG-NN}}\left(\hat{\mathbf{x}}_{k+i-1}^{\text{PIFG-NN}}, \mathbf{u}_{k+i-1}\right), \quad i = 1, 2, \ldots, 30 \tag{43}$$

Only the first step $i = 1$ uses true state feedback; all subsequent steps rely entirely on the model's own previous predictions, thus representing the most demanding open-loop evaluation scenario with no state updates within the MPC prediction horizon.

As shown in Figure 5, the PIFG-NN trajectories for all three state variables closely match the RK4 high-fidelity solution over 30 steps (1500 s) of autoregressive prediction without state updates, and no numerical divergence or physical violations such as negative mass are observed. This accuracy is attributable to two design choices: the embedded mass conservation ODE constraint keeps the error in gas-phase mass m_{ga}—which exhibits the strongest nonlinearity—within the physically feasible domain, while the skip connections mitigate gradient vanishing in the multi-timescale stiff system so that slow dynamics are fully captured. These results confirm that the open-loop long-horizon prediction accuracy of PIFG-NN meets the modeling requirements of the MPC prediction horizon (3 steps, 150 s).

https://doi.org/10.3390/pr14081240

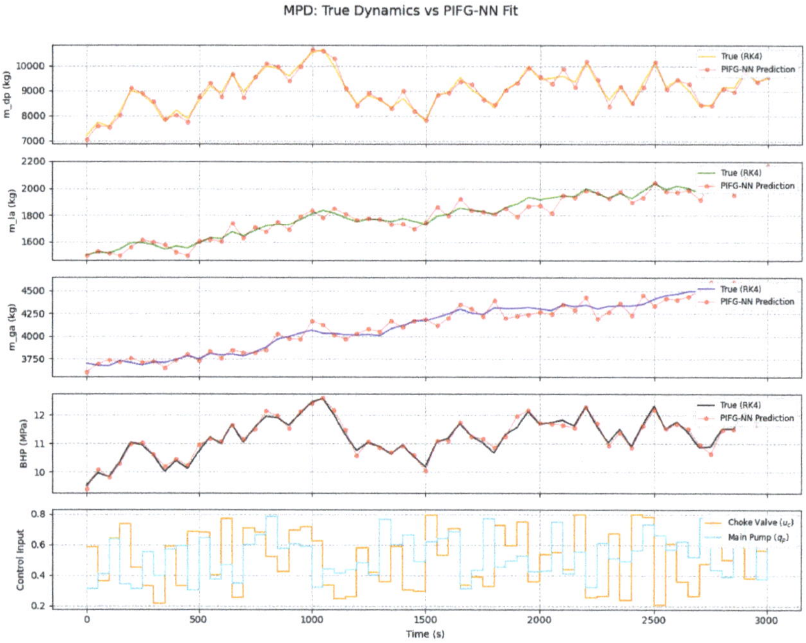

Figure 5. Comparison between RK4 numerical solution and PIFG-NN long-horizon autoregressive prediction (30 steps, 1500 s).

6. Simulation Results and Robustness Analysis

6.1. Baseline Performance Experiment (Noise-Free Ideal Conditions)

Under noise-free conditions, the simulation environment is initialized with standard parameters, and state feedback is taken directly from the RK4 integrator output. PINC-MPC, SLMPC, and ODE-NMPC (Section 4.5) are each run for a simulation duration of 5000 s with a control update period of 50 s, repeated five times with fixed random seeds 1–5, and the mean values are reported as deterministic baseline results.

Table 4 reveals that ODE-NMPC achieves the best tracking accuracy (IAE = 11.70 MPa·s) and serves as the theoretical upper bound, since its prediction model is identical to the plant. PINC-MPC closely approximates this upper bound (IAE = 12.90 MPa·s, only 10.3% higher), confirming that the PIFG-NN surrogate introduces minimal approximation error. SLMPC lags far behind (IAE = 63.40 MPa·s), showing the cost of linearization. Figure 6 shows the closed-loop control comparison.

This near-optimal accuracy comes with an 11.1× speedup: PINC-MPC's median solve time is 16.81 ms versus ODE-NMPC's 187.3 ms. Across the full 5000 s simulation (100 control steps), PINC-MPC accumulates a total solve time of only ∼1.68 s versus ∼18.73 s for ODE-NMPC—a 91% reduction. The computational bottleneck of ODE-NMPC lies in its need for $N_p \times$ steps_per_action $= 15$ full RK4 integration calls per cost function evaluation, compounded by the SLSQP optimizer's finite-difference gradient estimation requiring $2N_p + 1 = 7$ evaluations per iteration. PINC-MPC replaces the RK4 calls with a single neural network forward pass (approximately 0.1 ms per step), while CasADi/IPOPT provides exact gradients via symbolic differentiation, eliminating finite-difference overhead. These results confirm that PIFG-NN retains the accuracy of the full nonlinear ODE model while achieving real-time feasibility that the ODE model itself cannot provide.

https://doi.org/10.3390/pr14081240

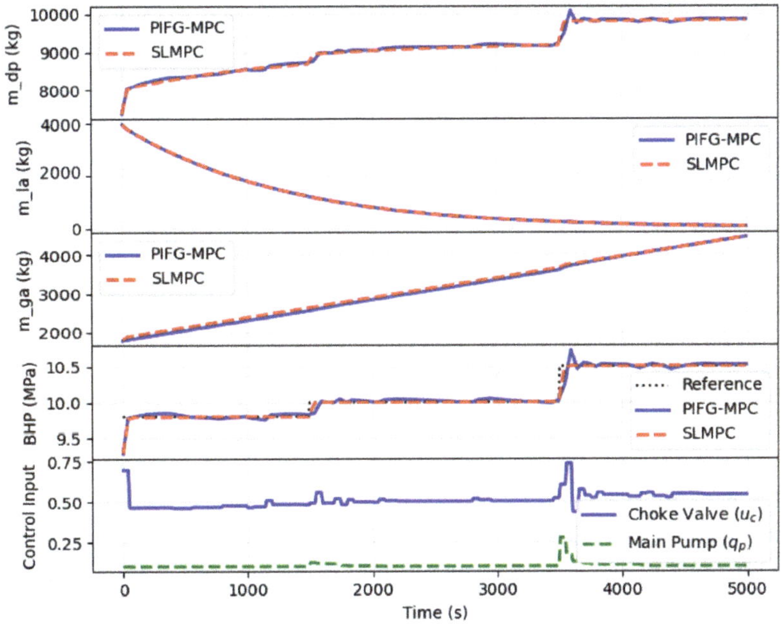

Figure 6. Closed-loop control comparison between PINC-MPC and SLMPC under noise-free conditions. The BHP reference trajectory consists of three step changes: $p_{\text{ref}} = 9.8$ MPa (0–1500 s), 10.0 MPa (1500–3500 s), and 10.5 MPa (3500–5000 s).

Table 4. Performance comparison under noise-free conditions (PINC-MPC vs. ODE-NMPC vs. SLMPC).

Controller	MAE (MPa)	RMSE (MPa)	IAE (MPa·s)	TV (u)	Solve (ms)
ODE-NMPC (upper bound)	0.0078	0.0412	11.70	1.2854	187.3
PINC-MPC	0.0086	0.0447	12.90	1.3127	16.81
SLMPC	0.0423	0.0747	63.40	8.3166	41.78

6.2. Measurement Noise Robustness Experiment

Measurement noise is modeled as additive white Gaussian noise:

$$\tilde{\mathbf{x}}_k = \mathbf{x}_k + \boldsymbol{\epsilon}_k, \quad \boldsymbol{\epsilon}_k \sim \mathcal{N}\Big(\mathbf{0}, \operatorname{diag}(\sigma_i^2)\Big) \tag{44}$$

The noise standard deviation is set as a percentage of each state variable's nominal value, $\sigma_i = \alpha \cdot x_{i,\text{nominal}}$, with $\alpha \in \{1\%, 3\%, 5\%\}$.

This additive white Gaussian noise model is a standard representation of downhole measurement uncertainty in MPD control studies [6,7]. In practice, downhole pressure sensors are subject to temperature drift, vibration-induced noise, and telemetry quantization errors; the 1–5% noise-to-signal range is consistent with reported measurement uncertainties of pressure-while-drilling (PWD) tools deployed in offshore wells [7]. Real sensor noise may exhibit non-Gaussian characteristics (e.g., spikes from shock events or mud-pulse interference), but the Gaussian model is a tractable and widely adopted baseline for evaluating controller robustness. For each noise level, 10 independent trials are conducted, and the mean and standard deviation of IAE are reported. An independent two-sample t-test

https://doi.org/10.3390/pr14081240

(two-tailed, significance level $\alpha = 0.05$) is used to compare the performance difference between the two controllers. Figure 7 shows the statistical distributions.

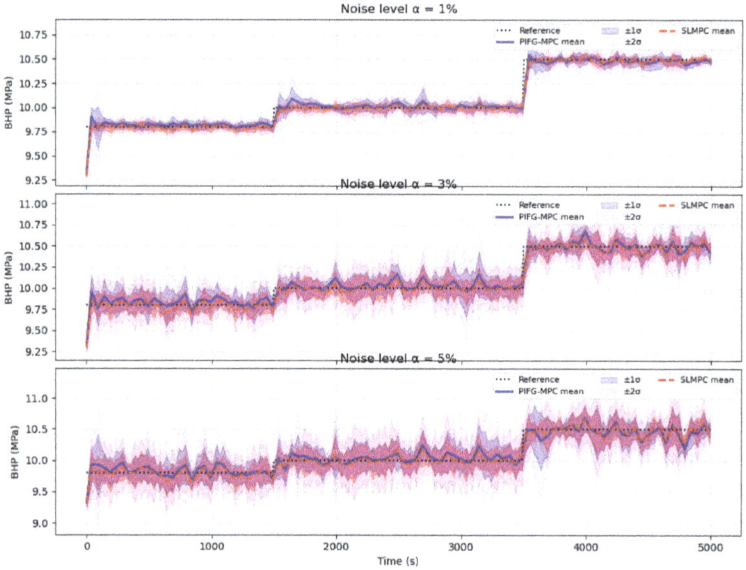

Figure 7. Statistical distributions of 10 repeated trials at different noise levels (mean ± confidence interval).

As shown in Table 5, the mean IAE of both controllers grows roughly linearly as noise increases from 1% to 5%; the growth factor for SLMPC (3.42×) is slightly higher than for PINC-MPC (3.25×). The more telling difference is in the standard deviation: at 5% noise, SLMPC's standard deviation (43.54) is 1.34 times that of PINC-MPC (32.43). The ODE loss term in PIFG-NN acts as a natural filter against states that violate mass conservation, giving MPC a more stable prediction basis. It should be noted that no state estimator or filter is included in the control loop; the mean IAE of both controllers therefore rises substantially after noise is introduced (PINC-MPC from 12.90 to 53.54 MPa·s), which is the expected behavior of an unfiltered closed-loop system. The advantage of PINC-MPC lies in its consistently lower IAE standard deviation, reflecting greater reliability under identical noise conditions. The noise-free results represent an ideal upper bound; the noisy results are more representative of practical conditions.

Table 5. IAE statistical comparison at different noise levels (mean ± standard deviation, 10 trials).

Noise Level α	PINC-MPC IAE (MPa·s)	SLMPC IAE (MPa·s)
1%	53.54 ± 6.55	77.08 ± 15.91
3%	140.34 ± 19.54	157.29 ± 22.02
5%	227.72 ± 32.43	263.68 ± 43.54

The repeated-trial noise experiments also provide an operational uncertainty quantification of the closed-loop system. The IAE standard deviation across 10 independent runs under identical noise conditions quantifies the stochastic spread of controller performance due to measurement uncertainty—a key concern in safety-critical MPD operations. As shown in Table 5, PINC-MPC has consistently lower IAE standard deviation than SLMPC at all noise levels (e.g., 32.43 vs. 43.54 at 5% noise). The mass conservation ODE constraint

https://doi.org/10.3390/pr14081240

embedded in PIFG-NN narrows the prediction distribution by rejecting physically implausible state trajectories: predictions that violate conservation laws are penalized during training, so the learned model concentrates its outputs on the physically feasible manifold.

However, this operational characterization does not constitute a formal probabilistic uncertainty quantification of the BHP prediction itself. PIFG-NN is a deterministic point-estimate model and does not output prediction intervals or confidence bounds. For safety-critical deployment where the engineer must verify that the predicted BHP remains within the pore pressure–fracture pressure window with a specified confidence level, more rigorous approaches—such as Monte Carlo Dropout [28], deep ensembles [29], or conformal prediction—would be required. These extensions are architecturally compatible with PIFG-NN (dropout can be inserted between hidden layers; ensembles require training multiple instances) and represent an important direction for future work.

6.3. Computational Real-Time Performance and Solver Efficiency Analysis

The MPC optimization solve time at each control step (control period 50 s) is recorded over the 5000 s simulation. Two solver backends are compared: CasADi/IPOPT constructs exact gradients via symbolic graph automatic differentiation and solves the nonlinear program using an interior-point method; SciPy/SLSQP relies on finite-difference numerical gradients and invokes a sequential quadratic programming solver.

As shown in Figure 8, the median single-step solve time for CasADi/IPOPT is 16.81 ms (maximum 32.40 ms), while that for SciPy/SLSQP is 41.78 ms (maximum 162.42 ms). The median ratio is 2.49× and the maximum ratio is 5.01×. Both approaches have maximum solve times well below the 50 s control period, satisfying real-time requirements. CasADi obtains exact gradients for all variables in a single forward pass through the symbolic graph, whereas SLSQP's finite-difference gradient estimation scales linearly with the number of decision variables—this is the primary cause of the timing difference.

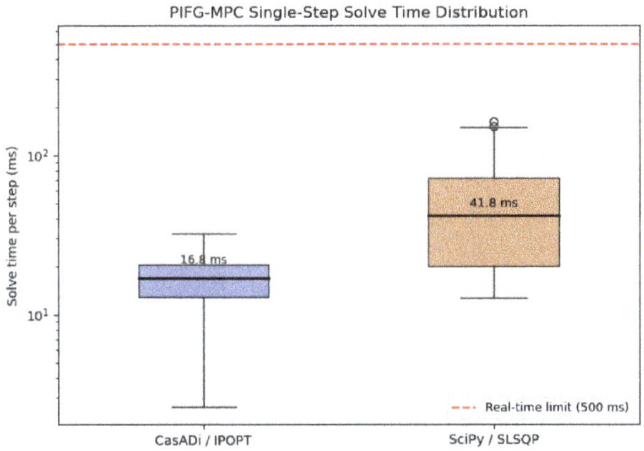

Figure 8. PINC-MPC single-step solve time distribution (CasADi/IPOPT vs. SciPy/SLSQP).

6.4. Pipe Connection Pump Stop Compensation Control

At $t = 2000$ s during the simulation, the main mud pump flow rate q_p is forced to drop to zero within 30 s, simulating pump shutdown during a pipe connection operation. The pump stop causes the annular frictional pressure drop Δp_{fric} to vanish instantaneously; without timely compensation, BHP drops sharply, posing a gas influx risk.

Pipe connection is one of the most frequent planned interruptions in MPD operations, typically occurring every 9–27 m of drilled interval and lasting 3–10 min [6,30]. The 30 s pump ramp-down profile used here matches the deceleration characteristics of modern

https://doi.org/10.3390/pr14081240

centrifugal mud pumps and is consistent with pump stop scenarios in published MPD simulation studies [30]. The primary hazard during pump stop is the instantaneous loss of annular friction pressure, which can reduce BHP by several MPa in deep wells, potentially underbalancing the formation and initiating gas influx. The choke valve must compensate for this loss by restricting annular flow—a response that requires anticipatory action due to finite valve actuation speed.

As shown in Figure 9, PINC-MPC rapidly closes the choke valve within 10 s, limiting the maximum BHP deviation to within 0.30 MPa. This response stems from the fact that PIFG-NN has learned the nonlinear mapping between the pump flow rate and frictional pressure drop, allowing it to correctly anticipate the frictional loss vanishing after pump stop within the prediction horizon and trigger compensatory action in advance. With $q_p = 0$ far from the normal linearization point, SLMPC produces severely inaccurate Jacobian estimates, resulting in a compensation delay of approximately 90 s and a maximum BHP deviation of 0.78 MPa, which risks triggering gas influx alarms.

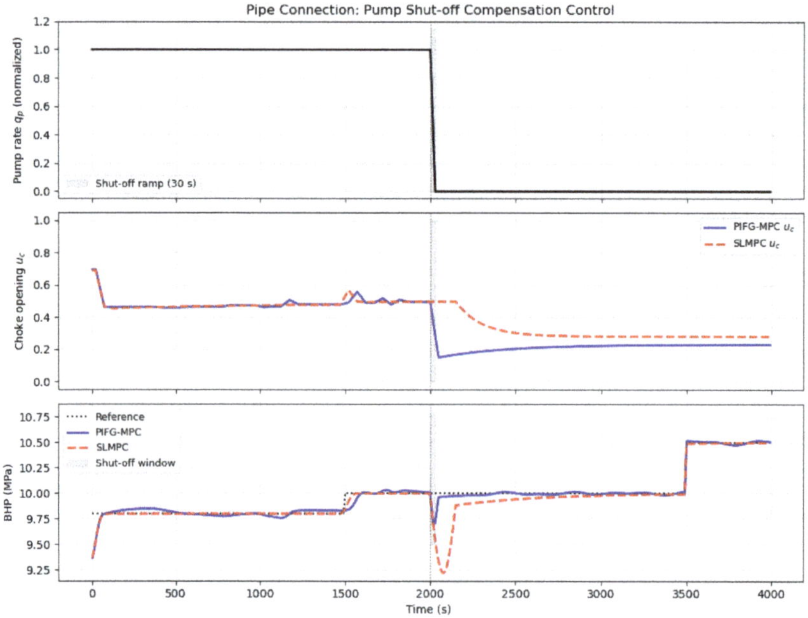

Figure 9. BHP compensation control response of PINC-MPC and SLMPC during pipe connection pump stop ($t = 2000$ s, BHP reference setpoint $p_{\text{ref}} = 10.0$ MPa).

6.5. Sudden Gas Influx Disturbance Suppression

During normal drilling ($t = 3000$ s), a pulse-shaped high-flow-rate disturbance lasting 100 s is applied to the formation gas influx term, causing the annular gas mass m_{ga} to surge, the mixed fluid density to decrease, and the hydrostatic column pressure to drop abruptly. This scenario simulates the early stage of a gas kick—one of the most hazardous well-control events encountered in the field.

The 100 s pulse duration represents the incipient stage of a gas kick, during which formation gas enters the wellbore before surface pit-gain alarms are triggered. Field studies report that early kick detection windows in MPD operations typically range from tens of seconds to a few minutes depending on formation permeability, differential pressure, and flow rate [4,31]. The pulse-shaped influx profile—rapid onset followed by decay as the near-wellbore pressure equalizes—is consistent with the transient inflow behavior modeled in MPD kick-detection literature [31]. This setup focuses the evaluation on the controller's early detection and rapid compensation capability. More severe scenarios, such as sustained influx or accelerating gas migration, remain important extensions for future work.

As shown in Figure 10, PINC-MPC identifies the abnormal growth of m_{ga} within 10 s, limits the BHP deviation to within 0.26 MPa by closing the choke valve, and recovers smoothly after the gas influx pulse subsides (approximately 300 s).

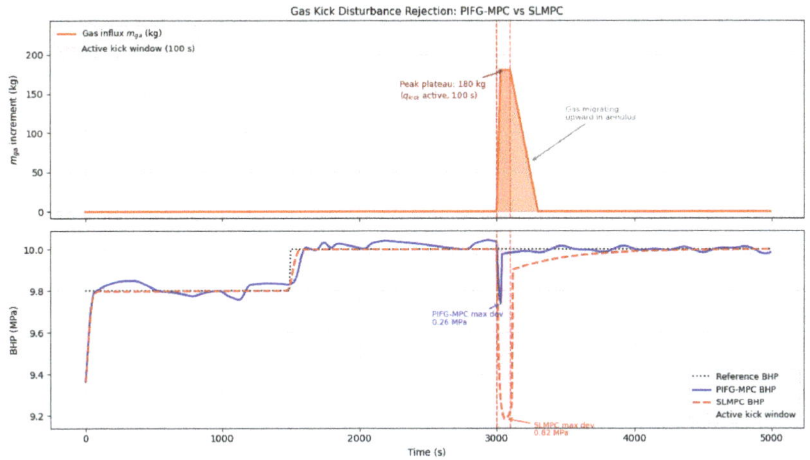

Figure 10. BHP suppression response of PINC-MPC and SLMPC under sudden gas influx disturbance ($t = 3000$ s, BHP reference setpoint $p_{\text{ref}} = 10.0$ MPa).

This anticipatory response originates from the mass conservation ODE embedded in PIFG-NN:

$$\frac{dm_{ga}}{dt} = q_{\text{kick}} - q_{\text{out,ga}} \tag{45}$$

The MPC can sense the impact of m_{ga} growth on the mixture density $\rho_{\text{mix}} = (m_{la} + m_{ga})/V_{\text{ann}}$ within the prediction horizon and apply compensation in advance. The linearized model of SLMPC fails to capture the nonlinear abrupt decrease in gas–liquid two-phase density coupling, resulting in a response delay of approximately 80 s and a maximum BHP deviation of 0.82 MPa (approximately 3.15 times that of PINC-MPC), which risks triggering an overflow alarm.

6.6. Control Weight Sensitivity Analysis

As noted in Section 4, PINC-MPC and SLMPC were each tuned independently and arrived at different optimal control variation weights ($w_u = 3.0$ vs. $w_u = 50.0$). To verify that PINC-MPC's performance advantage is not an artifact of this weight asymmetry, we evaluate PINC-MPC under $w_u = 50.0$—the same value used by SLMPC—and compare all four metrics against both the original PINC-MPC ($w_u = 3.0$) and SLMPC ($w_u = 50.0$) under noise-free conditions. Table 6 presents the weight sensitivity analysis results.

Even with w_u raised from 3.0 to 50.0, PINC-MPC still substantially outperforms SLMPC in IAE (29.10 vs. 63.40 MPa·s, a 54.1% reduction), confirming that the performance gap originates from prediction model quality rather than weight configuration. As expected, the higher w_u reduces TV(u) (from 1.3127 to 0.8815) at the cost of slower tracking (IAE increases from 12.90 to 29.10 MPa·s), illustrating the standard accuracy–smoothness trade-off in MPC cost function design. The independent-tuning protocol does not introduce systematic bias, and PINC-MPC's advantage holds across the w_u range tested.

Table 6. Control weight sensitivity analysis: PINC-MPC performance under matched ($w_u = 50$) and optimal ($w_u = 3$) weights, compared with SLMPC ($w_u = 50$).

Controller (w_u)	MAE (MPa)	RMSE (MPa)	IAE (MPa·s)	TV (u)
PINC-MPC ($w_u = 3$, optimal)	0.0086	0.0447	12.90	1.3127
PINC-MPC ($w_u = 50$, matched)	0.0194	0.0508	29.10	0.8815
SLMPC ($w_u = 50$, optimal)	0.0423	0.0747	63.40	8.3166

6.7. Prediction Horizon Sensitivity Analysis

The baseline configuration uses a prediction horizon of $N_p = 3$ steps (150 s). In real deep wells, gas slugs may take tens of minutes to migrate from the bottomhole to the surface, raising the concern that a 150 s window may be too short to capture such long-period dynamics. To investigate this, we vary $N_p \in \{3, 5, 10\}$ (corresponding to 150 s, 250 s, and 500 s, respectively) under noise-free conditions, with $N_c = N_p$ and all other parameters unchanged.

As shown in Table 7, extending the prediction horizon from $N_p = 3$ to $N_p = 10$ yields only a 25.6% reduction in IAE (12.90 to 9.60 MPa·s), with most of the gain already realized at $N_p = 5$. These diminishing returns are not evidence that longer horizons are unnecessary in general; they are a direct consequence of the lumped-parameter ODE model adopted in this study. In a lumped-parameter formulation, the entire wellbore is treated as a single control volume: a change in annular gas mass m_{ga} instantaneously alters the mixture density ρ_{mix} and thus the BHP—there is no spatial propagation delay. The dominant pressure dynamics therefore have time constants on the order of tens of seconds, and a 150 s window already captures the full response.

Table 7. Prediction horizon sensitivity analysis for PINC-MPC under noise-free conditions.

N_p	Window (s)	MAE (MPa)	RMSE (MPa)	IAE (MPa·s)	TV (u)	Solve (ms)
3	150	0.0086	0.0447	12.90	1.3127	16.81
5	250	0.0071	0.0389	10.65	1.2043	28.36
10	500	0.0064	0.0352	9.60	1.1580	53.92

This stands in contrast to real deep wells, where gas migration is a distributed process governed by the drift-flux or two-fluid equations with spatial extent. A gas slug that enters at 5000 m depth must travel upward through the annulus; as it rises, it expands under decreasing hydrostatic pressure, progressively accelerating the BHP reduction. This migration can take tens of minutes or longer, and a controller based on a distributed model would require a prediction horizon commensurate with the gas transit time to anticipate the delayed pressure wave.

The 150 s horizon is therefore sufficient for the lumped-parameter model used here, but may be inadequate for higher-fidelity distributed wellbore simulators. From a computational standpoint, extending the horizon is feasible for PINC-MPC: the PIFG-NN per-step inference cost is fixed, and the NLP solver complexity scales linearly with the number of decision variables ($2N_p$). At $N_p = 10$, the median solve time (53.92 ms) remains well within the 50 s control cycle. CasADi's symbolic automatic differentiation reduces the solver's sensitivity to horizon length compared with finite-difference approaches, making PINC-MPC more scalable to longer horizons than SLMPC. Future work incorporating the drift-flux model should re-evaluate the prediction horizon to match the gas slug transit timescale of the target well geometry.

https://doi.org/10.3390/pr1408124

Across all three robustness scenarios—measurement noise (1–5%), pipe connection pump stop, and sudden gas influx—PINC-MPC consistently outperforms SLMPC in both tracking accuracy and control smoothness. The physics-informed structure of PIFG-NN provides a natural filtering effect against noise, anticipatory compensation for mass conservation violations, and nonlinear density coupling that linearization-based methods cannot replicate. These results collectively demonstrate that the PINC-MPC framework is robust to the principal disturbance classes encountered in MPD field operations.

7. Conclusions and Future Work

This paper proposed the PINC-MPC framework for real-time BHP prediction and control in managed pressure drilling, addressing the dual challenge of maintaining physical consistency and meeting real-time computational constraints. The key results are as follows. PIFG-NN suppresses error accumulation over 30-step (1500 s) long-horizon autoregressive prediction, with predicted trajectories closely matching the RK4 high-fidelity solution. In closed-loop control, the IAE under noise-free conditions is reduced by approximately 79.7% compared with SLMPC (12.90 vs. 63.40 MPa·s), and $\mathrm{TV}(u)$ decreases by a factor of approximately 6.3. Under 1–5% noise, the IAE standard deviation remains consistently lower than that of SLMPC, reflecting the natural filtering effect of physics constraints on measurement noise. The median single-step solve time is 16.81 ms, representing an 11.1× speedup over the mechanistic model-based approach (187.3 ms); the total time for a 5000 s full-cycle closed-loop experiment is only 1.68 s compared with 18.73 s, an improvement of approximately 91%, meeting the real-time requirement of the 50 s control cycle. Under pipe connection pump stop and sudden gas influx conditions, the maximum BHP deviations are 0.30 MPa and 0.26 MPa, respectively, compared with 0.78 MPa and 0.82 MPa for SLMPC.

The main contributions of this paper are (1) applying the PINN framework to real-time BHP control in MPD via a PIFG-NN three-branch fusion architecture (static feature, temporal feature, and skip-connection sub-networks) that effectively suppresses error accumulation over 30-step long-horizon autoregressive prediction; (2) designing a multi-objective physics-constrained loss function embedding mass conservation ODE constraints, empirical flow rate constraints, and output feasibility constraints, with a gradient-balancing adaptive weighting strategy that resolves mutual suppression between physics and data-fitting losses; and (3) constructing a complete training–control–validation pipeline validated across closed-loop control, noise robustness, pump stop, and gas influx scenarios, demonstrating both real-time feasibility and robustness under the principal disturbance classes encountered in MPD operations.

It is important to note that all quantitative findings reported above are derived exclusively from a simulation environment rather than from physical hardware or field operations. While the simulation framework is grounded in real offshore drilling log data and validated against a high-fidelity RK4 solver, the gap between simulation and real-world deployment remains an open challenge. Transitioning the proposed framework to an operational MPD system would require additional validation against higher-fidelity multiphase flow simulators, multi-well field datasets, and hardware-in-the-loop experiments before any deployment decision can be justified.

This work has several limitations. A domain gap exists between training (real drilling log data) and validation (RK4 ODE simulation). The simplified lumped-parameter model does not explicitly capture gas expansion effects, potentially underestimating gas influx severity under real field conditions. Validation against higher-fidelity multiphase flow simulators or field data is recommended before deployment.

https://doi.org/10.3390/pr14081240

Regarding generalization, the pump stop and gas influx scenarios already constitute out-of-distribution tests, as the training set contains only normal drilling operations. The controller's ability to maintain BHP deviations within 0.30 MPa and 0.26 MPa demonstrates that the embedded physics constraints provide meaningful extrapolation beyond the training envelope. However, cross-well generalization—involving different formation properties, wellbore geometry, and fluid characteristics—has not been tested and will likely require transfer learning or domain adaptation strategies.

Future work will pursue four directions. First, the lumped-parameter ODE will be replaced with a drift-flux or two-fluid model to achieve higher physical fidelity and directly address the underestimation of gas expansion effects identified above. Second, the approach will be extended to diverse well conditions and operational environments: by incorporating multi-well field data through transfer learning, the pre-trained PIFG-NN can be fine-tuned to wells with different geometries (vertical, directional, or horizontal), formation properties (permeability, pore pressure gradient), and mud systems without retraining from scratch; online learning strategies can further enable incremental model adaptation as formation conditions evolve during a single drilling campaign, reducing reliance on offline pre-training. Third, online learning strategies will be explored for incremental adaptation to changing formation conditions. Fourth, hardware-in-the-loop testing will be conducted on embedded industrial platforms (e.g., PLCs or edge computing units), combined with model compression techniques such as quantization and pruning, to assess deployment feasibility on real-time control hardware.

Author Contributions: Conceptualization, C.Z., X.S. and M.Z.; Methodology, C.Z. and X.Y.; Software, L.W., Z.Z. and C.Z.; Validation, C.Z. and X.Y.; Formal analysis, G.L. and X.Y.; Investigation, X.S. and M.Z.; Resources, M.Z.; Data curation, Z.Z.; Writing—original draft, Z.Z.; Writing—review and editing, Z.Z., G.L. and X.Y.; Visualization, L.W.; Supervision, L.W., G.L. and X.S.; Project administration, L.W. and X.S.; Funding acquisition, L.W. All authors have read and agreed to the published version of the manuscript.

Funding: This research was funded by the National Science and Technology Major Project of the Ministry of Science and Technology of China (No. 2025ZD1403204).

Data Availability Statement: The data and code supporting the findings of this study are available from the corresponding author upon reasonable request.

Conflicts of Interest: The authors declare no conflicts of interest.

Abbreviations

The following abbreviations are used in this manuscript:

MPD	Managed Pressure Drilling
BHP	Bottomhole Pressure
PINN	Physics-Informed Neural Network
MPC	Model Predictive Control
PINC-MPC	Physics-Informed Fusion-Gated MPC
PIFG-NN	Physics-Informed Fusion-Gated Neural Network
SLMPC	Successive Linearization MPC
GRU	Gated Recurrent Unit
FCNN	Fully Connected Neural Network
ODE	Ordinary Differential Equation
IAE	Integral Absolute Error
TV	Total Variation

RK4	Fourth-Order Runge–Kutta
MAE	Mean Absolute Error
RMSE	Root Mean Square Error
PID	Proportional–Integral–Derivative
CFD	Computational Fluid Dynamics
NLP	Nonlinear Programming

Appendix A. Wellbore Physical Model Parameters

Table A1. Physical model parameters of the wellbore system.

Parameter	Symbol	Unit	Parameter	Symbol	Unit
Geometric parameters			**Two-phase flow parameters**		
Well depth	L	m	Gas volume fraction	α_g	—
Wellbore diameter	D	m	Liquid volume fraction	α_l	—
Drill pipe outer diameter	D_p	m	Gas velocity	v_g	m/s
Annular volume	V_a	m^3	Liquid velocity	v_l	m/s
Drill string internal volume	V_{dp}	m^3	Mixture velocity	v_m	m/s
Depth position	x	m	Gas slip velocity	v_{gr}	m/s
Spatial step size	Δx	m	Slip velocity coefficient	C_0	—
Time step size	Δt	s	Bubble correction factors	C_2, C_3, C_4	—
Fluid property parameters			**ODE model coefficients**		
Gravitational acceleration	g	m/s^2	Mud pump coefficient	k_{pump}	kg/s
Gas density	ρ_g	kg/m^3	Gas outflow coefficient	k_2	m^3/s
Liquid density	ρ_l	kg/m^3	Formation influx coefficient	k_{res}	m^3/s
Mixture density	ρ_m	kg/m^3	Bit nozzle flow coefficient	k_{bit}	$kg/(s \cdot MPa^{1/2})$
Liquid dynamic viscosity	μ_l	Pa·s	Liquid outflow coefficient	k_5	m^3/s
Wall roughness	ε	m			
Wellhead back-pressure	p_0	MPa	**System state variables**		
Friction factor	f_D	—	Drill string liquid mass	m_{dp}	kg
Reynolds number	Re	—	Annular liquid mass	m_{la}	kg
			Annular gas mass (gas influx)	m_{ga}	kg
Pressure and density variables					
Pressure at depth x	p_x	Pa	**Flow rate variables**		
Annular pressure	p_{la}	MPa	Pump injection mass flow rate	w_{pump}	kg/s
Drill string pressure	p_{dp}	MPa	Bit flow rate	w_{bit}	kg/s
Annular liquid density	ρ_{la}	kg/m^3	Formation gas influx rate	$w_{res,g}$	kg/s
Drill string liquid density	ρ_{dp}	kg/m^3	Formation liquid influx rate	$w_{res,l}$	kg/s
Momentum change term	Δp_m	Pa	Liquid outflow rate	$w_{out,l}$	kg/s
Wall friction force	F_f	Pa/m	Gas outflow rate	$w_{out,g}$	kg/s
			Control inputs		
			Choke valve opening	u_c	—
			Mud pump flow rate	q_p	m^3/s

Appendix B. Dataset Statistical Characteristics

Table A2. Statistical characteristics of the training dataset.

Parameter	Min	Max	Mean	Std
Bottomhole pressure P (MPa)	50.04	67.04	58.54	4.91
Gas volume fraction α_g ($\times 10^{-3}$)	0.12	8.74	1.53	1.07
Liquid velocity v_L (m/s)	0.38	0.92	0.62	0.16
Depth x (m)	4417.00	6003.00	5210.00	458.27
Time t (s)	13,860.00	1,111,680.00	516,240.64	319,720.97
Rate of penetration ROP (m/h)	0.90	30.00	6.67	3.72
Rotary speed R (RPM)	54	94	87	5.42
Weight on bit WOB (kN)	6	192	74	19
Surface torque ST (kN·m)	5.60	20.80	12.32	2.23

Table A2. *Cont.*

Parameter	Min	Max	Mean	Std
Flow rate Q (m^3/s)	0.0242	0.0295	0.0267	0.001
Mud density ρ_L (g/cm^3)	1.25	1.35	1.31	0.02
Mud viscosity μ_L (mPa·s)	13	20	16.40	1.61
Standpipe pressure SPP (MPa)	17.40	20.70	19.52	0.50
Mud temperature T_L (°C)	27.20	60.20	54.94	4.65
Mud pit volume MPV (m^3)	91.80	133.80	109.71	5.30

References

1. Zuber, N.; Findlay, J.A. Average Volumetric Concentration in Two-Phase Flow Systems. *J. Heat Transf.* **1965**, *87*, 453–468. [CrossRef]
2. Shi, H.; Holmes, J.A.; Durlofsky, L.J.; Aziz, K.; Diaz, L.R.; Alkaya, B.; Oddie, G. Drift-Flux Modeling of Two-Phase Flow in Wellbores. *SPE J.* **2005**, *10*, 24–33. [CrossRef]
3. Evje, S.; Fjelde, K.K. Hybrid Flux-Splitting Schemes for a Two-Phase Flow Model. *J. Comput. Phys.* **2002**, *175*, 674–701. [CrossRef
4. Zhou, J.; Nygaard, G.; Godhavn, J.M.; Breyholtz, Ø.; Vefring, E.H. Adaptive observer for kick detection and switched contro for bottomhole pressure regulation and kick attenuation during managed pressure drilling. In *Proceedings of the 2010 America Control Conference*; IEEE: New York, NY, USA, 2010; pp. 3765–3770. [CrossRef]
5. Nygaard, G.; Nævdal, G. Nonlinear model predictive control scheme for stabilizing annulus pressure during oil well drilling. *J. Process Control* **2006**, *16*, 719–732. [CrossRef]
6. Kaasa, G.O.; Stamnes, Ø.N.; Imsland, L.; Aamo, O.M. Simplified Hydraulics Model Used for Intelligent Estimation of Downhol Pressure for a Managed-Pressure-Drilling Control System. *SPE Drill. Complet.* **2012**, *27*, 127–138. [CrossRef]
7. Nandan, A.; Imtiaz, S. Nonlinear model predictive control of managed pressure drilling. *ISA Trans.* **2017**, *69*, 307–314. [CrossRef
8. Park, J.; Price, C.; Pixton, D.; Aghito, M.; Nybø, R.; Bjørkevoll, K.; Hedengren, J.D. Model predictive control and estimation o managed pressure drilling using a real-time high fidelity flow model. *ISA Trans.* **2020**, *105*, 256–268. [CrossRef]
9. Zhang, C.; Zhang, R.; Zhu, Z.; Song, X.; Su, Y.; Li, G.; Han, L. Bottom hole pressure prediction based on hybrid neural network and Bayesian optimization. *Pet. Sci.* **2023**, *20*, 3712–3722. [CrossRef]
10. Wang, Y.; Zhu, Z.; Song, X.; Liu, Z.; Zhang, C. Drilling Conditions Classification Model Based on an Expert-Knowledge-Fuse Convolutional Neural Network with a Multihead Attention Mechanism. *SPE J.* **2025**, *30*, 7539–7557. [CrossRef]
11. Wang, Z.; Song, X.; Kuru, E.; Li, H.; Zhu, Z.; Zhang, C.; Liu, Z. An explainable drilling status recognition method based o Attention-CNN-BiGRU with physics constraints. *Geoenergy Sci. Eng.* **2026**, *256*, 214088. [CrossRef]
12. Raissi, M.; Perdikaris, P.; Karniadakis, G.E. Physics-informed neural networks: A deep learning framework for solving forwar and inverse problems involving nonlinear partial differential equations. *J. Comput. Phys.* **2019**, *378*, 686–707. [CrossRef]
13. Cai, S.; Mao, Z.; Wang, Z.; Yin, M.; Karniadakis, G.E. Physics-informed neural networks (PINNs) for fluid mechanics: A review *Acta Mech. Sin.* **2021**, *37*, 1727–1738. [CrossRef]
14. Antonelo, E.A.; Camponogara, E.; Seman, L.O.; Jordanou, J.P.; de Souza, E.R.; Hübner, J.F. Physics-informed neural nets fo control of dynamical systems. *Neurocomputing* **2024**, *579*, 127419. [CrossRef]
15. Kittelsen, J.E.; Antonelo, E.A.; Camponogara, E.; Imsland, L.S. Physics-Informed Neural Networks with skip connections fo modeling and control of gas-lifted oil wells. *Appl. Soft Comput.* **2024**, *158*, 111603. [CrossRef]
16. Teng, T.; Wang, Y.; Chen, Y.; Li, Y.; Liu, X. Permeability and Energy Evolution of Coal under Unloading Confining Pressure. *Int. J. Geomech.* **2026**, *26*, 04025341. [CrossRef]
17. Gao, Y.; Ren, Z.; Chen, M.; Jiang, H.; Ding, S. Coupled geomechanical-thermal simulation for oil sand reservoirs with shal barriers under hot water injection in vertical well-assisted SAGD wells. *J. Pet. Sci. Eng.* **2022**, *208*, 109644. [CrossRef]
18. Li, X.; Zhu, L.; Liu, J.; Cao, Z.; Xue, Y.; Dang, F. Fracture evolution and mechanical deterioration of granite under cyclic therma and liquid nitrogen cryogenic impact. *Phys. Fluids* **2025**, *37*, 086652. [CrossRef]
19. Chen, Z.; Badrinarayanan, V.; Lee, C.Y.; Rabinovich, A. GradNorm: Gradient Normalization for Adaptive Loss Balancing in Dee Multitask Networks. *arXiv* **2018**, arXiv:1711.02257. [CrossRef]
20. Wang, S.; Yu, X.; Perdikaris, P. When and why PINNs fail to train: A neural tangent kernel perspective. *J. Comput. Phys.* **202** *449*, 110768. [CrossRef]
21. Ma, B.; Zhang, C.; Zhou, M.; Pan, T.; Song, X.; Zhu, Z. Physics-Informed Multitask Learning with Dynamic Weighting for Efficien Prediction of Extended-Reach Wellbore Cleanliness. *SPE J.* **2026**, *31*, 116–133. [CrossRef]
22. Camacho, E.F.; Bordons, C. *Model Predictive Control*, 2nd ed.; Springer: London, UK, 2004.
23. Rawlings, J.B.; Mayne, D.Q.; Diehl, M.M. *Model Predictive Control: Theory, Computation, and Design*, 2nd ed.; Nob Hill Publishing LLC: Santa Barbara, CA, USA, 2024.

https://doi.org/10.3390/pr1408124

. Mayne, D.Q.; Rawlings, J.B.; Rao, C.V.; Scokaert, P.O.M. Constrained model predictive control: Stability and optimality. *Automatica* **2000**, *36*, 789–814. [CrossRef]

. Qin, S.J.; Badgwell, T.A. A survey of industrial model predictive control technology. *Control Eng. Pract.* **2003**, *11*, 733–764. [CrossRef]

. Allgöwer, F.; Findeisen, R.; Nagy, Z.K. Nonlinear model predictive control: From theory to application. *J. Chin. Inst. Chem. Eng.* **2004**, *35*, 299–315.

. Ohtsuka, T. A continuation/GMRES method for fast computation of nonlinear receding horizon control. *Automatica* **2004**, *40*, 563–574. [CrossRef]

. Gal, Y.; Ghahramani, Z. Dropout as a Bayesian Approximation: Representing Model Uncertainty in Deep Learning. In *Proceedings of the 33rd International Conference on Machine Learning (ICML)*; PMLR: London, UK, 2016; pp. 1050–1059.

. Lakshminarayanan, B.; Pritzel, A.; Blundell, C. Simple and Scalable Predictive Uncertainty Estimation using Deep Ensembles. In Proceedings of the Advances in Neural Information Processing Systems (NeurIPS), Long Beach, CA, USA, 4–9 December 2017; Volume 30. [CrossRef]

. Godhavn, J.M. Control Requirements for Automatic Managed Pressure Drilling System. *SPE Drill. Complet.* **2010**, *25*, 336–345. [CrossRef]

. Hauge, E.; Aamo, O.M.; Godhavn, J.M. Model-based estimation and control of in/out-flux during drilling. In *Proceedings of the 2012 American Control Conference (ACC)*; IEEE: New York, NY, USA, 2012; pp. 4909–4914. [CrossRef]

https://doi.org/10.3390/pr14081240

processes

MDPI

Article

Real-Time Lexicographic MPC with Online Correction for Intelligent Drill-Bit Rotary Valves in Mud-Pulse Telemetry

Xuecheng Dong [1,2], Liangzhu Yan [1,2,*], Lingyun Wang [3], Zhiyuan Zhou [1,2], Youyan Jian [1,2] and Run Li [1,2]

[1] State Key Laboratory of Oil and Gas Reservoir Geology and Exploitation, Chengdu 610059, China; dongxc_cdut@cdut.edu.cn (X.D.)
[2] College of Energy Resources (Modern Industrial College of Shale Gas), Chengdu University of Technology, Chengdu 610059, China
[3] No. 10 Oil Production Plant of PetroChina Changqing Oilfield Company, Qingyang 745000, China; wanglingyunswpi@126.com
* Correspondence: ylzoffice@126.com

Abstract

Reliable front-end pressure-pulse generation is critical to mud-pulse telemetry because waveform distortion introduced at the rotary valve propagates through the telemetry chain and reduces downstream recoverability. This paper targets accurate and computationally tractable control of an intelligent drill-bit rotary valve under actuator limits, parameter drift, and downhole-like disturbances. A control-oriented electromechanical–hydraulic grey-box model is established, and a real-time lexicographic model predictive control (MPC) framework with candidate pre-screening, move blocking, and online correction/compensation is developed and compared with proportional–integral–derivative (PID) control and conventional MPC. Under a sampling period of $T_s = 20$ ms, the proposed controller reduces the step-tracking rise time from 2.18 s to 1.76 s and the steady-state pressure error from 0.1208 MPa to 0.0292 MPa relative to conventional MPC. In the pulse-output and mismatch–disturbance scenarios, it further maintains lower steady-state pressure error while reducing the cumulative input variation from 51.0 to 11.5 and from 121.5 to 19.5, respectively. The observed 99th-percentile and worst-case MATLAB workstation execution times remain below one sampling period, while supplementary mismatch–disturbance sensitivity maps indicate a favorable accuracy–timing compromise within the tested numerical envelope. These results support the proposed method as a simulation-validated candidate for low-complexity rotary-valve control and motivate subsequent bench/hardware-in-the-loop (HIL) validation rather than field-qualified deployment claims.

Keywords: intelligent drill-bit rotary valve; mud-pulse telemetry; grey-box modeling; lexicographic MPC; online disturbance compensation; embedded real-time control

1. Introduction

As unconventional oil and gas development extends toward long horizontal sections, high-temperature and high-pressure wells, and structurally complex reservoirs, drilling operations place increasingly stringent demands on stable downhole information acquisition, real-time transmission, and rapid response. Measurement while drilling (MWD) and logging while drilling (LWD) systems must deliver inclination, azimuth, tool-face angle, formation response, and key drilling parameters to the surface in a timely manner. Among available downhole telemetry technologies, mud-pulse transmission remains one of the most mature and widely deployed solutions.

In a mud-pulse telemetry chain, the intelligent drill-bit rotary valve is typically driven by an electric motor that rotates the valve core. By modulating the valve angle, the controller changes the throttling pressure drop and thereby generates identifiable pressure pulses in the drilling-fluid column. Open patent disclosures show that geometric changes in the rotary-valve opening directly modify the local pressure drop and the pulse-shaping process [1]. Mud-pulse channel-modeling studies further indicate that front-end waveform quality directly affects channel response and recoverability [2]. Fast simulation studies for downhole telemetry also show that chain modeling and controller design are tightly coupled [3]. Analyses of drill-string channel response similarly suggest that pressure-signal distortion degrades surface decoding quality [4]. Surface demodulation studies reach the same conclusion: front-end signal quality strongly influences downstream demodulation performance [5]. Therefore, the rotary valve is not merely an actuator; it is also a key waveform-shaping front-end in mud-pulse telemetry.

From an engineering perspective, the control performance of a rotary valve is commonly assessed through valve-angle tracking accuracy, pressure-pulse amplitude ΔP, carrier frequency f, and information coding rate R_b. Public patent disclosures indicate that the additional pressure-drop amplitude generated by a pulse generator can reach 100–300 psi, i.e., approximately 0.69–2.07 MPa; accordingly, the pressure-constraint window used in this paper is selected on the basis of that order of magnitude with an additional safety margin. Variable-forgetting-factor RLS equalization studies show that target-band recovery capability directly influences signal quality in time-varying telemetry channels [6]. Wavelet-network signal recovery results further show that multiscale band-energy distributions strongly affect the robust recovery of mud-pulse signals [7]. Fractional Fourier-transform equalization studies verify that phase and amplitude fidelity within a specific band are both important for decoding quality [8]. Similar evidence is provided by discrete Fourier-transform-based noise-suppression studies [9]. These observations imply that rotary-valve control errors are mapped onto pulse-waveform quality and waveform-level proxies associated with downstream recoverability rather than directly proving end-to-end decoding performance in the present study (Figure 1).

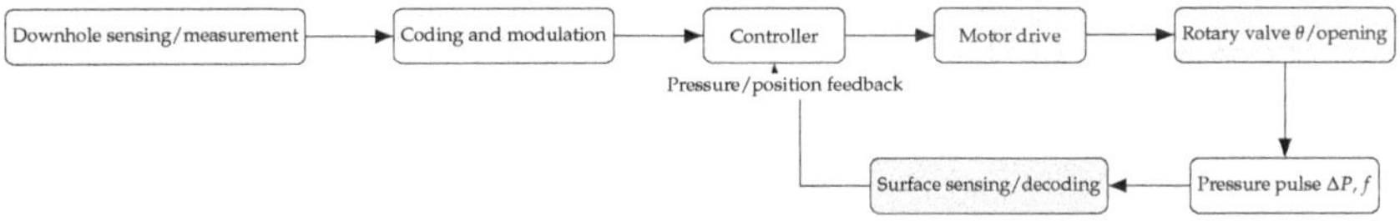

Figure 1. Closed-loop control chain of the rotary valve in mud-pulse telemetry.

Compared with conventional surface actuators, precise control of downhole rotary valves is subject to much harsher environmental constraints. Public MWD tool specifications indicate that downhole tools may operate at approximately 175 °C and 20,000 psi while also withstanding strong shocks and broadband vibration; under high-flow erosion and a high static-pressure background, wear of the valve-core surface, variation in friction coefficients, and clearance evolution continuously alter the valve–fluid coupling characteristics. More specifically, pump noise and mechanical noise reduce the signal-to-noise ratio and increase synchronization difficulty; fluid inertia and time-varying loads induce valve-position lag and waveform distortion; changes in flow rate, density, and local throttling characteristics magnify model mismatch; and limited downhole computation and power budgets shrink the real-time margin of complex optimization algorithms. These factors make rotary-valve control strongly coupled, strongly disturbed, and inherently time varying.

https://doi.org/10.3390/pr14101589

Existing control strategies for complex actuators and related motor-drive systems form a reasonably clear technical spectrum. PID/PI control is structurally simple and inexpensive to implement, but it tends to suffer from overshoot, oscillation, and steady-state degradation under strong disturbances and parameter drift. Fuzzy control reduces dependence on an accurate plant model, yet the design of rule bases and membership functions remains highly heuristic and often lacks cross-condition consistency. Adaptive control can improve environmental adaptability through online estimation, but its performance depends strongly on excitation conditions and measurement noise. By contrast, conventional model predictive control, including FCS-MPC, explicitly handles input, state, and switching constraints. Studies on model predictive power control in the stationary two-phase frame show that cost-function design directly affects dynamic behavior and steady-state ripple [10]. Dual-vector model predictive current control with parameter identification further demonstrates that model mismatch can significantly degrade prediction quality [11]. Enhanced model predictive control for marine propulsion systems also indicates that robustness and computational burden must be considered jointly [12]. In addition, two-vector sequential model predictive control shows that sequential optimization structures can alleviate the tuning difficulty associated with conventional single weighted costs [13].

Table 1 condenses the most relevant related-work streams, the remaining rotary-valve-specific gaps, and the corresponding study scope.

Table 1. Condensed related-work map and remaining rotary-valve-specific gaps.

Research Stream/Representative Refs.	Main Focus	Remaining Gap for Rotary-Valve Mud-Pulse Control	How This Study Responds
Mud-pulse channel and drill-string response	Channel modelling and propagation response	Does not model actuator-side valve dynamics or constrained front-end waveform generation.	Introduces a control-oriented rotary-valve actuation plant model and waveform-oriented pressure-pulse control metrics.
Surface demodulation and signal recovery	Denoising, recovery, equalization, and target-band fidelity	Does not close the loop from valve-control errors to telemetry decoding.	Limits the present claim to waveform-level proxies and identifies BER/synchronization metrics as future work.
Predictive motor/actuator control with identification [14]	MPC, parameter identification, and disturbance compensation	Does not include rotary-valve hydraulics, clearance/wear drift, and pressure safety windows.	Integrates grey-box hydraulic loading, wear/clearance terms, and online correction/compensation in one rotary-valve formulation.
Sequential and low-complexity MPC [15–17]	Reduced tuning burden, candidate/vector reduction, and online implementation	Is not tailored to finite candidate inputs and a 20 ms pressure-pulse control budget.	Proposes finite-candidate lexicographic MPC with candidate pre-screening, move blocking, and complexity reporting.

Existing studies have reported control-oriented actuator models, sequential MPC structures, and disturbance-compensated predictive control in related electromechanical systems. The novelty of the present work is therefore not claimed at the level of inventing grey-box modelling or lexicographic MPC in general. Instead, it lies in their rotary-valve-specific integration for mud-pulse telemetry: (i) a control-oriented electromechanical–hydraulic model that explicitly retains clearance/wear effects and pressure-coupled loading; (ii) a low-complexity lexicographic decision structure tailored to finite candidate inputs and a 20 ms sampling budget; and (iii) a joint validation protocol that examines tracking, pulse-shaping, robustness to drift and disturbance, and algorithmic complexity within a single framework.

Against that background, this work formulates three verifiable hypotheses.

https://doi.org/10.3390/pr1410158

Hypothesis 1 (H1). *Lexicographic sequential evaluation is expected to reduce reliance on a single aggregated weighted cost by making the priority order among control objectives explicit.*

Hypothesis 2 (H2). *Candidate pre-screening and move-blocking keep tail computation times below one sampling period under the adopted operating conditions.*

Hypothesis 3 (H3). *Online correction and disturbance compensation improve performance under the tested parameter-mismatch and disturbance cases.*

Accordingly, the paper contributes: (i) a control-oriented electromechanical–hydraulic grey-box prediction model with an explicit plant boundary; (ii) a real-time lexicographic MPC implementation with candidate pre-screening and reduced online search; and (iii) a simulation-based validation package covering tracking, pulse shaping, robustness proxies, and timing complexity. The remainder of this paper is organized as follows. Section 2 develops the grey-box model and the discrete prediction form. Section 3 presents the coupled estimation and decision layers of the improved MPC. Section 4 reports the simulation setup, evidence boundary, robustness discussion, and algorithmic implications. Section 5 concludes the paper and discusses the main limitations and future work.

2. Materials and Methods: Electromechanical–Hydraulic Grey-Box Modeling

2.1. Modeling Objective, Structural Composition, and Variable Definition

The modeling objective of this section is not a full high-fidelity reconstruction of the local downhole flow field, but a control-oriented grey-box description that remains interpretable, identifiable, and verifiable. To this end, the model retains the dominant mechanisms of mechanical motion, hydraulic throttling, hydrodynamic torque, and clearance-induced wear, while lumping unresolved high-frequency nonlinearities into bounded disturbance terms. Open patent disclosures indicate that geometric opening and wear can significantly affect the flow, leakage, and pressure-drop build-up process; therefore, the clearance variable g is explicitly introduced in the model .

The intelligent drill-bit rotary valve mainly consists of a drive motor, transmission components, a rotating valve core, a fixed valve seat, fluid passages, and sensing units. By regulating the motor output torque T_m, the controller drives the valve core to rotate, which changes the valve angle θ, modifies the effective flow area $A(\theta, g)$, and establishes an additional pressure drop p across the throttling pair. Here, p denotes the local additional pressure drop relative to the background circulation pressure, and its magnitude directly determines the mud-pulse amplitude. For controller design, the valve angle, angular speed, and additional pressure drop are selected as the primary state variables:

$$x(t) = \begin{bmatrix} \theta(t) \\ \omega(t) \\ p(t) \end{bmatrix}, \qquad u(t) = T_m(t), \qquad y(t) = \begin{bmatrix} \theta(t) \\ p(t) \end{bmatrix}. \tag{1}$$

If the actuator current is measurable, the relationship $T_m = K_t i$ can also be used to map current limits into equivalent torque constraints.

In this paper, the controlled object is the downhole rotary-valve actuation plant bounded by the motor torque input and the measured outputs $y = [\theta,\ p]^T$. It includes the motor–transmission–valve-core mechanics, the throttling/leakage hydraulics, and the slow wear-related clearance state g. The downstream mud channel and the surface demodulator are not included in the plant model in the present study; they are addressed only through waveform-based proxy metrics in the revised Results section. Accordingly, T_m (or ΔT_m)

https://doi.org/10.3390/pr14101589

is the manipulated variable, θ and p are the controlled outputs, ω is an internal state, q_{in}, d_τ, and d_p are exogenous disturbances, and g, C_d, B_{eq}, k_h, and T_p are slow-varying parameters updated or perturbed in the robustness analysis. The controlled-object boundary and main symbol definitions are summarized in Tables 2 and 3, respectively, and the electromechanical–hydraulic coupling schematic is shown in Figure 2.

Table 2. Controlled-object boundary and variables used in the MPC formulation.

Category	Symbol	Physical Meaning	Role in Control	Status
Manipulated variable	T_m, ΔT_m	Motor torque or torque increment	Control action generated by the controller	scheduled
Controlled outputs	θ, p	Valve angle and local pressure drop	Tracked outputs and safety-window variables	measured
Internal states	θ, ω, p	Valve angle, angular speed, and pressure drop	Predicted states in the corrected MPC model	estimated
External disturbances	q_{in}, d_τ, d_p	Inflow fluctuation and lumped disturbances	Exogenous uncertainty acting on the plant	disturbed
Slow-varying parameters	g, C_d, B_{eq}, k_h, T_p	Clearance, discharge coefficient, damping, coupling, and time constant	Scheduled or corrected plant quantities	scheduled
Inside plant model	–	Motor/drive mechanics, throttling leakage, pressure loading, and wear state	Included in the grey-box prediction model	inside
Outside plant model surface demodulator, telemetry receiver	Mud channel, Transmission and downstream decoding chain	Proxy-level discussion only	outside	

Table 3. Main symbols used in the plant and controller formulation.

Symbol	Meaning	Role/Unit
T_m, ΔT_m	Motor torque and torque increment	Manipulated variable
θ, ω, p	Valve angle, angular speed, additional pressure drop	States/outputs
$A(\theta, g)$	Equivalent throttling area	Geometry-derived plant nonlinearity
q_{in}, q_v, q_{leak}	Inflow, throttling flow, leakage flow	Hydraulic balance terms
g, C_d	Clearance/wear state and discharge coefficient	Slow-varying plant parameters
B_{eq}, k_h, T_p	Equivalent damping, hydrodynamic-torque coefficient, pressure time constant	Corrected prediction-model parameters
N_p, N_c, N_b	Prediction horizon, control horizon, move-blocking depth	MPC structural settings
$M(k)$, M_h, M_1, M_2, M_3	Candidate counts before and after filtering	Complexity-tracking quantities
ε_1, ε_2	Near-optimality tolerances for Stages 1 and 2	Lexicographic retention thresholds

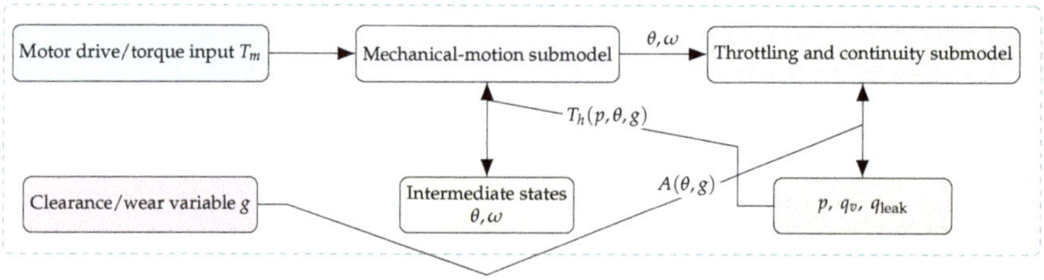

Figure 2. Electromechanical–hydraulic coupling and energy-flow schematic of the intelligent rotary valve. The dashed box denotes the controlled object used in this paper, namely the downhole actuation plant from the motor torque input T_m to the measured outputs (θ, p); the downstream mud channel and surface demodulator are outside this boundary.

2.2. Mechanical Submodel

The mechanical submodel satisfies the basic kinematic relation

$$\dot{\theta}(t) = \omega(t). \tag{2}$$

https://doi.org/10.3390/pr1410158

Considering the motor driving torque, friction torque, hydrodynamic torque, and lumped external disturbance, the mechanical dynamics of the rotary-valve core can be written as

$$J\dot{\omega}(t) = T_m(t) - T_f(t) - T_h(t) - d_\tau(t), \tag{3}$$

where J is the equivalent rotational inertia and $d_\tau(t)$ denotes unmodeled mechanical disturbance. To avoid oversimplifying friction as a purely viscous term, a grey-box friction expression better suited to the actuator layer is adopted:

$$T_f(t) = B\omega(t) + T_c \tanh\!\left(\frac{\omega(t)}{\varepsilon}\right) + k_p p(t)\,\mathrm{sgn}(\omega(t)) + d_{\tau,f}(t), \tag{4}$$

where B is the viscous damping coefficient, T_c is the equivalent Coulomb-friction amplitude, ε is the smoothing parameter, k_p characterizes the effect of pressure drop on the friction load, and $d_{\tau,f}(t)$ collects the remaining friction uncertainty. This expression captures low-speed nonsmooth behavior, pressure-load coupling, and unmodeled friction disturbance, and is therefore more representative of the downhole rotary-valve actuation layer than the simplified approximation $T_f \approx B\omega$.

Equation (4) means that the torque required to rotate the valve increases not only with speed-dependent damping but also with low-speed friction and pressure-induced loading. This reflects the practical fact that the valve is harder to move near reversals and under higher pressure drops.

2.3. Hydraulic Submodel

The hydraulic submodel is jointly determined by the throttling flow, leakage flow, and fluid compressibility. Let $p(t)$ denote the additional pressure drop across the rotary valve. The main throttling flow through the valve port is then written as

$$q_v(t) = C_d(t)\, A(\theta(t), g(t))\ \mathrm{sgn}(p(t))\sqrt{\frac{2|p(t)|}{\rho}}, \tag{5}$$

where $C_d(t)$ is the flow coefficient, ρ is the drilling-fluid density, and $A(\theta, g)$ is the equivalent throttling area jointly determined by the valve angle and the clearance. In the present study, $A(\theta, g)$ is implemented as a geometry-derived lookup surface over the admissible (θ, g) domain and evaluated online through bilinear interpolation. The corresponding static calibration route is illustrated later in Section 4.2. To account for leakage induced by long-term erosion and wear, a parametric leakage term is introduced as

To visualize the geometric relationship among valve angle, clearance, and equivalent throttling area in (5), Figure 3 presents a contour map of $A(\theta, g)$ over the operating range. This figure is intended to illustrate the direction of wear-induced equivalent-opening drift and the overall downward trend of flow capacity, rather than to replace the static calibration curves introduced later.

$$q_{\mathrm{leak}}(t) = q_{\mathrm{leak}}(p(t), g(t)), \tag{6}$$

where $g(t) = g_0 + \Delta g(t)$ denotes the slow evolution of the throttling-pair clearance. In the simulations, $q_{\mathrm{leak}}(p, g)$ is evaluated as a monotone pressure-driven leakage relation calibrated together with the nominal throttling map so that larger clearances produce both higher leakage and a reduced effective pressure build-up. Because wear significantly changes both the additional pressure drop and the leakage level, g is treated as a key slow-varying parameter in the pressure dynamics.

https://doi.org/10.3390/pr14101589

From the fluid-continuity relation, the local additional pressure drop satisfies

$$C_p \dot{p}(t) = q_{\text{in}}(t) - q_v(t) - q_{\text{leak}}(t) + d_p(t), \tag{7}$$

where $q_{\text{in}}(t)$ is the equivalent inflow entering the throttling control unit, $d_p(t)$ is the unmodeled pressure disturbance, and C_p is the equivalent compressibility coefficient of the hydraulic chamber. Furthermore,

$$C_p = \frac{V_e}{\beta_e}, \tag{8}$$

where V_e is the equivalent chamber volume and β_e is the equivalent bulk modulus. Compared with directly assuming a first-order pressure lag, (7) provides a more traceable derivation and is therefore better suited for later calibration and parameter identification.

Equation (7) states that the additional pressure rises when the equivalent inflow exceeds the throttling and leakage outflows, and falls otherwise. The model therefore keeps a direct physical connection between flow imbalance and pulse build-up.

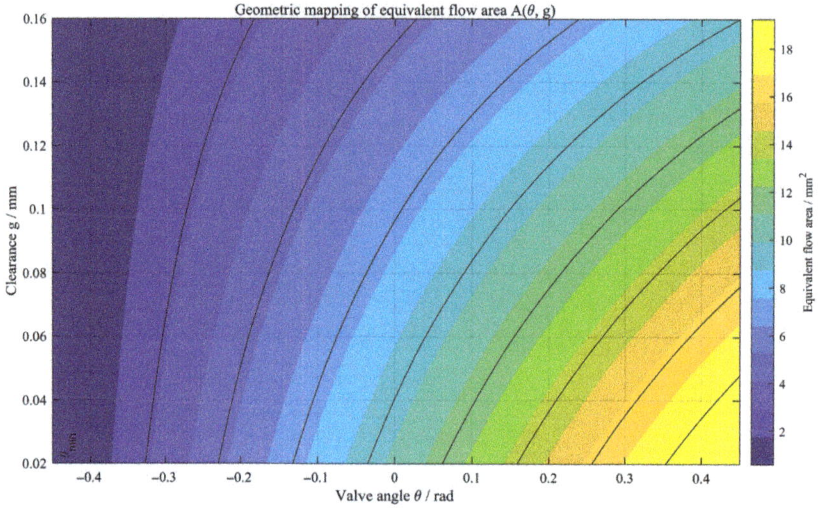

Figure 3. Model-derived contour map of the equivalent throttling area $A(\theta, g)$. Positive θ increases the effective opening, whereas larger clearance g represents wear-induced geometric drift and higher leakage tendency. The plot is used to explain the controlled plant geometry and is not a bench-measured calibration surface.

2.4. *Electromechanical Coupling and Unified Grey-Box Equations*

The hydrodynamic torque originates from the pressure distribution acting on the throttling pair and valve core. Its rigorous expression can be written as

$$T_h(t) = r_t \iint_{A_{\text{eff}}(\theta, g)} p_n \, \mathrm{d}A, \tag{9}$$

where r_t is the equivalent moment arm and p_n is the normal pressure distribution over the throttling pair. For online control and parameter-identification purposes, it can be approximated around the operating point as

$$T_h(t) \approx k_h p(t) + k_\theta (\theta(t) - \theta_0) + d_{\tau,h}(t), \tag{10}$$

https://doi.org/10.3390/pr1410158

where k_h is the equivalent coupling coefficient from pressure drop to hydrodynamic torque, k_θ is the additional hydrodynamic-torque coefficient induced by valve-angle deviation, and $d_{\tau,h}(t)$ is the residual hydrodynamic disturbance.

Equation (10) means that pressure build-up feeds back into the mechanical subsystem through an additional load torque, while the local valve position perturbs the same load around the operating point. This approximation keeps the dominant pressure–mechanics coupling visible without requiring a full CFD-resolution model during online control.

Combining the above terms, the unified continuous grey-box model of the intelligent drill-bit rotary valve can be written as

$$\dot{\theta}(t) = \omega(t), \tag{11}$$

$$J\dot{\omega}(t) = T_m(t) - T_f(t) - T_h(t) - d_\tau(t), \tag{12}$$

$$C_p\dot{p}(t) = q_{\text{in}}(t) - q_v(t) - q_{\text{leak}}(t) + d_p(t), \tag{13}$$

$$q_v(t) = C_d(t)\, A\big(\theta(t), g(t)\big)\ \text{sgn}\big(p(t)\big)\sqrt{\frac{2|p(t)|}{\rho}}. \tag{14}$$

This model jointly describes the coupling among mechanical motion, hydraulic throttling, leakage flow, hydrodynamic torque, and slowly varying clearance, and is therefore suitable for control-oriented modeling that combines physical mechanisms with parameter identification.

2.5. Linearization, State-Space Formulation, and Validity Range

To construct a model predictive controller, the grey-box model is linearized around the operating point $(\theta_0, \omega_0, p_0)$ using a small-signal approximation. Let

$$\Delta x = \begin{bmatrix} \Delta\theta \\ \Delta\omega \\ \Delta p \end{bmatrix}, \qquad \Delta u = \Delta T_m, \qquad \Delta\xi = \begin{bmatrix} \Delta q_{\text{in}} \\ \Delta g \\ \Delta C_d \end{bmatrix}, \tag{15}$$

Then the linearized model becomes

$$\Delta\dot{x}(t) = A_c\Delta x(t) + B_c\Delta u(t) + E_c\Delta\xi(t) + D_c d(t), \qquad y(t) = C\Delta x(t), \tag{16}$$

where

$$A_c = \begin{bmatrix} 0 & 1 & 0 \\ -\frac{k_\theta}{J} & -\frac{B_{\text{eq}}}{J} & -\frac{k_h^*}{J} \\ K_\theta & 0 & -\frac{1}{T_p} \end{bmatrix}, \qquad B_c = \begin{bmatrix} 0 \\ \frac{1}{J} \\ 0 \end{bmatrix}, \qquad C = \begin{bmatrix} 1 & 0 & 0 \\ 0 & 0 & 1 \end{bmatrix}. \tag{17}$$

Here, B_{eq} denotes the equivalent damping coefficient and k_h^* the linearized hydrodynamic-torque coefficient, while

$$T_p = \frac{C_p}{\left.\frac{\partial q_v}{\partial p}\right|_0 + \left.\frac{\partial q_{\text{leak}}}{\partial p}\right|_0}, \qquad K_\theta = -\frac{1}{C_p}\left.\frac{\partial q_v}{\partial \theta}\right|_0, \qquad K_q = \frac{1}{C_p}. \tag{18}$$

This shows that the commonly used first-order pressure model is not an independent assumption, but rather a natural result of linearizing the continuity equation. When the downhole controller is computationally constrained and the rotary valve mainly operates in the low- to mid-frequency pulse-modulation range, the pressure channel in (16) can be retained as a first-order reduced model for MPC. For large-amplitude modulation or wide-range operating analysis, however, the nonlinear form in (5) should be preserved and handled by piecewise linearization or gain scheduling.

https://doi.org/10.3390/pr14101589

The applicability bounds of the present model should be stated explicitly in five aspects. First, friction is not reduced to a single viscous term; Coulomb friction and pressure-load coupling are retained through (4). Second, the hydrodynamic torque is not empirically imposed, but first described mechanistically by (9) and then approximated in a control-oriented form around the operating point. Third, the pressure dynamics are derived from continuity and compressibility relations rather than being directly postulated as a first-order lag. Fourth, the clearance $g(t)$ and flow coefficient $C_d(t)$ are allowed to drift slowly so as to reflect erosion and wear. Fifth, under large-amplitude modulation, the dominant nonlinear throttling relationship should not be fully absorbed into a disturbance term, but instead retained and treated through segmentation or scheduling.

2.6. Discrete Prediction Model and Engineering Constraints

Considering the receding-horizon nature of MPC, (16) is discretized by a zero-order hold (ZOH) method, yielding

$$\Delta x(k+1) = A_d \Delta x(k) + B_d \Delta T_m(k) + E_d \Delta \xi(k) + D_d d(k), \qquad y(k) = C \Delta x(k). \tag{19}$$

where $A_d = e^{A_c T_s}$ and $B_d = \int_0^{T_s} e^{A_c \tau} B_c \mathrm{d}\tau$. Compared with directly optimizing the absolute torque, using the torque increment ΔT_m as the optimization variable is more effective in suppressing high-frequency switching and actuator shock.

In discrete time, the controller predicts how the current torque increment affects the next-step valve angle and pressure while enforcing actuator and safety limits. This is the model actually used by the receding-horizon decision layer at each 20 ms sample.

Consistent with the actuator-layer limits of the rotary valve, explicit constraints are imposed on the valve angle, angular speed, driving torque, torque increment, and additional pressure drop:

$$\theta_{\min} \le \theta(k) \le \theta_{\max}, \qquad |\omega(k)| \le \omega_{\max}, \tag{20}$$

$$T_{m,\min} \le T_m(k) \le T_{m,\max}, \qquad \Delta T_{m,\min} \le \Delta T_m(k) \le \Delta T_{m,\max}, \tag{21}$$

$$p_{\min} \le p(k) \le p_{\max}. \tag{22}$$

If the current is measurable, the following additional constraint can be imposed:

$$i_{\min} \le i(k) \le i_{\max}, \qquad T_m(k) = K_t i(k). \tag{23}$$

These constraints correspond, respectively, to the mechanical travel limits of the valve core, the avoidance of overspeed impact, the actuator output capability, switching smoothness, and the safety window of the pressure pulse.

2.7. Parameter Sources, Identification, and Validation Route

To avoid repeating the validation discussion in both the modeling and results sections, only a principle-level summary of parameter sources and identification requirements is provided here. The inertia J, damping B, friction parameter T_c, throttling-area function $A(\theta, g)$, flow coefficient C_d, and hydraulic compressibility parameters V_e, β_e all need to be obtained from CAD estimation, static calibration, free-decay identification, or step-response fitting. Details such as pressure-sensor placement, erosion-related measurement bias, the choice of online identification method, and the bench/swept-frequency validation procedure are deferred to Section 4.2, thereby forming a closed loop of modeling–control–validation.

3. Materials and Methods: Improved Real-Time Lexicographic MPC

The control input is the equivalent driving torque or torque increment u, the outputs are the valve angle θ and the local additional pressure drop p, and the controller consists

https://doi.org/10.3390/pr1410158

of an observer and a sequential (lexicographic) MPC. The estimation layer updates states, disturbances, and slowly varying parameters, while the decision layer performs receding-horizon optimization on the corrected discrete prediction model so as to simultaneously satisfy fast valve-angle tracking, low-ripple pressure-pulse shaping, and real-time feasibility for downhole implementation. Existing studies on sequential MPC indicate that such a structure reduces dependence on empirical weight tuning and is particularly suitable for industrial control problems with clearly prioritized objectives.

3.1. Control Architecture, Engineering Objectives, and Limitations of Standard MPC

For the engineering application of the rotary valve, the control targets are not expressed merely as abstract output-tracking goals, but are mapped to three verifiable classes of metrics:

1. Valve-position tracking performance: quantified by the steady-state angle error or the integral absolute error (IAE), requiring the valve core to reach the target accurately within a finite time;
2. Pulse-shaping quality: quantified by the additional-pressure amplitude error, RMS fluctuation, and peak-to-peak ripple, requiring the output pressure pulse to exhibit a better waveform-margin proxy associated with downstream decoding;
3. Actuation smoothness: quantified by the magnitudes of Δu and $\Delta^2 u$, requiring reduced high-frequency switching and mechanical shock.

The tracking errors and variation terms are defined as

$$e_\theta(k) = \theta^*(k) - \theta(k), \qquad e_p(k) = p^*(k) - p(k), \tag{24}$$

$$\Delta p(k) = p(k) - p(k-1), \qquad \Delta u(k) = u(k) - u(k-1), \qquad \Delta^2 u(k) = \Delta u(k) - \Delta u(k-1). \tag{25}$$

At each sampling instant, standard MPC predicts the future response over the horizon N_p based on (19) and solves a single weighted cost over the control horizon N_c. Although this method handles constraints explicitly, it still exhibits three shortcomings in the downhole rotary-valve scenario. First, valve tracking, pulse stability, and actuation smoothness typically rely on manually coordinated weights, which creates a substantial tuning burden. Second, multi-candidate enumeration or long prediction horizons can significantly increase the online computational cost. Third, when the model parameters drift or disturbances intensify, the prediction accuracy of a fixed model degrades rapidly. Studies on complex-vector disturbance observers with deadbeat predictive control show that online compensation can suppress harmonics and improve robustness. Weightless CCS model predictive control further shows that weakening empirical weight dependence can improve controller tunability [18]. Parameter-independent predictive current control highlights the need to reduce model dependence [19]. Low-complexity double-vector strategies demonstrate the direct real-time value of candidate-set reduction. Multi-voltage-vector model-free deadbeat control shows that updated reference vectors can improve dynamic quality. Enhanced model predictive control based on an adaptive super-twisting observer illustrates the advantage of online compensation under model mismatch. Together, these ideas provide transferable technical support for the two-layer control structure adopted here, as shown in Figure 4.

https://doi.org/10.3390/pr14101589

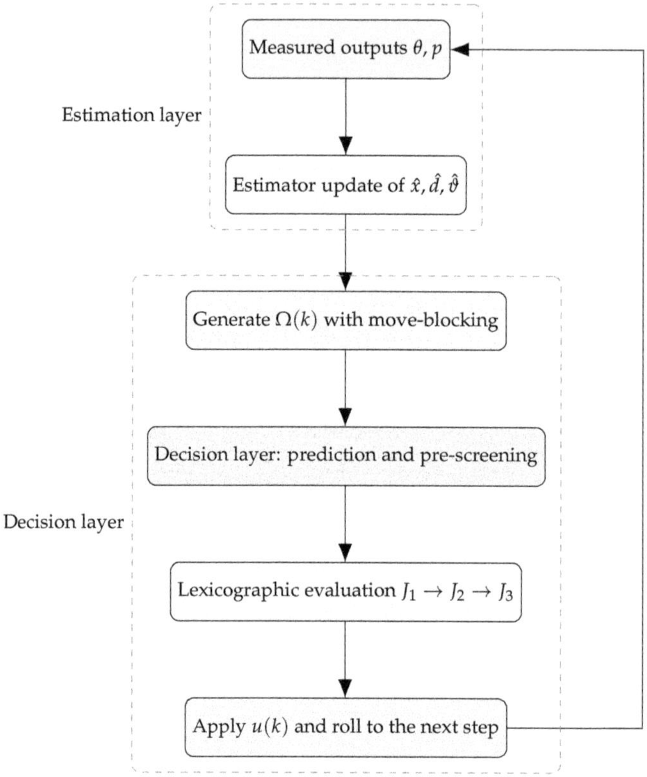

Figure 4. Integrated control architecture with an estimation layer and a sequential MPC decision layer.

3.2. Estimation Layer: Joint Update of States, Disturbances, and Parameters

Because the damping, hydrodynamic-torque coupling coefficient, and pressure-channel time constant may drift during long-term downhole operation, joint estimation of states, disturbances, and parameters is introduced around the controller. The parameter vector to be updated is defined as

$$\hat{\vartheta}(k) = \begin{bmatrix} \hat{B}_{\mathrm{eq}}(k) \\ \hat{k}_h(k) \\ \hat{T}_p(k) \end{bmatrix}, \tag{26}$$

The corrected prediction model is then written as

$$\hat{x}(k+1) = \hat{A}_d\big(\hat{\vartheta}(k)\big)\hat{x}(k) + \hat{B}_d\big(\hat{\vartheta}(k)\big)u(k) + E_d\hat{d}(k) + L\big(y(k) - \hat{y}(k)\big), \tag{27}$$

where $\hat{d}(k)$ is the equivalent disturbance estimate and L is the output-error feedback gain. The disturbance update is written as

$$\hat{d}(k+1) = \hat{d}(k) + L_d\big(y(k) - \hat{y}(k)\big). \tag{28}$$

To explicitly define the regressor and to prevent estimator divergence under transients and noisy measurements, a projection-based recursive least-squares (RLS) scheme is adopted for the slow-varying parameters:

$$\hat{\vartheta}_k = \Pi_\Theta\big\{\hat{\vartheta}_{k-1} + K_k\big(y_k - \hat{y}_k\big)\big\}, \tag{29}$$

https://doi.org/10.3390/pr1410158

$$K_k = \frac{P_{k-1}\phi_k}{\lambda + \phi_k^{\mathrm{T}} P_{k-1}\phi_k}, \qquad P_k = \lambda^{-1}\left(I - K_k\phi_k^{\mathrm{T}}\right)P_{k-1}, \tag{30}$$

where ϕ_k is the regressor assembled from the measured or estimated valve speed, the additional pressure drop, the previous input increment, and the one-step output mismatch, i.e.,

$$\phi_k = \left[\omega(k),\ p(k),\ \Delta u(k-1),\ y(k) - \hat{y}(k)\right]^{\mathrm{T}},$$

λ is the forgetting factor, and $\Pi_\Theta\{\cdot\}$ is the projection operator that enforces physically admissible parameter bounds. In the numerical implementation, the regressor is normalized before the update. This formulation keeps the update lightweight while preventing abrupt noise bursts or operating-point changes from driving the parameter estimates outside their feasible ranges.

From an implementation viewpoint, the estimation layer can also be extended to an augmented EKF or a sliding-mode observer (SMO), whose applicability is summarized in Table 4. Because the sampling period in this study is $T_s = 20$ ms and the outer-loop dynamics of the rotary valve are dominated by low- to mid-frequency pulse modulation, the simulations prioritize the combination of output-error observation and projection RLS to balance drift rejection against online computational burden.

Table 4. Comparison of disturbance and parameter compensation schemes.

Scheme	Core Idea	Advantages	Limitations	Inputs/Rate
Augmented EKF + feedforward compensation	Treats disturbance d as an augmented random-walk state and jointly estimates ϑ, then updates the prediction model and feedforward term	Good noise tolerance, can track slow drift, naturally coupled with MPC	Requires covariance tuning and heavier matrix operations	Sampling at T_s or faster; measurements θ, p
Sliding-mode observer + adaptive gain	Rapidly estimates matched disturbances online and compensates them, while adapting gains to reduce chattering	Strong robustness and relatively low computational cost	Boundary-layer and gain design strongly affect accuracy; anti-chattering treatment is required	Sampling at T_s; measurements θ, p

3.3. Decision Layer: Lexicographic Sequential MPC with Soft-Constraint Handling

After the estimation layer provides $\hat{x}(k)$, $\hat{d}(k)$, and $\hat{\vartheta}(k)$, the decision layer performs receding-horizon optimization using the corrected discrete model. For better engineering interpretability, the constraints are divided into hard and soft categories.

The hard constraints directly correspond to the mechanical and actuator limits:

$$\theta_{\min} \le \theta(k) \le \theta_{\max}, \qquad |\omega(k)| \le \omega_{\max}, \tag{31}$$

$$u_{\min} \le u(k) \le u_{\max}, \qquad \Delta u_{\min} \le \Delta u(k) \le \Delta u_{\max}. \tag{32}$$

The pressure safety window is also enforced as a hard constraint under normal conditions. When no candidate remains hard-feasible, a soft pressure constraint with slack variables is activated as a pragmatic feasibility-recovery mechanism:

$$p_{\min} - s_p(k+i) \le p(k+i|k) \le p_{\max} + s_p(k+i), \qquad s_p(k+i) \ge 0. \tag{33}$$

The corresponding slack penalty is written as

$$J_s^{(m)} = \rho_p \left\| s_p^{(m)} \right\|_2^2, \tag{34}$$

where ρ_p is the pressure soft-constraint penalty coefficient. This fallback limits the magnitude of pressure-window violation and avoids infeasibility-induced optimization failure in

https://doi.org/10.3390/pr14101589

the tested scenarios. It does not constitute a formal proof of recursive feasibility or closed-loop stability, and such a proof remains outside the scope of the present simulation study.

To support a reduced-weight-dependence claim rather than a fully weight-free claim, each stage objective is written in a dimensionless form normalized by the relevant physical constraint scale, instead of relying on a single strongly coupled set of empirical weights such as q_θ, q_p, and r_1. For the mth candidate sequence,

$$J_1^{(m)} = \sum_{i=1}^{N_p} \left[\left(\frac{e_{\theta,m}(k+i|k)}{\Delta\theta_{\max}} \right)^2 + \left(\frac{\omega_m(k+i|k)}{\omega_{\max}} \right)^2 \right], \tag{35}$$

$$J_2^{(m)} = \sum_{i=1}^{N_p} \left[\left(\frac{e_{p,m}(k+i|k)}{\Delta p_{\max}} \right)^2 + \left(\frac{\Delta p_m(k+i|k)}{\Delta p_{\max}} \right)^2 \right], \tag{36}$$

$$J_3^{(m)} = \sum_{j=0}^{N_c-1} \left[\left(\frac{\Delta u_m(k+j|k)}{\Delta u_{\max}} \right)^2 + \left(\frac{\Delta^2 u_m(k+j|k)}{\Delta u_{\max}} \right)^2 \right]. \tag{37}$$

where $\Delta\theta_{\max}$, $\Delta p_{\max}$, and $\Delta u_{\max}$ denote the physical scales of the valve-angle excursion, additional pressure-drop excursion, and input increment, respectively. In other words, the lexicographic structure reduces the tuning burden of a single global weighted cost by assigning priorities across stages, while still retaining a small number of intra-stage numerical scaling factors for stability.

From an engineering viewpoint, J_1 protects primary valve-motion quality, J_2 refines pressure-pulse quality once acceptable motion candidates have been retained, and J_3 finally suppresses aggressive actuator movement among the remaining candidates. The three-stage structure therefore makes the controller priorities explicit instead of burying them in a single aggregated weight set.

Once the candidate sequences satisfy either the hard constraints or the softened feasible set, the optimal solution is selected using a tolerance-based lexicographic rule. Define

$$J_{1,\min} = \min_{m\in\Omega(k)} J_1^{(m)}, \qquad J_{2,\min} = \min_{m\in\Omega_1(k)} J_2^{(m)}, \tag{38}$$

Then the first- and second-stage near-optimal candidate sets are

$$\Omega_1(k) = \left\{ m \in \Omega(k) \;\middle|\; J_1^{(m)} \le J_{1,\min} + \varepsilon_1 \right\}, \tag{39}$$

$$\Omega_2(k) = \left\{ m \in \Omega_1(k) \;\middle|\; J_2^{(m)} \le J_{2,\min} + \varepsilon_2 \right\}, \tag{40}$$

The final control sequence is determined by

$$m^* = \arg \min_{m\in\Omega_2(k)} \left(J_3^{(m)} + J_s^{(m)} \right) \tag{41}$$

This formulation is consistent with the standard expression of lexicographic optimization MPC and addresses two common methodological questions: whether pre-screening removes the optimum and whether the method still contains hidden empirical weighting. In the present implementation, $\varepsilon_1 = 0.03$ and $\varepsilon_2 = 0.02$ are used as empirical near-optimality thresholds to balance control performance against online computational effort. Supplementary Figure S4 reports retained-set tolerance proxies that visualize the observed accuracy–timing trade-off. A full tolerance sweep over $(\varepsilon_1, \varepsilon_2)$ is not included in the present simulation dataset; the added candidate-retention analysis is therefore interpreted as an empirical timing–retention proxy rather than a robustness-margin proof. The tolerances were selected empirically to balance near-optimality and online computation within the

https://doi.org/10.3390/pr1410158

tested numerical envelope, and a systematic robustness-margin-based selection remains future work.

3.4. Candidate-Set Design and Online Complexity-Reduction Strategy

For downhole controllers with limited onboard computation and a fixed sampling period, the continuous input space is not optimized globally at every sampling instant. Instead, a finite candidate set is generated locally around the previous input:

$$\Omega_u(k) = \{u(k-1) + \delta_i \mid \delta_i \in \{-\Delta u_{\max}, 0, \Delta u_{\max}\}\}. \tag{42}$$

To further reduce the search dimension, a move-blocking strategy is adopted over the prediction horizon: only the first N_b input moves are optimized, and the remaining moves are held at the last optimized value, i.e.,

$$\mathbf{U}_m(k) = [u_m(k|k), \cdots, u_m(k+N_b-1|k), u_m(k+N_b-1|k), \cdots, u_m(k+N_c-1|k)]. \tag{43}$$

Although the admissible control input is quantized into seven levels within $[u_{\min}, u_{\max}]$ in the numerical implementation, the online search is carried out locally through the three incremental moves $\{-\Delta u_{\max}, 0, \Delta u_{\max}\}$ around $u(k-1)$ and then clipped to the admissible seven-level grid. This strategy is especially suitable for a rotary valve with low-frequency dynamics, limited onboard resources, and explicit actuator constraints because it compresses the enumeration space without discarding hard feasibility checks. It should also be noted that FCS-type enumerative controllers remain sensitive to model mismatch, weight design, and search-space growth. Therefore, the present work adopts a compromise structure of "finite candidate set + sequential evaluation + move-blocking" to improve real-time implementability.

From an engineering implementation standpoint, the applicability of common low-complexity MPC variants is summarized in Table 5. Because the outer loop of the rotary valve is dominated by low- to mid-frequency pressure-pulse modulation and operates with a sampling period of about 20 ms, this work prioritizes a sequential lexicographic MPC with a finite candidate set rather than introducing more complex multi-vector duty allocation or an online QP solver.

Table 5. Comparison of low-complexity MPC variants for the rotary-valve scenario.

Variant	Advantages	Limitations	Complexity	Use Case
FCS-MPC (enumerative)	No QP solver is required and embedded implementation is straightforward	Enumeration grows with M and N_p and remains sensitive to model mismatch	Medium	Feasible at roughly $T_s = 20$ ms; suitable for finite candidate inputs
Multi-/double-vector sequential MPC	Lower output ripple and naturally compatible with sequential evaluation	Requires additional vector or duty-ratio allocation and is therefore more complex	Medium–high	More suitable for high-dynamic drive scenarios
Fast QP-MPC (linear MPC)	Smoother continuous control and access to embedded solvers	Relies on a linearized model and an online QP solver	Medium	Common in low-frequency outer loops around 50–200 Hz; can be combined with acados/FORCES

Because discretization error is another common implementation concern, it should be emphasized that A_d and B_d in (19) are computed offline exactly under the ZOH assumption by means of the matrix exponential, and are then accessed online through lookup or direct use of the precomputed matrices, rather than being approximated by a forward-Euler scheme.

https://doi.org/10.3390/pr14101589

3.5. Algorithmic Steps, Complexity, and Real-Time Implementation

Combining the estimation and decision layers, Figure 5 expands the online solution flow of candidate generation, constraint filtering, lexicographic evaluation, and soft-constraint fallback in greater detail. Its role is complementary to Figure 4: the latter emphasizes the two-layer control architecture, whereas the former focuses on the per-sample solution procedure and data flow.

For scalability discussion, let the local branching factor be $b = 3$ because the online search uses the three incremental moves $\{-\Delta u_{\max}, 0, \Delta u_{\max}\}$, and let N_b denote the move-blocking depth. The raw candidate upper bound is therefore

$$M_{\text{raw}} \le b^{N_b}. \tag{44}$$

For one candidate, one N_p-step rollout of an n_x-state prediction model with precomputed A_d and B_d costs $O(N_p n_x^2)$ in the present matrix-based implementation. The resulting upper-bound scaling is summarized as

$$\text{Conventional finite-set MPC}: \quad O\Big(M_{\text{raw}} N_p n_x^2\Big), \tag{45}$$

$$\text{Adaptive MPC target baseline}: \quad O\Big(n_\vartheta^2 + M_{\text{raw}} N_p n_x^2\Big), \tag{46}$$

$$\text{Proposed method}: \quad O\Big(n_\vartheta^2 + M_h N_p n_x^2 + M_1 N_p + M_2 N_p + M_3 N_c\Big), \tag{47}$$

with

$$M_3 \le M_2 \le M_1 \le M_h \le M_{\text{raw}}, \tag{48}$$

and memory complexity

$$O(N_p n_x + M_h). \tag{49}$$

These expressions are intended as architecture-independent upper-bound scaling relations rather than as formal complexity theorems. They highlight that the dominant online cost remains candidate rollout and stage-wise filtering, whereas the parameter-update cost scales only with the small correction vector $\vartheta = [B_{eq}, k_h, T_p]^T$. The controller-level complexity scaling used in this discussion is summarized in Table 6.

To make the adaptive-MPC reference formulation explicit rather than purely verbal, its single-objective formulation is written as

$$\begin{aligned} J_{\text{A-MPC}} = &\sum_{i=1}^{N_p} \Big(w_\theta e_\theta^2(k+i|k) + w_p e_p^2(k+i|k) + w_\omega \omega^2(k+i|k) \Big) \\ &+ \sum_{j=0}^{N_c-1} \Big(r_1 \Delta u^2(k+j|k) + r_2 \Delta^2 u^2(k+j|k) \Big). \end{aligned} \tag{50}$$

This reference formulation uses the same corrected prediction model and constraints as the proposed controller, but replaces the lexicographic decision layer with one weighted objective. It therefore separates the role of online correction from the role of staged decision filtering. It is retained as a reference formulation for complexity discussion, and it is excluded from the numerical KPI comparison because reproducible closed-loop trajectories under the same protocol are unavailable.

https://doi.org/10.3390/pr1410158

Table 6. Controller-level complexity scaling used in the complexity discussion.

Controller	Decision Structure	Complexity Upper Bound	Dependence on N_p, N_c, $M(k)$	Practical Implication
PID	Fixed feedback law	$O(1)$	Independent of N_p, N_c, and $M(k)$	Lowest online burden but no predictive constraint coordination.
Conventional MPC	Single weighted objective over all raw candidates	$O(M_{\text{raw}} N_p n_x^2)$	Grows directly with N_p and the full candidate count	Sensitive to candidate explosion when the horizon or branching depth increases.
Adaptive-MPC reference formulation	Corrected prediction model plus single weighted objective	$O(n_\vartheta^2 + M_{\text{raw}} N_p n_x^2)$	Adds parameter-update cost but still evaluates all raw candidates.	Retained only for complexity discussion because reproducible closed-loop trajectories under the same protocol are unavailable for numerical summary. It isolates online correction from staged decision filtering.
Proposed lexicographic MPC	Corrected prediction model plus staged filtering and final smoothness selection	$O(n_\vartheta^2 + M_h N_p n_x^2 + M_1 N_p + M_2 N_p + M_3 N_c)$	Depends on the retained candidate counts rather than the full raw set alone.	Candidate compression reduces tail-time risk while preserving explicit stage priorities.

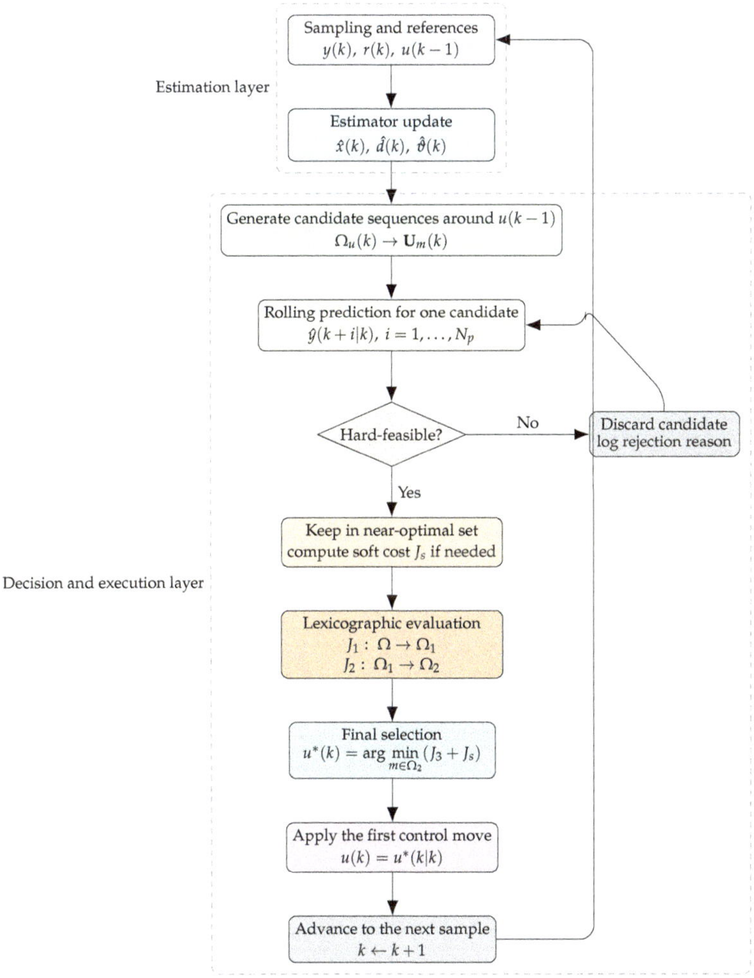

Figure 5. Detailed online solution workflow of the improved MPC.

https://doi.org/10.3390/pr14101589

Combining the estimation and decision layers, the control algorithm can be summarized as follows:

1. At sampling instant k, acquire the valve-angle and additional-pressure measurements, and update $\hat{x}(k)$, $\hat{d}(k)$, and $\hat{\vartheta}(k)$;
2. Construct the candidate action set $\Omega_u(k)$ around the previous input $u(k-1)$ and generate candidate control sequences through move-blocking;
3. Perform rolling prediction for each candidate sequence using $\hat{A}_d(\hat{\vartheta})$, $\hat{B}_d(\hat{\vartheta})$, and $\hat{d}(k)$;
4. Discard a candidate if hard constraints are violated; if no hard-feasible candidate exists, activate the soft constraint and compute the slack cost;
5. Compute $J_1^{(m)}$ and keep the near-optimal candidate set satisfying (39);
6. Compute $J_2^{(m)}$ within the near-optimal set and keep the secondary near-optimal set satisfying (40);
7. Compute $J_3^{(m)} + J_s^{(m)}$ for the remaining candidates and select the optimal one according to (41);
8. Apply the first control move of the optimal sequence to the rotary-valve system and advance to the next sampling instant.

To quantify real-time performance, the controller real-time load factor is defined as

$$\eta_{\text{rt}} = \frac{t_{\text{comp}}}{T_s}, \tag{51}$$

where t_{comp} is the single-step computation time. In addition to the average computation time, engineering evaluation should also report the worst-case execution time t_{wcet} and the statistical distribution of the candidate-set size $M(k)$. Section 4 therefore further reports the 99% percentile, WCET, and the evolution of $M(k)$ to characterize complexity fluctuation and the real-time implementation boundary.

A stronger theoretical guarantee of robust constraint satisfaction may be obtained by incorporating tube-MPC-based constraint tightening, in which the state and input sets are tightened as $\mathcal{X} \leftarrow \mathcal{X} \ominus E$ and $\mathcal{U} \leftarrow \mathcal{U} \ominus KE$, with E obtained from a disturbance bound. This extension is left for future work. Because the emphasis of this study is to construct a low-computation engineering control framework for the rotary-valve scenario, the primary implementation route adopted here is "disturbance observation + parameter correction + lexicographic MPC".

4. Results and Discussion

4.1. Simulation Setup, Benchmark Controllers, and Statistical Protocol

To avoid relying solely on qualitative curve inspection, the simulation setup is summarized before the scenario-level results are discussed. The evaluation protocol is organized into four reproducible KPI dimensions: dynamics, steady state, constraint satisfaction, and real-time performance.

1. Dynamic performance: the 10–90% rise time t_r, the ±2% settling time t_s, the steady-state error, and IAE/ITAE are used to evaluate valve-angle tracking;
2. Pulse-shaping quality: the steady-state additional-pressure error, peak-to-peak ripple, ripple RMS, and target-band energy are used to characterize pressure-pulse quality;
3. Constraint satisfaction and actuation smoothness: the number of constraint violations, the maximum violation magnitude, $\sum|\Delta u|$, and $\sum|\Delta^2 u|$ are used to evaluate actuator shock and constraint consistency;
4. Real-time performance: the mean computation time, the 99% percentile, the worst-case execution time (WCET), and the real-time load factor η_{rt} are used to assess online practicality within one sampling period.

https://doi.org/10.3390/pr1410158

The closed-loop rotary-valve simulation model is implemented in MATLAB R2022b. All verified controllers operate under the same sampling period $T_s = 20$ ms, the same state/input constraints, and the same reference trajectories, so that the reported numerical differences come from the control-law structure rather than from inconsistent operating conditions. The present simulation dataset contains verified trajectories for PID, conventional MPC, and the proposed lexicographic MPC. The adaptive-MPC formulation is retained as a reference formulation for complexity discussion because it shares the corrected prediction model but uses a single weighted objective. Since reproducible closed-loop trajectories under the same protocol are unavailable, it is excluded from the numerical KPI comparison. The KPI definitions and their engineering meanings are summarized in Table 7.

Table 7. Simulation KPIs and their engineering meanings.

Dimension	Main KPIs	Engineering Meaning
Dynamic performance	t_r, t_s, $e_{\theta,ss}$, IAE/ITAE	Characterizes how fast the rotary valve moves from command tracking to stable holding, together with the accumulated tracking deviation
Pulse-shaping quality	$e_{p,ss}$, peak-to-peak ripple, ripple RMS, target-band energy	Characterizes pressure-pulse amplitude stability and the waveform proxies associated with downstream decoding
Constraints and actuation smoothness	violation count, maximum violation magnitude, $\sum\lvert\Delta u\rvert$, $\sum\lvert\Delta^2 u\rvert$	Characterizes actuator shock, constraint consistency, and lifetime friendliness
Real-time performance	mean time, 99% percentile, WCET, η_{rt}	Characterizes the ability of the control algorithm to complete online optimization within one sampling period

To keep the statistical definitions used in figures and tables consistent across scenarios, Figure 6 summarizes the unified KPI-processing workflow from raw logs to pre-processing, metric calculation, and summary output. Metrics such as t_r, ripple RMS, target-band energy, the 99th percentile, and WCET are therefore computed using the same definitions for all controllers.

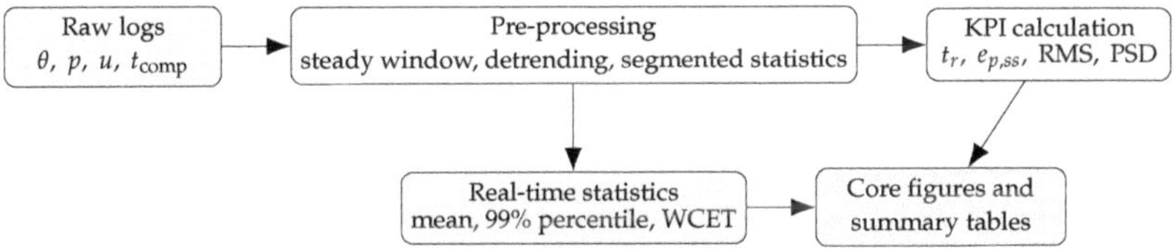

Figure 6. Unified KPI-processing workflow used for all controllers. Raw logs $\{\theta, p, u, t_{comp}\}$ are synchronized and pre-processed before time-domain, frequency-domain, constraint, and timing metrics are computed using the same definitions.

The simulation object adopts the discrete prediction model established in the previous section. The actuator-layer constraints are taken from the nominal parameter set: $\theta \in [\theta_{\min}, \theta_{\max}]$, $|\omega| \leq \omega_{\max}$, and $u \in [u_{\min}, u_{\max}]$, with an additional input-increment constraint applied to the proposed controller. Three representative scenarios are considered: a step valve-angle tracking scenario representing rapid downhole opening adjustment, a stable pressure-pulse output scenario representing continuous coded transmission, and a parameter-mismatch plus external-disturbance scenario representing common flow-rate fluctuation, structural wear, and load disturbance. Statistical post-processing is applied only to the existing verified trajectories. Only reproducible trajectory records in the present simulation dataset are reported.

The timing statistics are obtained on an Intel(R) Core(TM) i9-14900KF processor running MATLAB 9.13 (R2022b). The per-step computation time is measured by a unified timer at the entrance and exit of the controller solver, and then aggregated over all sampling steps into the mean, the 99% percentile, and the worst-case execution time. These timing results should be interpreted as indicators of algorithmic complexity distribution and tail risk rather than as direct evidence of WCET certification on a downhole embedded platform [20]. The controller settings and evidence status are summarized in Table 8.

Table 8. Controller settings and evidence status in the present study.

Controller	Decision Structure	Model Update, Constraints, and Evidence	Main Role
PID	No pre-screening and no sequential evaluation.	No online correction; same constraints and sampling period as the other verified controllers; reported numerically.	Low-complexity baseline reflecting the capability of conventional servo control.
Conventional MPC	Single weighted MPC cost without pre-screening or sequential evaluation.	No online correction; same constraints and sampling period as the proposed controller; reported numerically.	Baseline for predictive constrained control with one global objective.
Adaptive MPC	Same single weighted decision structure as conventional MPC.	Uses the corrected prediction model and the same constraints/sampling budget as the proposed controller, but is retained only as a reference formulation for complexity discussion because reproducible closed-loop trajectories under the same protocol are unavailable.	Reference formulation used to isolate the contribution of online correction from that of the lexicographic decision layer; excluded from the numerical KPI comparison.
Improved MPC	Candidate pre-screening followed by sequential lexicographic evaluation.	Includes online correction/compensation and shares the same constraints and sampling period; reported numerically.	Full method of this work, used to validate reduced search burden and mismatch robustness.

4.2. Parameter Sourcing, Identification Route, and Evidence Boundary

To reinforce the closure of the overall evidence chain, this section consolidates the parameter sources, identification means, and validation logic of both the model and controller. In general, rotary-valve studies should follow a four-level validation framework consisting of structural-parameter estimation, static calibration, dynamic identification, and statistical controller validation. In addition, drilling-fluid viscosity and hydraulic properties should incorporate temperature-, pressure-, and shear-rate-dependent corrections or uncertainty bounds [21]. The parameter identification routes are summarized in Table 9.

From the perspective of model validation, at least three steps should be carried out: (1) static Δp–θ and Δp–Q calibration to identify $A(\theta, g)$ and C_d; (2) valve-angle or torque step tests to fit key parameters such as J, B, and T_p; and (3) swept-frequency tests to obtain the magnitude and phase responses of the additional pressure drop, thereby marking the applicable frequency range of the model. In the present paper, these steps are described as the required identification route and as the evidence boundary of the current simulation study, not as already completed bench-validation tasks. For the downhole mud-pulse telemetry chain, pressure-sensor placement at the standpipe, sources of erosion-related error, and the interface to the decoding chain should also be documented in future bench/HIL work. The uncertainty ranges and sensitivity coverage are summarized in Table 10.

For online parameter identification, an augmented extended Kalman filter (EKF) can be used when field noise is significant and slow parameter drift must be tracked jointly with the states. When downhole computation is limited, recursive least squares (RLS) provides a lighter alternative for identifying the key parameters. The applicability of the two methods is summarized in Table 11.

Table 9. Parameter identification routes and current evidence status.

arameter	Nominal Status	Unit	Identification Route	Required Protocol	Evidence Status
	Internal nominal	$kg \cdot m^2$	CAD/inertia calculation	Rotor/drive inertia reconstruction or torque-step fit	Simulation only; no bench identification curve.
	Internal nominal	$N \cdot m \cdot s$ /rad	Free-decay identification	Free-spin or no-load decay test	Simulation only; no measured decay trace.
$_c$	Internal nominal	$N \cdot m$	Low-speed friction fit	Low-speed reversal/stiction test	Model term retained; no standalone bench curve.
$_h$	Internal nominal	–	Linearized fit/identification	Pressure-loading or torque-balance fit near the operating point	Corrected online in simulation; no bench fit.
$_d$	Internal nominal	–	Δp–Q calibration	Static flow/pressure calibration	Represented only by model-derived nominal baselines.
(θ, g)	Geometry-derived lookup surface	m^2	Structural geometry plus calibration logic	Static Δp–θ calibration over the admissible angle range	Model-derived nominal map only.
$_e$	Internal nominal	m^3	Chamber estimation	Geometry/volume estimate	Simulation only.
$_e$	Internal nominal	Pa	Fluid-property estimation	Mud-property characterization	Simulation only.
	Nominal clearance state around g_0	m	Manufacturing tolerance/wear estimate	Tolerance measurement or wear inspection	Slow-varying state only; no bench wear dataset.
$_p$	Internal nominal; not tabulated	s	Step-response fit/continuity-model fit	Pressure step or swept-frequency fit	Corrected online in simulation; no measured fit.

Table 10. Uncertainty ranges and sensitivity coverage.

arameter	Uncertainty/Sensitivity Coverage	Evidence Source	Remaining Gap
B, k_h, T_p	Included in the present mismatch proxy up to 30% with disturbance-intensity variation in Supplementary Figure S3.	Available qualitative sensitivity map plus corrected-model runs.	The present simulation dataset does not include a reproducible sweep script for regeneration or extension beyond the released grid.
$_c$	Not swept separately in the present simulation dataset.	Friction term retained in the grey-box model.	No dedicated low-speed friction identification or tolerance study is reported.
$_d$	Not swept separately; implicitly embedded in the nominal static baselines.	Model-derived Δp–θ and Δp–Q baselines in the static-calibration plot	Bench calibration data are unavailable, so the credibility boundary remains simulation based.
(θ, g) and g	Clearance and wear effects are represented qualitatively through the nominal map and robustness discussion.	Figure 3 plus the nominal geometric model.	No released numeric wear sweep or bench wear progression dataset is available.
$_e, \beta_e$	Not swept separately in the present simulation dataset.	Internal nominal simulation set.	Fluid-property uncertainty under realistic mud conditions remains to be characterized experimentally.
atic-calibration and FRF baselines	Used for operating-point selection, uncertainty-range design, and controller-tuning reference.	Model-derived nominal baselines, not bench-measured calibration curves.	These figures define the evidence boundary of the present simulation study rather than completed hardware calibration.

Table 11. Comparison of online identification methods for key rotary-valve parameters.

Method	Estimated Quantities	Advantages	Limitations	Complexity	Suitable Scenario
Augmented EKF	$x + \{B, k_h, T_p\}$	Stronger noise tolerance and able to track parameter drift	More complex tuning and covariance design	Medium–high	Significant noise and joint estimation required
RLS	$\{B, k_h, T_p\}$	Lower computation and suitable for online implementation	More sensitive to excitation and regressor conditions	Medium	Limited computation and slowly varying parameters

To further strengthen the evidence chain of parameter sourcing, calibration, identification, and control validation, Figure 7 presents the overall validation route of the model and controller. Figure 7 is intended to show what the present manuscript covers and what it does not yet cover: structural-parameter estimation, nominal calibration logic, dynamic-identification requirements, and simulation-based controller evaluation are discussed here, whereas bench/HIL deployment validation remains future work.

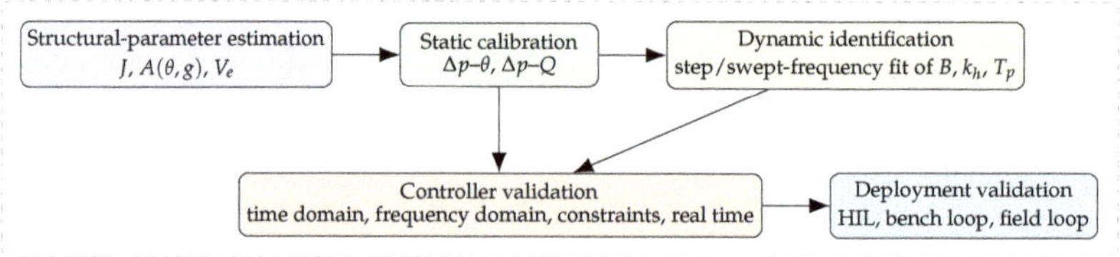

Figure 7. Validation route and evidence boundary of the present study. Structural estimation, nominal calibration, dynamic identification, and simulation-based controller evaluation are discussed here, whereas bench/HIL deployment validation remains future work.

In the absence of bench measurements, Figures 8 and 9 provide baseline plots generated from the nominal geometry, throttling equations, and linearized model. They are model-derived nominal baselines used for operating-point selection, uncertainty-range design, and controller-tuning reference; they are not bench-measured calibration or swept-frequency identification data. Their role is therefore to define the credibility boundary of the present simulation study rather than to claim completed hardware calibration.

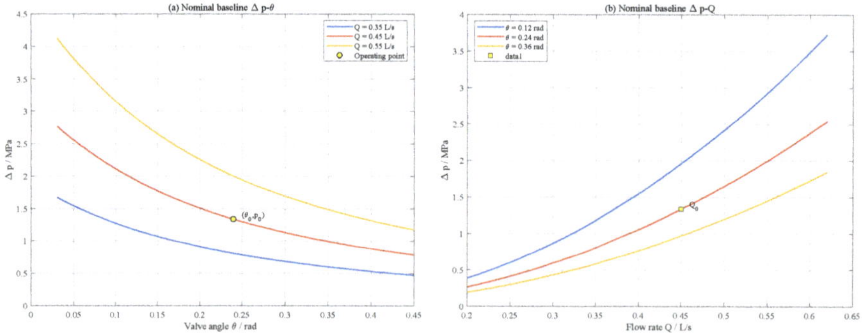

Figure 8. Model-derived nominal (**a**) Δp–θ and (**b**) Δp–Q baselines used for operating-point selection and uncertainty-range design; these curves are not bench-measured calibration data.

Figure 8a shows that, for a given flow level, the additional pressure drop decreases monotonically as the valve angle increases and the equivalent flow area becomes larger. Figure 8b shows that, for a fixed valve angle, the additional pressure drop grows approximately quadratically with the flow rate. These model-derived curves are used to select the nominal operating point (θ_0, p_0), to define admissible operating regions, and to motivate the uncertainty ranges later discussed in the robustness section; they are not treated as measured calibration evidence.

Figure 9 further gives the nominal magnitude and unwrapped-phase responses from the driving torque to the additional-pressure channel. The vertical dashed lines explicitly mark the target-band boundaries at 0.5 Hz and 3.0 Hz adopted in the pulse analysis. This model-derived baseline is used to define the controller target band and to explain the adopted small-signal validity range; it is not interpreted as a measured swept-frequency identification result. If future work extends the modulation band toward higher frequencies, the present baseline should be replaced by measured swept-frequency data so that the applicability boundary can be re-evaluated.

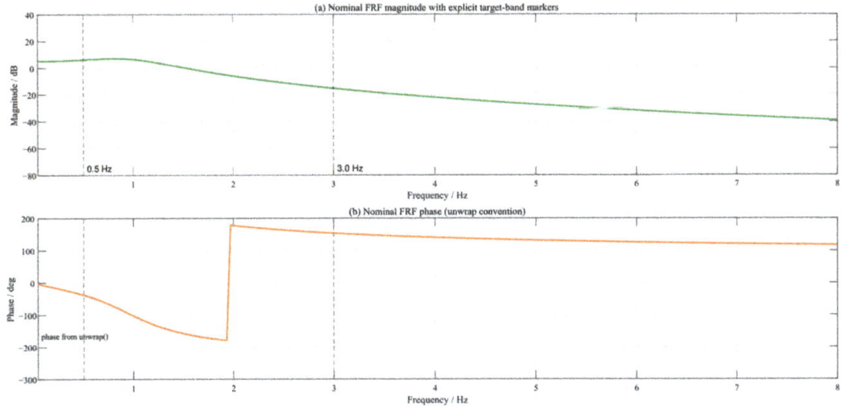

Figure 9. Model-derived nominal magnitude and unwrapped-phase frequency-response baseline used to define the controller target band; this figure is not a measured swept-frequency identification result.

4.3. *Step Valve-Angle Tracking Scenario*

In this scenario, the reference valve angle jumps from 0 to 0.22 rad at 0.5 s and is then further raised to 0.32 rad at 2.2 s. Figures 10 and 11 show that the improved MPC reduces the rise time from 2.18 s to 1.76 s relative to conventional MPC, lowers the steady-state pressure error from 0.1208 MPa to 0.0292 MPa, decreases the valve-angle IAE from 0.2424 to 0.1151, and cuts $\sum |\Delta u|$ from 113.0 to 17.0.

The step case is still nominal and therefore does not fully expose the advantage of online correction; PID retains a local steady-state advantage because the operating condition is simple and the constraint activity is mild. The main implication of this scenario is therefore not that the proposed controller dominates every scalar KPI, but that it achieves a better accuracy–smoothness compromise once pressure coordination and actuator restraint are considered jointly.

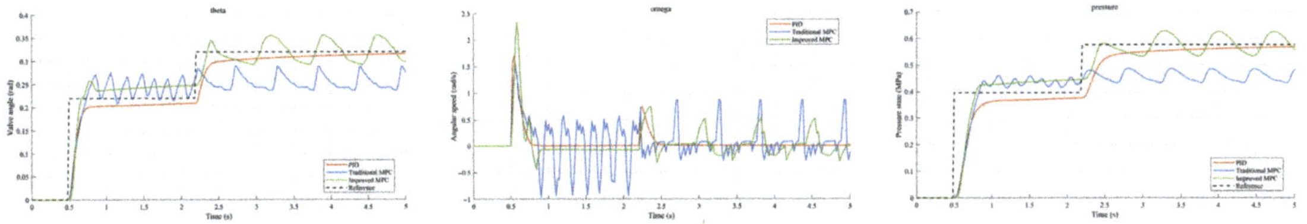

(**a**) Valve-angle response (**b**) Angular-speed response (**c**) Additional pressure-drop response

Figure 10. Main dynamic responses in the step valve-angle tracking scenario.

https://doi.org/10.3390/pr14101589

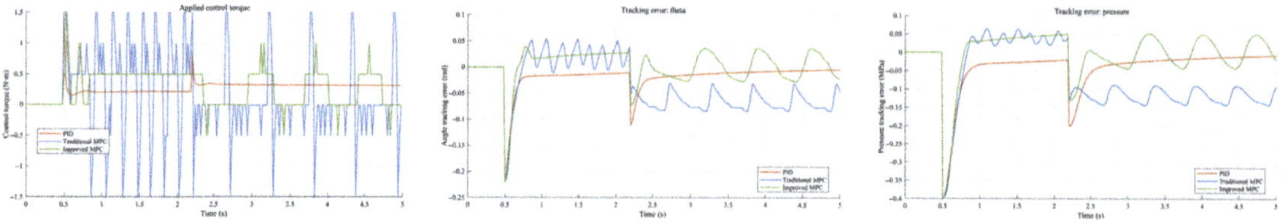

(**a**) Control input (**b**) Valve-angle error (**c**) Pressure-drop error

Figure 11. Control input and error comparison in the step valve-angle tracking scenario.

4.4. Stable Pressure-Pulse Output Scenario

In the stable pressure-pulse output scenario, the pressure-drop reference is composed of a bias term plus two sinusoidal components to emulate continuous mud-pulse modulation. Across Figures 12–16, the improved MPC shortens the valve-angle rise time to 0.18 s, lowers the steady-state pressure error to 0.0634 MPa, reduces the peak-to-peak ripple to 0.2541 MPa, and raises the target-band energy ratio to 98.11%. Relative to conventional MPC, the steady-state pressure error decreases by about 54.97%, the pressure-error IAE decreases from 0.8297 to 0.5473, and the in-band to out-of-band energy ratio increases from 12.49 to 17.90 dB.

The trade-off is that the ripple RMS of the improved MPC (0.0873 MPa) remains slightly higher than that of PID (0.0797 MPa), so the proposed method should be interpreted as improving the overall pulse-shaping compromise rather than minimizing every ripple metric simultaneously. The time–frequency concentration and leakage results are used here only as waveform-level proxies associated with downstream decoding: no full mud-channel attenuation model, surface demodulator dynamics, BER metric, or synchronization-jitter test is included in the available simulation evidence, and Supplementary Figure S6 is provided only to document that proxy mapping explicitly.

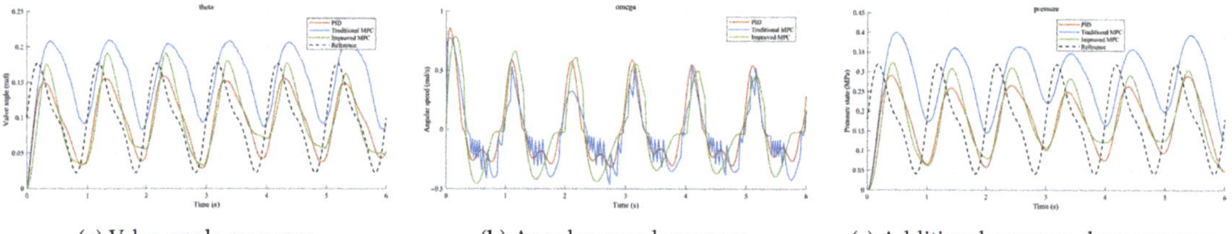

(**a**) Valve-angle response (**b**) Angular-speed response (**c**) Additional pressure-drop response

Figure 12. Main dynamic responses in the stable pressure-pulse output scenario.

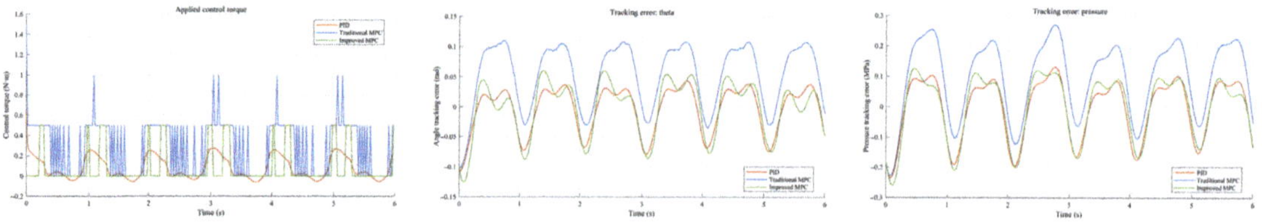

(**a**) Control input (**b**) Valve-angle error (**c**) Pressure-drop error

Figure 13. Control input and error comparison in the stable pressure-pulse output scenario.

https://doi.org/10.3390/pr1410158

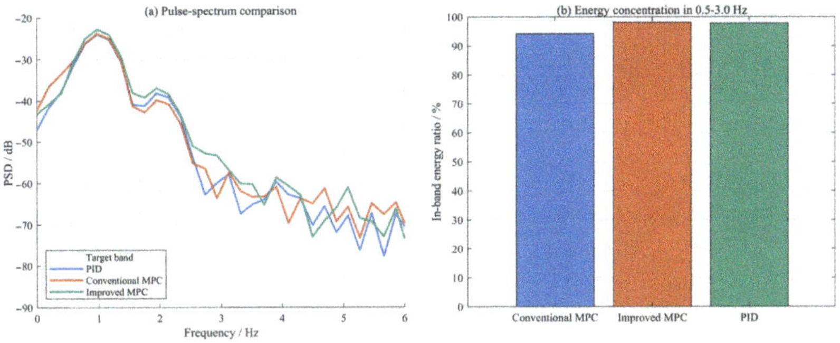

Figure 14. Spectrum and target-band (0.5–3.0 Hz) energy comparison in the pulse-output scenario.

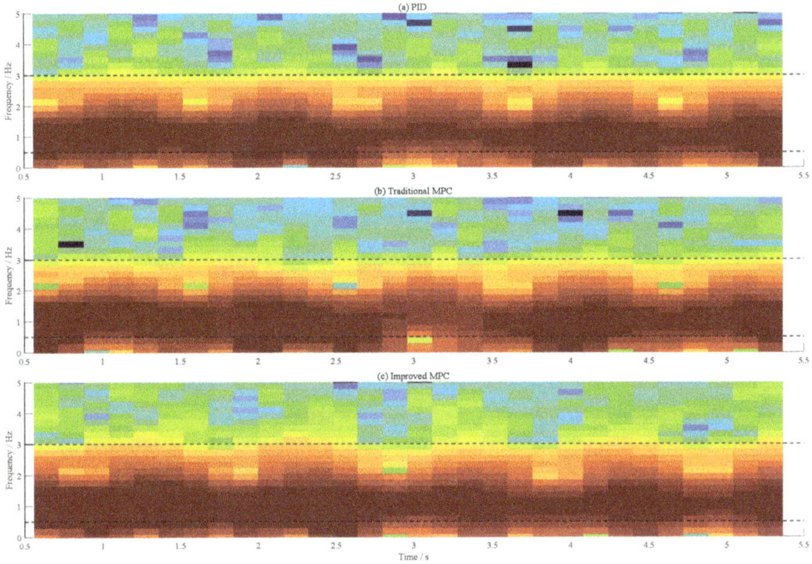

Figure 15. Time–frequency energy distribution in the pulse-output scenario. The label "Traditional MPC" in the legacy figure denotes the conventional MPC baseline.

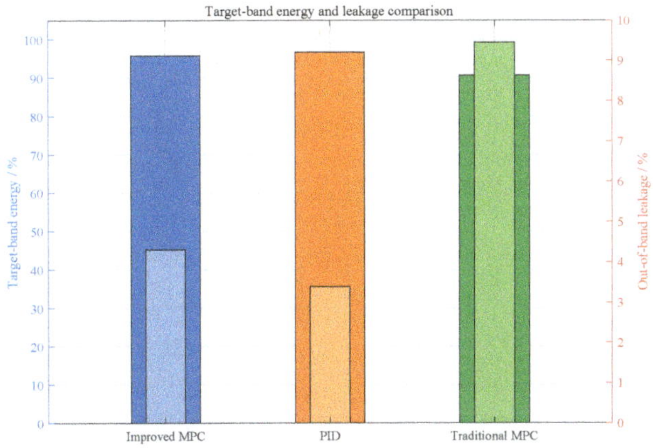

Figure 16. Out-of-band leakage and target-band energy ratio in the pulse-output scenario. The label "Traditional MPC" in the legacy figure denotes the conventional MPC baseline.

https://doi.org/10.3390/pr14101589

4.5. Robustness Evidence Under Severe Mismatch and Output-Noise Proxy Analysis

The available robustness trajectories correspond to a severe combined-mismatch case in which the effective inertia, damping coefficient, hydrodynamic-torque coefficient, and pressure-channel time constant are shifted from their nominal values while mechanical torque disturbance, inflow fluctuation, and an extra pressure disturbance are injected simultaneously. Under these tested conditions, the improved MPC shortens the rise time to 0.14 s, reduces the steady-state pressure error to 0.0606 MPa, lowers the pressure-error IAE to 0.5492, and cuts $\sum|\Delta u|$ to 19.5, which is only about 16.0% of the conventional-MPC value. Relative to conventional MPC, the steady-state pressure error decreases by about 48.83% even though the conventional controller retains a slightly smaller ripple RMS.

Figure 17 indicates that the controller continuity is maintained without repeated large pressure-window violations, while Figures 18, 19 and 20a,b show the dynamic responses, control/error comparison, correction factors, and disturbance estimates after disturbance injection. Supplementary Figure S3 adds the available two-dimensional sensitivity map over mismatch magnitude and disturbance intensity. Because only a qualitative sensitivity map is available in the present numerical dataset, that supplementary result is used for trend interpretation rather than for extracting quantitative robustness boundaries. Dedicated closed-loop measurement-noise injection over σ_p and σ_θ is outside the present trajectory dataset; therefore, the added noise-related evidence is limited to a descriptive pressure-envelope proxy rather than a closed-loop noise-injection test.

Dedicated closed-loop measurement-noise injection is outside the present trajectory dataset, but a derived descriptive proxy can still quantify how much pressure-error and ripple envelope is already occupied in the verified trajectories. For this purpose, the descriptive occupied envelope is defined as

$$\Delta p_{\mathrm{occ}} = |e_{p,ss}| + 3\,\mathrm{RMS}. \quad (52)$$

The resulting pressure-envelope proxy is summarized in Table 12, and the corresponding scenario-level statistics are summarized in Table 13. This derived metric supports descriptive interpretation only. It indicates how much output-side deviation is already present before hypothetical sensor noise is superposed. Smaller Δp_{occ} implies a larger residual measurement-noise headroom, but that headroom cannot be quantified exactly from the available summary tables because the corresponding pressure-window half-width is not tabulated.

Table 12. Pressure-envelope proxy under hypothetical output noise.

Scenario	Conventional MPC Δp_{occ}/MPa	Improved MPC Δp_{occ}/MPa	Relative Change	Interpretation
Step tracking	0.1760	0.1261	−28.35%	Smaller occupied envelope despite a larger RMS value, because the steady-state pressure bias is much lower.
Pulse output	0.3940	0.3253	−17.44%	Leaves more output-side envelope before hypothetical measurement noise is superposed or the available waveform.
Mismatch + disturbance	0.2525	0.2292	−9.23%	Shows a narrower pressure-error envelope even in the severe mismatch case.

https://doi.org/10.3390/pr1410158

Table 13. Summary statistics for numerically verified controllers across the three scenarios.

Controller	t_r/s	$e_{p,ss}$ MPa	RMS MPa	$\sum\lvert\Delta u\rvert$	Viol.	P99 ms	WCET ms
Step tracking							
PID	1.84	0.0102	0.0013	3.9991	0	0.2833	2.2015
Conventional MPC	2.18	0.1208	0.0184	113.0	0	2.4078	37.0310
Improved MPC	1.76	0.0292	0.0323	17.0	0	4.8981	10.5740
Pulse output							
PID	0.26	0.0729	0.0797	4.3565	0	0.0737	0.2172
Conventional MPC	1.34	0.1408	0.0844	51.0	0	1.8437	4.1552
Improved MPC	0.18	0.0634	0.0873	11.5	0	3.1568	3.3891
Mismatch + disturbance							
PID	0.20	0.0867	0.0596	4.6081	0	0.0491	0.1030
Conventional MPC	0.20	0.1184	0.0447	121.5	0	3.6926	3.7723
Improved MPC	0.14	0.0606	0.0562	19.5	0	2.1882	3.8136

The adaptive-MPC formulation is retained as a reference formulation for complexity discussion because it shares the corrected prediction model but uses a single weighted objective. Since reproducible closed-loop trajectories under the same protocol are unavailable, it is excluded from the numerical KPI comparison.

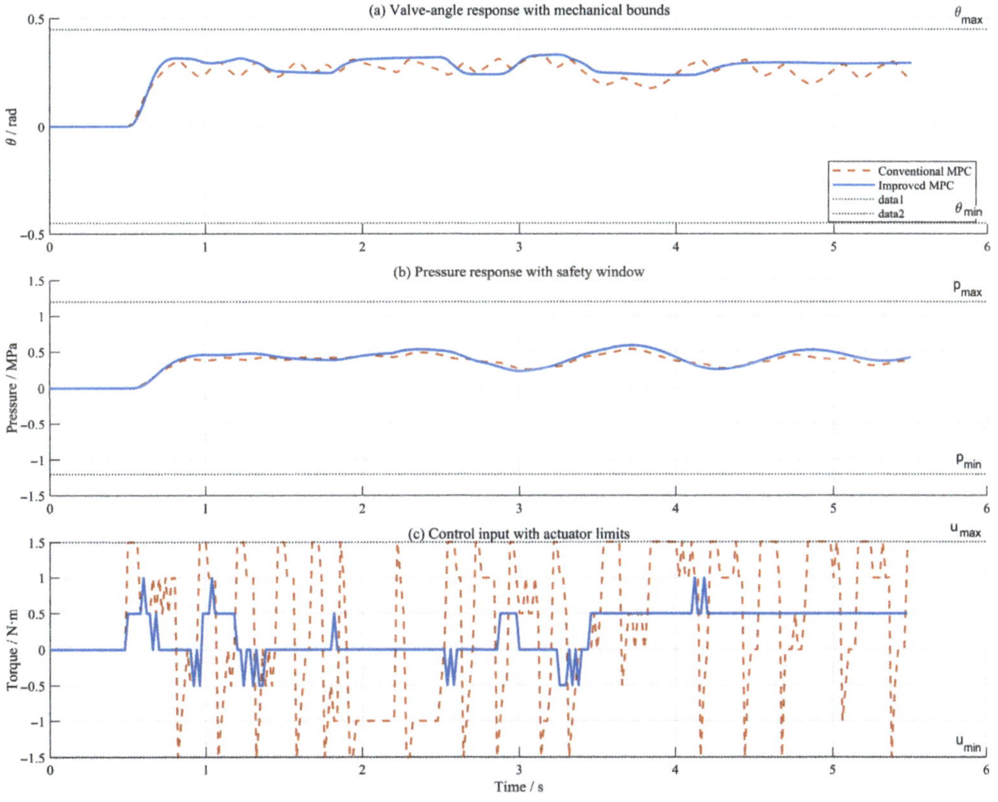

Figure 17. Overlay of key variables and constraint windows in the mismatch–disturbance scenario.

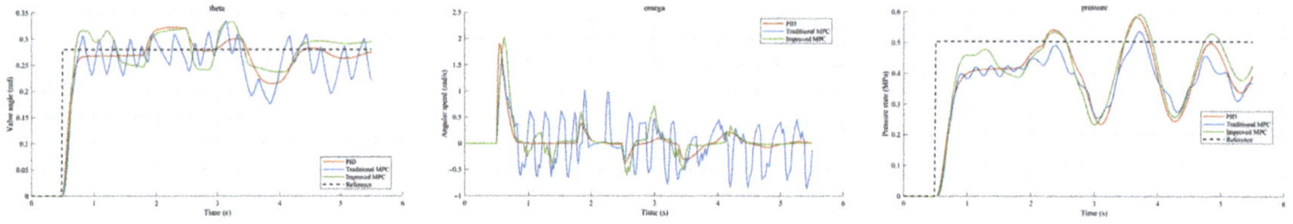

(**a**) Valve-angle response (**b**) Angular-speed response (**c**) Additional pressure-drop response

Figure 18. Main dynamic responses in the mismatch–disturbance scenario.

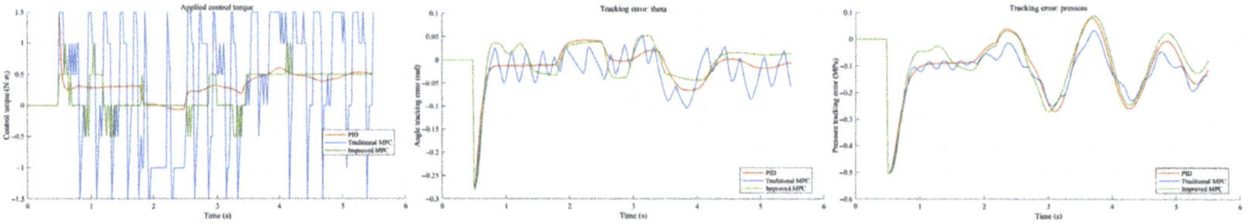

(**a**) Control input (**b**) Valve-angle error (**c**) Pressure-drop error

Figure 19. Control input and error comparison in the mismatch–disturbance scenario.

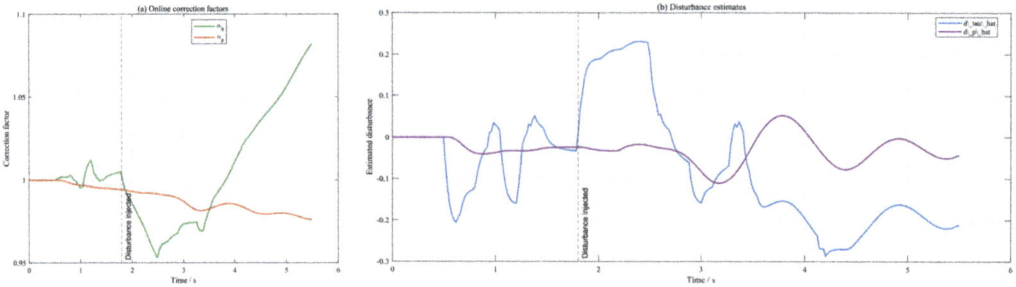

Figure 20. Trajectories of the online correction factors γ_h and γ_p and of the disturbance estimates $\hat{d}_\tau$ and $\hat{d}_p$ in the mismatch–disturbance scenario.

4.6. Real-Time Complexity and Scalability

Table 13 reports only the controllers whose trajectories were numerically verified from the present simulation dataset: PID, conventional MPC, and the improved MPC. The adaptive-MPC formulation is retained as a reference formulation for complexity discussion because it shares the corrected prediction model but uses a single weighted objective. Since reproducible closed-loop trajectories under the same protocol are unavailable, it is excluded from the numerical KPI comparison. The proposed controller is not uniformly best in every nominal scalar metric, yet it consistently reduces steady-state pressure error and input variation relative to conventional MPC while keeping the observed MATLAB workstation execution times below one sampling period in all verified scenarios.

The numerical implementation settings are listed in Tables 14–16. To reduce the scanning burden caused by dense multi-panel summaries, Figure 21 reorganizes the key improvements of the improved MPC relative to conventional MPC into a single heatmap. The figure shows that the most stable gains appear in steady-state pressure error and $\sum|\Delta u|$, whereas the ripple RMS or the 99th-percentile computation time is not always absolutely minimal in nominal scenarios. Monte-Carlo distributions, ablation views, retained-set tolerance proxies, and candidate-rejection statistics are kept in Supplementary Figures S1, S2, S4 and S5 so that the main text remains focused on the strongest supporting evidence.

https://doi.org/10.3390/pr1410158

Table 14. Sampling and horizon settings used in the numerical implementation.

Setting	Value Summary	Note
Sampling period T_s	PID/conventional MPC/adaptive-MPC reference/improved MPC: 20 ms	All controllers share the same sampling period.
Prediction horizon N_p	Conventional MPC/adaptive-MPC reference/improved MPC: 8	The MPC-type controllers are defined on the same prediction horizon.
Control horizon N_c	Conventional MPC/adaptive-MPC reference/improved MPC: 4	Shared control horizon in the implementation settings used in the simulations.
Admissible input levels	Conventional MPC/adaptive-MPC reference/improved MPC: 7 levels in $[u_{\min}, u_{\max}]$	The implementation uses the same quantized input grid for the three MPC-type controllers.

Table 15. Baseline controller gains and objective weights.

Setting/Controller	Value	Note
PID main-loop gains (PID)	$\{K_{p\theta}, K_{i\theta}, K_{d\theta}\} = (4.8, 2.4, 0.55)$	Low-complexity valve-angle reference loop.
PID pressure-loop gains (PID)	$\{K_{pp}, K_{ip}\} = (0.85, 0.45)$	Pressure-servo baseline.
Objective weights (conventional MPC)	$(w_\theta, w_\omega, w_p, w_{ps}, w_{\Delta u}, w_{\Delta^2 u}) = (120, 2.5, 50, 30, 0.10, 0.03)$	One weighted objective over the raw candidate set.
Objective weights (adaptive-MPC reference)	Defined reference formulation; to be re-tuned on the corrected model when scripts are available	No numerical values are reported because reproducible closed-loop trajectory data are unavailable for this comparator.
Intra-stage scales (improved MPC)	$(w_\theta, w_\omega, w_p, w_{ps}, w_{\Delta u}, w_{\Delta^2 u}) = (120, 2.0, 60, 35, 0.08, 0.02)$	Lexicographic prioritization keeps only a small set of intra-stage scales.

Table 16. Online correction and lexicographic settings.

Setting	Value	Note
Retained scales (improved MPC)	$(M_1, M_2, M_3) = (10, 5, 1)$	Only the improved MPC uses stage-wise candidate retention.
Lexicographic tolerances (improved MPC)	$(\varepsilon_1, \varepsilon_2) = (0.03, 0.02)$	Empirical near-optimality thresholds; a full tolerance sweep is not included in the present simulation dataset.
Correction gains (adaptive-MPC reference)	Same corrected-model update law as the proposed controller (defined only)	States the intended update mechanism of the adaptive-MPC reference formulation.
Correction gains (improved MPC)	$(\gamma_h, \gamma_p, l_\tau, l_p) = (0.04, 0.03, 0.28, 0.24)$	Hydrodynamic, pressure, and disturbance-estimation updates.

The available Monte-Carlo and ablation evidence is retained at the distribution level only. Supplementary Figure S1 shows that the improved MPC maintains a lower or near-lower distribution of steady-state pressure error than the other controllers shown in the supplementary distributions, while Supplementary Figure S2 shows that removing online correction most strongly worsens pressure regulation and removing pre-screening most strongly enlarges tail computation time. Because the present simulation dataset does not include the resampling scripts or raw trial tables needed for nonparametric hypothesis tests or confidence intervals, these observations are reported as distribution-level support rather than as formal statistical-significance claims.

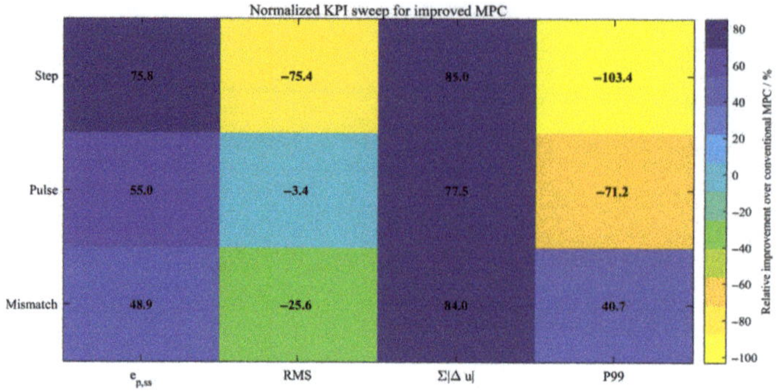

Figure 21. Normalized KPI-improvement heatmap of the improved MPC relative to conventional MPC.

Figure 22 summarizes the mean computation time, the 99% percentile, the WCET, and the mean candidate-set size at each stage for the improved MPC. The observed 99th-percentile times are 4.8981 ms, 3.1568 ms, and 2.1882 ms in the three verified scenarios, while the worst observed execution times are 10.5740 ms, 3.3891 ms, and 3.8136 ms. The mean number of Stage-1 candidates remains about 47–49 and is compressed to about 10 after Stage 1 filtering and to about 5 for the final selection stage, which is consistent with the upper-bound discussion of M_h, M_1, M_2, and M_3 in Section 3.5.

To complement the MATLAB workstation timing data, an architecture-independent operation-count interpretation is added here. With $n_x = 3$, local branching factor $b = 3$, and move-blocking depth N_b, the dominant online cost remains candidate rollout and stage-wise evaluation rather than parameter updating. However, these counts do not include fixed-point scaling, cache behaviour, task scheduling, memory contention, or RTOS overhead on a target downhole processor. Certified WCET therefore remains a target-platform task rather than a conclusion of the present workstation-based timing study, and a full N_p/N_c horizon sweep is not included in the present simulation dataset.

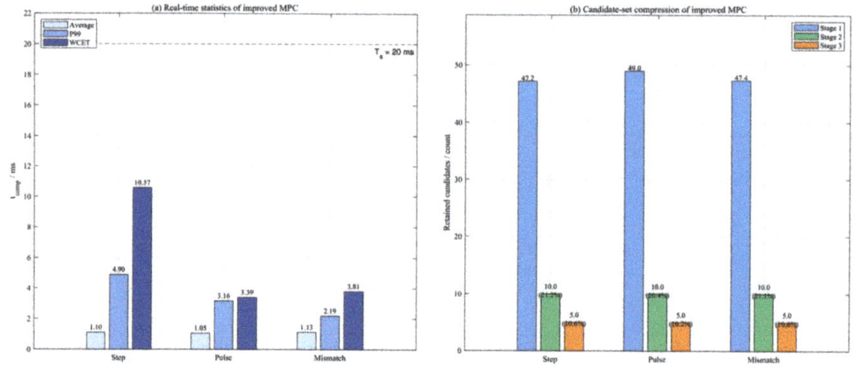

Figure 22. Real-time statistics and candidate-set compression of the improved MPC.

Because a full tolerance sweep over $(\varepsilon_1, \varepsilon_2)$ is not included in the present simulation dataset, the available candidate-compression data of Figure 22 are re-expressed here as a candidate-retention and timing proxy. The selected tolerances $\varepsilon_1 = 0.03$ and $\varepsilon_2 = 0.02$ are empirical near-optimality thresholds. Table 17 explains the retained-candidate ratios and timing behaviour for these selected values, but it does not establish global optimality or replace a full tolerance sensitivity sweep.

Table 17. Candidate-retention and timing proxy for the selected lexicographic tolerances.

Scenario	Observed M_h	Observed M_1	Observed M_2	M_1/M_h	M_2/M_h	P99/ms	WCET/ms
Step tracking	47.2	10.0	5.0	21.2%	10.6%	4.8981	10.5740
Pulse output	49.0	10.0	5.0	20.4%	10.2%	3.1568	3.3891
Mismatch + disturbance	47.4	10.0	5.0	21.1%	10.6%	2.1882	3.8136

The observed stage counts in Figure 22 correspond to M_h, M_1, and M_2; the final selected control is one element after Stage 3. This proxy is descriptive and does not replace a full tolerance sensitivity sweep.

The available scenario summaries support only a descriptive scenario-level consistency analysis rather than a formal statistical-significance test. To avoid leaving the comparison purely verbal, Table 18 summarizes the relative change of the proposed controller against conventional MPC over the three verified scenarios. Negative values denote reductions relative to conventional MPC.

Table 18. Scenario-level descriptive effect-size consistency relative to conventional MPC.

Metric	Step	Pulse	Mismatch	Median Relative Change	Direction Consistency
$e_{p,ss}$	−75.8%	−55.0%	−48.8%	−55.0%	3/3 lower
$\sum\lvert\Delta u\rvert$	−85.0%	−77.5%	−84.0%	−84.0%	3/3 lower
t_r	−19.3%	−86.6%	−30.0%	−30.0%	3/3 lower
RMS	+75.5%	+3.4%	+25.7%	+25.7%	3/3 higher

This is not a formal hypothesis test. It is a descriptive consistency analysis over the three verified scenarios and therefore does not provide p-values or confidence intervals.

4.7. Telemetry-Chain Proxy Analysis and Engineering Implications

For telemetry-oriented descriptive quantification, the available in-band/out-of-band ratio is re-expressed as the spectral-purity index

$$\mathrm{SPI} = 10\log_{10}\frac{E_{\mathrm{target}}}{E_{\mathrm{out}} + \epsilon_{\mathrm{num}}}, \tag{53}$$

where ϵ_{num} is a small numerical constant used only to avoid a zero denominator. In the available pulse-output record, SPI coincides with the reported in-band-to-out-of-band energy ratio of that scenario. These metrics are telemetry-oriented waveform proxies and should not be interpreted as BER, synchronization-error, or channel-capacity results. The telemetry-oriented waveform proxy summary is given in Table 19.

Table 19. Telemetry-oriented waveform proxy summary from available simulation outputs.

Metric	Available Numerical Proxy	Interpretation	Limitation
Steady-state pressure error $e_{p,ss}$	Pulse-output scenario: 0.1408 → 0.0634 MPa; mismatch scenario: 0.1184 → 0.0606 MPa	Smaller pressure bias leaves a larger waveform-amplitude reserve before any additional output perturbation is superposed.	No full channel attenuation or standpipe transfer model is included.
Ripple RMS/peak-to-peak ripple	Improved MPC pulse output: RMS = 0.0873 MPa, peak-to-peak ripple = 0.2541 MPa	Quantifies waveform fluctuation and template distortion at the valve outlet only.	No synchronization-error statistic is reported.
Target-band energy/SPI	Improved MPC target-band energy ratio = 98.11%; SPI = 17.90 dB versus 12.49 dB for conventional MPC	Indicates stronger concentration in the 0.5–3.0 Hz target band and a descriptive SPI gain of 5.41 dB in the available pulse outputs.	These metrics are not BER, synchronization-error, or channel-capacity results.
Input smoothness $\sum\lvert\Delta u\rvert$	Pulse-output scenario: 51.0 → 11.5	Lower input variation acts as a jitter-oriented actuator proxy for pulse-edge regularity.	No demodulator model or timing-jitter statistic is included.

These telemetry-oriented waveform proxies support descriptive interpretation only. They should not be interpreted as synchronization-error or channel-capacity results.

Taken together, the results of Section 4 indicate that the engineering value of the proposed controller does not lie in achieving the absolute best value of every KPI in every nominal case. Instead, it lies in obtaining a more balanced compromise across fast valve-angle tracking, pressure-pulse shaping, input smoothness, disturbance tolerance, and online solvability. In the telemetry context, lower front-end control error and lower out-of-band leakage are interpreted only as improvements in waveform-level proxies associated with downstream decoding rather than as direct proof of better decoding performance.

The present waveform-proxy chain remains intentionally narrow. Supplementary Figure S6 documents the qualitative mapping from control error to waveform quality and then to decode-related proxies, but the present simulation dataset does not include a full mud-channel attenuation model, additive coloured pump-noise surrogate, surface demodulator dynamics, or BER computation. BER, synchronization-error statistics, and channel-capacity metrics are beyond the evidence scope of the present waveform-proxy analysis.

This limitation is also the reason why the analysis stops at proxy-level interpretation. Pump noise, channel attenuation, and demodulator dynamics are acknowledged explicitly as missing subsystems, and the HIL/bench route in the following subsection is tied directly to that gap.

4.8. Limitations and Future Validation Route

At the same time, the parameter-source and validation chain still relies mainly on nominal geometric relations, control-oriented baseline plots, available simulation figures, and closed-loop numerical statistics. Figures 8 and 9 are still model-derived baselines rather than bench-measured curves, and the corresponding captions state that the key calibration baselines are not bench-measured calibration data. These derived metrics support descriptive interpretation only. They do not replace closed-loop noise-injection tests, full tolerance sweeps, BER-level telemetry validation, or target-platform WCET certification, which remain future work. The adaptive-MPC formulation is retained as a reference formulation for complexity discussion because it shares the corrected prediction model but uses a single weighted objective. Since reproducible closed-loop trajectories under the same protocol are unavailable, it is excluded from the numerical KPI comparison. The evidence should therefore be interpreted as simulation-based evidence for a rotary-valve-specific control integration, not as certification of hardware maturity.

From a reproducibility and deployment viewpoint, the KPI workflow and controller tables have been clarified, yet extrapolation to downhole hardware still requires caution. The reported mean time, 99% percentile, and WCET reflect algorithmic timing on the MATLAB workstation implementation, and the observed timing suggests, rather than certifies, embedded feasibility. Fixed-point realization, RTOS scheduling, peripheral communication, and processor-specific WCET certification remain future tasks. Pump noise, sampling delay, quantization error, channel attenuation, and surface demodulation should also be integrated into the next bench/HIL stage before any decoding-level claim is strengthened.

Figure 23 summarizes a deployment-oriented HIL/bench validation architecture. It places delay, quantization, pump-noise/channel modules, and WCET measurement points into the same closed-loop structure so that a unified interface is available for subsequent engineering validation.

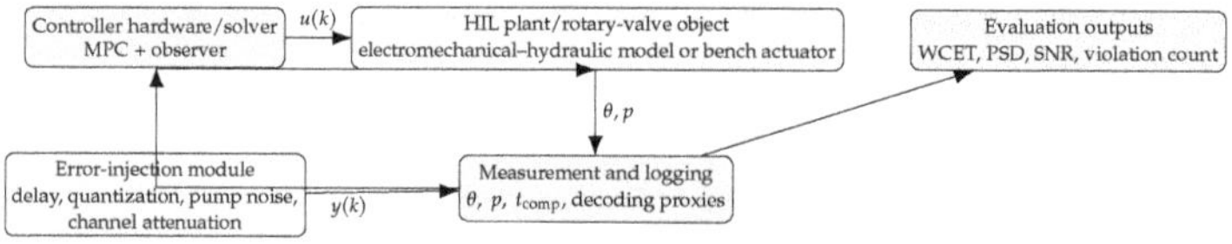

Figure 23. Planned HIL and bench-validation platform structure.

https://doi.org/10.3390/pr1410158

5. Conclusions and Outlook

This study investigated real-time constrained control of an intelligent drill-bit rotary valve for mud-pulse telemetry by combining a control-oriented electromechanical–hydraulic grey-box model with online correction and a lexicographic MPC architecture.

Across the step-tracking, pulse-output, and severe-mismatch scenarios, the proposed controller reduced the steady-state pressure error from 0.1208 to 0.0292 MPa, from 0.1408 to 0.0634 MPa, and from 0.1184 to 0.0606 MPa relative to conventional MPC, while also lowering cumulative input variation from 113.0 to 17.0, from 51.0 to 11.5, and from 121.5 to 19.5, respectively. The observed MATLAB workstation execution times remained below the 20 ms sampling period in all verified cases.

The main finding is not that the proposed method dominates every scalar KPI in every nominal condition, but that it provides a balanced compromise among pulse-shaping accuracy, pressure safety, disturbance tolerance, and online solvability within the tested numerical envelope. In addition to the verified scenario trajectories, the study includes descriptive analyses for pressure-envelope occupancy, candidate retention, cross-scenario effect-size consistency, and telemetry-oriented spectral purity.

The evidence remains simulation-based. Bench calibration curves, dedicated closed-loop measurement-noise injection, full tolerance sweeps, closed-loop adaptive-MPC trajectory benchmarking, target-processor WCET certification, a complete mud-pulse channel/demodulator model, BER-oriented validation, and formal recursive-feasibility/stability analysis remain future work.

Future work will therefore focus on bench/HIL validation, target-platform timing analysis, and telemetry-chain experiments using synchronization- and BER-oriented metrics.

Supplementary Materials: The supporting information can be downloaded at: https://www.mdpi.com/article/10.3390/pr14101589/s1, Figure S1: Monte-Carlo distributions; Figure S2: ablation comparison; Figure S3: mismatch–disturbance sensitivity map; Figure S4: retained-set tolerance proxy; Figure S5: candidate-rejection statistics; Figure S6: qualitative control-error-to-decoding proxy chain; Figure S7: data schema and reproducibility route.

Author Contributions: Conceptualization, X.D. and L.Y.; methodology, X.D. and Z.Z.; software, X.D.; validation, X.D., L.W., Z.Z., Y.J. and R.L.; formal analysis, X.D. and L.Y.; investigation, L.W., Z.Z., Y.J. and R.L.; resources, L.Y. and L.W.; data curation, X.D. and Z.Z.; writing—original draft preparation, X.D.; writing—review and editing, L.Y., L.W., Z.Z., Y.J. and R.L.; visualization, X.D.; supervision, L.Y.; project administration, L.Y.; funding acquisition, L.Y. and L.W. All authors have read and agreed to the published version of the manuscript.

Funding: This research was funded by the National Major Science and Technology Project for Deep Earth (DEEP) during the 14th Five-Year Plan Period of China (Grant No. 2024ZD1003504).

Data Availability Statement: The data and scripts supporting the reproducible figures and tables are available from the corresponding author upon reasonable request. Derived tables based on available simulation outputs are identified as descriptive analyses in the text. The current dataset does not include bench, HIL, or field measurements.

Acknowledgments: The authors acknowledge the support of the State Key Laboratory of Oil and Gas Reservoir Geology and Exploitation and the College of Energy Resources (Modern Industrial College of Shale Gas), Chengdu University of Technology.

Conflicts of Interest: Author Lingyun Wang was employed by the No. 10 Oil Production Plant of PetroChina Changqing Oilfield Company. The remaining authors declare that the research was conducted in the absence of any commercial or financial relationships that could be construed as a potential conflict of interest.

Abbreviations

The following abbreviations are used in this manuscript:

A-MPC	adaptive model predictive control
BER	bit error rate
CAD	computer-aided design
EKF	extended Kalman filter
FCS-MPC	finite-control-set model predictive control
HIL	hardware in the loop
IAE	integral absolute error
ITAE	integral time absolute error
KPI	key performance indicator
MPC	model predictive control
MWD	measurement while drilling
LWD	logging while drilling
PID	proportional–integral–derivative
PSD	power spectral density
QP	quadratic programming
RMS	root mean square
RLS	recursive least squares
RTOS	real-time operating system
SMO	sliding-mode observer
SNR	signal-to-noise ratio
WCET	worst-case execution time
ZOH	zero-order hold

References

1. Liu, J.; Xu, M.; Switzer, D.A.; Logan, A.W. Downhole Telemetry Signal Modulation Using Pressure Pulses of Multiple Puls Heights. U.S. Patent US20150330217A1, 19 November 2015.
2. Jia, M.; Geng, Y.; Yan, Z.; Zeng, Q.; Wang, W.; Yue, Y. Channel Modelling and Characterization for Mud Pulse Telemetry. *AEU-In J. Electron. Commun.* **2023**, *165*, 154654. [CrossRef]
3. Liang, H.; Yin, C.-C.; Su, Y.; Liu, Y.-H.; Li, J.; Gao, R.-Y.; Wang, L.-B. Fast Simulation of EM Telemetry in Vertical Drilling A Semi-Analytical Finite-Element Method with Virtual Layering Technique. *Pet. Sci.* **2025**, *22*, 3304–3314. [CrossRef]
4. Zhao, A.-S.; He, X.; Chen, H.; Wang, X.-M. Response Analyses on the Drill-String Channel for Logging While Drilling Telemetr *Pet. Sci.* **2023**, *20*, 2796–2808. [CrossRef]
5. Wang, W.; Yan, X. AI-Based Grey Wolf Chemical Optimization Approach for Energy-Efficient Downhole Telemetry Signa Demodulation toward Greener Subsurface Systems. *Microchem. J.* **2025**, *220*, 116394. [CrossRef]
6. Jia, M.; Geng, Y.; Zeng, Q.; Wang, W.; Yue, Y. Widely Linear RLS Equalizer with Variable Forgetting Factor and UD Factorizatio for Mud Pulse Telemetry. *AEU-Int. J. Electron. Commun.* **2024**, *183*, 155367. [CrossRef]
7. Zeng, Q.; Geng, Y.; Jiang, S.; Wang, W. WaveU-Net: Multi-Scale Wavelet Framework for Robust Recovery of Continuous Pressur Signals in Mud Pulse Telemetry. *Digit. Signal Process.* **2026**, *173*, 105853. [CrossRef]
8. Zhang, J.; Sha, Z.; Wan, L.; Su, Y.; Zhu, J.; Qu, F. A Fractional Fourier Transform-Based Channel Estimation and Equalizatio Algorithm for Mud Pulse Telemetry. *J. Mar. Sci. Eng.* **2025**, *13*, 1468. [CrossRef]
9. Zhang, J.; Sha, Z.; Tu, X.; Zhang, Z.; Zhu, J.; Wei, Y.; Qu, F. Noise Cancellation Method for Mud Pulse Telemetry Based on Discret Fourier Transform. *J. Mar. Sci. Eng.* **2025**, *13*, 75. [CrossRef]
10. Yao, X.; Huang, S.; Wang, J.; Ma, H.; Liu, T. Model Predictive Power Control of Permanent Magnet Synchronous Motor i Two-Phase Static Coordinate System. *Trans. China Electrotech. Soc.* **2021**, *36*, 60–67. [CrossRef]
11. Yao, X.; Huang, S.; Wang, J.; Ma, H.; Liu, T.; Zhang, G. A Two-Vector-Based Model Predictive Current Control with Onlin Parameter Identification for PMSM Drives. *Proc. Chin. Soc. Electr. Eng.* **2023**, *43*, 9319–9329.
12. Liu, T.; Yao, X.; Wang, J.; Kou, J. Enhanced Model Predictive Control for Induction Motor Drives in Marine Electric Powe Propulsion System. *J. Mar. Sci. Eng.* **2024**, *12*, 378. [CrossRef]
13. Liu, T.; Yao, X.; Wang, J.; Ma, C. Efficient Two-Vector-Based Sequential Model Predictive Control for IM Drives. *IEEE J. Emerg. Se Top. Power Electron.* **2024**, *12*, 903–912. [CrossRef]

https://doi.org/10.3390/pr1410158

Zhai, C.; Deng, Y.; Li, W.; Xu, L.; Kang, Y.; Cao, H.; Liu, X.; Zhang, Z.; Zhang, Y. Complex Vector Disturbance Observer-Based Deadbeat Predictive Current Controller for PMSM Current Harmonics Suppression. *IEEE Trans. Transp. Electrif.* 2026, *early access*. [CrossRef]

Wu, X.; Wang, Y.; Wang, N.; Xing, H.; Xie, W.; Lee, C.H.T. A Novel Double-Vector Model Predictive Current Control for PMSM with Low Computational Burden and Switching Frequency. *IEEE Trans. Power Electron.* **2025**, *40*, 11283–11295. [CrossRef]

Li, X.; Yang, Y.; Sun, J.; Xiao, Y.; Fan, M.; Ni, K.; Hu, J.; Wen, H.; Yang, H.; Rodriguez, J. Multiple-Voltage-Vector Model-Free Predictive Deadbeat Control with Updated Reference Voltage Vector for PMSM Drive. *IEEE Trans. Power Electron.* **2025**, *40*, 6492–6505. [CrossRef]

Zhang, Z.; Deng, Y.; Li, H.; Wang, J.; Liu, X.; Cao, H. Enhanced Model-Free Deadbeat Predictive Current Control for PMSM Drives Based on Generalized Adaptive Super-Twisting Observer. *Control Eng. Pract.* **2026**, *166*, 106611. [CrossRef]

Zhang, Z.; Zhang, J.; Wu, Y.; Zhu, Z.; Chang, S. A Weighted-Free CCS Model Predictive Current Control with Implicit Modulation for DTP-PMSMs. *IEEE Trans. Ind. Electron.* **2025**, *72*, 3335–3345. [CrossRef]

Zhang, X.; Cao, Y. A Simple Motor-Parameter-Free Model Predictive Current Control for PMSM Drive. *IEEE Trans. Ind. Electron.* **2025**, *72*, 3292–3302. [CrossRef]

Wilhelm, R.; Engblom, J.; Ermedahl, A.; Holsti, N.; Thesing, S.; Whalley, D.; Bernat, G.; Ferdinand, C.; Heckmann, R.; Mitra, T.; et al. The Worst-Case Execution Time Problem—Overview of Methods and Survey of Tools. *ACM Trans. Embed. Comput. Syst.* **2008**, *7*, 36. [CrossRef]

American Petroleum Institute. *Recommended Practice for the Rheology and Hydraulics of Oil-Well Systems: API Recommended Practice 13D*; API Publishing Services: Washington, DC, USA, 2017.

https://doi.org/10.3390/pr14101589

processes

Article

Research and Application of an Intelligent Cable-Controlled Injection–Production Integration and Control System

Jianhua Bai [1], Zheng Chen [1], Wei Zhang [1], Zhaochuan Zhou [2], Liu Wang [2], Yuande Xu [1], Shaojiu Jiang [1], Chengtao Zhu [2], Zhijun Liu [2], Le Zhang [1], Zechao Huang [1], Qiang Wang [1], Zhixiong Zhang [1], Chenwei Zou [1], Xiaodong Tang [1] and Yukun Du [3,4,*]

[1] CNOOC China Limited, Tianjin Branch, Tianjin 300459, China
[2] CNOOC EnerTech-Drilling & Production Co., Tianjin 300452, China
[3] State Key Laboratory of Deep Oil and Gas, China University of Petroleum (East China), Qingdao 266580, China
[4] School of Petroleum Engineering, China University of Petroleum (East China), Qingdao 266580, China
* Correspondence: duyukun@upc.edu.cn

Abstract

During offshore oilfield development, traditional injection–production processes commonly suffer from delayed regulation, low operational efficiency, and heavy reliance on manual intervention. Achieving real-time diagnosis of injection–production anomalies and dynamic optimization under complex geological conditions and harsh marine environments represents a core scientific challenge. This study presents the development and field deployment of an intelligent cable-controlled injection–production integrated management system. The work is positioned as an application- and system-oriented study, focusing on addressing practical challenges in offshore oilfield operations through the integration of established machine learning techniques into a cohesive operational platform. The system employs a cloud-native microservice architecture and integrates nine functional modules, enabling closed-loop management from data acquisition to intelligent decision making. Key methodological contributions include: (1) a weighted ensemble model combining Random Forest and SVM for blockage diagnosis, balancing global feature learning with boundary sample discrimination to achieve 92% diagnostic accuracy; (2) a Bayesian fusion framework that integrates static geological priors with dynamic sensitivity analysis for probabilistic quantification of injector–producer connectivity, achieving 85% identification accuracy with rigorous uncertainty propagation; and (3) a three-stage human–machine collaborative mechanism that substantially reduces anomaly response latency while ensuring field safety. Field application in Bohai oilfields demonstrates that the system shortens the injection–production response cycle by approximately 42%, reduces anomaly response time from over 72 h to less than 2 h (a 97% reduction), decreases water consumption per ton of oil by 27.6%, and increases injection–production uptime by 11.3 percentage points. This study provides an interpretable, extensible, and closed-loop technical solution for intelligent offshore oilfield development, with future directions including digital twin predictive simulation and reinforcement learning for real-time optimization.

Keywords: intelligent injection–production; machine learning; blockage diagnosis; connectivity analysis; microservice architecture; offshore oilfield

1. Introduction

With the deepening global energy transition and the rapid advancement of digital technologies, oil and gas field development is undergoing a significant shift from traditional to intelligent models [1]. The "1534" overall development strategy proposed by China National Offshore Oil Corporation (CNOOC) explicitly identifies digitalization and intelligence as critical directions for future oilfield development [2]. Within this context, traditional injection–production technologies face prominent issues, such as delayed regulation, low efficiency, and strong dependence on manual intervention, struggling to meet the demands for efficient offshore oilfield development. Particularly under complex geological conditions and harsh marine environments, achieving precise regulation and dynamic optimization of injection–production parameters becomes a key technical bottleneck constraining development effectiveness.

Intelligent injection–production completion systems can significantly enhance the overall recovery factor of oil and gas reservoirs through real-time monitoring and feedback control mechanisms [3]. According to the latest CNOOC statistics, by the end of 2023, cable-controlled intelligent injection–production technologies had been deployed in 421 wells across oilfields such as Liaodong, Bonan, and Qinhuangdao, with 181 wells in the Liaodong operating area and 110 in Bonan, forming a large-scale application demonstration [4]. The deployment of these intelligent tools provides a massive data foundation for building new intelligent oilfields characterized by unmanned operations, visualized wellbores, collaborative operations, and scientific decision making, also creating favorable conditions for this research study.

Scholars worldwide have conducted extensive frontier research in intelligent injection–production. Smith et al. [5] proposed a data-driven injection–production optimization method, using machine learning algorithms to analyze historical production data and establish parameter optimization models; Li et al. [6] developed a well blockage diagnosis model based on pressure transient analysis, enabling quantitative evaluation of blockage severity; Wang et al. [7] investigated connectivity analysis methods between injection and production wells based on dynamic production data, providing theoretical support for injection–production regulation. More recently, Faizah et al. [1] demonstrated the transformative potential of real-time edge intelligence for intelligent completion injectivity, showing that processes previously taking months can now be completed in minutes through automated systems. However, existing research often focuses on isolated technical problems, lacking systematic integrated solutions, particularly in terms of real-time capability, accuracy, and practicality.

This paper presents an application- and system-oriented intelligent cable-controlled injection–production integrated management system. Rather than proposing novel algorithmic methods, this work focuses on the integration of well-established machine learning techniques into a cohesive, field-deployable platform that addresses real-world operational challenges in offshore oilfields. The system innovatively combines machine learning algorithms with reservoir engineering theory, achieving closed-loop management encompassing data acquisition, anomaly diagnosis, and intelligent regulation. Compared with existing technologies, this system exhibits three significant advantages: (1) employing a microservice architecture to enable flexible expansion of functional modules [8]; (2) establishing multi-algorithm fusion intelligent diagnosis models to enhance analysis precision [9]; (3) developing standardized data interfaces for seamless integration with field equipment [10,11].

https://doi.org/10.3390/pr14081238

2. System Architecture Design

2.1. Overall Architecture Design

The system adopts the industry-leading cloud-native three-tier architecture (IAAS, PAAS, and SAAS) and microservice system to construct a highly modular and scalable technical framework (Figure 1). The overall architecture is divided into three main parts: data layer, service layer, and application layer. Each part is interconnected through microservice interfaces and supplemented with auxiliary service functions such as user permissions, security authentication, and microservice management. The architecture aligns with recent industry trends in cloud–edge collaboration for intelligent oilfield monitoring [12].

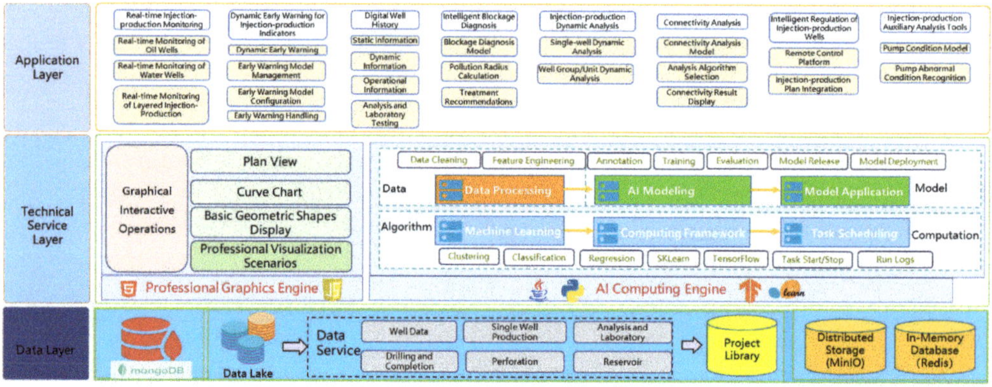

Figure 1. System three-tier architecture design diagram.

- Data Layer: The data engine provides storage capabilities for system-related data and offers a complete data processing workflow. It supports various data types, particularly the loading and parsing of data types relevant to the oil and gas industry, and delivers data-related services such as data collection, data querying, data preprocessing, and data loading. As the foundational support of the system, the data layer integrates multi-source heterogeneous data, including real-time monitoring data, static geological data, production dynamic data, and operational history data. This layer employs a distributed storage architecture, with structured data stored in the Kingbase database and unstructured data managed through the Minio object storage system. Specifically designed for the oil and gas industry, the data layer features a professional data processing engine that supports the parsing and processing of specialized data types such as well logging curves and well testing data, providing a complete Extract, Transform, Load (ETL) process.
- Technical Service Layer: This layer utilizes three public services from the oilfield cloud platform: user service, permission service, and log service. Serving as the intelligent core of the system, the technical service layer comprises two key components: first, the AI computing engine, which integrates various machine learning algorithms such as clustering, classification, regression, SKLearn, and TensorFlow within the algorithm model service; second, the professional graphics engine, which implements basic geometric visualization and specialized visual scenario business rules through the graphic interaction operation service.
- Application Layer: The presentation method is web-based, utilizing technologies such as JavaScript, Vue, and HTML5. The client requires Internet Explorer 10 or above, or a browser based on the Chrome kernel. This layer provides a user-friendly interface for end-users, ensuring a seamless experience across different ter-

https://doi.org/10.3390/pr1408123

minal devices. The application layer implements nine functional modules: real-time injection–production monitoring, dynamic pre-alarming, digital well history, blockage diagnosis, injection–production dynamic analysis, connectivity analysis, intelligent regulation, auxiliary analysis tools, and system management. Each functional module adopts a component-based design, allowing for flexible configuration based on user requirements.

To address key offshore operational constraints—such as limited on-site personnel, high communication latency, and intermittent network connectivity—the architecture incorporates several design features. The cloud-native microservice framework enables modular deployment and remote maintenance, reducing the need for on-site interventions. Local data caching and edge-compatible preprocessing modules ensure that core functions, including real-time monitoring and threshold-based alarming, remain operational during network interruptions. These design considerations ensure the system's reliability and usability under typical offshore working conditions.

Communication between the system layers is facilitated through well-defined interface protocols, with data exchange conducted in JSON format to ensure loose coupling and scalability. Additionally, the system architecture design fully considers the unique environment of offshore oilfields, ensuring the continued operation of core functionalities even under abnormal conditions such as network interruptions.

2.2. Technical Architecture Implementation

The system is implemented using a Spring Boot + Vue.js frontend/backend separation technology stack (Figure 2). Key technology selections are as follows:

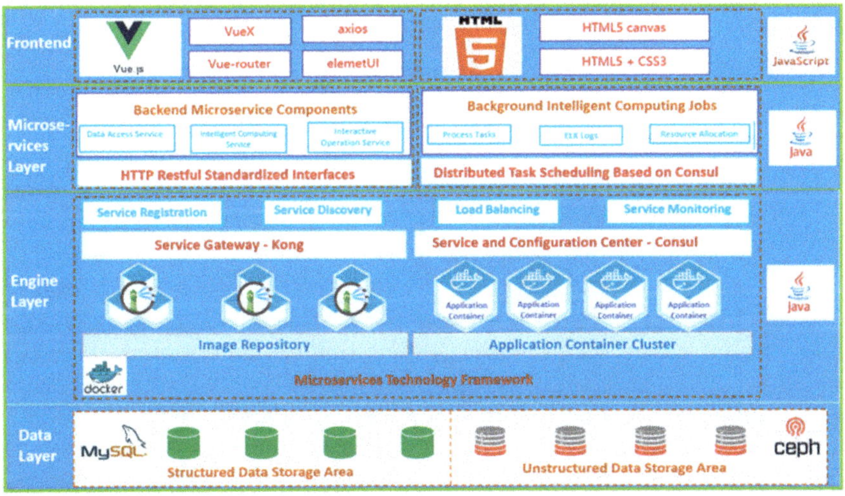

Figure 2. Spring Cloud + Vue.js technology stack.

- Development Language: Java JDK 1.8 (Oracle Corporation, Austin, TX, USA).
- Platform Frontend: The frontend uses the Vue development framework, specifically Vue 2, which provides reactive data binding and a component system, making frontend development more efficient and modular.
- Frontend UI Framework: Element UI (Version 2.1, Eleme Inc., Shanghai, China), a desktop-oriented component library based on Vue 2.
- Backend: The backend uses the Spring Boot framework (Version 2.7.1, VMware, Inc., Palo Alto, CA, USA), an open-source Java-based framework designed to create stan-

https://doi.org/10.3390/pr14081238

dalone, production-level Spring-based applications. It simplifies Spring application development by following the principle of "convention over configuration," allowing developers to quickly start and run Spring applications.

- Database: Renmin Jincang (Kingbase) (Version V8R6, Beijing Kingbase Data Technology Co., Ltd., Beijing, China) is a database management system that supports SQL standards, suitable for transactional processing, analytical processing, data warehousing, and other application scenarios.

Recent advancements in edge computing have demonstrated the value of integrating real-time data processing capabilities with cloud-based analytics. The system's architecture is designed to accommodate future edge computing modules that could enable downhole real-time decision making, following the paradigm established in recent intelligent completion deployments [1]. The development tools and version numbers used in this study are shown in Table 1.

Table 1. Tools and versions used.

No.	Tool Name	Purpose	Version
1	IntelliJ IDEA	Java development tool	Ver. 2018
2	Python	Provides the foundational environment for running machine learning algorithms	3.6.3
3	Renmin Jincang	Provides data storage services	9.5.21
4	Nginx	Load balancing, responsible for server forwarding	1.20.1
5	JDK	Provides the foundational runtime environment for backend applications	1.8
6	Minio	Used for storing large-capacity unstructured data	2020-12-29T23-29-29Z
7	Redis	Caching database for data caching operations	6.0.12

3. Key Technology Research

3.1. Intelligent Blockage Diagnosis Technology

3.1.1. Algorithm Model Research

Well blockage diagnosis is a core function of the intelligent injection–production system. Traditional diagnostic methods rely heavily on manual experience, suffering from subjectivity and inefficiency. This paper innovatively proposes a hybrid diagnostic algorithm integrating Random Forest and Support Vector Machine (SVM), significantly improving diagnostic accuracy.

The Random Forest algorithm avoids overfitting inherent in single decision trees through ensemble learning by constructing multiple trees. It uses information gain as the feature selection criterion:

$$g(D,A) = H(D) - H(D|A) \quad (1)$$

where $H(D)$ is the empirical entropy representing the dataset's disorder and $H(D|A)$ is the conditional entropy representing the disorder given feature A [13]. By recursively selecting features with the highest information gain for node splitting, a highly generalizable classification model is ultimately constructed.

The SVM algorithm maps nonlinear problems in low-dimensional space to linearly separable problems in high-dimensional space by using the kernel trick. This paper employs the Radial Basis Function (RBF) kernel:

$$K(x_i, x_j) = \exp\left(-\gamma \,||x_i - x_j||^2\right) \tag{2}$$

Balancing model complexity and generalization performance is achieved by adjusting the penalty coefficient C and kernel parameter γ [14]. SVM is particularly suitable for small-sample, high-dimensional classification problems, complementing Random Forest's strengths.

The model ensemble strategy uses weighted voting, fusing the predictions of Random Forest and SVM. Weight coefficients were determined via grid search and cross-validation, set to 0.6 for Random Forest and 0.4 for SVM. This specific weighting was determined via grid search and cross-validation, and it was found to be optimal, as it leverages Random Forest's strength in capturing global feature interactions while still incorporating SVM's discriminative power on boundary samples to correct edge-case errors. This approach preserves Random Forest's ability to capture global features [9] while leveraging SVM's discriminative power on boundary samples.

3.1.2. Feature Engineering Processing

High-quality feature engineering is fundamental to model performance. Three categories of features were extracted from raw production data:

- B Basic Features: They include directly monitored indicators such as tubing pressure, casing pressure, daily water injection volume, and daily liquid production rate. These reflect the well's fundamental production status and are primary inputs for blockage diagnosis.
- Derived Features: Obtained through specialized calculations, including apparent water absorption index, productivity index, pressure derivative, etc. the Apparent Water Absorption Index (AWAI), is defined as:

$$AWAI = \frac{Q_w}{P_t - P_c + \epsilon} \tag{3}$$

where: Q_w is daily water injection Volume, P_t is tubing pressure and P_c is casing pressure. ϵ is a small positive constant to avoid division by zero. These features incorporate reservoir engineering expertise, providing deeper insights into changes in formation flow characteristics.

- Temporal Features: Generated via sliding window statistics, including 3-day moving averages, week-over-week change rates, etc. These capture dynamic trends in production parameters, crucial to early blockage warning.

All features underwent rigorous preprocessing before model input: missing values were filled using linear interpolation; outliers were identified and corrected using the 3σ principle; numerical features were normalized using Min–Max scaling as

$$x' = \frac{x - \min(x)}{\max(x) - \min(x)} \tag{4}$$

and categorical features were converted into numerical form via one-hot encoding.

3.1.3. Model Training and Optimization

Model training employed a 10-fold cross-validation strategy: the dataset was randomly split into 10 subsets; models were trained on 9 subsets and validated on the remaining

https://doi.org/10.3390/pr14081238

1; this process was repeated 10 times; average performance across all folds was used as the evaluation metric. This strategy maximizes the use of limited data while ensuring reliable assessment.

Hyperparameter optimization used Bayesian Optimization, building a probabilistic model of the parameters to guide the search direction, significantly improving efficiency. Key hyperparameters included: Random Forest (n_estimators, max_depth); SVM (C, γ).

Model performance was evaluated using multiple metrics: accuracy reflects overall classification effectiveness; precision measures the reliability of positive predictions; recall assesses the ability to identify positive samples; F1-score, the harmonic mean of precision and recall, is suitable for imbalanced classification problems. Final model performance on the test set is shown in Table 2. As shown in Table 2, the ensemble model achieved the highest F1-score of 0.92, demonstrating its superior balance between precision and recall compared with the individual algorithms.

Table 2. Blockage diagnosis model performance evaluation results.

Algorithm	Accuracy	Precision	Recall	F1-Score
Random Forest	0.89	0.91	0.87	0.89
SVM	0.90	0.88	0.92	0.90
Ensemble Model	0.92	0.93	0.91	0.92

3.1.4. Field Application Validation

The data were labeled according to the method shown in the wellbore blockage reference (Figure 3): Taking the oil pressure index as an example, mark the declining phase and the low-level fluctuation phase as label 0 (normal production); mark the rising phase as label 1 (mild blockage); and mark the peak phase and the stable phase after the peak as label 2 (severe blockage).

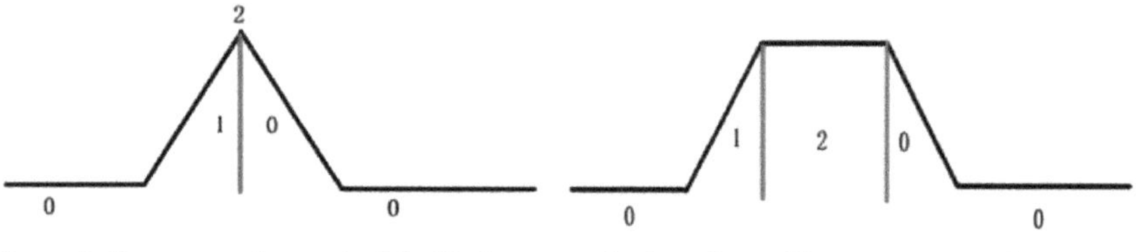

Figure 3. Data annotation method for blockage warning based on tubing pressure trends.

Taking Well B23 as an example (Figure 4), comparing model results with field operations, the model successfully predicted the blockage progression. Field operations performed acid stimulation on 10 November 2022. The blockage severity prediction chart (where the blue line represents the actual operating parameters, namely oil pressure, daily water injection volume, and visual water absorption index, and the red line represents the predicted values of these parameters, consistent with the method in Figure 3. The blockage severity can be marked as normal production, mile blockage, and severe blockage based on the shape of the curve) shows that, excluding misjudgments caused by shut-ins due to other operations, key instances of slight and severe blockage could be identified by the model, which consequently generated an alarm, verifying its accuracy.

https://doi.org/10.3390/pr14081238

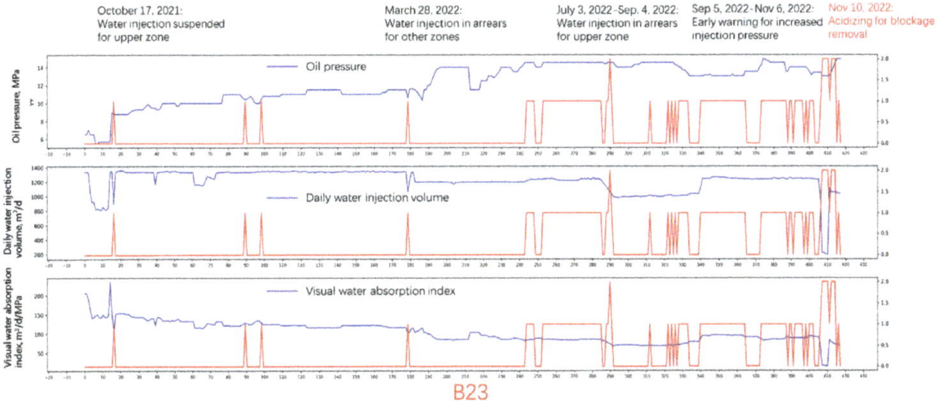

Figure 4. Well B23 blockage prediction chart.

The annotation method based on tubing pressure trends served as the initial coarse label generation stage. This approach originates from the empirical knowledge of field engineers: when wellbore blockage occurs, the injection pressure typically exhibits a "rise–peak–stabilization" pattern. However, this was merely the starting point of label generation, not the final stage. To transform these preliminary labels into "ground truth labels" suitable for supervised learning, we conducted rigorous cross-validation based primarily on the following two categories of hard data: direct evidence from downhole operation records and skin factor inversion. Through the aforementioned comparison, the labels generated solely by the rules shown in Figure 3 achieved approximately 80% consistency with operational records. For samples where the rules conflicted with the records, we first eliminated operational interferences and then incorporated expert experience combined with other dynamic data such as injection volume and water cut to conduct final arbitration for the disputed periods.

It is important to acknowledge the potential impact of labeling uncertainty on model performance. The initial labels generated based on tubing pressure trends are subject to uncertainty arising from operational interferences (e.g., well shut-ins) and ambiguous blockage conditions. To mitigate this uncertainty, we cross-validated the labels against two independent data sources: downhole operation records and skin factor inversion. Approximately 80% of the rule-based labels were consistent with operation records. For conflicting samples, expert arbitration was conducted using additional dynamic data (e.g., injection volume, water cut) to resolve discrepancies. This multi-source validation process enhances the reliability of the training labels and, consequently, the robustness of the trained model.

The model also provides quantitative evaluation of blockage severity and treatment recommendations. For wells with slight blockage (damage degree < 30%), pressure increase/enhanced injection is recommended; for severe blockage (damage degree $\geq$ 30%), acidizing is suggested. This functionality provides a scientific basis for field decisions, significantly improving treatment effectiveness.

3.2. Pollution Radius Calculation Model

3.2.1. Theoretical Basis

Pollution radius is a key parameter for evaluating formation damage, directly influencing the design and effectiveness prediction of stimulation treatments. This paper establishes a pollution radius calculation model based on the Dupuit formula and composite reservoir theory.

https://doi.org/10.3390/pr14081238

For an ideal well, production is given by the Dupuit formula:

$$Q = \frac{2\pi kh(p_e - p_w)}{\mu \ln\left(\frac{r_e}{r_w}\right)} \quad (5)$$

When formation damage occurs, the additional flow resistance is characterized by the skin factor S, and the production rate becomes:

$$Q' = \frac{2\pi kh(p_e - p_w)}{\mu \ln\left(\frac{r_e}{r_w}\right) + S} \quad (6)$$

Treating the damaged zone as the inner region of a composite reservoir, the skin factor can be expressed as:

$$S = (\frac{k}{k_s} - 1)\ln(\frac{r_s}{r_w}) \quad (7)$$

By solving these equations simultaneously, the pollution radius r_s and damaged zone permeability k_s can be determined [15].

3.2.2. Numerical Simulation Validation

Numerical simulations were conducted for reservoirs with different permeabilities (Table 3) to validate model reliability. Simulation results show the following: The lower the reservoir permeability, the more significant the impact of damage on production. For the same permeability, the production decline exhibits an approximately linear relationship with the damage degree.

Table 3. Simulated production decline under different damage degrees.

Permeability (mD)	Damage Degree	Production Decline (%)
2000	0.3	12.5
2000	0.5	20.8
2000	0.7	29.2
1000	0.3	15.6
1000	0.5	26.3
1000	0.7	37.5
500	0.3	18.9
500	0.5	31.4
500	0.7	44.8

Based on the simulation results, a pollution radius correction formula was established:

$$r_s = r_{sc}\exp(\frac{RQ_a}{RQ_r} - 1) \quad (8)$$

where r_{sc} is the reference pollution radius (20 m), RQ_a is the actual production decline, and RQ_r is the theoretical production decline. This correction accounts for reservoir heterogeneity and fluid properties, making the results more consistent with real conditions.

3.2.3. Application Case Analysis

Taking Well A in a Bohai oilfield as an example, its production declined from 85 m^3/d to 62 m^3/d while the bottomhole flowing pressure was kept constant. Model calculations yielded: skin factor S = 3.2, pollution radius rs = 15.8 m, and damage degree Ds = 0.42.

Based on the calculation results, a small-scale acidizing treatment was recommended and implemented for the well. After the operation, the production of the well was restored

https://doi.org/10.3390/pr14081238

to 80 m^3/d, close to the initial production capacity, verifying the model's reliability in assessing pollution degree and guiding treatment design.

3.3. Connectivity Analysis Technology

3.3.1. Static Connectivity Analysis

Static connectivity analysis is based on geological and reservoir static data, including the following:

(1) Sandbody Distribution Analysis: Determining sandbody connectivity between injectors and producers through depositional facies studies and sandbody correlation.
(2) Fault Seal Evaluation: Analyzing fault impact on fluid flow using methods like Shale Gouge Ratio (SGR).
(3) Well Trajectory Analysis: Considering the spatial positional relationships of horizontal well sections within the reservoir.

Based on static and dynamic data (well locations, deviations, layers, sandbodies, production dynamics, and well histories), built-in algorithms automatically determine static injector–producer relationships using predefined rules. Expert knowledge is then used to calibrate these results. Subsequently, big data techniques deeply mine patterns within injection–production dynamic data to build dynamic correlation models [16], ultimately achieving quantitative characterization of injector–producer relationships driven by the fusion of geological and injection–production data (Figure 5).

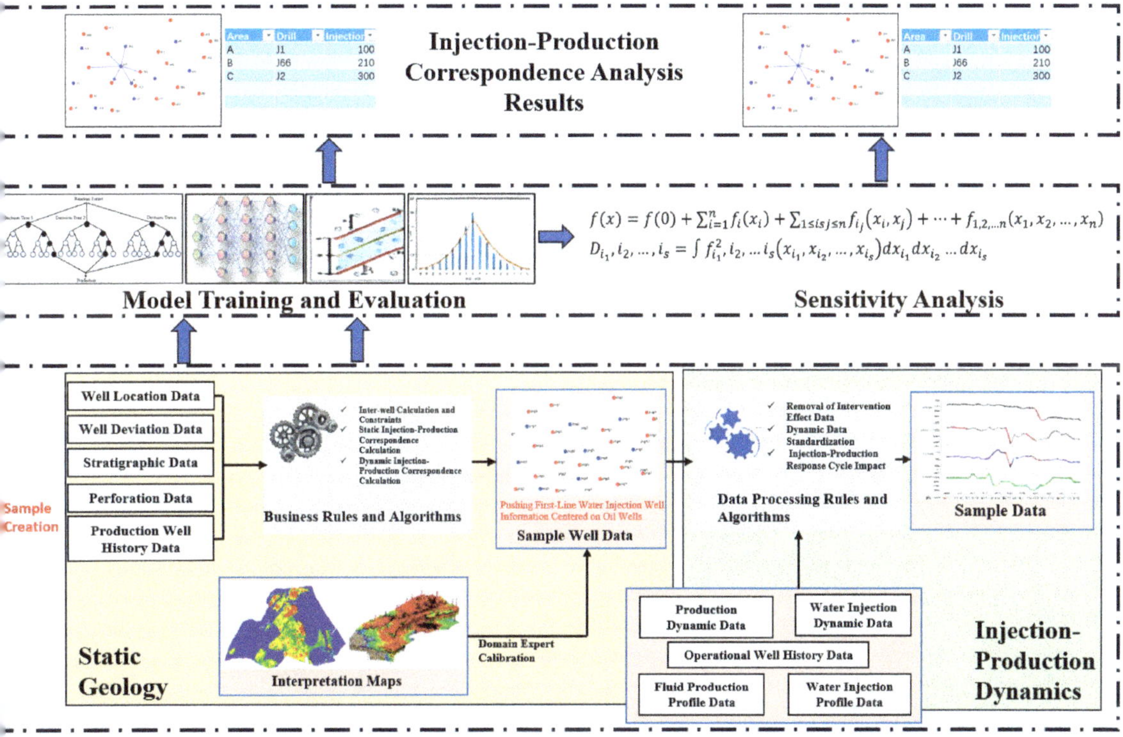

Figure 5. Technical approach for injector–producer connectivity analysis.

3.3.2. Dynamic Connectivity Analysis

Dynamic connectivity analysis utilizes production dynamic data and employs the Sobol Global Sensitivity Analysis method [17]. This method is a time-series model that requires using lagged injection volumes as inputs (for example, employing the NARX model—Nonlinear Autoregressive with Exogenous Inputs model). This method decomposes the output variance of a model into contributions from individual input variables and their interactions:

$$f(x) = f_0 + \sum_i f_i(x_i) + \sum_{i<j} f_{ij}(x_i, x_j) + \ldots + f_{1,2,\ldots,n}(x_1, \ldots, x_n) \tag{9}$$

The connectivity coefficient between injector i and producer j is defined as:

$$S_{ij} = D_{ij}/D \tag{10}$$

where D denotes the total variance and D_{ij} represents the contributed variance from injection well i to production well j.

Implementation steps:

(1) Construct a Multi-Layer Perceptron (MLP) neural network model with injector rates as inputs and producer liquid rates as output;
(2) Generate input samples via Monte Carlo sampling;
(3) Calculate the main effect index and total effect index for each injector;
(4) Obtain standardized connectivity coefficients through normalization.

3.3.3. Integrated Static–Dynamic Analysis Method

An innovatively proposed comprehensive evaluation method integrates static and dynamic analyses:

(1) Static analysis provides geological constraints, excluding obviously non-connected well pairs;
(2) Dynamic analysis quantifies connection strength, reflecting actual flow characteristics;
(3) Results are fused using a Bayesian framework to obtain posterior connectivity probability:

$$P(C_{ij}|D) = \frac{P(D|C_{ij})P(C_{ij})}{P(D)} \tag{11}$$

where $P(C_{ij})$ is the prior connectivity probability derived from static geological analysis, and $P(D|C_{ij})$ is the likelihood function inferred from dynamic production data [17]. This integrated approach aligns with emerging hybrid modeling techniques that combine physics-based understanding with data-driven insights [16].

Bayesian updating and propagation process:

(1) On the Numerically Stable Quantification of $P(C_{ij})$

We agree that converting geological interpretations (e.g., fault seal capacity) into a stable numerical probability is challenging. To minimize subjectivity, we employed a multi-attribute scoring system combined with a frequentist statistical approach.

(2) On the Propagation of Geological Uncertainty through the Bayesian Update

The Bayesian framework is inherently designed to propagate uncertainty. This is handled in our model as follows:

- Prior as a Distribution: The static prior $P(C_{ij})$ is not implemented as a single scalar but as a probability distribution. High uncertainty in the geological interpretation (e.g., a fault with sparse calibration data) is represented by a prior distribution with a larger variance (e.g., a Beta distribution with appropriate parameters).

https://doi.org/10.3390/pr1408123

- Dominance of the Likelihood: The dynamic likelihood $P(D|C_{ij})$ acts as an "uncertainty filter:"
    - If the dynamic data are strong (e.g., clear interference test results), the likelihood function is sharp. This dominant evidence will override the prior uncertainty, causing the posterior distribution to converge towards the dynamic indication. Strong flow data correct geological ambiguity.
    - If the dynamic data are also weak, the posterior distribution retains the wide uncertainty from the prior, effectively warning the user that the conclusion is based on weak evidence.
- Sensitivity Analysis: We conducted sensitivity tests on the most subjective prior parameters (e.g., fault seal probability). As shown in Table 4, the final identification accuracy (85%) remained robust across a reasonable range of prior values. This suggests that in our specific case study, the dynamic likelihood dominated the updating process, further mitigating concerns about prior subjectivity.

Table 4. Results of integrated static–dynamic connectivity analysis.

Well Pair	Static Probability	Dynamic Coefficient	Fused Probability	Actual Result	Conclusion
B01	0.85	0.92	0.91	Connected	Correct
B02	0.92	0.15	0.33	Not Connected	Correct
B03	0.78	0.87	0.85	Connected	Correct
B04	0.95	0.05	0.22	Not Connected	Correct
...	...	...	...	...	...
B32	0.95	0.88	0.93	Connected	Correct

Our workflow ensures that the static prior is quantifiable and traceable. The Bayesian framework then rigorously propagates geological uncertainty into the final posterior, providing a more honest assessment of connectivity risk.

Application examples (Table 4) show that this method correctly identified connectivity for 27 out of 32 total well pairs in Oilfield B, achieving an identification accuracy of 85%, significantly higher than single-method analyses.

4. System Implementation and Applications

4.1. Implementation of Core Functions

4.1.1. Real-Time Monitoring Module

The Real-Time Monitoring Module integrates monitoring data from four main categories: oil wells, water injection wells, electrical submersible pumps (ESPs), and zonal injection–production tools (Figure 6). Key features include the following:

(1) High-Frequency Data Acquisition: Key parameters are sampled at 1 Hz to ensure capture of dynamic features.
(2) Multi-Dimensional Visualization: It supports curves, dashboards, 3D wellbore views, and other formats.
(3) Anomaly Detection: It implements real-time anomaly identification based on Statistical Process Control (SPC) methods.

Data Downsampling and Aggregation: High-frequency raw data (e.g., once per second) are aggregated over fixed time windows and converted into low-frequency feature data for use by upper-level intelligent modules.

https://doi.org/10.3390/pr14081238

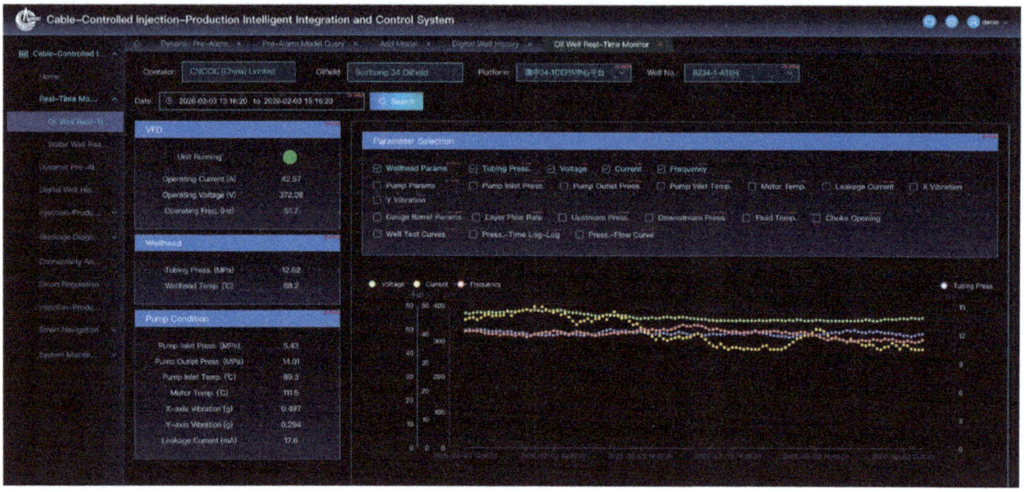

Figure 6. Real-time monitoring interface (oil well/water well/ESP).

(1) Mean/Extreme Value Aggregation: Calculate the average, maximum, or minimum values per minute, hour, or day. For example, convert 3600 s level pressure points into a single hourly average.
(2) Sampling Storage: Retain data points only at fixed intervals (e.g., save one point every 5 min), and discard the intermediate points.
(3) Interval Statistics: In addition to the mean, store statistical features such as variance and rate of change within the time period to preserve the fluctuation characteristics of the data.

The module is connected to over 200 wells, processing more than 20 million data points daily, providing the data foundation for intelligent analysis [18].

4.1.2. Dynamic Early Warning Module

The Dynamic Early Warning Module employs a multi-level warning mechanism (Figure 7):

(1) Threshold Warning: It sets upper/lower limits for single parameters.
(2) Trend Warning: It identifies abnormal trends using time-series analysis.
(3) Composite Warning: It combines multiple parameters using logical rules.

Warning rules support graphical configuration, allowing users to flexibly define various warning scenarios. The system also provides a warning analysis dashboard displaying key metrics like warning statistics and handling progress.

4.1.3. Digital Well History Module

The Digital Well History Module (Figure 8) manages full lifecycle data:

(1) Dynamic Information: production curves, operational history, etc.
(2) Static Information: well configuration, log interpretations, etc.
(3) Monitoring Data: production profiles, injection profiles, etc.
(4) Laboratory Data: fluid properties, core analysis, etc.

Adopting a "Well–Layer–Parameter" three-tier architecture, the module supports multi-dimensional data correlation analysis, significantly improving historical data utilization.

https://doi.org/10.3390/pr1408123

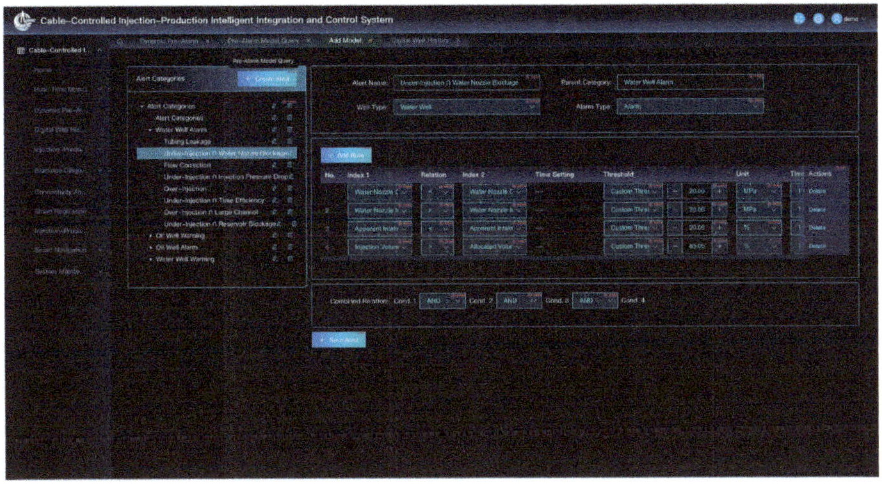

Figure 7. Multi-level warning configuration interface.

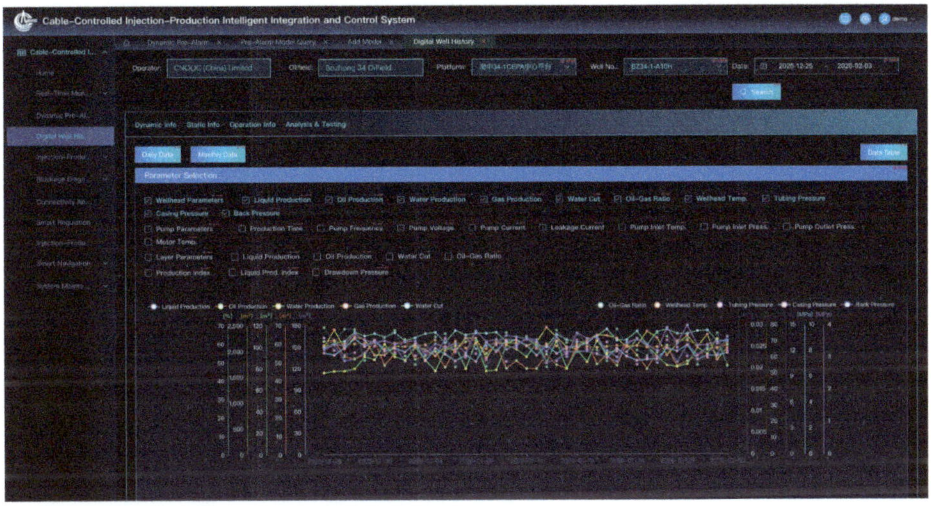

Figure 8. Digital Well History Data Architecture.

4.2. Three-Stage Human–Machine Collaboration

We have reduced the response time from >72 h to <2 h precisely through the following standardized workflow.

Stage 1—Automated Alerting and Diagnosis (Duration: <1 min). System Action: Real-time data streams feed into the model. When the blockage probability exceeds a threshold (e.g., 90%), the system automatically triggers a red alert on the interface and generates the diagnostic recommendation panel mentioned above.

Human Status: The engineer receives a notification on their mobile device or workstation but does not take immediate action, waiting instead for the system to complete its preliminary diagnosis.

Stage 2—Manual Review and Confirmation (Duration: <30 min). Human Action: Upon seeing the red alert, the duty engineer first reviews the diagnostic basis at Level 3. If SHAP indicates that the primary contributing factor is "pressure derivative upwarping" and the engineer confirms this by retrieving the pressure buildup curve, they click "Confirm Diagnosis."

Collaboration Mechanism: If the basis provided by the system conflicts with the engineer's experience (for example, if the system mistakenly identifies pressure gauge drift as blockage), the engineer can click "Reject." The sample is then flagged and returned to the data lake for subsequent model retraining. This step is critical to ensuring field safety, as humans retain ultimate decision-making authority.

Stage 3—Instruction Issuance and Closure (Duration: <1.5 h). System Action: Once the diagnosis is confirmed, the system automatically displays contingency plans based on the severity of the blockage: for mild blockage, it recommends "pressurized water injection + monitoring"; for severe blockage, it recommends "acidizing operation" and pushes an operation design template.

Human Action: The shift manager reviews the plan and issues instructions with an electronic signature. In cases of severe blockage, the operation team begins preparing equipment; for mild blockage, the control room remotely adjusts the injection pressure.

Result Feedback: After the operation is completed, personnel manually enter the operation record into the system. This record serves as new labeling data, forming a closed loop of "alert–diagnosis–operation–feedback."

Before the system was implemented, engineers had to manually retrieve data, plot curves, and convene meetings for discussion, a process that typically took 1–3 days. The system handles "monitoring" and "preliminary diagnosis," compressing 72 h of data analysis time into just 1 min. Humans retain "decision making" and "review," streamlining the previously meeting-intensive discussion process into a simple "confirm" or "reject" action on a unified interface, typically completed within 2 h.

4.3. Field Application Results

The system was deployed in multiple oilfields in the Bohai Sea, achieving significant results:

(1) Injection–Production Efficiency Improvement: Water injection utilization increased from 68% to 78.5% through intelligent regulation.
(2) Anomaly Diagnosis Timeliness: Diagnosis time reduced from an average of 3 days to 2 h.
(3) Treatment Success Rate: Success rate of treatments recommended by the model increased to 92%.
(4) Labor Cost Savings: Reduced field inspection frequency by over 30%.
(5) Typical Application Cases:

- Case 1—L Oilfield (12 Well Patterns): After the full application of intelligent regulation in 2023, the response time for water injection well groups was shortened by approximately 42% (Table 5). Connectivity analysis was used to optimize injection–production correspondence, effectively reducing ineffective circulation. Precision de-plugging based on blockage diagnosis models significantly improved single-well oil production [10]. These improvements align with recent industry reports on digital transformation benefits in offshore operations [12].

Table 5. Injection response time comparison in L Oilfield.

Stage	Traditional (Pre-Application)	Intelligent (Post-Application)
2021	85 days	-
2022	78 days	48 days (System Pilot)
2023	72 days	42 days (Full Deployment)

https://doi.org/10.3390/pr14081238

- Case 2—Q Oilfield: The application of the Dynamic Early Warning Module enables proactive prediction of equipment failures (e.g., electric submersible pumps), while the pollution radius model optimizes treatment design. The integrated use of these technologies led to a marked improvement in overall operational efficiency: anomaly response time was reduced from an average of over 72 h to less than 2 h, a decrease of over 97%; single-well regulation frequency decreased by 72%; the injection–production time rate increased by 11.3%; and water consumption per ton of oil dropped by 27.6% (Table 6) [15]. The 97% reduction in response time parallels the 90% improvement reported in recent edge intelligence deployments for similar offshore applications.

Table 6. Key application performance indicators in Q Oilfield.

Evaluation Metric	Pre-Application	Post-Application	Improvement	Calculation Method
Injection–Production Uptime	83.2%	94.5%	+11.3%	(Production Time/Calendar Time) × 100%
Anomaly Response Time	>72 h	<2 h	−97%	Mean Time from Detection to Action
Regulation Frequency per Well	4.3 times/year	1.2 times/year	−72%	Total Annual Regulation Commands/Well Count
Water Consumption per Ton of Oil	5.8 m^3/t	4.2 m^3/t	−27.6%	Annual Water Injection (m^3)/Annual Oil Production (t)

5. Conclusions

This study presents an application- and system-oriented intelligent cable-controlled injection–production integrated management system developed for offshore oilfield operations. Rather than proposing novel algorithmic methods, this work focuses on the integration of well-established machine learning techniques—including ensemble learning, Bayesian inference, and SHAP-based interpretability—into a cohesive, field-deployable platform that addresses real-world operational challenges. Based on a "data-driven + knowledge-driven" integrated approach, an intelligent cable-controlled injection–production integrated management system was constructed, yielding three new insights:

(1) Multi-source feature fusion combined with ensemble learning significantly enhances blockage diagnosis accuracy. To overcome the limitations of single algorithms in global feature extraction and boundary sample discrimination, a weighted ensemble model integrating Random Forest and Support Vector Machine was constructed. Through grid search and cross-validation, the weight coefficients were optimized (RF:SVM = 0.6:0.4), enabling effective capture of global feature interactions while maintaining precise discrimination of boundary samples. Field validation demonstrates that the ensemble model achieves a diagnostic accuracy of 92% and an F1-score of 0.92, significantly outperforming single algorithms. The model also enables quantitative evaluation of blockage severity and intelligent recommendation of treatment strategies, providing a scientific basis for precise de-plugging operations.

(2) The static–dynamic integrated Bayesian framework enables probabilistic quantitative characterization of injector–producer connectivity. To address the dual challenges of subjectivity in static geological interpretation and uncertainty in dynamic production responses, a connectivity quantitative evaluation method integrating Sobol global sensitivity analysis with Bayesian inference was established. Static geological information—such as fault seal capacity and sandbody connectivity—is quantified as probabilistic prior distributions, while dynamic production responses serve as the

https://doi.org/10.3390/pr14081238

likelihood function. Through Bayesian updating, rigorous propagation and fusion of uncertainty are achieved. Sensitivity analysis shows that when dynamic evidence is strong, the posterior probability converges toward the dynamic indication, effectively overcoming prior subjectivity; when dynamic evidence is weak, the posterior retains a wide distribution, providing users with a quantified risk warning. This method achieved an identification accuracy of 85% across 32 well pairs in Oilfield B, significantly outperforming single-method analyses.

(3) The human–machine collaborative closed-loop mechanism substantially reduces anomaly response latency. To address the trust barrier and responsibility allocation challenges faced by intelligent models in industrial field applications, a three-stage human–machine collaborative decision-making framework was designed and validated. Stage 1 enables automated alerting and preliminary diagnosis, compressing data response time to within 1 min. Stage 2 introduces SHAP explainability analysis as the basis for manual review, allowing engineers to verify the model's diagnostic logic based on physical mechanisms, thereby establishing a human–machine trust bridge. Stage 3 enables instruction issuance and operational feedback, forming a closed loop of "alert–diagnosis–operation–feedback." This framework reduces anomaly response time from over 72 h to less than 2 h—a reduction of 97%—while ensuring field safety, significantly improving the timeliness and accuracy of anomaly management.

Through the above technical integration and mechanism innovation, systematic field application validation was conducted in Bohai oilfields. Application results demonstrate that the injection–production response cycle was shortened by approximately 42%, the injection–production uptime increased by 11.3 percentage points, water consumption per ton of oil decreased by 27.6%, and the regulation frequency per well decreased by 72%. This system provides an interpretable, extensible, and closed-loop technical solution for the intelligent development of offshore oilfields, effectively addressing the critical challenge of transitioning from "algorithm accuracy" to "field practicality."

Future Directions: Although this study has achieved significant application results in Bohai oilfields, the generalization capability of the AI models under different geological and fluid conditions requires further validation. Future research will focus on the following directions: (1) integrating digital twin technology to construct virtual injection–production systems for real-time simulation of pressure and saturation distributions, enabling predictive "what-if" analysis for regulation strategies [19]; (2) exploring reinforcement learning for real-time optimization of injection–production parameters to achieve adaptive dynamic control; (3) developing edge computing modules to support downhole real-time decision making, reducing dependence on network transmission [20]; and (4) constructing cross-regional intelligent injection–production cloud platforms to facilitate large-scale replication and promotion of the technical achievements [21–23].

Author Contributions: Conceptualization, J.B., Z.C. and W.Z.; methodology, Z.Z. (Zhaochuan Zhou) and L.W.; software, C.Z. (Chengtao Zhu), L.Z. and Z.L.; validation, Y.X., L.Z., Q.W. and Z.H.; formal analysis, Z.Z. (Zhixiong Zhang); investigation, C.Z. (Chenwei Zou); resources, W.Z.; data curation, X.T.; writing—original draft preparation, Z.C., S.J. and Y.D.; writing—review and editing, Y.D.; visualization, Y.X.; supervision, J.B.; project administration, L.W.; funding acquisition, Y.D. All authors have read and agreed to the published version of the manuscript.

Funding: This research study was funded by the Natural Science Foundation of Shandong Province. (ZR2023ME119 and ZR2022ME077).

Data Availability Statement: The original contributions presented in this study are included in the article. Further inquiries can be directed to the corresponding author.

Conflicts of Interest: Authors J.B., Z.C., W.Z., Y.X., S.J., L.Z., Z.H., Q.W., Z.Z. (Zhixiong Zhang), C.Z. (Chenwei Zou) and X.T. were employed by Tianjin Branch of CNOOC China Limited. Authors Z.Z. (Zhaochuan Zhou), L.W., C.Z. (Chengtao Zhu) and Z.L. were employed by CNOOC EnerTech-Drilling & Production Co. The remaining authors declare that the research was conducted in the absence of any commercial or financial relationships that could be construed as potential conflicts of interest.

eferences

Faizah, N.; Tan, C.; Tan, A.S.; Fah, G.G.K.; Kalidas, S.; Musayev, R.; Barretto, G. Real-Time Edge Intelligence and Remotely Controlled Optimization of Intelligent Completion Injectivity. In Proceedings of the SPE/IADC Asia Pacific Drilling Technology Conference and Exhibition, Bangkok, Thailand, 7–9 May 2024; SPE-219272-MS.

CNOOC. *Intelligent Oilfield Construction White Paper*; CNOOC: Beijing, China, 2023.

Zhang, W.; Liu, H.; Wang, D.; Li, Y.; Chen, S.; Zhao, L. Application of Intelligent Completion Technology in Offshore Oilfields. *Pet. Explor. Dev.* **2021**, *48*, 621–628.

National Energy Administration. *China Oil and Gas Field Intelligent Development Report*; NEA: Beijing, China, 2024.

Du, S.; Zhao, X.; Xie, C.; Zhu, J.; Wang, J.; Yang, J.; Song, H. Data-driven Production Optimization using Particle Swarm Algorithm Based on the Ensemble-learning Proxy Model. *Pet Sci.* **2023**, *20*, 2951–2966. [CrossRef]

Gurav, S.; Kumar, P.; Ramshankar, G.; Mohapatra, P.K.; Srinivasan, B. Machine Learning Approach for Blockage Detection and Localization using Pressure Transients. In Proceedings of the 2020 IEEE International Conference on Computing, Power and Communication Technologies (GUCON), Greater Noida, India, 2–4 October 2020; pp. 189–193.

Zhang, L.; Liao, X.; Dong, P.; Hou, S.; Li, B.; Chen, Z. An Efficient Method for Identifying Inter-Well Connectivity Using AP Clustering and Graphical Lasso: Validation with Tracer Test Results. *Processes* **2024**, *12*, 2143. [CrossRef]

Siddiqui, H.; Khendek, F.; Toeroe, M. Microservices Based Architectures for IoT Systems—State-of-the-Art Review. *Internet Things* **2023**, *23*, 100854. [CrossRef]

Chen, T.; Guestrin, C. XGBoost: A Scalable Tree Boosting System. In Proceedings of the 22nd ACM SIGKDD International Conference on Knowledge Discovery and Data Mining, San Francisco, CA, USA, 13–17 August 2016; pp. 785–794.

Valestrand, R.; Khrulenko, A.; Hatzignatiou, D.G. Smart Wells for Improved Water Management in the Presence of Geological Uncertainty. In Proceedings of the SPE Bergen One Day Seminar, Bergen, Norway, 2 April 2014; SPE-169223-MS.

Mustafa, M.H.; Aliraani, B.; Kazi, S. Fiber Optics Reservoir Monitoring System Sustainability in Extreme Gas Well Conditions. In Proceedings of the Middle East Oil, Gas and Geosciences Show (MEOS GEO), Manama, Bahrain, 16–18 September 2025; D031S121R004.

Lang, H.; Zhang, Z.; Yang, Q.; Zhao, Q. Design and Implementation of Intelligent Oilfield Monitoring and Data Transmission System Based on Cloud-Edge Collaboration Technology. *J. Comput.* **2024**, *35*, 109–122. [CrossRef]

Quinlan, J.R. Induction of Decision Trees. *Mach. Learn.* **1986**, *1*, 81–106. [CrossRef]

Cortes, C.; Vapnik, V. Support-Vector Networks. *Mach. Learn.* **1995**, *20*, 273–297. [CrossRef]

Economides, M.J.; Hill, A.D.; Ehlig-Economides, C.; Zhu, D. *Petroleum Production Systems*, 2nd ed.; Prentice Hall: Upper Saddle River, NJ, USA, 2013; pp. 287–291.

Saihood, A.; Saihood, T.; Jebur, S.A.; Ehlig-Economides, C.; Alzubaidi, L.; Gu, Y. Artificial Intelligence Based-Improving Reservoir Management: An Attention-Guided Fusion Model for Predicting Injector-Producer Connectivity. *Eng. Appl. Artif. Intell.* **2025**, *146*, 110205. [CrossRef]

Saltelli, A.; Ratto, M.; Andres, T.; Campolongo, F.; Cariboni, J.; Gatelli, D.; Saisana, M.; Tarantola, S. *Global Sensitivity Analysis: The Primer*; Wiley: Hoboken, NJ, USA, 2008; pp. 45–68.

Faizah, N.; Tan, C.; Tan, A.S.; Fah, G.G.K.; Kalidas, S.; Musayev, R. Real-Time Edge Intelligence and Remote Smart Completion for Oil and Gas Producers. In Proceedings of the SPE/IADC Asia Pacific Drilling Technology Conference and Exhibition, Bangkok, Thailand, 7–8 August 2024; D021S005R008.

Ilyushin, Y.; Nosova, V.; Krauze, A. Application of Systems Analysis Methods to Construct a Virtual Model of the Field. *Energies* **2025**, *18*, 1012. [CrossRef]

Xin, G.; Zhang, K.; Wang, H.; Sun, Z.; Zhang, L.; Liu, P.; Wang, Y.; Yao, J. Dual-Reward Reinforcement Learning with Intrinsic Exploration Mechanisms for Real-Time Reservoir Management. *SPE J.* **2025**, *30*, 4501–4517. [CrossRef]

Allahloh, A.S.; Sarfraz, M.; Ghaleb, A.M.; Dabwan, A.; Ahmed, A.A.; Al-Shayea, A. Integration of Industrial Internet of Things (IIoT) and Digital Twin Technology for Intelligent Multi-Loop Oil-and-Gas Process Control. *Machines* **2025**, *13*, 940. [CrossRef]

https://doi.org/10.3390/pr14081238

22. Saputelli, L.; Nikolaou, M.; Economides, M.J. Self-Learning Reservoir Management. In Proceedings of the SPE Annual Technica Conference and Exhibition, Denver, CO, USA, 5–8 October 2003; SPE-84064-MS.
23. Hao, J.; You, Q.; Peng, Z.; Ma, D.; Tian, Y. A Model-Free Toolface Control Strategy for Cross-Well Intelligent Directional Drillin *Eng. Appl. Artif. Intell.* **2024**, *133*, 108272. [CrossRef]

https://doi.org/10.3390/pr1408123

processes

MDPI

rticle

eismic Attribute Fusion and Reservoir Prediction Using Multiscale Convolutional Neural Networks and Self-Attention: A Case Study of the B Gas Field, South Sumatra Basin

iyun Cheng [1], Wensong Huang [1], Xiaoling Zhang [1], Zhanxiang Lei [1], Guoliang Hong [1], Wenwen Wang [1], lengyang Zhang [2,3,*], Linze Li [2,3] and Jian Li [1]

1 PetroChina Research Institute of Petroleum Exploration & Development, Beijing 100083, China; ziyuncheng@petrochina.com.cn (Z.C.); hwshws6@petrochina.com.cn (W.H.); zhangxiaoling-yt@petrochina.com.cn (X.Z.); leizhanxiang@petrochina.com.cn (Z.L.); hongguoliang@petrochina.com.cn (G.H.); wangww21@petrochina.com.cn (W.W.); lijianlijian@petrochina.com.cn (J.L.)
2 National Key Laboratory of Petroleum Resources and Engineering, China University of Petroleum (Beijing), Beijing 102249, China; llzllz1021@163.com
3 College of Geosciences, China University of Petroleum (Beijing), Beijing 102249, China
* Correspondence: zmyocean999@gmail.com; Tel.: +86-13319070919

Abstract

Strong heterogeneity and ambiguous seismic responses hinder reliable sandstone thickness prediction when using a single seismic attribute in the lower sandstone interval of the Talang Akar Formation (hereafter abbreviated as the LTAF interval) in the B gas field, South Sumatra Basin. To address this challenge, we propose a seismic attribute fusion and reservoir sweet-spot prediction framework based on a multiscale convolutional neural network (CNN) integrated with a self-attention module. Multiple seismic attribute volumes are organized as multi-channel 2D attribute slices, and parallel convolutions with kernel sizes of 3×3, 5×5, and 7×7 are employed to capture spatial features ranging from thin-bed boundaries and channel morphology to sand-body assemblage distribution. The self-attention module explicitly models inter-attribute dependencies and performs adaptive weighted fusion to suppress noise and emphasize informative attributes. The network adopts a dual-output design, producing (i) a sandstone thickness prediction map at the same spatial resolution as the input and (ii) attribute importance scores for quantitative attribute selection and geological interpretation. Using 3D seismic data and well-constrained thickness labels, the proposed model achieves an R^2 of 0.8954, outperforming linear regression (R^2 = 0.8281) and random forest regression ($R^2 \approx 0.8453$). The learned importance scores indicate that amplitude-related attributes (e.g., RMS amplitude and maximum amplitude) contribute most to thickness prediction, whereas frequency- and energy-related attributes show relatively lower contributions, which is consistent with bandwidth-limited resolution effects. Overall, the proposed framework unifies attribute fusion, thickness prediction, and interpretability within a single model, providing practical support for fine reservoir characterization and development optimization in heterogeneous sandstone reservoirs.

Keywords: South Sumatra Basin; multiscale convolutional neural network; self-attention mechanism; seismic attribute fusion; sandstone thickness prediction

1. Introduction

Seismic exploration is an important technique for reservoir identification and prediction in petroleum exploration and development, which has been widely used in the exploration and development of oil and gas resources worldwide. By analyzing the characteristics of seismic wave propagation in underground strata, seismic exploration can provide critical geological data for reservoir identification, prediction, and development [1–3]. In hydrocarbon exploration, there are multiple challenges for the prediction of sandstone reservoirs. The study of deep tight sandstones in the Qigu Formation of the Junggar Basin conducted by Tian Lei et al. (2024) suggests that deposition controls the distribution and composition of sandstone bodies, which is the basic contributor to reservoir heterogeneity, while diagenesis is a key factor that affects the heterogeneity of deep tight sandstones by modifying sandstone pores in terms of structure and type [4]. The study of Guo Song et al. (2013) on the beach-bar sandstones of the Boxing Oilfield in the Bohai Bay Basin shows that these sandstone reservoirs are highly heterogeneous due to the development of thinly interbedded sandstone and mudstone, their hydrocarbon-bearing properties vary significantly in different well areas and sandstone formations, their quality is closely correlated with their hydrocarbon-bearing properties, and the degree of hydrocarbon filling and hydrocarbon distribution are directly controlled by reservoir heterogeneity [5]. Typically, sandstone reservoirs are highly heterogeneous, with complex variations in porosity and permeability distributions, and they are affected by multiple factors such as depositional environments and tectonic deformation. These factors make it very difficult to predict sandstone reservoirs.

In the selection of favorable seismic attributes, traditional methods often employ techniques such as correlation coefficient analysis, principal component analysis, and multiple regression, which are standardized, easy to implement, and computationally efficient. They are suitable for intuitive interpretation based on geological knowledge and can effectively support reservoir prediction when the number of seismic attributes is small and the seismic response characteristics are relatively simple. However, with the significant increase in seismic attribute types and data dimensions, these methods overly rely on linear correlation and human experience, making it difficult to characterize the complex inter-attribute relationships. Furthermore, due to collinearity and redundancy among seismic attributes, the results of attribute selection are easily affected by changes in samples and parameters, and the stability and applicability of these methods are limited [6,7]. Therefore, traditional methods for seismic attribute selection can barely meet the needs of accurate reservoir prediction, and more intelligent, data-driven attribute selection and fusion techniques need to be introduced. In addition, the statistical results for frequency attributes show that the dominant frequency of seismic data is usually in the range of 20–40 Hz (peak frequency: about 25–30 Hz), leading to weakened relationships between traditional seismic attributes and sandstone reservoirs and limited accuracy of reservoir prediction based on a single seismic attribute [8,9]. The relatively low contribution of frequency-related attributes is likely influenced by the limited seismic bandwidth (dominant frequency of ~25–30 Hz) and the corresponding vertical resolution, which may reduce their sensitivity to subtle thickness variations compared with amplitude-related attributes. Therefore, the lower weights reflect data-dependent sensitivity rather than inherent geological irrelevance. More importantly, the seismic responses of sandstone reservoirs often exhibit high ambiguity. Different seismic attributes may lead to different interpretations of the same reservoir. Due to this reason, traditional seismic attribute analysis methods have significant limitations in terms of accuracy and reliability, making it difficult to meet the needs of accurate reservoir prediction [10–12].

https://doi.org/10.3390/pr1406098

Therefore, accurately capturing and interpreting reservoir features using more advanced seismic exploration techniques has become a key factor in improving the efficiency of hydrocarbon exploration and development. In recent years, the rapidly developing deep learning techniques have provided new solutions, and in particular, the application of convolutional neural networks (CNNs) and attention mechanisms has achieved breakthroughs in many fields. According to Zhang Chenjia et al. (2021), CNNs integrated with attention mechanisms can adaptively weight features at different scales such as channel and space and have significantly enhanced feature representation capabilities that enable them to achieve better results than traditional methods in complex tasks such as image recognition, fully demonstrating the advantages of deep learning techniques in automatic feature extraction and nonlinear modeling [13]. CNNs can automatically extract multiscale spatial features from raw seismic data [14], while attention mechanisms can highlight key features and suppress irrelevant information by dynamically weighting input features, significantly improving the performance of CNN models in complex tasks [15]. These deep learning techniques have shown great potential in image processing in the field of seismic exploration.

In this paper, a seismic attribute fusion method based on a multiscale CNN and a self-attention mechanism is proposed to address the accuracy and reliability issues in sandstone reservoir prediction. This method can improve the accuracy of reservoir prediction by extracting multiscale features from seismic data through the multiscale CNN and by performing weighted fusion of features through the attention mechanism. It enables more effective extraction of key features from seismic data, providing more reliable technical support for accurately predicting sandstone reservoirs.

In summary, existing works on the prediction of complex sandstone reservoirs still have several deficiencies. First, they often use a single seismic attribute or a small number of seismic attributes for linear modeling, making it difficult to fully explore the nonlinear relationships between multi-source seismic attributes and resulting in limited accuracy of reservoir prediction. Second, they rely primarily on manual experience or simplified statistical indicators for the selection of favorable seismic attributes, lacking a unified framework for quantitative characterization of contributions from seismic attributes. Third, most of the studies on the sandstones in the Lower Talang Akar Formation (LTAF) of the B field in the South Sumatra Basin focus on geological understanding and conventional seismic interpretation, lacking deep learning-driven practices in sandstone thickness prediction. Therefore, an intelligent seismic attribute characterization method integrating a multiscale CNN and a self-attention mechanism was proposed to perform sandstone thickness prediction and seismic attribute selection for the LTAF sandstone reservoir in the B field, and its effectiveness was verified using actual production data.

The main contributions of this study are threefold. (1) We design a multiscale CNN backbone with parallel receptive fields to capture seismic–geological features at multiple spatial scales, which is critical for heterogeneous channelized sandstone reservoirs. (2) We incorporate a self-attention module to adaptively fuse multi-attribute features and mitigate redundancy/noise among seismic attributes. (3) We propose a dual-output strategy that outputs both a thickness prediction map and quantitative attribute importance scores, enabling joint prediction and interpretation within a unified framework. In comparison with existing research, our method offers substantial improvements in both prediction accuracy and interpretability. Recent studies have demonstrated the advantages of integrating seismic attributes with machine learning for reservoir prediction, yet limitations remain. For example, multi-attribute regression frameworks have been successfully applied to predict porosity and shale volume in complex deltaic settings, and feature importance analysis has been used to interpret influential seismic attributes [16,17]. However, these approaches do not fully exploit multiscale seismic information or attention mechanisms. More recently, deep learning-driven multi-frequency seismic inversion methods that combine multiscale convolutional modules with self-attention have shown improved characterization of thin-layer stratigraphy, illustrating the benefits of capturing contextual relationships in seismic data [18–20]. Building on these advances, our framework incorporates multiscale learning and self-attention to effectively fuse diverse seismic attributes and enhances interpretability through a dual-output design, providing both thickness prediction maps and attribute importance scores—features that are often lacking in conventional methods.

2. Geological Setting and Overview of the Study Area

2.1. Regional Geological/Tectonic Setting and Location of the Study Area

Located in southeastern Sumatra, the South Sumatra Basin is one of the most important hydrocarbon-bearing basins in Indonesia and one of the regions with the greatest hydrocarbon potential in the world [21]. The formation of this basin is closely related to its unique tectonic setting, and it has undergone multiple stages of tectonic evolution. Its oil and gas resources are mainly concentrated in medium-deep sandstone reservoirs, with great potential and value for exploration and development [22]. The abundant oil and gas resources in the basin are closely related to its complex geological structure, depositional environment, and petrophysical properties. Therefore, systematically studying the geological characteristics of the basin is of great significance for hydrocarbon exploration and development.

The B field is situated in the core area of the South Sumatra Basin. The basin's tectonic evolution process can be roughly divided into three stages, namely, the rift stage, the post-rift and back-arc subsidence stage, and the inversion and uplift stage. During the rift stage, the subduction of the West Sumatra Trench caused the basin to extend, creating a series of half-grabens. During the post-rift stage, tectonic activity weakened, the subsidence of the basin intensified, the sea level rose, and the basin entered the marine transgression stage. During the inversion and uplift stage, the basin experienced multistage structural inversion accompanied by the uplift of the Barisan Mountains and the formation of compressional-torsional folds. These structural inversion events facilitated the formation and adjustment of various types of reservoirs, providing favorable conditions for hydrocarbon accumulation and preservation [23].

2.2. Stratigraphic and Sedimentary Characteristics

The South Sumatra Basin has a diverse and complex geological framework, and the evolution of its depositional environment and tectonic activity have profoundly influenced the formation and distribution of reservoirs. The depositional environment of the basin has evolved from a lacustrine environment to a shallow sea environment and then to a deep-sea environment. In particular, multistage marine transgressions have significantly affected the spatiotemporal distribution of the basin's strata and the development of reservoirs in the region.

The main sedimentary units of the basin include the pre-rift basement/bedrock, the rift-stage Lahat, Lemat, and Talang Akar formations, and post-rift formations such as Batu Raja, Gumai, Air Benakat, and Muara Enim. The significant differences in sediment type and tectonic setting among these formations directly control the degree of source rock development and the spatiotemporal distribution of reservoirs. The Talang Akar Formation and the Gumai Formation differ significantly in sediment characteristics and lithology. The Gumai Formation is dominated by marine sediments and volcaniclastic rocks, with well-developed secondary pores providing ample space for hydrocarbon storage. In contrast, the depositional system of the Talang Akar Formation (the main stratigraphic unit studied

https://doi.org/10.3390/pr1406098

herein) is dominated by deltaic and fluvial facies, with favorable facies zones such as delta fronts and channel deposits. The sandstones of the Talang Akar Formation generally have high porosity and high permeability, forming a series of high-quality sandstone reservoirs in zones with delta fronts and channel deposits [24]. Due to differences between these two formations in sedimentary facies and lithology, there are multiple types of reservoirs and various hydrocarbon accumulation patterns in the study area, creating great resource potential and broad prospect for subsequent hydrocarbon exploration and development.

In this study, the target interval is the lower sandstone interval of the Talang Akar Formation in the B gas field (hereafter abbreviated as the LTAF interval). This abbreviation is used for convenience within the study area. It is mainly distributed in the rift zone and the slopes on both sides of the rift zone. In terms of lithology, it is dominated by medium- to fine-grained sandstone interbedded with a small number of thin conglomerate and mudstone layers, and channel sandstone is the most prevalent type of reservoir rock. In terms of depositional environment, LTAF was formed in a lacustrine-deltaic depositional system with multiple sedimentary facies, including braided channels, meandering channels, crevasse splays, and floodplains. The delta front sandstones and channel-filling sandstones generally have high porosity and high permeability, making them favorable facies zones for hydrocarbon accumulation. The thickness of LTAF is notably characterized by coupled effects of tectonic-depositional processes. It is relatively thick in the central depression zone, reaching tens of meters to nearly 100 m, and it gradually thins towards structural highs and basin margins until it pinches out. Controlled by both sedimentary facies distribution and faulting, the LTAF sandstone reservoir is highly heterogeneous in both horizontally and vertically, with significant variations in hydrocarbon charge degrees and gas-bearing properties in different sections. Therefore, it is necessary to accurately predict the distribution and thickness of the LTAF sandstone body based on seismic attributes using intelligent algorithms in order to provide support for adjusting the mid- to late-stage development strategy of the B field (Figure 1).

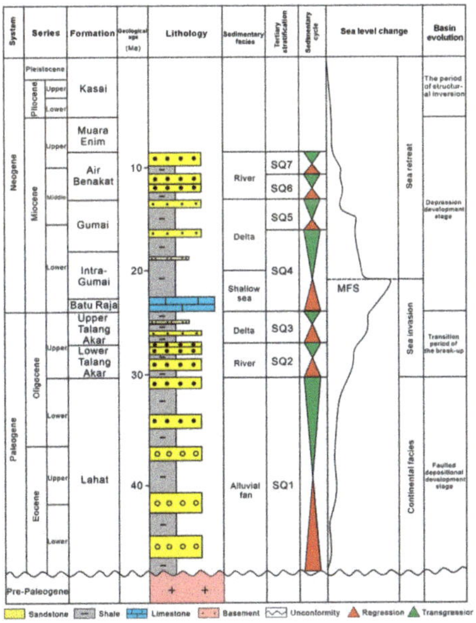

Figure 1. Composite stratigraphic column of the South Sumatra Basin showing lithology, depositional facies, sequence stratigraphy, and sea-level changes; different dot patterns in sandstone indicate grain-size variations.

2.3. Research Data Overview

Reservoir prediction in this study was performed using 3D seismic data and well logs from the study area. The 3D seismic volume covers approximately 80 km^2, with a trace spacing of 20 m × 20 m and a sampling interval of 2 ms, which satisfies the resolution requirements for reservoir-scale geological analysis. Formation of 2D input slices: For each seismic attribute volume, 2D slices were extracted within the target interval bounded by the picked top and base horizons. The selected 18 seismic attributes were stacked along the channel dimension to form multi-channel inputs. Each full 2D slice is defined on a regular grid with a spatial sampling of 20 m × 20 m and a size of H × W = 876 × 712 grid nodes, where H = 876 and W = 712 correspond to the numbers of nodes in the J- and I-directions, respectively. For model training, we constructed well-centered 2D patches cropped from these slices and stacked the same 18 seismic attributes to form patch tensors of shape h × w × 18 (where h and w denote the patch dimensions). Each attribute channel was normalized using z-score normalization based on training set statistics. The label for each patch was the log-interpreted sandstone thickness at the corresponding well location, which was used to supervise thickness prediction. The seismic data contain stable and effective frequency information within the target interval. Sandstone thickness labels interpreted from logs in 113 wells (out of 237 wells in the study area) were used as supervised-learning targets.

During data processing, the seismic volume was first preprocessed and calibrated using conventional workflows. Multiple seismic attributes (e.g., average negative amplitude, average positive amplitude, RMS amplitude) were extracted within the time window of the target interval.

To construct training and validation samples, well markers were tied to the seismic time domain, and multi-attribute seismic responses were matched with the log-derived sandstone thickness at the corresponding well locations. In addition, to ensure comparable magnitudes across different attributes and improve training stability, each attribute channel was normalized using statistics computed from the training set. The dataset was divided into training and validation sets at a ratio of 70%:30% for model training and evaluation.

3. Multi-Source Seismic Attribute Intelligent Fusion and Deep Learning-Based Modeling

In this study, a seismic attribute selection and fusion model based on a multiscale CNN integrated with a self-attention mechanism was fully built, considering the characteristics of the responses of seismic attributes to geological structures and lithofacies at different scales. This model takes multiple seismic attributes as input, extracts local to global features through multiscale convolutions, and then simulates inter-attribute correlations through the self-attention mechanism, achieving a unified process of reservoir prediction and seismic attribute selection (Figure 2).

3.1. Multiscale Convolutional Feature Input and Extraction

The core idea of the multiscale CNN architecture is to extract multilevel/multiscale features using convolution kernels of different sizes, thus capturing multiscale information in the input data more comprehensively [25–27]. To capture the details presented by different seismic responses at different scales, multiple convolution kernels of different sizes were employed to extract multilevel/multiscale features from seismic data. The detailed process is described below.

https://doi.org/10.3390/pr1406098

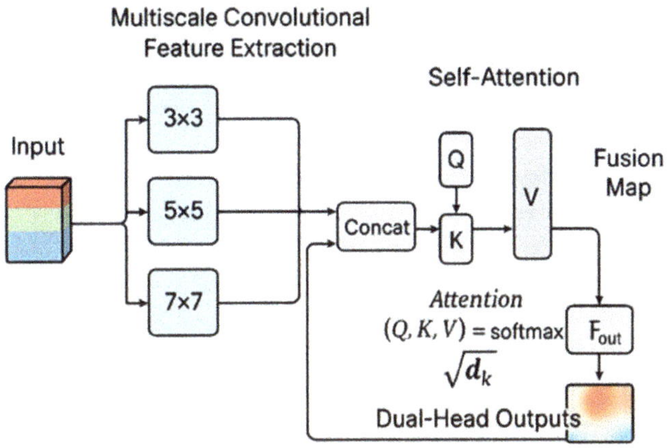

Figure 2. Schematic of the proposed multiscale CNN self-attention framework. Multiscale convolutions (kernel sizes 3 × 3, 5 × 5, and 7 × 7) extract spatial features at different scales, which are fused by a self-attention module. The network produces dual outputs: a sandstone thickness prediction map and attribute importance scores for interpretability.

First, two-dimensional (2D) seismic attribute slices are input using Equation (1).

$$X = [A_1, A_2, \ldots, A_N], A_i \in R^{(H \times W)} \quad (1)$$

where H and W represent the spatial dimensions of the seismic profile, and A_i represents the ith seismic attribute (amplitude, phase, instantaneous frequency, or impedance).

Following this, multiple convolutional layers with convolution kernels of different sizes are used for feature extraction. The 3 × 3 convolution kernel extracts texture information, amplitude micro-variations, and the boundaries of thin sandstone layers. The 5 × 5 convolution kernel extracts feature related to the morphology, distribution, direction, and lateral extension of channel sandstone bodies. The 7 × 7 convolution kernel captures features related to the overall morphology of channel groups or sandstone groups and the continuity of sandstone bodies.

The output of convolutions at the aforementioned three scales can be expressed as:

$$F_k = f_{k \times k}(X), k \in \{3,5,7\} \quad (2)$$

where $f_{k \times k}(\cdot)$ represents convolution operations with a kernel size of $^{k \times k}$.

Finally, the extracted multiscale features are spliced and fused using Equation (3) to provide the basis for inter-attribute correlation analysis.

$$F = \text{Concat}(F_3, F_5, F_7) \quad (3)$$

3.2. Feature Enhancement Through Self-Attention Mechanism

Self-attention is an attention mechanism that has been widely used in deep learning in recent years. It can dynamically adjust the weights of input features based on inter-feature correlations, thereby strengthening the learning of important features and suppressing irrelevant features [28,29]. To optimize the feature fusion process of the multiscale CNN, a self-attention mechanism was introduced in this study, obtaining a weight matrix by calculating the correlations between input features. The greater the weight of a feature is, the more important the feature is to the final prediction results. In the seismic attribute fusion process, the weights of seismic attributes can be dynamically adjusted based on inter-attribute correlations to enable the network model to focus more on important features

https://doi.org/10.3390/pr14060981

while suppressing noise and irrelevant information. The features extracted by the multiscale CNN in the subsequent feature extraction process are input into the self-attention module. The self-attention module calculates the weight of each feature and performs weighted fusion of features. In this way, the network model can focus on the most important seismic attributes, thereby improving the accuracy of reservoir prediction.

First, the feature map F is linearly transformed into query (Q), key (K), and value (V) matrices.

$$Q = FW_Q, K = FW_K, V = FW_V \tag{4}$$

The attention weight matrix is calculated as follows:

$$\text{Attention}(Q, K, \text{V}) = \text{softmax}\left(\frac{QK^T}{\sqrt{d_k}}\right)V \tag{5}$$

An enhanced feature is obtained through residual connections and layer normalization as follows:

$$F_{out} = \text{LayerNorm}(F + \text{Attention}(Q, K, V)) \tag{6}$$

3.3. Dual-Output Design

To obtain and represent the results of seismic attribute fusion more clearly, a heatmap showing the distribution of attribute importance weights (Figure 3) and a fused map of predicted sandstone thickness values (Figure 4) were created based on multiscale convolutional feature extraction and self-attention feature fusion for seismic attribute selection and feature importance interpretation.

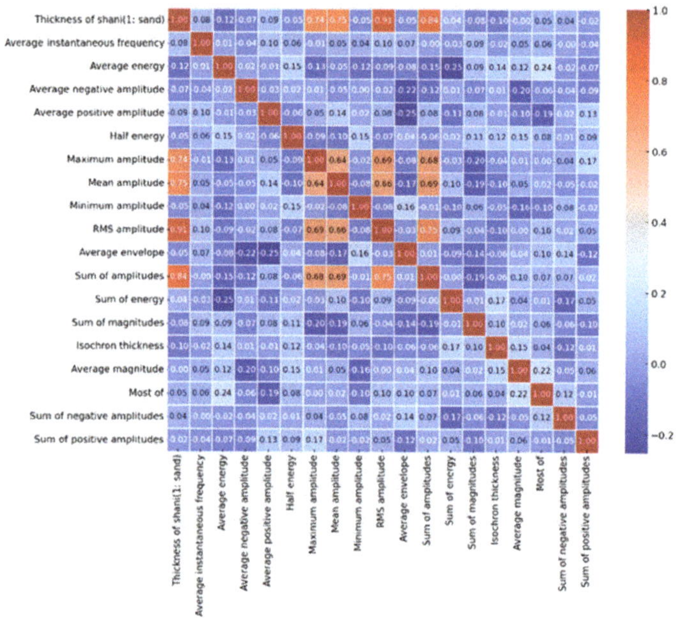

Figure 3. Seismic attribute contribution matrix (based on attention weight statistics).

https://doi.org/10.3390/pr1406098

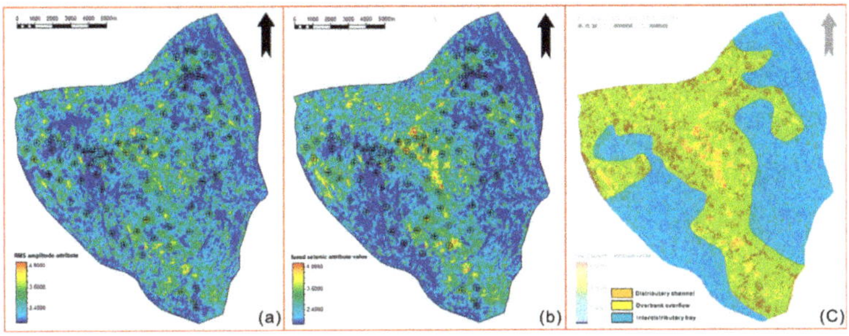

Figure 4. Map comparison for seismic attributes, fused prediction, and facies interpretation within the target interval of the B gas field. (**a**) RMS amplitude attribute map. (**b**) Fused seismic attribute response and predicted sandstone thickness distribution obtained from the proposed multiscale CNN with self-attention; validation wells are marked by the well symbols in the map to enable well-constrained consistency checking. (**c**) Interpreted facies zoning (distributary channel, overbank overflow, and interdistributary bay), used to support geological interpretation and to assess the spatial consistency between predicted thick-sand zones and sedimentary facies architecture. The scale bar is shown at the top of each panel, and the arrow indicates the north direction. Well locations are marked by circle-with-cross symbols, representing validation wells used for consistency checking.

3.3.1. Feature Map Obtained by Attribute Fusion

The feature tensor F_{out} output by the self-attention module is mapped through the 1×1 convolutional layer or up-sampling layer to a sandstone thickness prediction map with the same resolution as the input to represent the spatial distribution of sandstone thickness values corresponding to favorable sandstone bodies. The calculation process can be expressed as follows:

$$Y_{\text{pred}} = f_{\text{conv}}(F_{\text{out}}) \tag{7}$$

where $f_{\text{conv}}(\cdot)$ is the convolutional mapping function, and Y_{pred} is the final spatial distribution predicted by the model.

3.3.2. Attribute Importance Analysis

To provide an interpretable measure of attribute contribution, we compute the average response values of various channels (corresponding to seismic attributes) by performing global average pooling (GAP) on the fused feature map in the spatial dimension. This process generates channel-wise statistics, which are then normalized using a SoftMax function, allowing the resulting weights to be interpreted consistently across attributes. These importance scores are used to (i) rank seismic attributes for the target interval and (ii) support the geological interpretation of why the model emphasizes certain seismic responses.

$$w_c = \frac{1}{H'W'} \sum_{h=1}^{H'} \sum_{w=1}^{W'} F_{\text{out}}(h, w, c) \tag{8}$$

where H' and W' are the height and width of the feature map, respectively, and $F_{\text{out}}(h, w, c)$ represents the response value of the cth channel at the spatial location (h, w). Then, the SoftMax function is applied for normalization so that the weights of seismic attributes can be directly interpreted as probability distributions.

$$\widetilde{w}_c = \frac{e^{w_c}}{\sum_{j=1}^{C} e^{w_j}} \tag{9}$$

https://doi.org/10.3390/pr14060981

Finally, the importance weight ($\widetilde{w}_c$) of each seismic attribute/channel is obtained, which can be used for seismic attribute selection and physical importance analysis. A larger $\widetilde{w}_c$ value indicates that the attribute contributes more significantly to reservoir identification and has higher geological sensitivity.

3.4. Loss Function and Training Strategy

During model training, the measured sandstone thickness curve or equivalent sandstone thickness data were used as labeled data for supervised learning, and the fused feature was mapped to a sandstone thickness prediction map of the same size as the input profile. In this paper, the mean squared error (MSE) is used as the loss function.

$$L_{\text{pred}} = \frac{1}{N}\sum_{i=1}^{N}\left(h_i^{\text{pred}} - h_i^{\text{obs}}\right)^2 \tag{10}$$

where h_i^{pred} and h_i^{obs} are the predicted and measured sandstone thickness values at the *ith* well or sampling point, and N is the number of sample points.

To suppress abnormal weight amplification and improve training stability, the L2 regularization term is introduced to form a joint loss function as follows:

$$L = L_{\text{pred}} + \lambda \parallel \mathbf{w} \parallel_2^2 \tag{11}$$

where $\mathbf{w}$ is the weight vector of the attribute channel, and λ is the balance coefficient. The training data is divided into training and validation sets at a ratio of 70%:30%. The labeled samples were split into training and validation subsets at a ratio of 70%/30%. This split is commonly adopted in supervised learning to balance model fitting and unbiased performance assessment, especially under a limited number of labeled wells. We used the Adam optimizer and early stopping (training terminates when the validation loss does not decrease for a preset number of epochs) to reduce the risk of overfitting. The Adam optimization algorithm is used for backpropagation. To reduce the risk of overfitting, model training is terminated when the validation loss no longer decreases within a certain number of consecutive iterations.

4. Results of Seismic Attribute Fusion and Model Validation

4.1. Seismic Attribute Contribution Matrix and Sandstone Thickness Prediction Results

Based on the trained model, we obtain (i) an attribute importance matrix (Figure 3) by summarizing attention-related statistics and (ii) a fused sandstone thickness prediction map (Figure 4b). Figure 3 indicates that amplitude-related attributes (e.g., RMS amplitude and maximum amplitude) contribute more strongly to thickness prediction, whereas frequency- and energy-related attributes contribute less. This provides a quantitative basis for attribute selection in the LTAF interval.

The matrix in Figure 3 shows the relative contributions of different seismic attributes. Amplitude-related attributes (such as the RMS amplitude and maximum amplitude) contribute more significantly to sandstone thickness prediction, while frequency and energy-related attributes contribute relatively less. This provides a basis for attribute selection. This observation is consistent with the bandwidth-limited resolution of the available seismic data, which may weaken the response of some frequency-derived measures to fine-scale thickness variations. Hence, the lower weights should be interpreted as reduced sensitivity under the present data conditions rather than inherent geological irrelevance.

Compared with (Figure 4a), the fused prediction (Figure 4b) exhibits clearer geological patterns and improved continuity, suggesting that multiscale feature extraction helps capture channel morphology and lateral connectivity, while adaptive fusion suppresses scattered noise and enhances spatially consistent sand-body signals.

The importance of each seismic attribute was quantitatively evaluated using the seismic attribute contribution matrix, and weights were dynamically assigned to seismic attributes based on sandstone thickness prediction. Then, multiple attribute information was deeply mined and synergistically fused using the CNN in combination with the self-attention mechanism to improve the continuity and clarify of the spatial changes in amplitude, frequency, and energy features, significantly improving the accuracy of sandstone thickness prediction and providing reliable reference for subsequent geological modeling and the fine characterization of sandstone reservoirs.

In Figure 4, compared with the original seismic data map that has not been processed by deep learning, the fused map shows clearer key geological features that have been effectively enhanced by multiscale convolutional feature extraction through the CNN and the weighted fusion of extracted features through the self-attention mechanism. In particular, these feature extraction and fusion processes have enhanced the model's ability to identify sandstone bodies. The amplitude-related attributes (taking the RMS amplitude as an example) and sandstone thickness prediction results obtained after seismic attribute fusion show a clearer trend of accumulation in the spatial dimension, which is difficult to observe in the original seismic data map.

Before attributing fusion, the seismic attributes in the original map are often sparsely distributed and characterized by high noise levels, making it difficult to reveal reservoir complexity and detail. The CNN model with self-attention can effectively suppress noise and irrelevant information while enhancing the representation of key features. In this process, the weights of different seismic attributes are dynamically adjusted by the self-attention mechanism to further highlight areas where sandstone bodies are predicted to exist.

Compared with the original map, the fused thickness prediction map shows the relationship between reservoir features and seismic attributes more clearly, which facilitates interpretation.

4.2. Model Validation

To evaluate the predictive accuracy of the proposed multiscale CNN with self-attention, we performed well-constrained validation and compared the results with baseline methods. The coefficient of determination (R^2) was used as the primary metric, and additional residual-based analysis was conducted to inspect deviation patterns.

4.2.1. Validation Metrics

For the purpose of this study, the coefficient of determination (R^2) is used as the main metric to verify the model's accuracy in sandstone thickness prediction. R^2 measures the goodness of fit between the model's predictions and actual observations, representing the proportion of actual sandstone thickness variations that the model can interpret. R^2 is calculated as follows:

$$R^2 = 1 - \frac{\sum_{i=1}^{n} (y_i - \hat{y}_i)^2}{\sum_{i=1}^{n} (y_i - \overline{y})^2} \quad (12)$$

where y denotes the measured (actual) sandstone thickness, $\hat{y}$ denotes the corresponding model prediction, and $\overline{y}$ is the mean of y. A larger R^2 indicates a better agreement between predictions and measurements.

https://doi.org/10.3390/pr14060981

4.2.2. Benchmark Comparison with Baseline Methods

Figure 5 shows the cross-plot between the sandstone thickness predicted by the proposed model and the measured thickness at wells. The validation result indicates a strong correlation, with $R^2 = 0.8954$, suggesting that the proposed model can explain most of the variation in sandstone thickness in the study area.

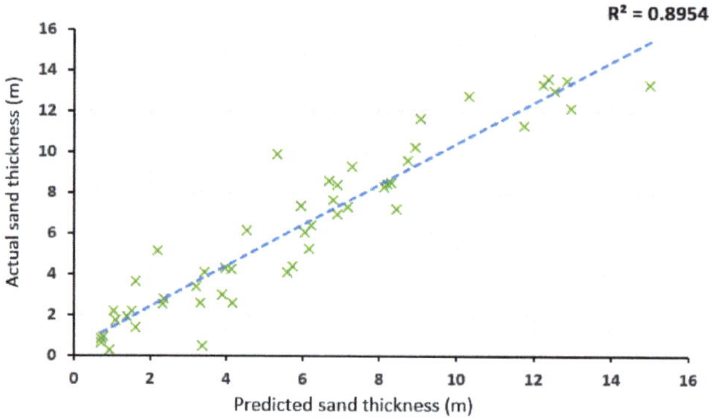

Figure 5. Cross-plot of sandstone thickness predicted by the proposed model versus well-measured thickness (m). Each × symbol represents a well data point. The dashed line denotes the fitted regression, and the coefficient of determination is $R^2 = 0.8954$.

For comparison, we further evaluated two commonly used baselines. First, a linear regression model was built using a representative single seismic attribute (RMS amplitude) as the predictor. The corresponding cross-plot is shown in Figure 6, where the coefficient of determination is $R^2 = 0.8281$, which is notably lower than that of the proposed model. Second, a random forest regression model was used to represent a nonlinear baseline in Figure 7. Its performance is $R^2 \approx 0.8453$, which also remains lower than that of the proposed multiscale CNN with self-attention.

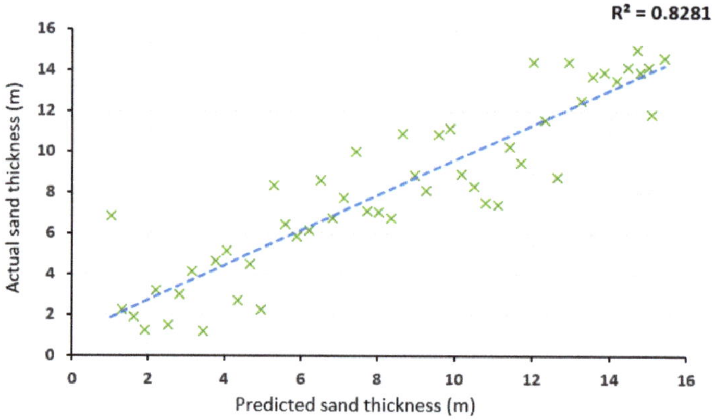

Figure 6. Cross-plot of sandstone thickness predicted by the linear regression model versus well-measured thickness (m). Each × symbol represents a well data point. The dashed line denotes the fitted regression, and the coefficient of determination is $R^2 = 0.8281$.

https://doi.org/10.3390/pr14060981

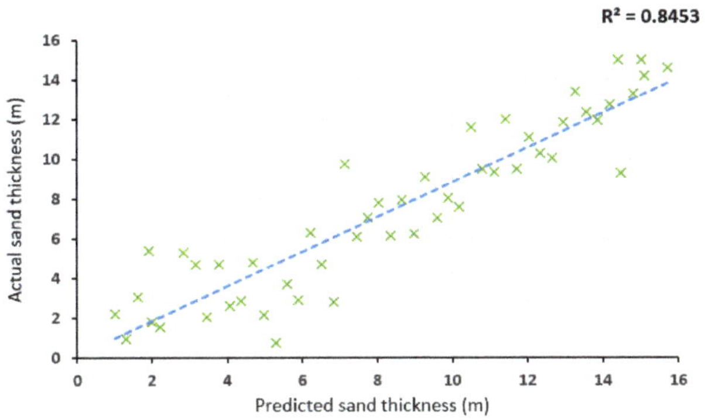

Figure 7. Random Forest predictions versus measured sandstone thickness (m). Each × symbol represents a well data point. The dashed line denotes the fitted regression ($R^2 \approx 0.8453$).

Overall, the proposed model achieves the highest R^2 among the tested methods, indicating improved nonlinear regression capability and better generalization for thickness prediction in complex heterogeneous sandstone reservoirs.

4.2.3. Ablation Study

To quantify the contribution of the self-attention mechanism to attribute fusion, we conducted an ablation study by removing the self-attention module while keeping the multiscale CNN backbone, input attributes, and training strategy unchanged. In addition, we included a single-scale CNN baseline (without multiscale kernels and without attention) to evaluate the benefit of multiscale feature extraction. All models were trained and evaluated using the same dataset split and the same evaluation metrics for fair comparison. The ablation experiments have been designed to quantify the contributions of multiscale feature extraction and the self-attention mechanism under identical data split and training settings. The quantitative results for the ablated models will be reported in Table 1 after completing the additional runs.

Table 1. Ablation study design for quantifying the effects of multiscale convolution and the self-attention module (results to be filled).

Model	Multiscale	Self-Attention	R^2
Single-scale CNN	No	No	TBF
Multiscale CNN (w/o attention)	Yes	No	TBF
Proposed (Multiscale + attention)	Yes	Yes	0.8954

Note: TBF (to be filled) denotes the quantitative results of the additional ablation runs under the same data split and training settings.

This ablation setup enables a controlled evaluation of how multiscale convolutions and self-attention contribute to thickness prediction performance and interpretability, and the finalized statistics will be included in the revised version. The ablation results provide direct evidence on how the self-attention module improves performance by explicitly modeling inter-attribute correlations and enabling adaptive weighting during the fusion process.

4.2.4. Deviations and Uncertainty Analysis

Residuals between predicted and measured sandstone thickness are analyzed using a residual scatter plot (Figure 8) and a residual histogram (Figure 9), where the resid-

https://doi.org/10.3390/pr14060981

ual is defined as Predicted − Actual (m). As shown in Figure 8, residuals are generally centered around the zero line across the prediction range with a slight negative bias (mean = −0.41 m; median = −0.21 m), indicating mild underestimation on average. Quantitatively, the overall error level is MAE = 1.01 m and RMSE = 1.37 m. Figure 9 shows that most residuals concentrate near zero, while the tails and a few outliers indicate localized uncertainty. These larger deviations are likely related to (i) low signal-to-noise ratio and local acquisition/processing artifacts, (ii) thin-bed tuning/interference effects, (iii) rapid lateral facies transitions, and/or (iv) structural complexity (faulting or steep dips). Therefore, zones with larger absolute residuals should be treated as higher-risk targets and prioritized for additional verification, such as updated well ties or local seismic conditioning. Overall, the residual analysis suggests stable predictions with meter-level typical errors and limited localized uncertainty.

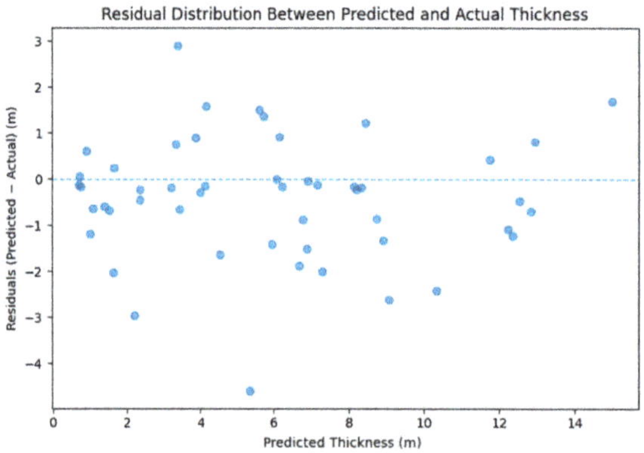

Figure 8. Residual scatter plot of sandstone thickness prediction, where residuals are defined as Predicted − Actual (m). The dashed line indicates zero residual; points above/below the line represent over-/under-prediction, and larger absolute residuals indicate higher prediction uncertainty.

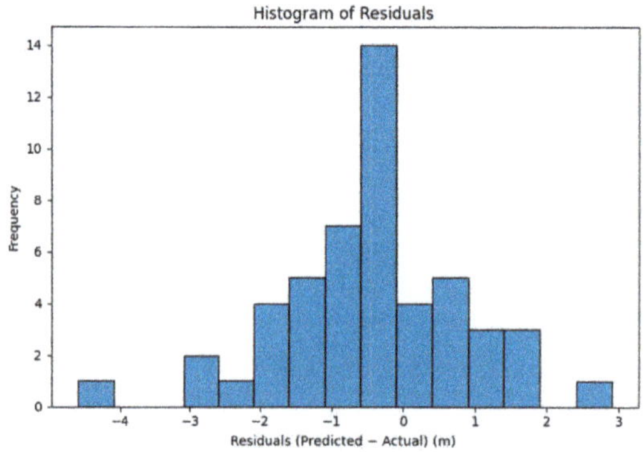

Figure 9. Histogram of residuals (Predicted − Actual, m) for sandstone thickness prediction. The peak near zero indicates that most predictions have small errors, whereas the spread and tails reflect high-deviation cases that may require further validation.

4.3. Geological and Engineering Implications

The predicted sandstone bodies in the LTAF interval are mainly distributed along braided channels and delta-front fairways in the central depression zone of the B field, which aligns with the established sedimentary facies understanding. Several favorable

https://doi.org/10.3390/pr14060981

zones extend toward structural lows and slope areas, indicating potential targets that warrant further evaluation.

From an engineering perspective, the thickness prediction map can be translated into an actionable decision-support workflow (Figure 10). First, it can be used to screen and rank candidate locations for new wells by applying thickness thresholds and connectivity criteria within the predicted sand-body fairways, while avoiding isolated anomalies likely caused by noise. Second, for existing wells located near predicted thickness gradients, the results provide a quantitative basis to prioritize intervention options (e.g., sidetracking or infill drilling) by jointly considering predicted thickness, sand-body continuity, and production dynamics. Third, the attribute importance output offers an interpretable link between predictions and seismic responses, supporting risk management by identifying which attribute groups drive model decisions in different blocks. Overall, the proposed framework provides a feasible technical route for mid- to late-stage development optimization in complex sandstone reservoirs.

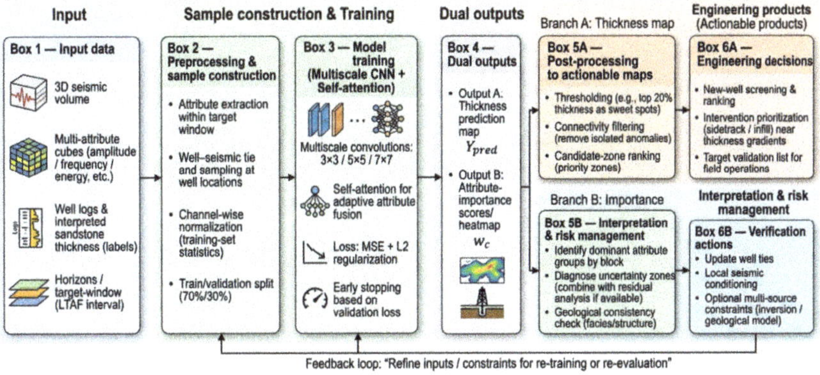

Figure 10. Engineering decision-support workflow based on the proposed framework. Multi-attribute seismic data and well-constrained thickness labels are used to construct training/validation samples and train the multiscale CNN integrated with a self-attention module. The model produces dual outputs: (i) a sandstone thickness prediction map and (ii) attribute importance scores for interpretability. The thickness map is further converted into actionable products via thresholding and connectivity-based filtering to support candidate-zone ranking for new wells and intervention planning, while the attribute importance output supports interpretation and risk management. Different colored boxes represent major stages of the workflow, and the numbered boxes (Box 1–Box 6) indicate sequential processing steps from data input to engineering decision-making.

5. Conclusions

(1) In this study, an intelligent seismic attribute representation model integrating a multi-scale CNN and a self-attention mechanism was built, achieving simultaneous seismic attribute fusion, reservoir thickness prediction, and attribute importance analysis under the same network architecture while ensuring both predictive accuracy and interpretability.

(2) The multiscale convolutional feature extraction module effectively extracts multi-level/multiscale spatial features from thin-layer boundaries and channel morphology to sandstone distribution, while the self-attention mechanism models inter-attribute correlations and performs adaptive weighted fusion of features, enabling key seismic attributes to present more concentrated and continuous spatial responses in zones with thick sandstones. The results of well-constrained model validation show that the

https://doi.org/10.3390/pr14060981

model's R^2 value for sandstone thickness prediction reaches 0.8954, indicating that the model outperforms baseline methods such as linear regression and random forest.

(3) The results of attribute weight analysis show that amplitude-related attributes such as the RMS amplitude and maximum amplitude are most sensitive to sandstone thickness, while some frequency and energy-related attributes contribute less to sandstone thickness prediction. These results provide a quantitative basis for the selection of favorable seismic attributes for the study area.

(4) The model proposed in this paper has achieved good results in the LTAF sandstone reservoir of the B field in the South Sumatra Basin, and it has the potential for extended application in similar zones with medium- to high-frequency seismic data. However, the model's performance still depends on the quantity and quality of well constraints and the quality of seismic attributes, and its applicability in areas with low signal-to-noise ratio (SNR), very few wells, and/or higher heterogeneity needs to be further evaluated. In the future, we will attempt to introduce joint constraints of 3D convolution/temporal attention and multi-source data (e.g., well logs, geological models, inversion volumes) and carry out uncertainty quantification to further improve the model's reliability and engineering applicability.

Author Contributions: Conceptualization, X.Z.; methodology, Z.C.; software, Z.C., Z.L., G.H., W.W., M.Z. and L.L.; validation, Z.C. and X.Z.; formal analysis, Z.C., W.H., Z.L. and J.L.; investigation, Z.L., L.L. and J.L.; resources, Z.C. and X.Z.; data curation, Z.C., W.H. and G.H.; writing—original draft, Z.C.; writing—review & editing, M.Z.; visualization, W.H. and W.W. All authors have read and agreed to the published version of the manuscript.

Funding: This research received no external funding.

Data Availability Statement: The data presented in this study are available on request from the corresponding author. The seismic and well data used in this study are proprietary and confidential to the data owner/field operator and are not publicly available due to legal and commercial restrictions. The data may be made available from the corresponding author upon reasonable request and with the permission of the data owner.

Conflicts of Interest: Ziyun Cheng, Wensong Huang, Xiaoling Zhang, Zhanxiang Lei, Guoliang Hong, Wenwen Wang and Jian Li were employed by PetroChina Research Institute of Petroleum Exploration & Development. The remaining authors declare that the research was conducted in the absence of any commercial or financial relationships that could be construed as a potential conflict of interest.

References

1. Liu, J.F. Application and development of seismic exploration technology in petroleum and geological exploration. *Petrochem. Technol.* **2020**, *27*, 137–138.
2. Xiao, P. Study of the development status and trend of petroleum exploration technology. *Petrochem. Technol.* **2020**, *27*, 285–286.
3. Cao, H.; Dai, S.H.; Li, J.F.; Zheng, X.F.; Liu, H.S.; Zhou, K. Seismic exploration technology for tight oil and gas reservoirs in China and its development direction. *Acta Pet. Sin.* **2025**, *46*, 236–254. [CrossRef]
4. Tian, L.; Song, F.; Zhang, S.Q.; Yang, Y. Heterogeneity of deep tight sandstone reservoirs and controlling factors—A case study of the Jurassic Qigu Formation in Yongjin Area of the Junggar Basin. *J. Northeast. Pet. Univ.* **2024**, *48*, 29–40, 136–137.
5. Guo, S.; Tan, L.J.; Lin, C.Y.; Li, H.N.; Lü, X.X. Effect of reservoir heterogeneity on hydrocarbon accumulation—A case study of the beach-bar sandstones of the upper sub-section of the 4th member of the Shahejie Formation in Boxing Oilfield. *Pet. Geol. Exp.* **2013**, *35*, 534–538.
6. Deng, Y.; Liu, X.H.; Wang, L.L.; Zhang, H.L.; Liu, Z.X.; Hao, J.J. Study and application of thin reservoir prediction methods based on broadband seismic data. *Comput. Tech. Geophys. Geochem. Explor.* **2025**, *47*, 209–221.
7. Jiang, Z.H. Research on reservoir prediction based on intelligent frequency-divided inversion—A case study of the southern exploration area of the Heiyupao Depression, Daqing. *Chin. J. Geol.* **2021**, *56*, 1052–1061.
8. Cui, Y.Z.; Yao, R.; Pan, H.H.; Cui, G.Z.; Guo, J.L.; Wang, X.H.; Liu, Y.Y.; Xu, C.Y.; Liu, S. Application of multi-filter combination in seismic frequency expanding. *Geophys. Prospect. Pet.* **2025**, *64*, 1060–1071. [CrossRef]

https://doi.org/10.3390/pr1406098

Wang, J.; Ren, X.F.; Zhang, J.H.; Wang, X.F.; Zhang, P.; Zhang, D. Thickness prediction method for favorable sandstone and conglomerate reservoirs based on PCA-RF. *CT Theory Appl.* **2021**, *29*, 311–318. [CrossRef]

Wang, H.X.; Feng, S.Y.; Liu, Z.L. A geological structure identification model based on an improved random forest algorithm. *Int. J. Coal Sci. Technol.* **2023**, *10*, 25.

Liu, W.X.; Wang, H.; Wan, Q.H.; Heng, L.Q.; Yin, N.X.; Pan, S.J. Fine-scale analysis of the configuration of braided river delta reservoirs based on frequency-divided RGB fusion technology. *J. Earth Sci. Environ.* **2022**, *44*, 765–774.

Li, T.T.; Wang, Z.; Ma, S.Z.; Wang, Z.; Yuan, Z.L. A review of seismic attribute fusion methods. *Prog. Geophys.* **2015**, *30*, 378–385.

Zhang, C.J.; Zhu, L.; Yu, L. A review of attention mechanisms in convolutional neural networks. *Comput. Eng. Appl.* **2021**, *57*, 64–72.

Wang, S.H.; Yu, H.Z.; Tan, S.Q.; Chen, X.G. Research on seismic facies identification technology based on deep convolutional neural networks. *Comput. Tech. Geophys. Geochem. Explor.* **2020**, *42*, 475–480.

Zhao, M.; Zhao, Y.; Shen, D.H.; Wang, J.; Dai, X. High-resolution processing of seismic data using the U-net adaptive attention mechanism. *Oil Geophys. Prospect.* **2024**, *59*, 675–683.

Ejaita, E.; Omonefe, F. Multi-attribute predictions of volume of shale and porosity from seismic data over Akaso Field using machine learning regression. *Discov. Geosci.* **2026**, *4*, 80. [CrossRef]

Tan, P.; Fu, S.; Huang, L.; Chen, Z.; Cao, J. Effects of orthogonal cleat structures on hydraulic fracture evolution behavior. *Geoenergy Sci. Eng.* **2024**, *241*, 213119. [CrossRef]

Zhang, J.; Shan, X.; Huo, S.; Huang, L.; Zheng, W.; Zhou, X.; Liu, E. Deep learning-driven multi-frequency seismic inversion for enhanced thin-layer stratigraphic characterization. *J. Appl. Geophys.* **2025**, *239*, 105749. [CrossRef]

Huang, L.; Liu, J.; Zhang, F.; Dontsov, E.; Damjanac, B. Exploring the influence of rock inherent heterogeneity and grain size on hydraulic fracturing using discrete element modeling. *Int. J. Solids Struct.* **2019**, *176*, 207–220. [CrossRef]

Ilyushin, Y.V.; Novozhilov, I.M. Temperature field control of a metal oil-well tubing for producing of high-paraffin oil. In *Proceedings of the 2020 XXIII International Conference on Soft Computing and Measurements (SCM), Online, 27–29 May 2020*; IEEE: New York, NY, USA, 2020; pp. 149–152.

Li, M.Y.; Zhu, R.K.; Hu, S.Y. Geological characteristics and resource potential of overseas terrestrial shale oil. *Lithol. Reserv.* **2022**, *34*, 163–174. [CrossRef]

Hong, G.L.; Wang, H.J.; Zhu, H.Q.; Bai, Z.H.; Wang, W.W. Hydrocarbon accumulation conditions and favorable zones of the Miocene Gumai lithologic reservoirs in Block J of the South Sumatra Basin. *Lithol. Reserv.* **2023**, *35*, 138–146.

Hao, R.L.; Huang, W.H.; Hu, G.C.; Liu, J.Y.; Zhu, H.L. Superposition pattern of Miocene progradational deltaic sandstone bodies in Block J of the South Sumatra Basin. *Pet. Geol. Recovery Effic.* **2020**, *27*, 52–62.

Chen, W.; Shi, K. Multi-scale Attention Convolutional Neural Network for time series classification. *Neural Netw.* **2021**, *136*, 126–140. [CrossRef] [PubMed]

Yang, X.; Zhu, Y.; Guo, Y.; Zhou, D. An image super-resolution network based on multi-scale convolution fusion. *Vis. Comput.* **2022**, *38*, 4307–4317. [CrossRef]

Liu, X.; Ng, A.H.-M.; Lei, F.; Ren, J.; Liao, X.; Ge, L. Hyperspectral Image Classification Using a Multi-Scale CNN Architecture with Asymmetric Convolutions from Small to Large Kernels. *Remote Sens.* **2025**, *17*, 1461. [CrossRef]

Vaswani, A.; Shazeer, N.; Parmar, N.; Uszkoreit, J.; Jones, L.; Gomez, A.N.; Kaiser, Ł.; Polosukhin, I. Attention Is All You Need. In Proceedings of the Advances in Neural Information Processing Systems (NeurIPS), Long Beach, CA, USA, 4–9 December 2017; pp. 5998–6008. [CrossRef]

Liu, J.W.; Liu, J.W.; Luo, X.L. Research progress in attention mechanism in deep learning. *Chin. J. Eng.* **2021**, *43*, 1499–1511. [CrossRef]

Chaudhari, S.; Mithal, V.; Polatkan, G.; Ramanath, R. An Attentive Survey of Attention Models. *ACM Trans. Intell. Syst. Technol.* **2021**, *12*, 1–32. [CrossRef]

https://doi.org/10.3390/pr14060981

processes

MDPI

Article

Numerical Simulation and Optimization of Polyacrylamide Solution Flow in a Polymer Injector Using an Improved Viscosity Constitutive Model

Qin Qian [1], Tengyu Li [2], Congkun Ren [1,*], Yantao Zhou [2], Chuanrui Che [1], Xuemei Zhang [1], Jiaxing Ma [1], Pengxu An [2] and Qiuyang Zhao [2,*]

[1] Research Institute of Petroleum Engineering Technology, SINOPEC Shengli Oilfield Company, Dongying 257000, China
[2] State Key Laboratory of Multiphase Flow in Power Engineering, Xi'an Jiaotong University, Xi'an 710049, China
* Correspondence: rencongkun.slyt@sinopec.com (C.R.); qyzhao@xjtu.edu.cn (Q.Z.)

Abstract

Previous numerical simulations of polymer injectors often rely on fixed-viscosity models, which fail to accurately capture the severe shear degradation of non-Newtonian fluids under high-shear throttling conditions. To address this limitation and enhance polymer flooding efficiency, this study proposes an improved Carreau–Yasuda viscosity constitutive model to precisely simulate the flow behavior of polyacrylamide (HPAM) solutions. A comprehensive computational fluid dynamics (CFD) model was developed and validated, showing a viscosity prediction error of less than 8.6% across a wide shear rate range (0.1–10,000 s^{-1}). Based on this dynamic rheological model, the internal flow channel of the injector was optimized, resulting in a novel spindle-type throttling unit. Simulation and field validation results demonstrate that the optimized structure achieves a significant pressure drop of 6.03 MPa at an injection flow rate of 96 m^3/d—representing a 65% improvement over traditional designs—while successfully maintaining a viscosity retention rate above 85%. This research overcomes the traditional design conflict between high pressure reduction and viscosity preservation, providing an accurate numerical framework and practical guidance for engineering high-flow, robust-throttling polymer injectors.

Keywords: polymer injector; polyacrylamide solution; non-Newtonian fluid; Carreau–Yasuda model; numerical simulation; structural optimization

1. Introduction

Polymer flooding has emerged as one of the most widely applied and effective chemical Enhanced Oil Recovery (EOR) techniques in mature oilfields [1,2]. By increasing the viscosity of the injected fluid, polymer solutions significantly improve the mobility ratio and volumetric sweep efficiency [3]. However, in highly heterogeneous reservoirs, conventional continuous injection often leads to severe fingering effects [4,5]. To mitigate these interlayer conflicts and ensure precise fluid distribution across strata with varying permeabilities, layered polymer injection technology has been aggressively developed and deployed [6,7].

The polymer injector serves as the core equipment in this layered injection process. Its fundamental principle relies on utilizing the annular gap between the throttling core and the mandrel to achieve staged throttling and pressure reduction for high-viscosity polymer fluids. Despite its widespread application, a critical engineering bottleneck persists: to achieve the high pressure drops required for deep formations, traditional injector designs inevitably induce extreme shear rates within the narrow throttling zones [8,9]. This excessive mechanical shear leads to the irreversible scission of polyacrylamide (HPAM) molecular chains [10,11], causing a dramatic loss in apparent viscosity and directly impairing the ultimate oil displacement efficiency. Extensive research has been conducted to optimize the throttling tooth profile to balance pressure reduction and shear intensity. However, previous numerical simulations of polymer injectors predominantly relied on fixed-viscosity models or simplified non-Newtonian frameworks (such as the standard power-law model) [12–14].

These conventional models fail to accurately capture the highly nonlinear, dynamic evolution of polymer apparent viscosity under extreme shear rate variations, particularly when shear rates exceed 1000 s^{-1}. Consequently, the existing studies have been unable to effectively resolve the core conflict between "achieving a high pressure drop" and "maintaining high viscosity retention", as they lack a reliable theoretical basis to precisely regulate the local shear field.

To address the shortcomings of previous research, this study introduces a comprehensive structural optimization approach grounded in an advanced rheological framework. The primary innovation of this research lies in the establishment of an improved Carreau–Yasuda viscosity constitutive equation, derived from transient rheological experiments covering an exceptionally wide shear rate range (10^{-1} to 10^4 s^{-1}). By dynamically coupling this accurate constitutive model into computational fluid dynamics (CFD) simulations, we effectively capture the real-time response between shear rate and viscosity. Ultimately, this study proposes a novel spindle-shaped throttling unit, aiming to provide a robust, scientifically grounded solution that simultaneously maximizes pressure-reducing performance and preserves polymer viscosity, thereby offering significant practical value for EOR engineering applications.

2. Materials and Methods

2.1. Rheological Property Testing

In the experiment, the polymer used was partially hydrolyzed polyacrylamide (HPAM), appearing as white powdery crystals. To facilitate the reproducibility of this study, it is specified that the HPAM was supplied by SNF Floerger (Andrezieux, France), a widely recognized manufacturer for chemical EOR applications. The polymer has a weight-average molecular weight of 20×10^6 Da and a degree of hydrolysis of 25%.

A specific mass of deionized water was accurately weighed in a 300 mL beaker. Furthermore, it should be noted that deionized water was intentionally selected as the solvent for solution preparation in this fundamental phase of the study. While field applications typically involve formation water, the use of deionized water effectively isolates the baseline rheological behavior of the HPAM solution, avoiding the complex electrostatic shielding effects and viscosity fluctuations introduced by multivalent ions (e.g., Ca^{2+}, Mg^{2+}) found in actual or simulated formation water. This controlled approach ensures that the fitting of the improved Carreau–Yasuda model strictly reflects the purely shear-dependent viscosity evolution. The coupled effects of salinity and formation water chemistry on the fluid's mechanical degradation represent a recognized limitation of the current model and will be a primary focus of our future optimization research.

To systematically evaluate the rheological properties, we prepared 10 different mass concentrations of the HPAM aqueous solution, ranging from 0.05 wt% to 0.50 wt% with an interval of 0.05 wt%. Following widely accepted international practices for enhanced oil recovery (EOR) fluid preparation [15,16], a specific mass of deionized water was accurately weighed in a 300 mL beaker. The sample was slowly added along the vortex wall over 30 s while switching on the digital slurry constant-speed stirrer at (400 ± 20) r/min, and then stirred at (700 ± 20) r/min for 1 h [17]. The solution was allowed to stand for 24 h until the polyacrylamide was fully dissolved and the flow characteristics stabilized. The rheological tests were carried out in a Haake rotational rheometer (model Haake Mars60, Thermo Fisher Scientific, Waltham, MA, USA) with a double slit cylindrical rotor CC27 DG, with the shear rate varying linearly in the range of 0.1~10,000 s^{-1}. For each concentration, the rheometer recorded continuous data points across the entire shear rate sweep to ensure robust data fitting.

2.2. Geometric Model and Simulation Setup

Figure 1a illustrates the operational assembly of the injector, highlighting a narrow conduit known as the flow channel through which fluid is directed at the core of the device. Utilizing Solidworks software (Version 30.0, Dassault Systèmes, Vélizy-Villacoublay, France), a three-dimensional geometric model of the injector's flow channel has been crafted. The working cylinder, characterized by a diameter of 28.375 mm, is depicted in the model. A detailed view of the flow channel for an individual throttle unit is presented in Figure 1b, with the entire flow channel being composed of 16 such units. Upon completion of the geometric model, it is subsequently imported into the ANSYS Fluent 2022 R1 (Version 22.1, ANSYS, Inc., Canonsburg, PA, USA) workbench for mesh generation and computational fluid dynamics (CFD) analysis (Figure 2). As indicated in the flowchart (Figure 2), the "Adjust Mesh and Conditions" step refers to the essential iterative process of conducting grid independence tests and fine-tuning CFD solver settings (such as under-relaxation factors and near-wall treatments) to ensure the numerical pressure drop matches the physical loop experimental validation data within an acceptable error margin.

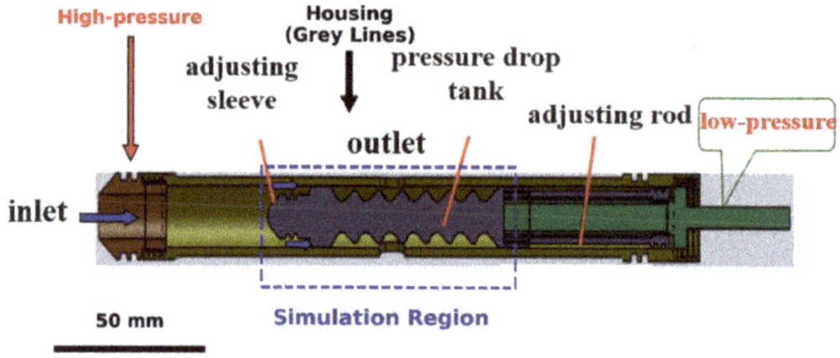

(a)

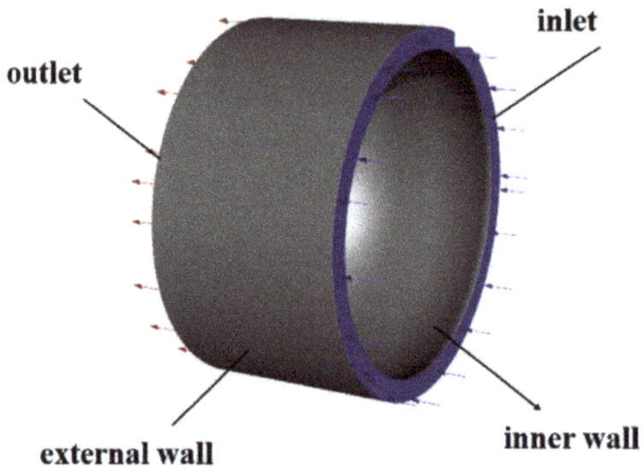

(b)

Figure 1. Geometry of the polymer injector: (**a**) 2D schematic with simulation region (blue dashed box) and high-pressure inlet; (**b**) 3D view of the assembly. The scale bar indicates 50 mm. The grey lines in (**a**) represent the housing structure.

https://doi.org/10.3390/pr1406088

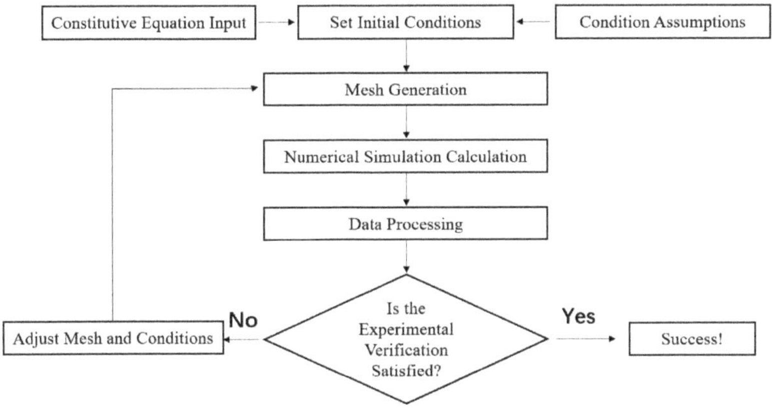

Figure 2. Flow Numerical Simulation Flowchart.

2.3. Numerical Simulation Method

To address the flow characteristics of non-Newtonian fluid within the polymer injector, the numerical simulation is conducted based on the following assumptions: (1) the fluid is incompressible with constant density; (2) the flow is steady-state, neglecting transient effects; and (3) heat transfer is neglected. The flow of the incompressible non-Newtonian fluid under isothermal conditions is governed by the continuity equation:

$$\nabla\cdot(\rho u) = 0 \quad (1)$$

and the Navier–Stokes momentum equations:

$$\rho(u\cdot\nabla) = -\nabla p + \nabla\cdot\tau \quad (2)$$

where ρ is the fluid density, u is the velocity vector, p is the static pressure, and τ is the extra stress tensor.

Given the high flow velocities of the polymer fluid (>10 m/s), the standard $k-\epsilon$ turbulence model was selected to effectively capture the turbulent flow field [18,19]. To accurately represent the physical flow loop, the computational domain was configured with a mass-flow inlet and a pressure outlet boundary condition. No-slip boundary conditions were applied to all solid walls. The numerical solution was achieved using the SIMPLEC algorithm for pressure–velocity coupling, with second-order upwind schemes for spatial discretization.

During the numerical simulation process, Fluent dynamically calculates the fluid viscosity based on the Carreau–Yasuda model defined in the User-Defined Function (UDF). This model was specifically chosen over the standard power-law model because the fluid in the injector transitions rapidly from a low-velocity Newtonian region to an extreme-shear throttling gap (>1000 s^{-1}). The Carreau–Yasuda model is uniquely capable of smoothly capturing both the zero-shear and infinite-shear viscosity plateaus at these extremes. The model characterizes the viscous behavior as follows:

$$\eta(\dot{\gamma}) = \eta_\infty + (\eta_0 - \eta_\infty)\left[1 + (\lambda\dot{\gamma})^a\right]^{\frac{n-1}{a}} \quad (3)$$

where η_0 is the zero-shear viscosity, η_∞ is the infinite-shear viscosity, λ is the relaxation time, n is the power-law index, and a is the Carreau constant [20–22]. It is critically important to note that polyacrylamide (HPAM) is a typical shear-thinning (pseudoplastic) fluid. There-

https://doi.org/10.3390/pr14060883

fore, its theoretical power-law index (n) must inherently be less than 1. Cases exhibiting a power-law index above 1 represent shear-thickening behavior, which contradicts the physical reality of the HPAM solution and were thus strictly excluded from this study. The viscosity retention rate was reconstructed using a weighted approach based on shear rate distribution.

To ensure the accuracy of the numerical results, a grid independence test was conducted. As shown in Figure 3, the total pressure drop stabilizes when the number of grid elements exceeds 1.5 million. Therefore, a mesh size of approximately 1.54 million elements was selected for all subsequent simulations.

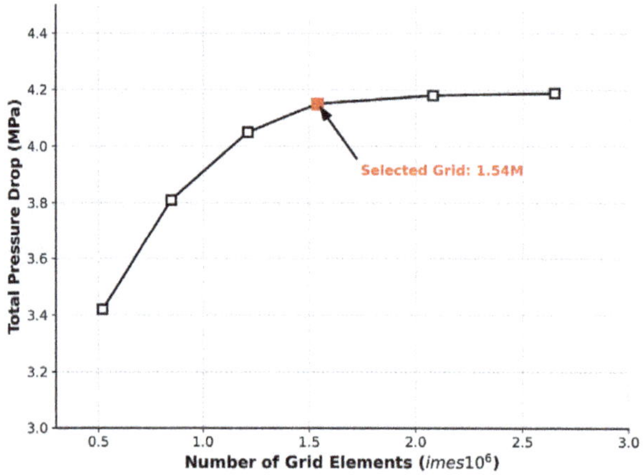

Figure 3. Grid independence test results.

3. Results and Discussion

3.1. Rheological Characteristics and Constitutive Equation

For polymeric fluids, viscosity is the most crucial parameter characterizing their viscous flow behavior. As shown in Figure 4, its rheological curve encompasses five distinct regions: the zero-shear plateau, the pseudoplastic region, the infinite-shear plateau, the viscoelastic region, and the degradation region. In the initial stage of shearing, the solution viscosity decreases rapidly ("shear thinning"). When the shear rate reaches approximately 7300 s^{-1}, the rheological properties undergo significant changes, identified as the critical shear rate [5,23,24].

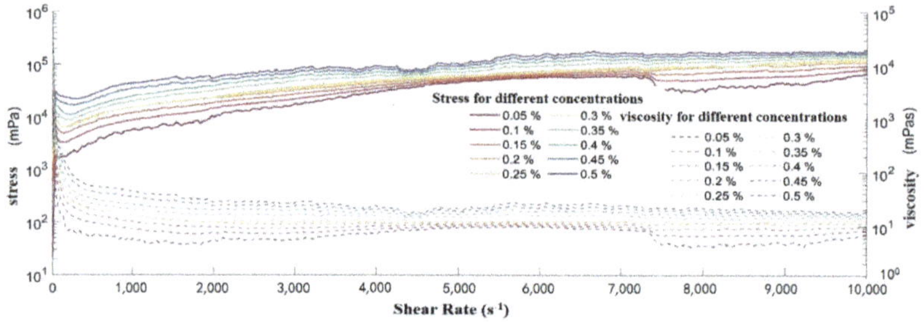

Figure 4. Variation curves of shear stress and viscosity with shear rate for aqueous solutions of different concentrations.

Based on the experimental data, the key parameters of the Carreau–Yasuda model (a, n, λ) were fitted using the least squares method (Table 1). The fitting accuracy (R^2) exceeded 0.977 and percentage errors were below 8.6% for all concentrations. As shown in Figure 5,

https://doi.org/10.3390/pr14060883

the regression curves exhibit excellent agreement with the experimental values (using 0.2 wt% concentration as a representative example), validating the model's applicability.

Table 1. Carreau–Yasuda model fitting results.

Mass Fraction (wt%)	Relaxation Time (λ)	a	n	Fitting Accuracy (R^2)	Percentage Error (%)
0.05	0.02156	0.006245	0.3264	0.9979	6.1
0.1	0.02204	0.008342	0.3499	0.9982	5.1
0.15	0.02578	0.01564	0.2515	0.9895	6.5
0.2	0.02794	0.01103	0.2560	0.9774	7.3
0.25	0.01964	0.01452	0.2123	0.9851	8.1
0.3	0.02097	0.01261	0.2883	0.9967	7.8
0.35	0.00921	0.1594	0.2434	0.9851	7.4
0.4	0.004072	0.1974	0.5045	0.9948	6.2
0.45	0.015611	0.2666	0.4256	0.9914	7.6
0.5	0.006026	0.1594	0.2739	0.9923	8.5

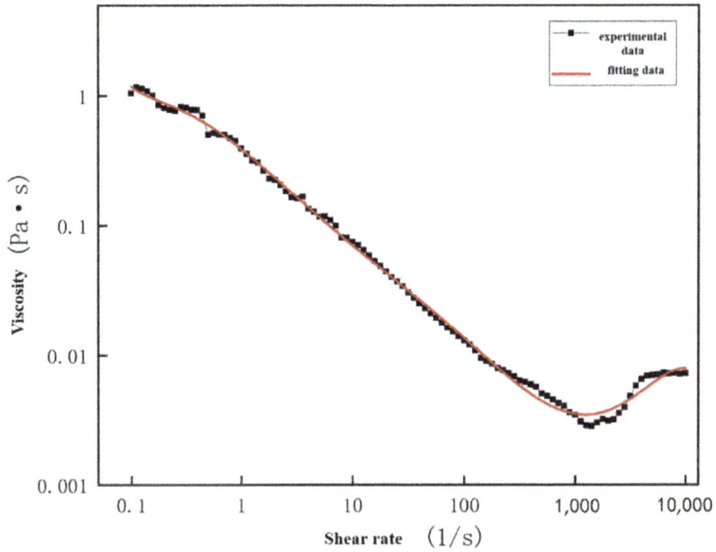

Figure 5. Comparison of experimental data and Carreau–Yasuda model fitting results for 0.2 wt% HPAM solution.

3.2. Model Verification

To verify the accuracy of the numerical model, a closed-loop flow experimental system was constructed. The HPAM solution was injected into the test section using a progressive cavity pump, specifically chosen to minimize initial shear degradation. High-precision pressure transmitters were installed immediately upstream and downstream of the test channel to record the differential pressure. In this study, Case 1 is defined as the physical model with a 129 mm clearance, and Case 2 corresponds to the model with a 94 mm clearance [6,7,10].

Verification was performed by comparing the simulated and measured pressure drops. Figure 6 illustrates the static pressure distribution contours for both cases. It is important to note that the negative pressure values displayed in the color scale (e.g., -3.58×10^3 Pa in Case 1) represent the gauge pressure relative to the reference operating pressure. As the fluid enters the narrow annular gap, the sudden cross-sectional contraction causes a drastic increase in flow velocity. According to Bernoulli's principle (the Venturi effect), this sharp increase in dynamic pressure leads to a corresponding drop in local static pressure.

https://doi.org/10.3390/pr14060883

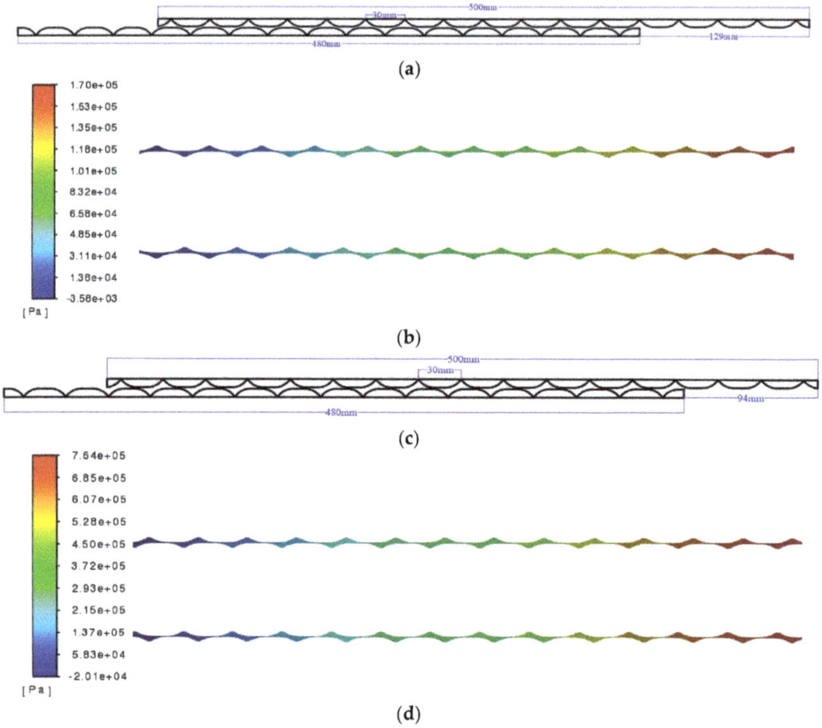

Figure 6. Experimental simulation comparison chart for working Case 1/2: (**a**) Case 1 fit dimension drawing; (**b**) Case 1 Pressure Distribution Contour; (**c**) Case 2 fit dimension drawing; (**d**) Case 2 Pressure Distribution Contour.

As shown in Table 2, for a 129 mm clearance, the simulated pressure drop deviated from the measured value by −0.03 MPa; for a 94 mm clearance, the deviation was +0.06 MPa. Under both conditions, deviations were within 15%, demonstrating that the established model correctly simulates the flow behavior and pressure drop variations.

Table 2. Specifications of the test section.

Groups	Volumes (m^3/d)	Pressure Drop (MPa)	Deviation (MPa)
Case 1 test	96	0.2	
Case 1 simulation	96	0.17	−0.03
Case 2 test	96	0.7	
Case 2 simulation	96	0.76	+0.06

3.3. Structural Optimization of Throttling Unit

This study systematically analyzed the influence of the upstream/downstream contour ratio and contour type on the throttling pressure drop. The dimensions for six optimization schemes are illustrated in Figure 7 (labeled Scheme I to Scheme VI). Note that although Scheme I and Scheme IV share the same D/L ratio of 20/30, the "streamlined" downstream profile in Scheme IV refers to a spline curve design intended to minimize abrupt flow separation compared to the linear profile.

https://doi.org/10.3390/pr1406088

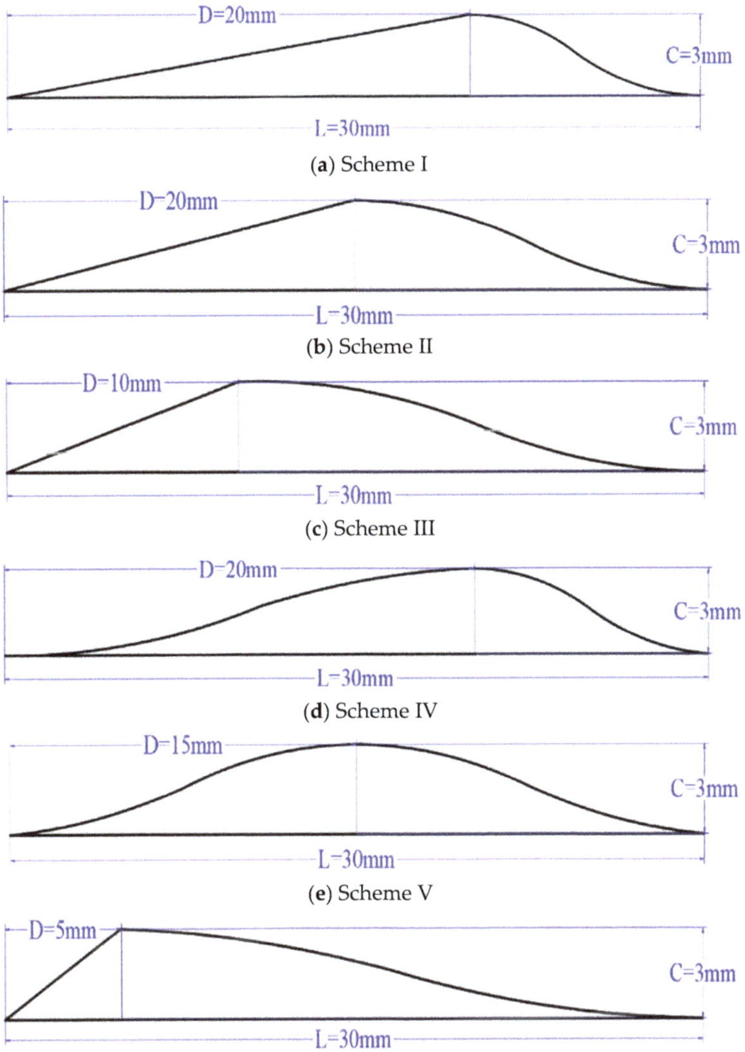

Figure 7. Schematic diagrams of six structural optimization schemes: (**a**) Scheme I (D/L = 20/30, Linear); (**b**) Scheme II (D/L = 15/30, Linear); (**c**) Scheme III (D/L = 10/30, Linear); (**d**) Scheme IV (D/L = 20/30, Streamline); (**e**) Scheme V (D/L = 15/30, Streamline); (**f**) Scheme VI (D/L = 5/30, Linear, Optimal Spindle-shaped Design).

The comprehensive working pressure drop data are visually compared in Figure 8.

To explain this, we analyzed the strain rate distribution. As shown in Figure 9, Scheme III generates a much larger high-shear zone at the throat entrance compared to Scheme I. The optimization of the downstream profile helps mitigate flow separation and stabilizes the boundary layer, a mechanism consistent with findings in other complex fluid systems [3,14,25,26].

https://doi.org/10.3390/pr14060883

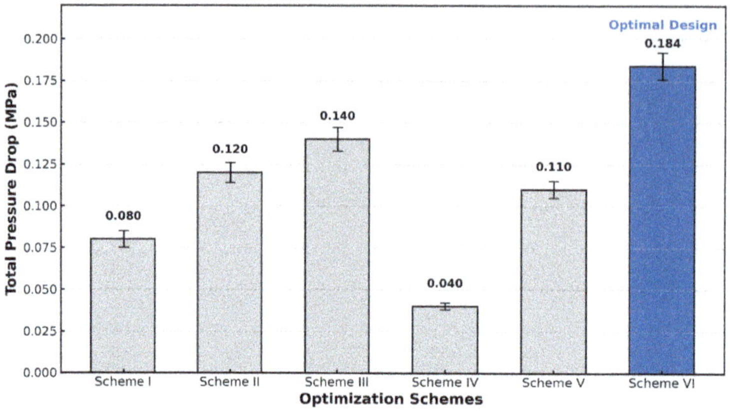

Figure 8. Comparison of total pressure drop across different optimization schemes. Scheme VI exhibits the highest pressure drop performance (0.184 MPa), significantly outperforming other designs.

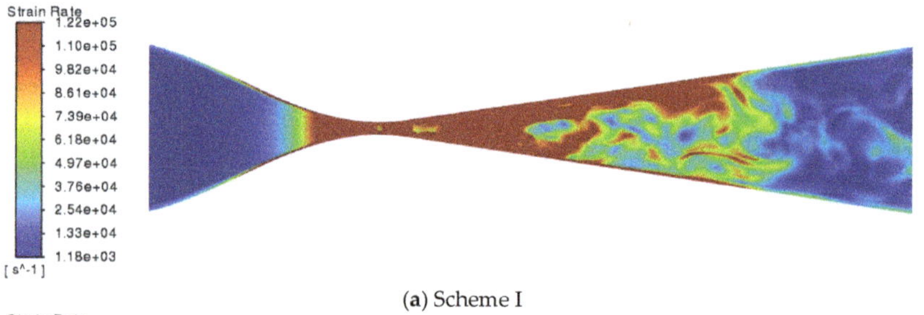

Figure 9. Comparison of strain rate distribution contours (unit: s^{-1}) between: (**a**) Scheme I (Original Linear Design), showing high-shear regions (red) at the throat entrance; and (**b**) Scheme III, demonstrating a smoother flow transition and significantly reduced shear intensity.

The contour clearly shows that Scheme III generates a much larger high-shear zone at the throat entrance compared to Scheme I, leading to greater energy dissipation. Second, while a streamlined downstream design (Scheme IV) reduces shear intensity, it significantly weakens the pressure drop (50% decrease compared to Scheme I). This confirms that the linear downstream design (used in Schemes I, II, III, VI) plays a crucial compensatory role by maintaining necessary shear perturbation.

Based on these findings, a novel "spindle-shaped" optimization scheme (Scheme VI) was proposed. The detailed flow field of the optimal scheme, including the explicitly controlled shear rate distribution, is presented in Figure 10. Furthermore, the successful control of shear forces and pressure drop in Scheme VI demonstrates the potential applicability of

https://doi.org/10.3390/pr1406088

such geometric optimizations to other functional polymer systems, such as hydrogels and nanoparticle-enhanced fluids [27–30].

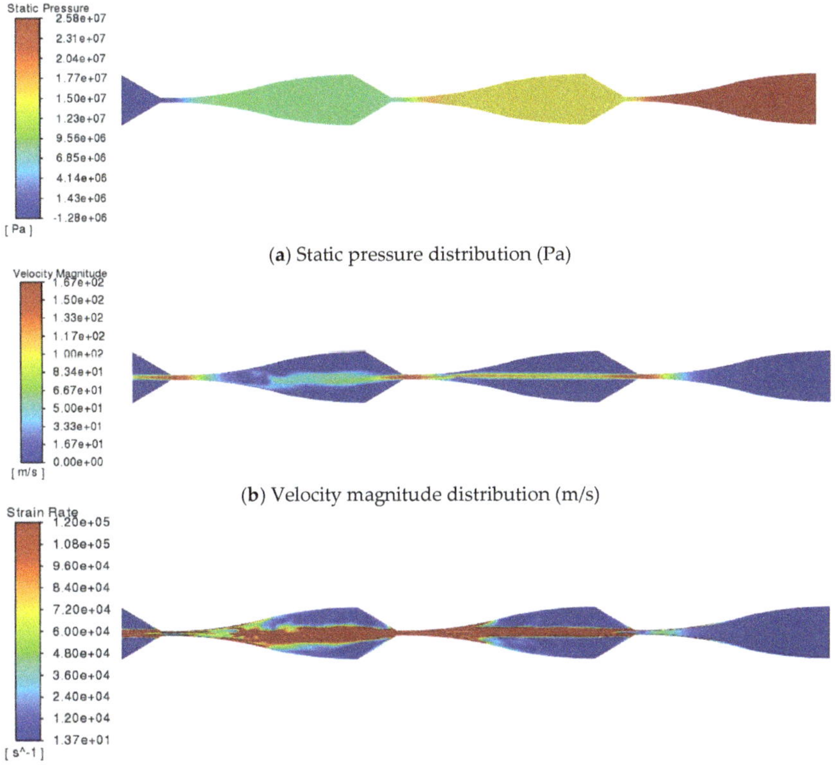

(**a**) Static pressure distribution (Pa)

(**b**) Velocity magnitude distribution (m/s)

(**c**) Shear rate distribution (s^{-1})

Figure 10. Local flow field distributions of the optimal Scheme VI (Spindle-shaped) showing 3 representative throttling units: (**a**) static pressure distribution (Pa); (**b**) velocity magnitude distribution (m/s); (**c**) shear rate distribution (s^{-1}).

The fundamental mechanism underlying the superior performance of the spindle-shaped design (Scheme VI) lies in its optimized spatial distribution of kinetic energy and strain rates(Table 3). Traditional linear designs typically force the fluid through an abrupt cross-sectional contraction, generating a highly localized, extreme-shear spike at the throat entrance that inevitably exceeds the critical shear rate for polymer chain scission. In contrast, the gradual upstream contraction in the spindle design pre-conditions the flow, distributing the velocity gradient more evenly over a longer axial distance. Subsequently, the controlled downstream expansion creates a stable flow separation zone that maximizes turbulent energy dissipation (contributing to the high pressure drop) without subjecting the bulk polymer chains to destructive mechanical forces.

When translating these findings to subsequent engineering optimizations, special attention must be paid to the coupled effects of bottom-hole temperature variations and multiphase flow conditions. Elevated temperatures in deep reservoirs may alter the polymer's relaxation time (λ) and shift the critical shear threshold, potentially requiring dynamic adjustments to the D/L ratio of the throttling unit. To visually summarize this flow mechanism, a conceptual schematic detailing the velocity gradient and polymer chain integrity within the optimized throttling unit is presented in Figure 11.

https://doi.org/10.3390/pr14060883

Table 3. Comparison data table of working pressure drop for six optimization schemes.

Schemes	Downstream Ratio D/L	Downstream Profile	Pressure Drop (MPa)
I	20:30	linear	0.08
II	15:30	linear	0.12
III	10:30	linear	0.14
IV	20:30	streamlined	0.04
V	15:30	streamlined	0.11
VI	5:30	linear	0.184

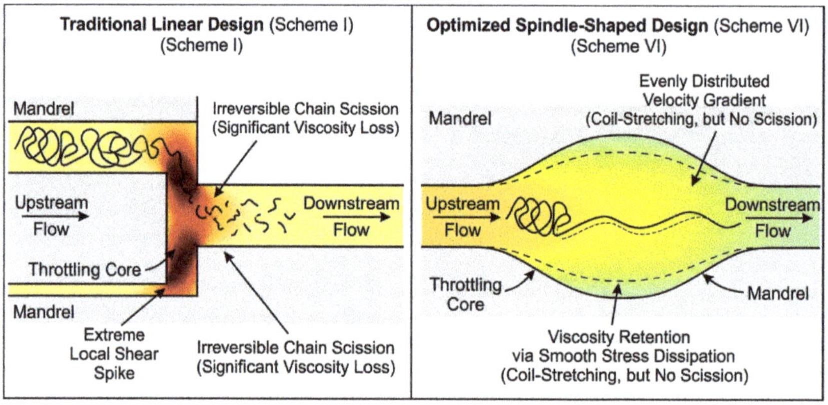

Figure 11. Conceptual schematic illustrating the mechanism of viscosity retention: comparison of highly localized shear and irreversible polymer chain scission in traditional abrupt contractions (left, Scheme I) versus evenly distributed velocity gradients and chain structural preservation in the optimized spindle-shaped design (right, Scheme VI).

4. Conclusions

(1) An improved Carreau–Yasuda viscosity constitutive model was established and validated. By incorporating the zero-shear and infinite-shear viscosity plateaus, this model accurately predicts the rheological behavior of HPAM solutions across a wide shear rate range (0.1–10,000 s^{-1}) with a prediction error of less than 8.6%, providing a reliable theoretical basis for injector simulation.

(2) Numerical optimization revealed the physical mechanism of the throttling unit: the upstream contraction ratio dominates the pressure drop generation by controlling the maximum velocity gradient, while the linear downstream profile is essential for maintaining shear perturbation to enhance energy dissipation.

(3) The novel spindle-shaped throttling unit (Scheme VI) resolves the conflict between high pressure drop and viscosity retention. Field applications confirm that the optimized injector achieves a 6.03 MPa pressure drop at 96 m^3/d (a 65% improvement over existing designs) while maintaining a viscosity retention rate above 85%. This design offers a robust solution for deep-well layered polymer injection in heterogeneous reservoirs.

Author Contributions: Methodology, J.M.; software, Y.Z. and X.Z.; validation, P.A.; formal analysis, Q.Q., C.R., X.Z. and Q.Z.; investigation, C.R.; resources, C.C.; data curation, Q.Q. and T.L.; writing—original draft, T.L.; writing—review and editing, Q.Z.; visualization, P.A.; supervision, J.M.; project administration, Y.Z. and C.C.; funding acquisition, T.L. All authors have read and agreed to the published version of the manuscript.

https://doi.org/10.3390/pr1406088

Funding: This research received no external funding.

Data Availability Statement: The original contributions presented in this study are included in the article. Further inquiries can be directed to the corresponding authors.

Conflicts of Interest: Qin Qian, Congkun Ren, Chuanrui Che, Xuemei Zhang and Jiaxing Ma were employed by SINOPEC Shengli Oilfield Company. The remaining authors declare that the research was conducted in the absence of any commercial or financial relationships that could be construed as a potential conflict of interest.

.eferences

Seright, R.S.; Wang, D. Polymer flooding: Current status and future directions. *Pet. Sci.* **2023**, *20*, 1248–1259. [CrossRef]

Song, K.; Tao, J.; Li, X.; Xu, Y.; Liu, S. Recent Advances in Polymer Flooding in China. *Molecules* **2022**, *27*, 6978. [CrossRef] [PubMed]

Salem, K.G.; Tantawy, M.A.; Gawish, A.A.; Gomaa, S.; El-hoshoudy, A.N. Nanoparticles assisted polymer flooding: Comprehensive assessment and empirical correlation. *Geoenergy Sci. Eng.* **2023**, *226*, 211753. [CrossRef]

Sorbie, K.S. *Polymer-Improved Oil Recovery*; CRC Press: Boca Raton, FL, USA, 1991.

Standnes, D.C.; Skjevrak, I. Literature review of implemented polymer field projects. *J. Pet. Sci. Eng.* **2014**, *122*, 761–775. [CrossRef]

Zaitoun, A.; Makakou, P.; Blin, N.; Al-Maamari, R.S.; Al-Hashmi, A.R.; Abdel-Goad, M.; Al-Sharji, H.H. Shear stability of EOR polymers. *SPE J.* **2012**, *17*, 335–339. [CrossRef]

Jung, M.; Shakeri, P.; Seemann, R. Effect of viscoelasticity on displacement processes in porous media. *Front. Phys.* **2023**, *11*, 1099073. [CrossRef]

Husveg, T.; Stokka, M.; Jouenne, S.; Husveg, R. The Development of a Low-Shear Valve Suitable for Polymer Flooding. *SPE J.* **2020**, *25*, 2632–2647. [CrossRef]

Al Hashmi, A.R.; Al Maamari, R.S.; Al Shabibi, I.S.; Mansoor, A.M.; Zaitoun, A.; Al Sharji, H.H. Rheology and mechanical degradation of high-molecular-weight partially hydrolyzed polyacrylamide during flow through capillaries. *J. Pet. Sci. Eng.* **2013**, *105*, 100–106. [CrossRef]

10. Jouenne, S.; Chakibi, H.; Levitt, D. Polymer stability after successive mechanical-degradation events. *SPE J.* **2018**, *23*, 18–33. [CrossRef]
11. Morel, D.; Jouenne, S.; Vert, M.; Nahas, E. Polymer Injection in Deep Offshore Field: The Dalia Angola Case. In *SPE Annual Technical Conference and Exhibition*; Society of Petroleum Engineers: Houston, TX, USA, 2015. [CrossRef]
12. Bird, R.B.; Armstrong, R.C.; Hassager, O. *Dynamics of Polymeric Liquids, Volume 1: Fluid Mechanics*, 2nd ed.; Wiley-Interscience: New York, NY, USA, 1987.
13. Sochi, T. Non-Newtonian Flow in Porous Media. *Polymer* **2010**, *51*, 5007–5023. [CrossRef]
14. Salam, A.H.; Alsaif, B.; Hussain, S.M.S.; Khan, S.; Kamal, M.S.; Patil, S.; Al-Shalabi, E.W.; Hassan, A.M. Advances in Understanding Polymer Retention in Reservoir Rocks: A Comprehensive Review. *Polym. Rev.* **2024**, *211*, 1387–1413. [CrossRef]
15. Zhong, H.; Shi, B.; Bi, Y.; Shen, W.; Xu, Y.; Yin, H.; Xia, H.; Zhao, X. Flow mechanism of viscoelastic polymer flooding: State of the art review and outlook. *Chin. J. Theor. Appl. Mech.* **2024**, *56*, 847–861. [CrossRef]
16. Zhao, Y.; Dordzie, G.; Huh, C.; Balhoff, M.; Lu, Y. Rheology, Phase Stability, and Microstructure of Nanosilica-Assisted Partially Hydrolyzed Polyacrylamide. *Energy Fuels* **2024**, *38*, 2145–2158. [CrossRef]
17. Seright, R.S.; Fan, T.; Wavrik, K.; Balaban, R.C. New insights into polymer rheology in porous media. *SPE J.* **2011**, *16*, 35–42. [CrossRef]
18. Versteeg, H.K.; Malalasekera, W. *An Introduction to Computational Fluid Dynamics: The Finite Volume Method*, 2nd ed.; Pearson Education: Harlow, UK, 2007.
19. Launder, B.E.; Spalding, D.B. The numerical computation of turbulent flows. *Comput. Methods Appl. Mech. Eng.* **1974**, *3*, 269–289. [CrossRef]
20. Yasuda, K.; Armstrong, R.C.; Cohen, R.E. Shear flow properties of concentrated solutions of linear and star branched polystyrenes. *Rheol. Acta* **1981**, *20*, 163–178. [CrossRef]
21. Morrison, F.A. *Understanding Rheology*; Oxford University Press: New York, NY, USA, 2001.
22. Macosko, C.W. *Rheology: Principles, Measurements, and Applications*; Wiley-VCH: New York, NY, USA, 1994.
23. Zhang, H.; Lou, T.; Wang, X. Synthesis and properties of xanthan gum-acrylamide-carboxymethyl cellulose ternary anionic flocculants. *J. Water Process Eng.* **2023**, *53*, 104455. [CrossRef]
24. He, W.; Luo, J.; Wu, Y.; Luo, T.; Tang, C. Differently charged polyacrylamides (PAMs) significantly affect adsorption affinity and associated floc growth behaviors. *Sep. Purif. Technol.* **2023**, *329*, 125279. [CrossRef]
25. Grabsch, A.F.; Fawell, P.D.; Davies, M.G. Flocculating fine cuttings particles suspended within partially hydrolysed polyacrylamide (PHPA) solutions. *Miner. Eng.* **2024**, *206*, 108510. [CrossRef]

https://doi.org/10.3390/pr14060883

26. Al-Shakry, B.; Skauge, T.; Shaker Shiran, B.; Skauge, A. Impact of Mechanical Degradation on Polymer Injectivity in Porous Media. *Polymers* **2018**, *10*, 742. [CrossRef]
27. Husveg, R.; Stokka, M.; Husveg, T.; Albustin, R.; Jouenne, S. Field Testing a Low Shear Valve Suitable for Polymer Flooding in a Mother Solution Injection Scheme. In *Proceedings of the IOR 2021—21st European Symposium on Improved Oil Recovery, Vienna, Austria, 19–22 April 2021*; European Association of Geoscientists & Engineers (EAGE): Houten, The Netherlands, 2021. [CrossRef]
28. Åsen, S.M.; Stavland, A.; Strand, D.; Hiorth, A. An Experimental Investigation of Polymer Mechanical Degradation at cm and m Scale. In Proceedings of the SPE Improved Oil Recovery Conference, Tulsa, OK, USA, 14–18 April 2018. [CrossRef]
29. Huang, B.-Q.; Cui, H.-G.; Feng, T.-Y.; Yuan, C.-L.; Chang, A.-L.; Shao, M.-L.; Zhu, C.-F. Thin film composite polyamide nanofiltration membranes with interlayer constructed with core-shell structured polystyrene-polyacrylamide nanospheres. *J. Water Process Eng.* **2022**, *48*, 104844. [CrossRef]
30. Su, Z.; Xue, B.; Xu, C.; Dong, X. Mussel-Inspired Calcium Alginate/Polyacrylamide Dual Network Hydrogel. *Polymers* **2023**, *15*, 4498. [CrossRef]

https://doi.org/10.3390/pr14060883

processes

MDPI

rticle

aleoenvironmental Evolution and Its Dominant Controls on)rganic Matter Enrichment: Insights from the Lower Cambrian)iongzhusi Formation Shale

hengkai Lan [1,2], Jianhong He [3,*], Fangfang Chen [4], Tao Yang [5] and Lei Wang [6]

1 College of Petroleum Engineering, Xi'an Shiyou University, Xi'an 710065, China; lanzhengkai@cug.edu.cn
2 Research Institute of Marginal Oil and Gas Development Technology, Xi'an Shiyou University, Xi'an 710065, China
3 No. 3 Oil Production Plant, PetroChina Qinghai Oilfield Company, Mangya 817500, China
4 Research Institute of Exploration and Development, Tarim Oilfield Company, PetroChina, Korla 841000, China; chenfangfang@petrochina.com.cn
5 Oilfield Technology Development Company, PetroChina Liaohe Oilfield Company, Panjin 124010, China; yangtao@petrochina.com.cn
6 School of Mathematics and Physics, China University of Geosciences, Wuhan 430074, China
* Correspondence: hjhbyqh@petrochina.com.cn

Abstract

Based on data from core observations, thin-section petrography, scanning electron microscopy, whole-rock X-ray diffraction, organic geochemical analysis, and element analysis, in this study, we characterized the mineralogical–petrological features and sedimentary environment of the Lower Cambrian Qiongzhusi Formation shale in Western Hubei Province, and we clarified their relationships with organic matter enrichment. The results are as follows: (1) Five dominant rock types were identified in the Qiongzhusi Formation, namely, siliceous shale, argillaceous–siliceous mixed shale, argillaceous–calcareous shale, calcareous–siliceous shale, and calcareous shale. Vertically, the lithofacies transition follows the sequence of siliceous shale facies → mixed shale facies → calcareous shale facies. Laterally, from the marine trough to the trough margin, the thicknesses of the siliceous shale, argillaceous–siliceous mixed shale, and calcareous–siliceous mixed shale gradually decrease, whereas the thickness of the argillaceous–calcareous mixed shale increases progressively. (2) From the early to late sedimentary periods of the Qiongzhusi Formation and from the marine trough to the trough margin, a consistent evolutionary trend can be observed: gradual shallowing of the water depth, intensified hydrodynamic conditions, increased dissolved oxygen content of the bottom water, weakened upwelling currents, reduced paleoproductivity in the surface water, enhanced water mass stagnation, increased terrigenous input, and a corresponding gradual decrease in the total organic carbon (TOC) content. (3) The formation of the late-stage organic-rich shale was comprehensively controlled by the terrigenous input, redox conditions, paleoproductivity, water mass stagnation, and upwelling activity. Among these factors, the redox conditions, water mass stagnation, and paleoproductivity were the primary drivers responsible for the difference in the TOC contents in the Western Hubei marine trough and its margin, while the terrigenous input played a secondary role.

Keywords: Qiongzhusi Formation; shale; mineralogical–petrological characteristics; sedimentary environment; organic matter enrichment; controlling factors

1. Introduction

The Lower Cambrian Qiongzhusi Formation represents a principal stratigraphic unit for marine-shale gas exploration and development in southern China [1]. It is characterized by a wide geographic distribution, substantial sedimentary thickness, high organic-matter abundance, and advanced thermal maturation of its source rocks. At present, shale gas drilling in this formation is mainly concentrated in the Upper Yangtze Sichuan Basin, and wells such as We201, Jinye1, and Jinshi1 in the Weiyuan-Qianwei area have produced industrial gas flows [2], representing a major breakthrough in exploration [3]. However, exploration of the Qiongzhusi Formation in the Central Yangtze Western Hubei region remains in the early stage. Nevertheless, recent flows reported for wells such as Eyangye1 and Eyiye1 [4] indicate promising prospects for shale gas exploration in western Hubei. Prior investigations of the Qiongzhusi Formation in Western Hubei have emphasized the geochemical characteristics, shale gas pore structure and gas storage capacity, and the geological features of shale gas accumulation; however, studies addressing the linkage between paleoenvironmental evolution and organic-matter enrichment remain limited. The conditions of the sedimentary environment fundamentally regulate the formation and preservation of organic matter in marine shales; therefore, reconstructing the depositional environments provides a scientific basis for predicting the distribution of high-quality shale intervals and identifying exploration sweet spots.

Previous studies indicate that the variability in the mineralogy, sedimentary structures, organic-matter content, fossil assemblages, and lithological abundances reflects the combined influences of the basin architecture, hydrodynamics, terrigenous input, climatic variability, and eustatic sea-level changes during deposition [5,6]. Classifying shale lithotypes and reconstructing their depositional environments are therefore fundamental to understanding the sedimentation of black shale and identifying shale gas sweet spots.

Therefore, in this study, we focused on the Qiongzhusi Formation shale in Western Hubei by selecting representative outcrop profiles and wells in the region. We integrated core samples, thin sections, whole-rock X-ray diffraction, organic geochemical analysis, and element geochemical analysis. We analyzed the development characteristics of the shale facies, reconstructed the sedimentary paleoenvironment, investigated its relationship with organic matter enrichment, and established a sedimentary evolution model. The results provide geological constraints for predicting and mapping the distribution of high-quality shale reservoirs in the Qiongzhusi Formation in Western Hubei.

2. Regional Geological Overview

The Sinian–Cambrian transition constitutes a pivotal geological interval in Earth's evolution, coinciding with a suite of major global and regional geological events, including the breakup of the Rodinia Supercontinent, the Cambrian Explosion, Snowball Earth episodes, and global anoxic events [7,8]. At the end of the Late Sinian Dengying Period, the Middle Yangtze Platform underwent tectonic uplift associated with the Huiting Movement. The central Hubei region was elevated, forming the Central Hubei Old Land, which is predominantly composed of carbonate rocks and distinctly differs from the clastic-dominated old land in the Upper Yangtze Platform [9–11].

During the Early Cambrian, inheriting the Late Sinian paleogeographic framework and controlled by regional extensional rifting, rapid global sea-level rise occurred. Against this tectono-sedimentary backdrop, a widespread sequence of organic-rich black shales was deposited across the Middle Yangtze Platform [12,13]. The stratigraphic nomenclature for this Early Cambrian black shale succession remains inconsistent across different subregions of the Middle Yangtze Platform. For the sake of research consistency, this study collectively refers to it as the Qiongzhusi Formation. The Qiongzhusi Formation in the study area was

https://doi.org/10.3390/pr1406088[illegible]

deposited primarily in a shelf depositional environment. Proximal to the Central Hubei Old Land, the sedimentary facies are dominated by shoreline–shallow-water shelf deposits. Toward the southwest, the water depth gradually increases, and the sedimentary facies transition sequentially to semi-deep-water shelf–deep-water shelf facies (Figure 1).

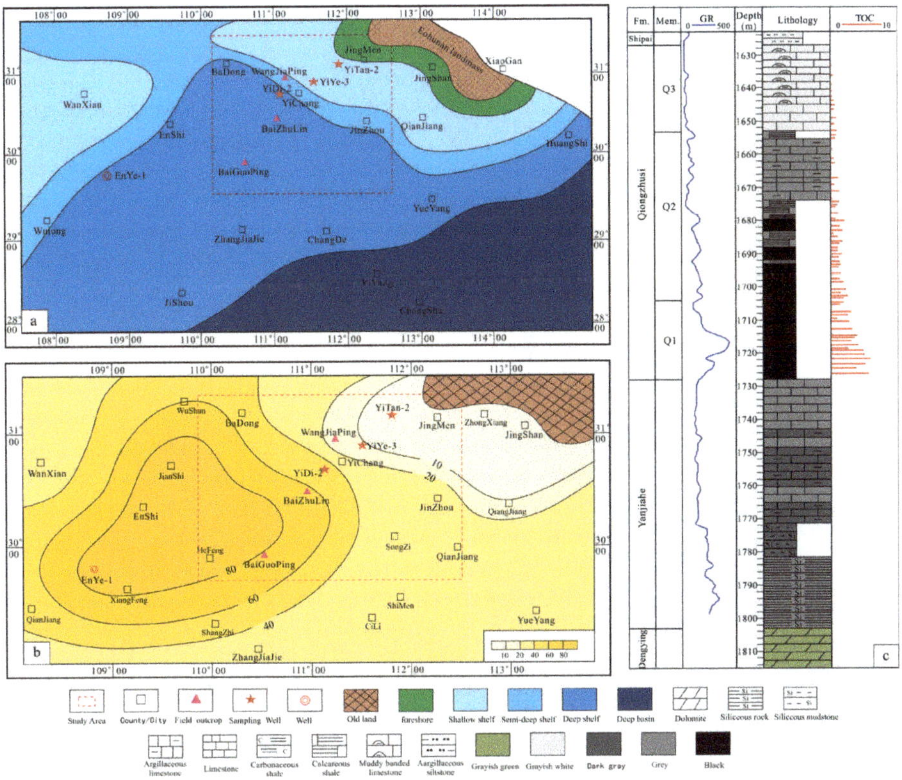

Figure 1. (**a**) Lithofacies paleogeography of the Lower Cambrian Qiongzhusi Formation in the Central Yangtze region. (**b**) Planar distribution of the thickness of the organic-rich shale in the study area. (**c**) Stratigraphic column of the Qiongzhusi Formation in Well Yidi 2.

The stratigraphic succession in the study area, from bottom to top, comprises the Upper Sinian Dengying Formation, Lower Cambrian Yanjiahe Formation, Qiongzhusi Formation, and Shipai Formation. The Dengying Formation is dominated by gray to light gray dolomite and is in unconformable contact with the overlying Yanjiahe Formation. The Yanjiahe Formation has a restricted distribution, primarily occurring on the southern and western flanks of the Huangling Anticline and in the core of the Changyang Anticline [14, 15], and its extent was strongly controlled by rifting. Vertically, the Yanjiahe Formation can be divided into two members: the lower member mainly consists of grayish-black siliceous mudstone and dark gray siliceous rock, while the upper member is dominated by grayish-black argillaceous limestone and gray limestone. The Qiongzhusi Formation overlies the Dengying Formation paraconformably or the Yanjiahe Formation conformably, and it is in conformable contact with the overlying silt-bearing mudstone of the Shipai Formation. The sedimentary thickness of the Qiongzhusi Formation varies significantly laterally across the study area. In the Enshi-Xianfeng-Sangzhi area, which is dominated by the Western Hubei Trough, the thickness of high-quality organic-rich shale ranges from 60 to 100 m. In

https://doi.org/10.3390/pr14060882

contrast, in the Jingmen-Zhongxiang-Jingshan area, which is strongly influenced by the Central Hubei Old Land, the thickness of the organic-rich shale is relatively thin, generally less than 20 m. Overall, the thickness of the organic-rich shale in the Qiongzhusi Formation exhibits a gradual thinning trend from southwest to northeast (Figure 1).

Based on lithological observations and logging response characteristics, the Qiongzhusi Formation in the study area can be vertically subdivided into three lithological members from bottom to top (Figure 1):

- Member 1 (Qiong-1) is dominated by black siliceous shale and is characterized by high natural gamma (GR) values (150–600 API, average of 349 API), making it the interval with the highest GR values among the three members of the Qiongzhusi Formation.
- The base of Member 2 (Qiong-2) is dominated by black carbonaceous shale, which transitions upward to calcareous shale interbedded with argillaceous rocks of unequal thickness. The GR values are significantly lower than those of Member 1, showing a medium-low amplitude funnel-shaped logging curve (96.2–280.5 API, average of 146.7 API).
- Member 3 (Qiong-3) is mainly composed of limestone or calcareous mudstone, and its GR values rapidly decrease and exhibit a weakly serrated box-shaped logging curve (17.2–100 API, average of 79.8 API).

3. Materials and Methods

In this study, a total of 155 samples were collected from the black shale section of the Qiongzhusi Formation, including 52 from well Baizhuling, 67 from well YiDi2, 23 from YiYe3, and 13 from YiTan2. The samples were analyzed to determine their organic carbon contents, mineral compositions, and lithological characteristics and micropaleontological identification was conducted. Based on the analysis results, 13 and 17, of the samples from well YiYe3 and Baizhuling, respectively, were selected for elemental analysis.

All of the analyses conducted in this study were performed in the Geological Laboratory of the PetroChina Southwest Oil and Gas Field Company. The organic carbon content was measured using a CS-344 carbon and sulfur analyzer (LECO Corporation, St. Joseph, MI, USA), with an analytical accuracy of ±0.5%. The thin sections of the samples were stained with a mixture of alizarin red and potassium ferricyanide (Aladdin, Shanghai, China). The X-ray diffraction analysis was performed using a Phillips Xpert-MPD X-ray diffractometer (Malvern Panalytical, Almelo, The Netherlands). The major elements were analyzed using a Rigaku 100e wavelength dispersive X-ray fluorescence spectrometer (XRF) (Rigaku, Nagano, Japan) after sample treatment using the alkali fusion method. The trace and rare earth elements were analyzed using a Thermo Scientific Element XR inductively coupled plasma mass spectrometer (Thermo Fisher Scientific, Wilmington, DE, USA), with an accuracy better than 5%.

4. Lithofacies Development and Distribution Characteristics

4.1. Lithofacies Classification

Based on whole-rock X-ray diffraction data, the shale lithofacies in the study area were classified with reference to the ternary diagram classification scheme for siliceous minerals (quartz + feldspar)–carbonate minerals–clay minerals [16]. The classification system is defined as follows:

- Siliceous shale facies (siliceous mineral content > 50%, both carbonate and clay mineral contents < 50%);
- Argillaceous shale facies (clay mineral content > 50%, both siliceous and carbonate mineral contents < 50%);

https://doi.org/10.3390/pr1406088

- Calcareous shale facies (carbonate mineral content > 50%, both siliceous and clay mineral contents < 50%);
- Mixed shale facies (all three mineral groups range from 25% to 50%), further subdivided into argillaceous–siliceous mixed shale facies, calcareous–siliceous mixed shale facies, and argillaceous–calcareous mixed shale facies.

Using 25%, 50%, and 75% as the critical mineral content thresholds, the aforementioned lithofacies were further refined into nine specific shale lithofacies types. The details of the classification criteria are provided in Figure 2 and Table 1.

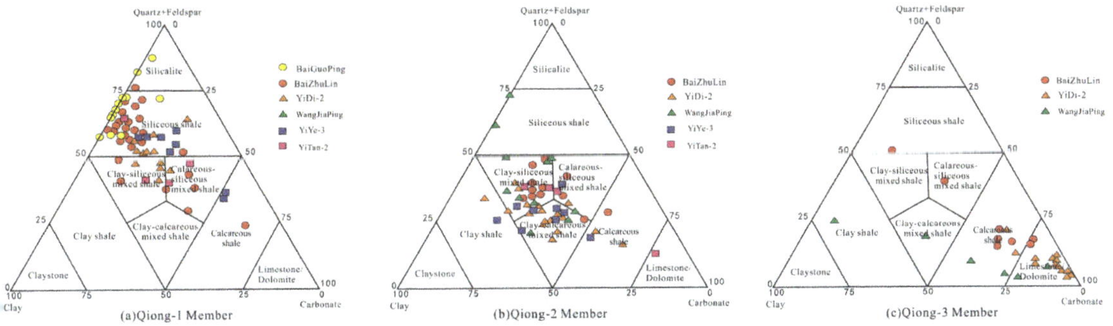

Figure 2. Ternary diagram for shale lithofacies classification of the Qiongzhusi Formation in the study area.

Table 1. Classification scheme of the marine shale lithofacies types.

Lithofacies Type	Mass Fraction of Rock Mineral Components (%)		
	Quartz + Feldspar	Calcite + Dolomite	Clay
Siliceous shale facies	50–75	<30	10–50
Clay shale facies	25–50	<30	50–75
Calcareous shale facies	<30	50–75	25–50
Clay like siliceous mixed shale facies	30–50	<33	30–50
Clay calcareous mixed shale facies	<33	30–50	30–50
Calcareous siliceous mixed shale facies	30–50	30–50	<33

4.2. Development Characteristics of the First Section of the Qiongzhusi Formation

The first member of the Qiongzhusi Formation (hereafter referred to as the Qiong-1 Member) is dominated by deep-water shelf subfacies, and the siliceous shale facies are the dominant lithofacies. Core observations and field outcrop surveys reveal that this siliceous shale facies occurs as thin- to medium-bedded intervals and is predominantly grayish-black to black. The rocks are rigid and brittle and are characterized by well-developed horizontal bedding, pyrite bands, and siliceous concretions (Figure 3a–c). Locally, black calcareous mudstone is present in the Well Yiye-3 block at the trough margin.

Microscopic analyses indicate that the Qiong-1 Member siliceous shale in the study area is characterized by high organic matter abundance and poor light transmittance. The matrix particles are dominated by flocculent microspheres and biogenic framework grains, which are mostly rounded pellets with a particle size range of 9–75 μm, dispersed within the clay matrix (Figure 3d). It is enriched in siliceous fossils, primarily consisting of sponge spicules and radiolarians, which have relatively simple morphologies (predominantly spherical or ellipsoidal; Figure 3e,f). Comprehensive investigations revealed that the quartz in the Qiong-1 Member siliceous shale is predominantly of biogenic siliceous origin, indicative of deposition in a deep-water, low-energy, and stable reducing environment.

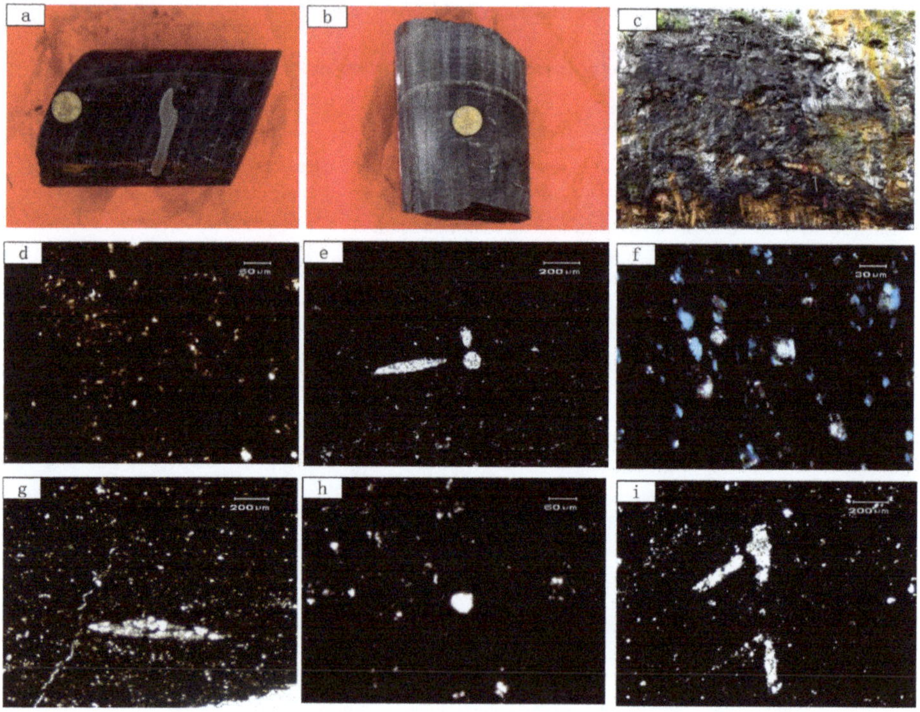

Figure 3. Characteristics of lithofacies in the first section of the Qiong Formation. (**a**) Gray black siliceous shale with horizontal bedding and pyrite clusters, suitable for exploration in two wells, 5056 m. (**b**) Gray black siliceous shale with pyrite bands, suitable for exploration in two wells, 5059.1 m. (**c**) Black carbonaceous shale, with a large number of siliceous nodules visible at the top of three layers, Baizhuling. (**d**) Siliceous shale, developed siliceous organisms, Baizhuling. (**e**) Siliceous shale with visible sponge bone needles, high TOC content, Baizhuling. (**f**) Siliceous shale with visible radiolarians, Yiye 3 well, 3048.8 m. (**g**) Siliceous shale, siliceous spicules replaced by calcite, well Yiye 3, 3052.8 m. (**h**) Siliceous shale with high content of siliceous organisms, suitable for exploration in two wells, 3059.95 m. (**i**) Siliceous shale, visible bone needles, Baiguoping.

The shales at the center of the trough (Baizhuling and Baiguoping areas) are enriched in siliceous minerals (quartz + feldspar), with an average content of 65%. Specifically, the average quartz content is 39.2%, the average feldspar content is 8%, clay minerals account for 23.3% on average, the average carbonate mineral content is 18%, and pyrite is the least abundant (average of 3%). The total organic carbon (TOC) content ranges from 1.5% to 15.1%, with a mean value of 4%. In contrast, in the trough margin area (well blocks Yidi-2, Yitan-2, and Yiye-3), the average siliceous mineral content decreases to 40%, the carbonate mineral content increases to 30% on average, and the mean TOC content is 3%—exhibiting a vertical trend of gradual reduction from bottom to top (Figure 4). These features illustrate that the siliceous mineral (quartz + feldspar) and TOC contents of the siliceous shale facies are significantly higher at the center of the trough than in the trough margin (Central Hubei Old Land) area, whereas the carbonate mineral content is notably lower. These lateral differences in the mineral composition and TOC content are strongly governed by variations in the sedimentary environments and are consistent with the material distribution patterns of deep-water sedimentary systems.

https://doi.org/10.3390/pr1406088

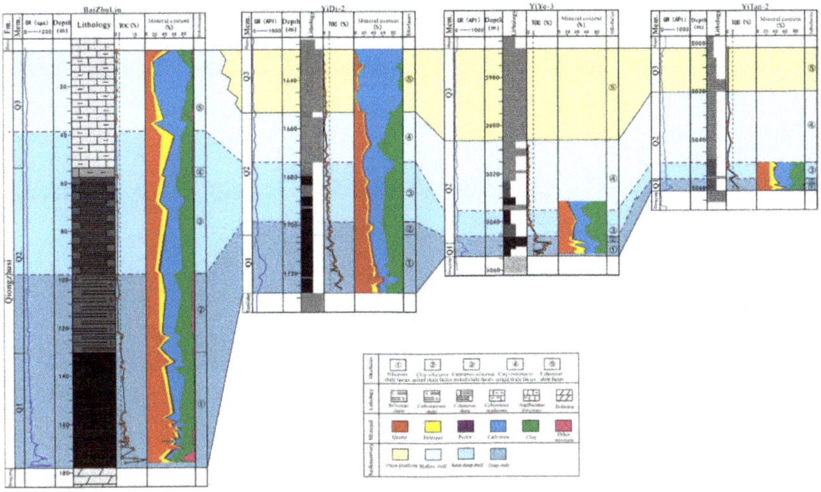

Figure 4. Cross-well correlation profile of the Qiongzhusi Formation along the trough-to-old land transect: Changyang Baizhuling–Well Yidi-2–Well Yiye-3–Well Yitan-2.

4.3. Development Characteristics of the Second Section of the Qiongzhusi Formation

The lower interval of the second member of the Qiongzhusi Formation (hereafter designated as the Qiong-2 Member) is dominated by deep-water shelf subfacies, and the lithology predominantly consists of grayish-black carbonaceous shale, locally interbedded with calcareous shale interlayers. Vertically, it exhibits a continuous depositional transition from deep-water shelf to semi-deep-water shelf and shallow-water shelf subfacies upward (Figure 4), and the dominant lithofacies shifts to mixed shale facies, including argillaceous–siliceous mixed shale and argillaceous–calcareous mixed shale. Overall, the depositional conditions were less favorable during the deposition of the Qiong-2 member than during the deposition of the Qiong-1 Member.

Field outcrop investigations revealed that the mixed shale of the Qiong-2 Member is characterized by spheroidal weathering (Figure 5a). The rocks are rigid and exhibit intense effervescence upon contact with dilute hydrochloric acid, indicative of an elevated calcareous component content. Microscopic observations revealed the occurrence of well-developed horizontal laminations, manifested as alternating low-transmissivity organic-rich mudstone and low-transmissivity calcareous layers. The degree of development of the lamination in the Qiong-2 Member is notably greater than that in the Qiong-1 Member (Figure 5b). In comparison to the Qiong-1 Member, the Qiong-2 Member has a substantially lower biogenic silica content, with grains showing irregular morphologies and smaller individual dimensions. The calcareous components are significantly enriched and occur as granular aggregates scattered sporadically within the claystone matrix. Siliceous biogenic fossils (e.g., radiolarians) are scarce (Figure 5c–f).

Comprehensive analysis of the lithological, textural, and biotic fossil characteristics indicates that the water depth was significantly shallower during the deposition of the Qiong-2 Member than during the deposition of the Qiong-1 Member. The depositional environmental stability decreased, and the water column oxygenation increased. These features are consistent with the evolutionary trajectory of shelf facies deposits under a background of eustatic sea-level fall.

Vertical correlation analysis revealed the occurrence of substantial differences in the shale mineral assemblages between the Qiong-2 Member and the Qiong-1 Member. Specifically, the average quartz content declines to 28.4%, exhibiting a pronounced downward

https://doi.org/10.3390/pr14060882

trend relative to the Qiong-1 Member. In contrast, the average feldspar content (7.4%) and pyrite content (2%) do not exhibit statistically significant deviations from those of the Qiong-1 Member. In addition, the average clay mineral content increases to 28.3%, while the average carbonate mineral content increases to 31.3%, with both showing marked enrichment compared to the Qiong-1 Member (Figure 4).

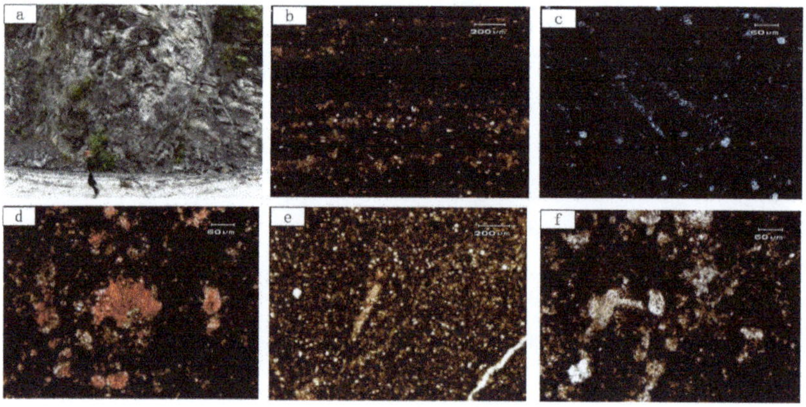

Figure 5. Characteristics of lithofacies in the second section of the Qiong formation: (**a**) Gray calcareous siliceous mixed shale, locally interbedded with calcareous shale, developed spherical weathering, Baizhuling. (**b**) Calcareous siliceous mixed shale, with calcareous fine silt and calcareous source rocks developed along the bedding planes, Baizhuling. (**c**) Clay mixed with siliceous shale, with poor transparency and visible siliceous organisms, suitable for well 3, 3041.3 m. (**d**) Limestone mudstone, sporadic calcareous debris, Yiye 3 well, 3046 m. (**e**) Clay calcareous mixed shale with scattered calcareous debris and low content of siliceous organisms, suitable for well 3, 3040.09 m. (**f**) Clay calcareous mixed shale, low content of siliceous organisms, irregular solidification resembling bone needles, suitable for well 3, 3046 m.

These systematic variations in the mineral composition were primarily governed by the evolution of the regional dynamic sedimentary regime during the deposition of the Qiong-2 Member. During this interval, gradual eustatic sea-level fall triggered a progressive forced regression. This tectono-sedimentary event induced southwestward contraction of the sedimentary center within the Western Hubei Trough and sustained shallowing of the depositional water column [13–15]. Against the backdrop of this regression, the paleotopography of the Central Hubei Old Land underwent relative uplift, leading to a significant increase in the flux of terrigenous clastic materials, including a substantial increase in the carbonate-bearing terrigenous input. These observations align with the typical sedimentary response pattern of regressive sequences, characterized by reductions in the siliceous components and enrichment of terrigenous clays and carbonates, thereby reflecting the dominant control of the evolution of the depositional environment on the mineralogical composition of the shale.

4.4. Development Characteristics of the Third Section of the Qiongzhusi Formation

The dominant lithofacies of the third member of the Qiongzhusi Formation (hereafter referred to as the Qiong-3 Member) are calcareous shale facies and limestone facies, and there is pronounced lateral heterogeneity of the sedimentary facies zonation. The central part of the Western Hubei Trough is predominantly composed of shallow-water shelf subfacies, which are underlain by a basal interval of semi-deep-water shelf subfacies. In contrast, the trough margin is dominated by open platform facies, which is stratigraphically characterized by light gray argillaceous limestone at the base, interbedded argillaceous

https://doi.org/10.3390/pr14060882

limestone and dolomitic limestone in the middle section, and light gray limestone in the upper part section (Figure 6a–c). Petrographically, the limestone is primarily classified as muddy grainstone according to the Dunham carbonate texture classification system, and the grain content ranges from 65% to 80%. The grain fraction is dominated by bioclasts, and intense effervescence occurs upon contact with dilute hydrochloric acid, indicative of enrichment of the significant calcareous component.

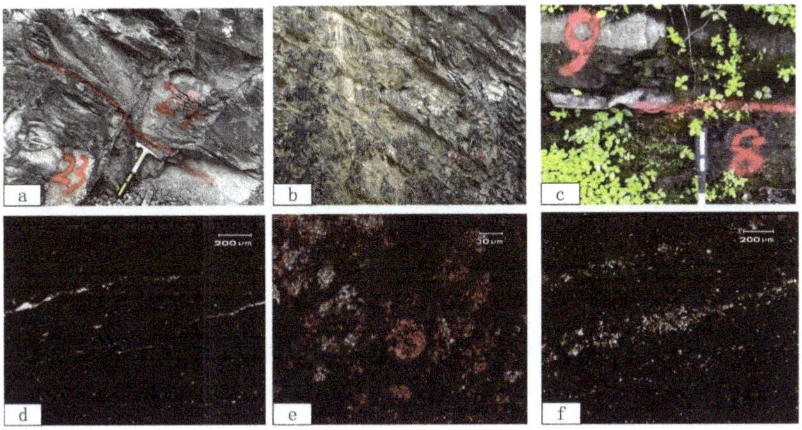

Figure 6. Characteristics of lithofacies in the third section of the Qiong formation: (**a**) Gray white calcareous shale with high calcium content, acid dripping and foaming, Baizhuling. (**b**) Gray white mudstone with high calcium content, acid dripping and foaming, Baizhuling. (**c**) Gray white middle layer mudstone, interbedded with a small amount of siliceous shale, Wangjiaping. (**d**) Calcareous shale, calcareous fine silt and calcareous sediment along the page, Baizhuling. (**e**) Calcareous shale, calcareous fine silt and calcareous cohesive source along the bedding plane, Baizhuling. (**f**) Gray mudstone, gray sediment with good translucency interbedded with clay rock with poor translucency, developed horizontal bedding, Baizhuling.

Microscopic petrographic analysis revealed that the Qiong-3 Member rocks exhibit excellent light transmittance and well-developed horizontal bedding, and the amount of silty components is negligible. Relative to the Qiong-2 Member, the content of calcareous bioclasts is substantially higher, whereas siliceous fossils (e.g., radiolarians) are virtually absent (Figure 6d–f). The geochemical data indicate that the TOC content of the Qiong-3 Member shale is generally less than 1%, and it locally declines to below 0.5%. Comprehensive analysis of the sedimentary facies, lithological, and geochemical characteristics revealed that the Qiong-3 Member was deposited in an overall shallow-water shelf–open platform depositional setting. Markedly enhanced hydrodynamic conditions and improved water mass circulation resulted in elevated paleo-oxygenation levels. This oxic depositional environment was unfavorable for the preservation of organic matter, ultimately leading to the notably low TOC content of this member.

Vertical comparative analysis of the mineral composition revealed the existence of distinct evolutionary trends in the shale mineral assemblages of the Qiong-3 Member. The average quartz content is 19.6%, and the average clay mineral content is 19.7%, both showing significant decreasing trend compared to the Qiong-2 Member. In contrast, the average feldspar content (8.7%) and pyrite content (1.2%) are not statistically significant from those of the Qiong-2 Member. Conversely, the average carbonate mineral content increases sharply to 49.1%, representing substantial enrichment relative to the Qiong-2 Member (Figure 4). Driven by sustained global eustatic sea-level fall, the Western Hubei Trough entered a phase of progressive regression, characterized by continuous shallowing of the depositional water depth and a transition in the sedimentary environment from a deep-water, low-energy setting to a shallow-water, relatively high-energy regime. During this regressive episode, the paleotopography of the Central Hubei Old Land provenance underwent relative uplift, which significantly enhanced the flux of terrigenous clastic materials—particularly a substantial increase in the carbonate-bearing terrigenous input. This process dominated the evolutionary trajectory of the shale mineral assemblages, which is highly consistent with the typical sedimentary response pattern of regressive sequences, namely, carbonate enrichment coupled with attenuation of the siliceous and clay components.

5. Sedimentary Environment Analysis

5.1. Changes in Ancient Sea Level

Previous studies have well established that the cerium (Ce) anomaly is a robust geochemical proxy for reconstructing depositional water redox conditions and paleo-sea-level fluctuations [17,18]. The underlying response mechanism is as follows: under oxic conditions, Ce (III) is oxidized to Ce (IV), which readily precipitates due to its low solubility, resulting in relative depletion of Ce in sediments and the manifestation of a distinct negative Ce anomaly. In contrast, under reducing conditions, Ce predominantly exists in the soluble Ce (III) state and thus undergoes mobilization and enrichment. This process weakens the negative Ce anomaly or even changes it to a positive anomaly. As the paleo-water depth varies, the redox interface within the water column migrates synchronously. Consequently, the Ce anomaly can be employed to indirectly infer the processes of paleo-sea-level transgression and regression.

The geochemical analysis results for the Qiongzhusi Formation samples from Well Yiye-3 in the Western Hubei region reveal the following characteristics (Table 2): the δCe values of the Qiong-1 Member range from 0.66 to 0.81, demonstrating a significant negative anomaly. This geochemical signature indicates that the depositional environment was characterized by anoxic bottom water during this period. Regionally, this interval was dominated by rapid transgression, accompanied by a marked rise in sea level, which caused the redox interface to synchronously migrate upward and become decoupled from the sediment-water interface. Upon transitioning to the Qiong-2 Member, the negative Ce anomaly weakened significantly and stabilized, with δCe values distributed between 0.85 and 0.94. This transition reflects a gradual sea-level fall trend, but the overall sea level remained at a relatively high stand (Figure 7).

Table 2. Statistical Table of Rare Earth Element Content in Yiye-3 Well.

Sample No.	Depth (m)	La (μg/g)	Ce (μg/g)	Nd (μg/g)	δCe
YY3-29	3034.9	52.646	102.632	43.428	0.92
YY3-27	3035.7	53.460	104.303	44.980	0.92
YY3-25	3037	50.182	94.021	40.698	0.89
YY3-21	3038.7	50.952	99.734	39.999	0.94
YY3-19	3040.09	35.843	69.173	30.847	0.91
YY3-17	3041.3	44.857	83.061	34.569	0.89
YY3-13	3046	37.056	65.643	27.941	0.85
YY3-11	3047.3	38.762	75.646	33.235	0.92
YY3-14	3048	34.901	63.427	27.794	0.87
YY3-8	3048.8	28.766	45.721	21.357	0.77
YY3-6	3049.68	25.527	44.729	24.435	0.81
YY3-4	3051.4	34.641	53.252	27.161	0.74
YY3-2	3052.8	35.124	46.567	20.866	0.66

Note: $\delta Ce = Ce/Ce^* = 3CeN/(2La_N + Nd_N)$.

https://doi.org/10.3390/pr1406088

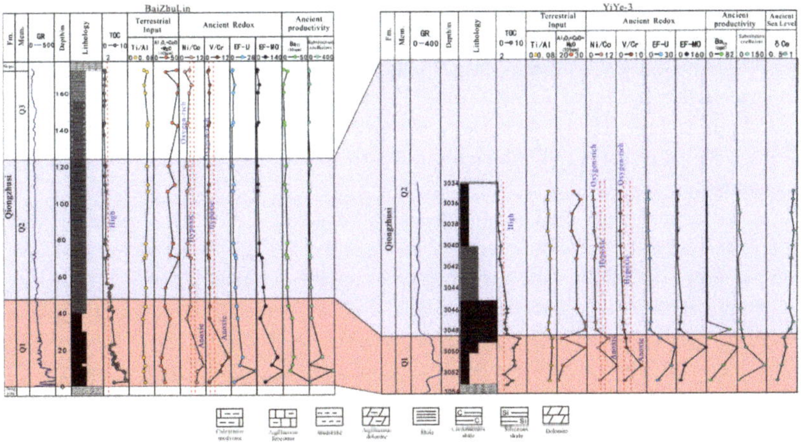

Figure 7. Standard stratigraphic column of depositional environments for the Qiongzhusi Formation, from the Baizhuling section (Changyang) to Well Yiye-3. The red dashed line is the threshold value for classification.

The rapid sea-level rise during the deposition of the Qiong-1 Member exerted two key controls on the depositional system. First, upwelling currents transported substantial fluxes of nutrients (e.g., phosphate and sulfate) into the basin, providing a fundamental material basis for the explosive proliferation of plankton and algae. This biological bloom substantially enhanced the marine primary productivity, which is a critical prerequisite for organic matter accumulation. Second, the massive mortality of thriving organisms in the surface water column led to continuous sedimentation of organic matter on the seafloor. The subsequent decomposition of this organic matter consumed large quantities of dissolved oxygen from the bottom water, ultimately giving rise to a stratified, anoxic, or even sulfidic reducing environment at the seafloor. This reducing setting provided favorable conditions for the preservation of sedimentary organic matter.

5.2. Terrestrial Input

Aluminum (Al), potassium (K), and rubidium (Rb) are transported to sedimentary basins via fluvial or aeolian processes and are primarily associated with aluminosilicate minerals. Due to their inherent chemical stability and resistance to post-depositional alteration, these elements are widely recognized as robust proxies for quantifying the terrigenous input to sedimentary systems [17,18]. The terrigenous input, typically manifested as the clastic flux, is predominantly governed by relative sea-level fluctuations: during transgressive phases, clastic depocenters migrate landward, whereas regressive episodes drive progradation of these depocenters toward the interior of the basin [19]. Therefore, the Al_2O_3 content and Ti/Al ratio are commonly used to assess terrigenous input into sediments. Specifically, Al_2O_3 content is positively correlated with the degree of terrigenous influence, whereas the Ti/Al ratio exhibits an inverse relationship [19–21].

Within the study area, the Baizhuling section (Changyang) and Well Yiye-3 are proximal to the Central Hubei Old Land, and the terrigenous material supply was dominated by carbonate detritus. The geochemical data for the Qiong-1 Member reveal that it has low $(Al_2O_3 + CaO + MgO) \times 10^4$ values (averages of 19.31 for the Baizhuling section and 23.06 for Well Yiye-3) and relatively high Ti/Al ratios (averages of 0.048 for the Baizhuling section and 0.062 for Well Yiye-3). These signatures collectively indicate the occurrence of a low-intensity terrigenous input regime, which is attributed to eustatic sea-level rise driven by seafloor spreading and consequent deepening of the depositional water column.

https://doi.org/10.3390/pr14060882

During the deposition of the Qiong-2 Member, the intensified tectonic activity triggered a gradual eustatic sea-level fall. Corresponding geochemical proxies record a notable increase in the average (Al_2O_3 + CaO + MgO) × 10^4 values (33.64 for the Baizhuling section and 26.66 for Well Yiye-3) and a consistent Ti/Al ratio of 0.056 for both localities. This transition signifies a substantial increase in the terrigenous input relative to the underlying Qiong-1 Member, reflecting the tectonically driven reorganization of the sediment routing systems.

During the deposition of the Qiong-3 Member, the tectonic activity intensified further, and this was accompanied by sustained shallowing of the depositional environment. The Baizhuling section exhibits a markedly high average (Al_2O_3 + CaO + MgO) × 10^4 value of 39.99, indicative of a substantial terrigenous flux. Overall, the terrigenous input intensity was low during the deposition of the Qiong-1 Member in the western Hunan-Hubei region, and it gradually increased upward through the Qiong-2 and Qiong-3 Members, reflecting a long-term regressive trend (Figure 7).

Extensive research has demonstrated that the terrigenous input exerts three primary modulating effects on organic matter enrichment: (1) dilution of organic matter by terrigenous clastic particles, directly reducing its concentration in the sedimentary record; (2) attenuation of organic matter sedimentation rates, prolonging exposure to oxidative degradation in the water column; and (3) weakening of the organic matter burial efficiency, disrupting the sequestration of organic carbon in seafloor sediments. Geochemical correlation analyses within the study area revealed that the TOC content is generally positively correlated with the Ti/Al ratio (Figure 8a) and is strongly negatively correlated with both the Al_2O_3 content and (Al_2O_3 + CaO + MgO) × 10^4 values (Figure 8b). These relationships indicate that the intervals in the Qiong-1 Member that are characterized by a low terrigenous input are associated with elevated TOC contents. Conversely, the intensified terrigenous input during the deposition of the Qiong-2 Member resulted in extensive organic matter dilution, leading to a discernible decrease in the TOC content compared to the Qiong-1 Member.

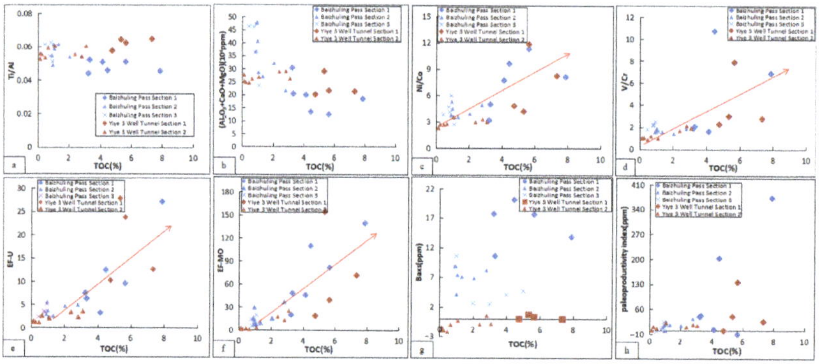

Figure 8. Correlation diagram of sedimentary environment indicators: (**a**,**b**) are the correlations between terrestrial input indicators and TOC, (**c**–**f**) are the correlations between paleoredox indicators and TOC, and (**g**,**h**) are the correlations between paleoproductivity indicators and TOC. The red arrow indicates a positive correlation.

5.3. Ancient Redox Conditions

Redox-sensitive elements (RSEs) such as uranium (U) and molybdenum (Mo) are widely employed as proxies for reconstructing marine redox conditions [20]. Generally, the concentrations of U and Mo exhibit positive correlations with the anoxia intensity. In oxic environments, U and Mo occur in high-valence states and remain dissolved in the water

https://doi.org/10.3390/pr14060882

column, leading to low accumulation in sediments. Conversely, under reducing conditions, these elements precipitate as low-valence authigenic minerals and become enriched in sediments [21]. In this study, three diagnostic indices were adopted: the Ni/Co, V/Cr ratio and the terrigenous input-corrected enrichment factor (X_{EF}), which was calculated as follows: (X/Al) sample/(X/Al) PAAS [22]. Specifically, in anaerobic environments, Ni/Co is greater than 7, V/Cr is greater than 4.25, whereas in oxygen-rich environments, Ni/Co is less than 5 and V/Cr is less than 2.0 [22–24].

The vertical variation patterns of the Ni/Co and V/Cr ratios in the study area are highly consistent. During the deposition of Member 1 of the Qiongzhusi Formation, the average values of Ni/Co, V/Cr, the uranium enrichment factor (U_{EF}), and the molybdenum enrichment factor (Mo_{EF}) in the Baizhuling section were 7.52, 5.19, 11.11, and 74.46, respectively, while those in Well Yeye-3 were 7.29, 4.00, 18.63, and 71.00, respectively. The maximal enrichment of U and Mo confirms that the water column was anaerobic during this period, which favored the accumulation of redox-sensitive U and Mo. During the deposition of Member 2, the Ni/Co ratios in Baizhuling were predominantly < 5 (with only one exception of 5.31 at the base), all of the V/Cr ratios were <2, and the average U_{EF} and Mo_{EF} values were 3.91 and 18.10, respectively. In Well Yeye-3, all of the Ni/Co ratios were <5, all of the V/Cr ratios were <2, and the average U_{EF} and Mo_{EF} values were 2.50 and 7.68, respectively. The significant reductions in U and Mo enrichment compared to Member 1 reflect gradual weakening of the reducibility of the bottom water from the Exi Trough to the Exi Old Land, exhibiting a redox transition sequence of dysoxic–oxic (Baizhuling) → oxic (Well Yeye-3). During the deposition of Member 3, the Ni/Co ratios in Baizhuling were mostly <5, the V/Cr ratios ranged from 1.57 to 2.50 (average of 2.06), and the enrichment factors of U and Mo were 3.37 and 11.75, respectively. The continuous decreasing trends of the U and Mo enrichments indicate that the water column was persistently oxic throughout this interval.

The redox proxies (U_{EF}, Mo_{EF}, Ni/Co, and V/Cr) exhibit strong positive correlations with the TOC content (Figure 8c–f), suggesting that the redox conditions exerted a significant control on the organic matter preservation. Consequently, (1) Member 1 in the study area was characterized by an anaerobic environment, which facilitated excellent organic matter preservation and resulted in relatively high TOC contents; (2) at the base of Member 2, the environment gradually transitioned to dysoxic, with the middle-upper part shifting to oxic conditions, leading to a notable decrease in the TOC content; and (3) during the deposition of Member 3, the study area was dominated by oxic conditions, which were unfavorable for U and Mo enrichment, thus resulting in low TOC contents.

5.4. Degree of Water Retention

The molybdenum/total organic carbon (Mo/TOC) ratio serves as a reliable proxy for evaluating water mass stagnation in anoxic marine basins [23,24]. Low Mo/TOC ratios indicate a strong water retaining environment, and vice versa (Figure 9). However, this method is only suitable for anaerobic environments with certain water parameter constraints [25–28]. All of the samples from the second member of the Qiongzhusi Formation in Well Yiye-3 were deposited under oxic conditions, so these data were only utilized as a reference for the variation trend of the water mass stagnation (Figure 9).

Molybdenum (Mo) and uranium (U) exhibit high mobility under aquatic anoxic conditions, which facilitates their enrichment in sediments. The covariation pattern of the Mo enrichment factor (Mo_{EF}) and U enrichment factor (U_{EF}) can also be employed to assess the degree of water mass stagnation [28]. In non-stagnant, relatively dysoxic (weakly reducing) environments, U enrichment precedes Mo enrichment, with Mo_{EF}/U_{EF} ratios ranging from 0.1 to 0.3 times that of normal seawater. As the reducing conditions intensify, the rate of Mo enrichment surpasses that of U enrichment, and the Mo_{EF}/U_{EF} ratio increases to 1–3 times the value of normal seawater. In anaerobic, weakly stagnant environments, Mo enrichment is further enhanced, leading to Mo_{EF}/U_{EF} ratios as high as 3–10 times that of normal seawater. In strongly stagnant environments, enhanced water mass confinement limits the supplies of Mo and U. Under anoxic conditions, the rate of Mo uptake and consumption exceeds that of U, resulting in Mo_{EF}/U_{EF} ratios below that of normal seawater (i.e., <1 times) [29,30].

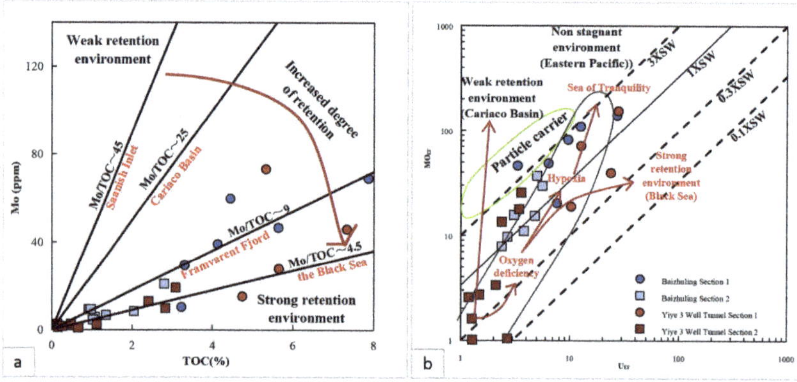

Figure 9. Mo/TOC relationship diagram and Mo_{EF}/U_{EF} covariance diagram in western Hubei region: (**a**) The solid line in the Mo/TOC relationship graph represents the Mo/TOC ratio of four modern ocean basins. (**b**) The solid line in the Mo_{EF}-U_{EF} covariance plot represents the Mo/U molar concentration ratio (SW = 7.9) in seawater, while the three dashed lines represent Mo/U ratios that are 0.1 times, 0.3 times, and 3 times higher than seawater, respectively.

In the western Hubei region, the Mo/TOC ratios generally range from 4.5 to 25 (average of 6.94), which is close to the semi-stagnant environment threshold. This infers that the Qiongzhusi Formation was deposited in a semi-stagnant marine basin (Figure 9a). For the Baizhuling Section, the average Mo/TOC ratios are 8.67 for the first member of the Qiongzhusi Formation (Qiongzhusi 1 Member) and 6.63 for the second member (Qiongzhusi 2 Member). For Well Yiye-3, the average Mo/TOC ratios are 7.03 for the Qiongzhusi 1 Member and 6.29 for the Qiongzhusi 2 Member, indicating that the degree of water mass stagnation was stronger during the deposition of the Qiongzhusi 2 Member than during the deposition of the Qiongzhusi 1 Member.

As shown in Figure 9b, the Mo_{EF}/U_{EF} ratios in the western Hunan-western Hubei region range from 0.3× to 3× that of seawater. With increases in these enrichment factors, the Mo_{EF}/U_{EF} ratio remains essentially constant. This differs from the continuous increase in the Mo_{EF}/U_{EF} ratio in open marine basins, thus deviating from the non-stagnant marine range. The Mo/U ratios of the Qiongzhusi Formation shale are only 0.3×–3× those of seawater, which are lower than the ratios affected by Mo adsorption onto manganese-iron oxide particles (Mo/U ratios: 3×–10× that of seawater). For high enrichment factors, there is no trend of decreasing Mo/U ratios with increasing enrichment factors, indicating that it does not belong to the strong stagnation basin model. Therefore, the Qiongzhusi Formation was deposited in a semi-stagnant marine basin. This setting resulted in low Mo concentrations in seawater, limiting the continuous increase in the Mo/U ratio, as observed in non-stagnant basins. In addition, the incomplete isolation of the basin allowed a certain amount of replenishment of U and Mo in the seawater, unlike the situation in a strong stagnation environment where the Mo in seawater was transferred to the sediments without subsequent replenishment, leading to a decrease in the Mo/U ratio. This inference

https://doi.org/10.3390/pr1406088

that the organic-rich shale of the Qiongzhusi Formation formed in a semi-stagnant marine basin is consistent with the conclusion derived from the Mo/TOC ratios. Furthermore, the Mo_{EF}/U_{EF} covariation diagram for the western Hubei region shows that the Mo_{EF}/U_{EF} ratios of most of the samples from the Qiongzhusi 2 Member are less than $1 \times$ SW, further confirming that the degree of water mass stagnation was significantly stronger during the deposition of the Qiongzhusi 2 Member than during the deposition of the Qiongzhusi 1 Member.

The degree of water mass stagnation is closely related to tectonic activities and sea-level fluctuations, which in turn affect the exchange of nutrients in the water column and play vital roles in organic matter enrichment [19]. Overall, the western Hubei region was dominated by a semi-stagnant environment during the deposition of the Qiongzhusi Formation. During the deposition of the Qiongzhusi 1 Member, the tectonic activity was stable, the sea level was rising, the sedimentary water depth was relatively large, and the degree water mass stagnation was weak. After entering the deposition period of the Qiongzhusi 2 Member, the tectonic activity intensified, the sea level fell, the sedimentary water depth became shallower, and the degree of water mass stagnation increased. This led to weakened seawater exchange, which affected the redox conditions and paleoproductivity, resulting in a slight decrease in the TOC content.

5.5. *Upwelling*

Upwelling currents are typically anoxic, nutrient-rich, silica-enriched marine currents, and they can fuel intense biological productivity in surface waters. Elevated nutrient fluxes and subsequent organic matter degradation consume substantial amounts of dissolved oxygen, thereby establishing conditions of high organic matter productivity and facilitating the accumulation and preservation of organic matter [20]. According to Sweere et al. [21], cobalt (Co) and manganese (Mn) exhibit analogous distribution patterns but show pronounced variability across distinct depositional environments. In upwelling systems, deep-water-sourced nutrient supplies are characteristically depleted in Co and Mn, and the authigenic enrichment of these elements is limited by insufficient availability. In contrast, within restricted basin settings, Co and Mn inputs are governed by Co- and Mn-rich fluvial discharge, making it improbable for the authigenic enrichment to become supply-limited. Thus, the geochemical proxy Al—$CO_{EF} \times Mn_{EF}$ can be employed to discriminate between upwelling systems and restricted environments. In this study, the cobalt enrichment factor (CO_{EF}) and manganese enrichment factor (Mn_{EF}) were utilized to negate the effects of clastic input and dilution. As a general discriminant criterion: $CO_{EF} \times Mn_{EF}$ values > 2 indicate a restricted environment; values < 0.5 correspond to an open/upwelling environment; and values of 0.5–2 indicate an unstable seasonal upwelling environment [31].

In Figure 10a, the samples from the Qiongzhusi 1 Member in the western Hubei region—characterized by weak watermass restriction—cluster around a Co × Mn value of 0.4, which approximates the discriminant boundary between restricted and open/upwelling environments. The robust positive correlation between Co × Mn and aluminum (Al) implies that the Co × Mn values are predominantly controlled by terrigenous clastic minerals, and these samples are therefore inferred to have been influenced by seasonal upwelling. This weak water mass restriction also reflects a certain degree of connectivity between the western Hubei region and open marine realms. The samples from the Qiongzhusi 2 Member and Qiongzhusi 3 Member in the Baizhuling Section and Well Yiye-3—proximal to the Central Hubei Old Land—mostly plot within the restricted environment field (with the exception of three samples from the Baizhuling Section), indicating an increase in the degree of water mass restriction. In Figure 10b, the Co and

https://doi.org/10.3390/pr14060882

Mn enrichment factors were calculated to effectively eliminate the confounding effects of clastic input and dilution. Consistent with the observations in Figure 10a, the samples from the Qiongzhusi 1 Member in the Baizhuling Section and Well Yiye-3 were primarily influenced by open/upwelling conditions, whereas the Qiongzhusi 2 Member and Qiongzhusi 3 Member reflect intensified water mass restriction. The samples with $Co_{EF} \times Mn_{EF}$ values of 0.5–2.0 (predominantly from the Qiongzhusi 2 Member and Qiongzhusi 3 Member in the Baizhuling Section and Well Yiye-3) could not be precisely delineated into a specific water mass environment. This ambiguity is presumably attributed to the influence of unstable seasonal upwelling.

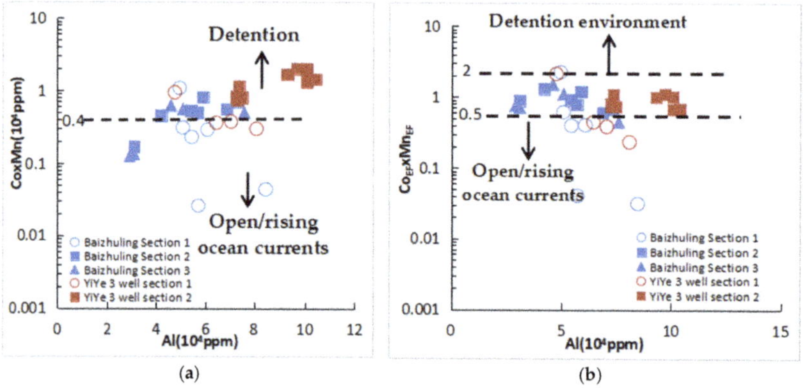

Figure 10. Plots of Al vs. Co × Mn and Al vs. $Co_{EF} \times Mn_{EF}$: (**a**) Plot of Al vs. Co × Mn: Co × Mn values > 0.4 indicate a restricted environment, while values < 0.4 indicate an open/upwelling environment. (**b**) Plot of Al vs. $Co_{EF} \times Mn_{EF}$: Values > 2 correspond to a restricted environment, values < 0.5 denote an open/upwelling environment, and values of 0.5–2 represent an unstable seasonal upwelling environment.

Collectively, these geochemical results demonstrate that the Qiongzhusi 1 Member in the western Hubei region was influenced by open/upwelling currents, which sustained a high level of paleoproductivity. In contrast, the Qiongzhusi 2 Member and Qiongzhusi 3 Member experienced enhanced water mass restriction, diminished open/upwelling influence, reduced paleoproductivity, and a consequent decrease in organic matter content. This interpretation is consistent with inferences based on the Mo/TOC ratios and the Mo_{EF}/U_{EF} covariation pattern, providing consistent evidence of the evolution of the depositional environment and its control on the organic matter enrichment.

5.6. Ancient Productivity

Organic matter enrichment in shales requires sustained high productivity in seawater to provide a robust source of organic carbon. Paleoproductivity variations are typically inferred using proxies such as excess barium (Baxs), the Ba/Al ratio, and the composite paleoproductivity index (Y) [32,33]. The excess barium is calculated with aluminum (Al) correction via the formula Baxs = Basample − Alsample × $(Ba/Al)_{PAAS}$. It should be noted that sulfate reduction and iron oxide dissolution under anoxic conditions will lead to the decomposition of $BaSO_4$ in organic matter, respectively, which distorts the Ba indicators [32–35].

In contrast, copper (Cu), nickel (Ni), and zinc (Zn) exhibit more reliable behavior, Chen et al. [35] developed a composite paleoproductivity index (Y) using these elements: Y = 0.593955 × Cuxs + 0.523396 × Nixs + 0.610962 × Znxs. Excess values (Cuxs, Nixs, and

https://doi.org/10.3390/pr14060882

Znxs) are calculated relative to the Post-Archean Australian Shale (PAAS) standard and with titanium (Ti) correction: Elementxs = Elementsample − Tisample × $(Element/Ti)_{PAAS}$.

During deposition of the Qiongzhusi 1 Member in western Hubei, elevated proxy values were recorded in both the Baizhuling Section (average Baxs = 3381.30; average Y = 108.65) and Well Yiye-3 (average Baxs = 34.86; average Y = 50.64), indicating high paleoproductivity. For the Qiongzhusi 2 Member, a significant decline occurred: the Baxs (416.19–893.48) and Y (average of 17.03) values in the Baizhuling section and the Baxs (0–58.79) and Y (average of 6.92) values in Well Yiye-3 are substantially lower than those in the Qiongzhusi 1 Member. Similarly, the Qiongzhusi 3 Member in the Baizhuling section exhibits even lower values (average Baxs = 495.25; average Y = 9.80). Overall, the paleoproductivity was the highest during the deposition of Qiongzhusi 1 Member, and it progressively weakened upward through the stratigraphic succession.

The results of this study confirm that paleoproductivity exerts a critical control on organic matter enrichment. This is supported by strong positive correlations between the TOC and both Baxs and Y in the study area (Figure 8g,h). From the Qiongzhusi 1 Member to the Qiongzhusi 3 Member, the consistent decline in paleoproductivity—accompanied by decreasing nutrient availability in the surface waters—resulted in a gradual reduction in organic matter content.

6. Sedimentary Environment Evolution Model

Early Depositional Stage of the Qiongzhusi Formation (Qiongzhusi 1 Member): This stage corresponds to a period of rapid sea-level rise followed by an early phase of sea-level fall. It was characterized by relatively deep depositional water depths and the development of marine water column stratification. Owing to inadequate oxygen circulation, anoxic to hypoxic conditions prevailed in the bottom waters of the western Hubei Trough, as evidenced by the intense enrichment of RSEs. The Mo/TOC ratios and Mo_{EF}/U_{EF} values indicate that the study area was characterized by a semi-confined environment during this interval, which was influenced by seasonal upwelling currents. These currents transported nutrients from the Paleo-South China Ocean to the study area, enriching the surface waters with nutrients and sustaining a high level of paleoproductivity. Furthermore, sea-level rise displaced the Yangtze Basin farther from its provenance areas, resulting in reduced terrigenous input. The reduced dilution effect of the terrigenous detritus on the organic matter, coupled with the high organic matter abundance, created favorable conditions for the enrichment and preservation of organic matter (Figure 11a).

Middle Depositional Stage of the Qiongzhusi Formation (Qiongzhusi 2 Member): With the uplift of the ancient lands surrounding the Middle-Upper Yangtze region, the extensional intensity of the rift basin decreased and the sea level fell. This transition led to an increase in the dissolved oxygen content of the water column, disrupting the anoxic bottom-water conditions and establishing a predominantly hypoxic environment. Deposited in this paleoenvironmental setting, the Qiongzhusi 2 Member in the western Hubei Trough is dominated by semi-deep-water shelf facies deposits. From the western Hubei Trough to the Central Hubei Old Land, the facies transition from semi-deep-water shelf to shallow-water shelf is associated with a gradual decrease in the reducibility of the bottom-water, shifting from hypoxic to oxic conditions. Concurrently, the degree of water mass restriction intensified, the influence of open/upwelling currents waned, the surface water nutrient availability declined progressively, the terrigenous input increased, the paleoproductivity decreased, and the TOC content decreased accordingly (Figure 11b).

Late Depositional Stage of the Qiongzhusi Formation (Qiongzhusi 3 Member): During the late depositional stage of the Qiongzhusi Formation, the ancient lands surrounding the Middle-Upper Yangtze region underwent sustained uplift, which further accelerated

https://doi.org/10.3390/pr14060882

the infilling rate of the rift basin. The western Hubei area was dominated by a shallow-water, oxic depositional environment. There was a continued increase in the terrigenous clastic input, which intensified the dilution effect on the organic matter. Concurrently, the paleoproductivity level persistently decreased, culminating in the lowest organic matter content across the entire Qiongzhusi Formation stratigraphic succession.

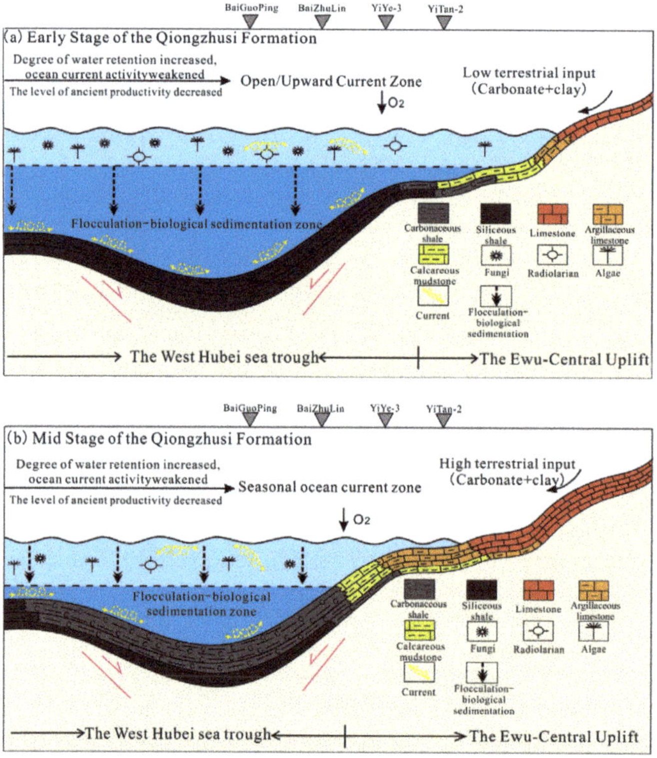

Figure 11. Schematic diagram showing the shale formation and evolution during the Early Cambrian Qiongzhusi Period: Western Hubei Trough–Central Hubei Old Land.

7. Conclusions

(1) The TOC content in the western Hubei area exhibits a consistent decreasing trend both vertically and laterally from the center of the trough to its margins. Specifically, the TOC content of the Qiongzhusi 1 Member generally exceeds 3% (regarded as an excellent source rock quality), is primarily between 1.13% and 3% in the Qiongzhusi 2 Member, and is mostly less than 1% in the Qiongzhusi 3 Member, reflecting a progressive deterioration of the organic matter enrichment potential upward through the stratigraphic sequence.

(2) The Qiongzhusi Formation in the western Hubei area is characterized by strong vertical heterogeneity, with a distinct lithofacies transition sequence from siliceous shale facies → mixed shale facies → calcareous shale facies from bottom to top. Laterally and areally driven by the sustained uplift of the ancient lands surrounding the Middle-Upper Yangtze region, the thicknesses of the siliceous shale, argillaceous–siliceous mixed shale, and calcareous–siliceous mixed shale gradually thin toward the northeast. In contrast, the thickness of the argillaceous–calcareous mixed shale progressively

increases, which is closely linked to the intensified terrigenous clastic input from the northeast paleo-provenance.

(3) From the early to late depositional stages of the Qiongzhusi Formation in the western Hubei area, a systematic paleoenvironmental transition occurred: the water depth gradually shallowed, the hydrodynamic energy intensified, the dissolved oxygen content of the bottom waters increased progressively, the upwelling current intensity weakened continuously, the surface-water paleoproductivity declined steadily, the degree of water mass restriction increased gradually, and the terrigenous clastic input increased consistently. This suite of coeval environmental changes collectively governed the vertical evolution of the organic matter enrichment conditions.

(4) The formation of the organic-rich shale in the study area was predominantly controlled by the synergistic effects of the terrigenous input, redox conditions, paleoproductivity, water mass restriction, and upwelling currents. Favorable conditions for organic matter enrichment include a low terrigenous clastic input (minimizing dilution), anoxic bottom-water environments (enhancing preservation), high paleoproductivity (increasing organic matter supply), semi-confined water mass, and active upwelling currents (transporting nutrients). Among these factors, the redox conditions, degree of water mass restriction, and paleoproductivity were the primary drivers responsible for the differences in the TOC contents at the center and margins of the western Hubei Trough, while the terrigenous input was the secondary controlling factor.

Author Contributions: Conceptualization, Z.L., J.H., F.C., T.Y. and L.W.; methodology, Z.L., F.C., T.Y. and L.W.; formal analysis, L.W.; investigation, F.C.; resources, J.H. and T.Y.; data curation, J.H.; writing—original draft, Z.L.; supervision, L.W. All authors have read and agreed to the published version of the manuscript.

Funding: This research received no external funding.

Data Availability Statement: The original contributions presented in this study are included in the article. Further inquiries can be directed to the corresponding author.

Conflicts of Interest: JianHong He was employed by No. 3 Oil Production Plant, PetroChina Qinghai Oilfield Company. Fangfang Chen was employed by Research Institute of Exploration and Development, Tarim Oilfield Company. Tao Yang was employed by Oilfield Technology Development Company, PetroChina Liaohe Oilfield Company The remaining authors declare that the research was conducted in the absence of any commercial or financial relationships that could be construed as a potential conflict of interest.

eferences

Wang, P.W.; Zou, C.; Li, X.J.; Ma, L.Q.; Jiang, L.W.; Jia, D.; Mei, J.; Huang, L.; Chen, X.Y.; Li, Q.F. Geochemical characteristics and paleoenvironmental significance of the Qiongzhusi Formation in the Dianqianbei area. *J. China Univ. Pet. Ed. Nat. Sci.* **2021**, *45*, 51–62.

Zou, C.N.; Dong, D.Z.; Wang, S.J.; Li, J.; Li, X.; Wang, Y.; Li, D.; Cheng, K. Formation mechanism, geological characteristics and resource potential of shale gas in China. *Pet. Explor. Dev.* **2010**, *37*, 641–653. [CrossRef]

Nie, H.K.; Zhang, J.C.; Li, Y.X. Accumulation conditions of Lower Cambrian shale gas in the Sichuan Basin and its peripheral areas. *Acta Pet. Sin.* **2011**, *32*, 959–967.

Luo, S.Y.; Chen, X.H.; Liu, A.; Li, H. Geochemical characteristics and genesis of shale gas from the Lower Cambrian Shuijingtuo Formation in Yichang area, Central Yangtze Block. *Oil Gas Geol.* **2019**, *40*, 999–1010.

Wu, S.; Hu, Z.G.; Qiu, X.S.; Li, S.L.; Li, H.D.; Liu, S.F. Genesis and significance of mineral composition of organic-rich shale from the Lower Cambrian Niutitang Formation in Tongshan area, eastern Hubei. *Sci. Technol. Eng.* **2020**, *20*, 11867–11874.

Wan, M.M.; Feng, M.S.; Meng, W.B.; Liu, S.B.; Lu, Y.F.; Nie, Y.; Zhao, Z.; Guo, M. Geochemical characteristics and geological significance of black shales from the Lower Cambrian Shuijingtuo Formation in Yichang area. *Bull. Mineral. Petrol. Geochem.* **2021**, *40*, 946–957.

https://doi.org/10.3390/pr14060882

7. Wu, Y.; Fan, T.L.; Ding, H.Y. Lithofacies types and sedimentary model of Lower Cambrian marine shale in the Upper Yangtze area. *Geoscience* **2017**, *31*, 1223–1232.
8. Och, L.M.; Shields-Zhou, G.A.; Poulton, S.W.; Manning, C.; Thirlwall, M.F.; Li, D.; Chen, X.; Ling, H.; Osborn, T.; Cremonese, L. Redox changes in Early Cambrian black shales at Xiaotan section, Yunnan Province, South China. *Precambrian Res.* **2013**, *225*, 166–189. [CrossRef]
9. Wang, Z.J.; Wang, J.; Zhuo, J.W.; Yang, P.; Liu, J.H.; Xie, S.K. Crustal extension during the Sinian-Cambrian transition in the Yangtze Block: Evidence from sedimentary sequences and geochemistry. *Geol. Rev.* **2011**, *57*, 731–742.
10. Zhou, M.Z.; Luo, T.Y.; Huang, Z.L.; Liu, S.R. Records of Early Cambrian volcanic activity in Guizhou and their geological significance. *Acta Mineral. Sin.* **2011**, *31*, 453–461.
11. Qiu, X.S.; Hu, M.Y.; Hu, Z.G. Lithofacies palaeogeography and shale accumulation conditions of the Lower Cambrian in the Central Yangtze area. *J. Cent. South Univ. Sci. Technol.* **2014**, *45*, 3174–3185.
12. Li, W.Z.; Zhang, J.Y.; Li, H.H.; Wang, X.F.; Zou, Q.; Jiang, H.; Fu, X.D.; Wang, P.W.; Xu, Z.; Ma, L.Q.Y. Distribution characteristics and petroleum exploration significance of intracratonic rifts in Western Hubei-Eastern Chongqing area. *Nat. Gas Geosci.* **2020**, *31*, 675–685.
13. Li, Z.X.; Lu, Y.C.; Wang, J.; Duan, T.Z.; Gao, Y.H. Sedimentary characteristics and lithofacies palaeogeography of the Late Sinian-Early Cambrian in the Central Yangtze area. *J. Palaeogeogr. Chin. Ed.* **2004**, *6*, 151–162.
14. Hu, Y.; Chen, X.H. Geochemical characteristics and environmental significance of black shales during the Precambrian-Cambrian transition in the Three Gorges area. *Geol. Sci. Technol. Inf.* **2017**, *36*, 61–71.
15. Wei, K.; Chen, X.H.; Wang, C.S.; Liu, A.; Zeng, X.W.; Li, Z.H. Genesis of siliceous rocks from the Late Ediacaran-Early Cambrian in Western Hunan-Hubei area and their shale gas geological significance. *Bull. Geol. Sci. Technol.* **2020**, *39*, 20–30.
16. Wang, Y.M.; Wang, S.F.; Dong, D.Z.; Li, X.Z. Lithofacies characterization of Lower Silurian Longmaxi Formation shale in southern Sichuan Basin. *Earth Sci. Front.* **2016**, *23*, 119–133.
17. German, C.R.; Elderfield, H. Application of the Ce anomaly as a paleoredox indicator: The ground rules. *Palaeoceanography* **1990**, *5*, 823–833. [CrossRef]
18. Wilde, P.; Mary, S.; Erdtmann, B.D. The whole-rock Cerium anomaly: A potential indicator of eustatic sea-level changes in shales of the anoxic facies. *Sediment. Geol.* **1996**, *101*, 43–53. [CrossRef]
19. Tribovillard, N.; Algeo, T.J.; Baudin, F.; Riboulleau, A. Analysis of marine environmental conditions based on molybdenum-uranium covariation—Applications to Mesozoic paleoceanography. *Chem. Geol.* **2012**, *324–325*, 46–58. [CrossRef]
20. Li, Y.F.; Zhang, T.; Ellis, G.S.; Shao, D. Depositional environment and organic matter accumulation of Upper Ordovician-Lower Silurian marine shale in the Upper Yangtze platform, South China. *Palaeogeogr. Palaeoclimatol. Palaeoecol.* **2017**, *466*, 252–264. [CrossRef]
21. Lu, Y.B.; Jiang, S.; Lu, Y.C.; Xu, S.; Shu, Y.; Wang, Y.X. Productivity or preservation? The factors controlling the organic matter accumulation in the late Katian through Hirnantian Wufeng organic-rich shale, South China. *Mar. Pet. Geol.* **2019**, *109*, 22–35. [CrossRef]
22. Li, Y.F. Geochemical Characteristics, Organic Matter Enrichment and Paleoenvironmental Significance of Late Ordovician-Early Silurian Shale in the Upper Yangtze Area. Ph.D. Thesis, Lanzhou University, Lanzhou, China, 2017.
23. Algeo, T.J.; Maynard, J.B. Trace-element behavior and redox facies in core shales of Upper Pennsylvanian Kansas-type cyclothems. *Chem. Geol.* **2004**, *206*, 289–318. [CrossRef]
24. Wang, T.L.; Hao, A.S.; Chen, Q.; Li, C.; Wang, Q.T.; Lu, H.; Liu, D.Y. Main controlling factors of shale development in the Wufeng Formation and Longmaxi Formation in Yichang area, Central Yangtze Block. *Nat. Gas Geosci.* **2018**, *29*, 616–631.
25. Rowe, H.D.; Loucks, R.G.; Ruppel, S.C.; Rimmer, S.M. Mississippian Barnett Formation, Fort Worth Basin, Texas: Bulk geochemical inferences and Mo-TOC constraints on the severity of hydrographic restriction. *Chem. Geol.* **2008**, *257*, 16–25. [CrossRef]
26. Algeo, T.J.; Lyons, T.W. Mo-total organic carbon covariation in modern anoxic marine environments: Implications for analysis of paleoredox and paleohydrographic conditions. *Palaeoceanography* **2006**, *21*, PA2004. [CrossRef]
27. He, L. Organic Matter Enrichment Mechanism and Sedimentary Environment Evolution of Wufeng-Longmaxi Formation Shale in the Southeastern Margin of the Sichuan Basin. Ph.D. Thesis, University of Chinese Academy of Sciences (Guangzhou Institute of Geochemistry, Chinese Academy of Sciences), Guangzhou, China, 2020.
28. Xiao, B.; Liu, S.G.; Ran, B.; Yang, D.; Han, Y.Y. Application of discriminant indicators for organic matter enrichment factors in marine sedimentary rocks based on elements Mn, Co, Cd and Mo in the northern margin of the Sichuan Basin. *Geol. Rev.* **2019**, *65*, 1317–1330.
29. Huang, Z.Q.; Fang, C.G.; Li, J.Q.; Zhang, C.C.; Wang, Y.J.; Liu, L.X.; Zhao, S. U-Mo covariation pattern and paleobasin watermass restriction degree of Wufeng-Gaojiabian Formation shale in Ningzhen area. *J. Chengdu Univ. Technol. Sci. Technol. Ed.* **2020**, *47*, 443–450.
30. Algeo, T.J.; Tribovillard, N. Environmental analysis of paleoceanographic systems based on molybdenum-uranium covariation. *Chem. Geol.* **2009**, *268*, 211–225. [CrossRef]

https://doi.org/10.3390/pr14060882

. Sweere, T.; van den Boorn, S.; Dickson, A.J.; Reichart, G.-J. Definition of new trace-metal proxies for the controls on organic matter enrichment in marine sediments based on Mn, Co, Mo and Cd concentrations. *Chem. Geol.* **2016**, *441*, 235–245. [CrossRef]
. Qiu, Z.; Wang, Q.C. Geochemical characteristics and sedimentary setting of Middle-Upper Permian siliceous rocks in Xiangqiangui area. *Acta Petrol. Sin.* **2010**, *26*, 3612–3628.
. Gao, Q.; Wang, X.Z.; Zhu, Y.Q.; Zhao, S.X.; Zhang, R.; Xiao, Z.Y. Elemental geochemical characteristics and main controlling factors of organic matter enrichment of Longmaxi Formation in southern Sichuan Basin. *Lithol. Reserv.* **2019**, *31*, 72–84.
. Dymond, J.; Suess, E.; Lyle, M. Barium in deep-sea sediment: A geochemical proxy for paleoproductivity. *Palaeoceanography* **1992**, *7*, 163–181. [CrossRef]
. Chen, H.; Xie, X.N.; Li, H.J.; Su, M.; Peng, W.; Hu, C.Y. Evaluation of Permian marine hydrocarbon source rocks in the Shangsi section, Sichuan Basin using paleoredox and paleoproductivity proxies. *J. Palaeogeogr. Chin. Ed.* **2010**, *12*, 76–85.

https://doi.org/10.3390/pr14060882

processes

MDPI

Article

In Situ Self-Assembled Particle-Enhanced Foam System for Profile Control and Enhanced Oil Recovery in Offshore Heterogeneous Reservoirs

Mengsheng Jiang [1,2,3,4], Shanfa Tang [1,2,3,4,*] and Yu Xia [1,2,3,4,*]

1 School of Petroleum Engineering, Yangtze University, Wuhan 430100, China; jms2026@126.com
2 Hubei Key Laboratory of Oil and Gas Drilling and Production Engineering, Yangtze University, Wuhan 430100, China
3 National Engineering Research Center for Oil & Gas Drilling and Completion Technology, School of Petroleum Engineering, Yangtze University, Wuhan 430100, China
4 State Key Laboratory of Low Carbon Catalysis and Carbon Dioxide Utilization, Wuhan 430100, China
* Correspondence: tangsf2005@126.com (S.T.); xydsb1997@163.com (Y.X.)

Abstract

Severe reservoir heterogeneity in offshore oilfields often leads to dominant flow channels, high water cut, and low sweep efficiency during long-term water flooding. In this study, an in situ self-assembled composite foam system combining soft polymer particles with a low-interfacial-tension foaming agent was developed for profile control and enhanced oil recovery (EOR) in offshore heterogeneous reservoirs. The self-assembly characteristics and physicochemical properties of different particle systems were evaluated to optimize the composite foam structure. Static and dynamic experiments were conducted to assess foam stability, plugging performance, injectivity behavior, and oil displacement efficiency. Results show that the optimized composite foam undergoes in situ self-assembly under reservoir conditions, forming a stable particle–foam structure that enhances selective plugging and mobility control. Core flooding experiments demonstrate that the system increases oil recovery by up to 27.2% across a wide permeability range. Field application further confirms its effectiveness in regulating interlayer water absorption, stabilizing injection pressure, and reducing water cut. These results indicate that the proposed in situ self-assembled composite foam is a promising technique for integrated profile control and enhanced oil recovery in offshore heterogeneous reservoirs.

Keywords: self-assembled foam; profile control and flooding; stabilizing oil and water control

1. Introduction

Offshore oilfields play an important role in oil and gas production [1–3], yet many have entered the middle and late stages of development and are increasingly challenged by high water cut and declining oil production. Due to loose cementation and strong reservoir heterogeneity [4], long-term water and polymer injection commonly lead to the formation of high-permeability flow channels [5,6], resulting in inefficient water circulation and poor sweep efficiency [7–9].

Chemical profile control and enhanced oil recovery (EOR) technologies, such as polymer flooding, gel treatments, particle plugging, and foam flooding, have been widely applied to mitigate water channeling [10]. However, conventional polymer and gel systems often suffer from limited stability and short effective periods, while rigid particle systems

 https://doi.org/10.3390/pr1403041

show poor adaptability to complex pore–throat structures [11]. Foam flooding offers effective mobility control, but its performance in offshore reservoirs is frequently constrained by insufficient stability under high salinity and crude oil conditions [12].

To improve foam stability and profile control performance, particle-stabilized foam systems have attracted increasing attention [13]. Nevertheless, most existing systems rely on pre-formed particle–foam mixtures, in which particles mainly act as external stabilizers and lack adaptability to dynamic reservoir conditions [14]. In contrast, soft polymer particles capable of in situ self-assembly can undergo deformation and aggregation under reservoir conditions, enabling dynamic reconstruction and selective plugging at the pore scale [15].

Unlike conventional rigid particle systems that rely on fixed mechanical jamming, the adaptability of the proposed system is defined as the ability of soft polymer particles to undergo in situ softening and viscoelastic deformation under reservoir temperature and salinity conditions, allowing them to migrate through narrow pore throats and achieve deep profile control. Based on this concept, an in situ self-assembled composite foam system consisting of soft nano/micron-sized polymer particles and a low-interfacial-tension foaming agent is developed for offshore heterogeneous reservoirs. Under reservoir conditions, the system forms a dynamic particle–foam structure that enables simultaneous mobility control and profile modification. Experimental results from static foam tests, core flooding, and field applications indicate that particle morphology and surface charge dominate the balance between foam stability and plugging performance, while particle size mainly affects transport behavior. Particles with loose layered or three-dimensional network morphologies and moderate negative surface charges exhibit superior dispersion stability and foam strength, whereas positively charged particles tend to agglomerate, enhancing plugging but reducing injectivity. Accordingly, the HK system with a nanoscale layered morphology and moderate electronegativity is identified as the optimal formulation.

2. Experimental Section

2.1. Experimental Materials and Conditions

2.1.1. Main Reagents

The primary chemicals used in this study included acrylamide (AM, Meryer, Shanghai, China), 2-acrylamido-2-methylpropane sulfonic acid (AMPS, Meryer, Shanghai, China), *N,N′*-Methylenebisacrylamide (MBA, Meryer, Shanghai, China), sorbitan monolaurate (Span 20, J&K Scientific, Beijing, China), polyethylene glycol mono-octylphenyl ether (OP-10, J&K Scientific, Beijing, China), sodium hydroxide (NaOH, Energy Chemical, Shanghai, China), ammonium persulfate (APS, Energy Chemical, Shanghai, China), sodium bisulfite ($NaHSO_3$, Meryer, Shanghai, China), anhydrous ethanol, a dodecyl-based foaming agent (ZHDB-5, Meryer, Shanghai, China), and polyacrylamide (PAM, J&K Scientific, Beijing, China) with an average molecular weight of approximately 12 million.

The main experimental instruments employed were a constant-temperature water bath (JingCheng Instrument, QingDao, China), a JJ-1B digital constant-speed stirrer (JingCheng Instrument, QingDao, China), an EVOS optical microscope (KEYENCE, Osaka, Japan), a Fourier transform infrared (FTIR) spectrometer (JingCheng Instrument, QingDao, China), a JEM-2100F transmission electron microscope (TEM, KEYENCE, Osaka, Japan), a Phenom ProX scanning electron microscope (SEM, Phenom, Shanghai, China), a Zetasizer Nano ZS90 dynamic/static light scattering analyzer (Malvern Panalytical, Great Malvern, UK), a Mastersizer 3000 laser particle size analyzer (Malvern Panalytical, Great Malvern, UK), and a Teclis Foamscan foam analyzer (TECLIS, Lyon, France).

https://doi.org/10.3390/pr14030411

2.1.2. Sand-Packed Tube Model

A rectangular sand-packed tube model with dimensions of 30.0 cm × 4.5 cm × 4.5 cm was used in this study. The model was packed with quartz sand and encapsulated in resin to ensure structural integrity. Three pressure measurement ports were evenly distributed along the main flow direction of the sand-packed core. By adjusting the sand packing conditions, the model was designed to represent three levels of formation permeability: 500 mD, 1000 mD, and 3000 mD.

2.1.3. Experimental Conditions

Crude oil used in this study was obtained from the J oilfield. Particle size analysis of the mechanical impurities in the raw crude oil revealed a wide distribution, with a median diameter (D50) of 12.5 μm and a D90 of 58.2 μm. To eliminate experimental errors caused by end-face plugging from macro-aggregates while retaining the oil's native emulsification characteristics, the crude oil was filtered through a stainless-steel screen with a mesh size of 0.045 mm at 60 °C. The filtered oil was subsequently dehydrated at temperatures below 120 °C for more than 1 h, until the water content was reduced to less than 0.5%. The experimental water was field injection water collected from the J oilfield. The schematic of the experimental procedure is presented in Figure 1.

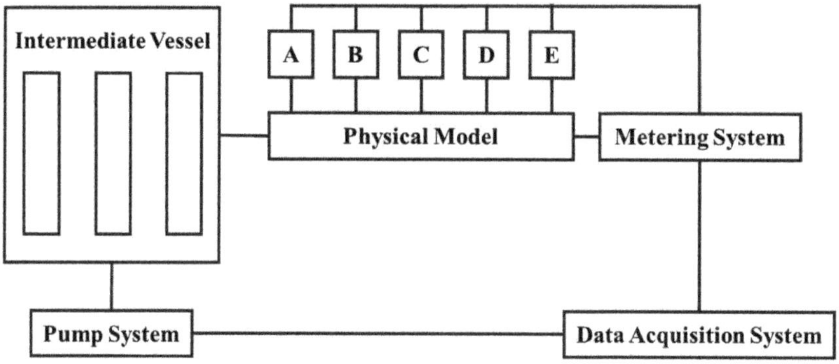

Figure 1. Dynamic performance evaluation experimental process diagram.

2.2. Experimental Procedures

2.2.1. Preparation of Self-Assembled Systems

Nano-sized polyacrylamide microspheres were synthesized via a reverse microemulsion polymerization method under strictly controlled conditions to ensure reproducibility. A composite emulsifier composed of sorbitan monolaurate and polyethylene glycol mono-octylphenyl ether was first prepared, followed by the addition of acrylamide and 2-acrylamido-2-methylpropane sulfonic acid at predetermined molar ratios, with a total monomer concentration fixed at 15 wt%. The crosslinking agent N and initiator were introduced at 0.05 wt% and 0.3 wt% of the total monomer content, respectively. Polymerization was carried out at 55–60 °C for 4 h under continuous stirring. After completion of the reaction, anhydrous ethanol was added to the emulsion at a volume ratio of 1:14, and the mixture was stirred and filtered to remove residual surfactants and unreacted species, yielding polyacrylamide nanospheres. The obtained microspheres were subsequently surface-modified to form functionalized particles, which are capable of softening and bonding with one another under reservoir conditions, leading to in situ self-assembly.

In this study, six self-assembled systems with different particle sizes, pore–throat adaptability, and surface charge characteristics were prepared and designated as PMB-1

https://doi.org/10.3390/pr14030411

(nanoscale), PMB-2 (micron scale), PMB-3, PMB-4, PMB-5, and HK. Six self-assembled particle–foam systems were rationally designed to investigate the effects of particle scale, surface charge, and morphology on profile control performance. A multiscale pore-matching strategy was adopted by tuning particle sizes from the nanometer scale (PMB-1, HK) to the micron scale (PMB-2, PMB-5) to evaluate transportability across heterogeneous reservoirs (500–3000 mD). Most systems (HK, PMB-1/2/4/5) were engineered with negative zeta potentials to enhance dispersion stability, while PMB-3 was functionalized with a positive surface charge to assess electrostatic hetero-coagulation as a mechanism for intensified plugging in high-permeability channels. In addition, discrete spherical particles (PMB-1, PMB-5) were designed for deep migration, whereas layered or three-dimensional network structures (PMB-2, HK) were intended to enhance foam film stability and viscoelasticity, enabling systematic correlation between particle characteristics and profile control functions.

2.2.2. Characterization and Performance Evaluation of Self-Assembled Systems

Prior to characterization, the self-assembled systems were pretreated through extraction, filtration, drying, and grinding. The morphological characteristics of the treated samples were then examined using scanning electron microscopy (SEM, Phenom, Shanghai, China).

The particle size, particle size distribution, and zeta potential of the self-assembled systems were measured using a dynamic/static laser light scattering analyzer and a laser particle size analyzer. Based on the pore–throat characteristics of the target oilfield in the Bohai Sea, the optimal self-assembled system was determined.

The dynamic performance of the self-assembled foam drive systems was evaluated in terms of sealing performance, pressure reduction, and injectivity behavior, as well as displacement efficiency.

3. Experimental Results and Discussion

3.1. Morphological Characteristics of Self-Assembled Particles

The morphologies of the self-assembled systems HK, PMB-1, PMB-2, PMB-3, PMB-4, and PMB-5 were examined using a Phenom ProX scanning electron microscope, and the corresponding images are presented in Figure 2. As shown in Figure 2, the HK sample is uniformly dispersed and exhibits a loose layered structure. After extraction with ethanol and subsequent drying, the PMB-1 sample appears as discrete spherical particles with an average particle size of approximately 200 nm.

In contrast, the PMB-2 sample is well dispersed and displays a three-dimensional network structure with uniformly distributed pores. The PMB-3 and PMB-4 samples show a pronounced tendency to agglomerate, indicating reduced dispersion stability. The PMB-5 sample consists of irregular, small spherical particles with uniformly dispersed microspheres.

These distinct morphological characteristics directly govern the pore-scale transport and selective plugging mechanisms of the system. The discrete spherical structures (PMB-1, PMB-5) exhibit low flow resistance and superior dispersibility, facilitating deep migration into tight pore throats. In contrast, the loose layered and 3D network structures (HK, PMB-2) provide enhanced viscoelasticity and steric hindrance; they effectively adsorb at gas–liquid interfaces to stabilize foam lamellae while retaining the ability to deform and pass through narrow constrictions. Conversely, the agglomerated structures (PMB-3, PMB-4) tend to form bridge plugging at pore throats, which enhances localized blockage but may limit deep penetration. Therefore, the HK system's deformable layered morphology offers an optimal physical architecture for balancing deep propagation with effective mobility control.

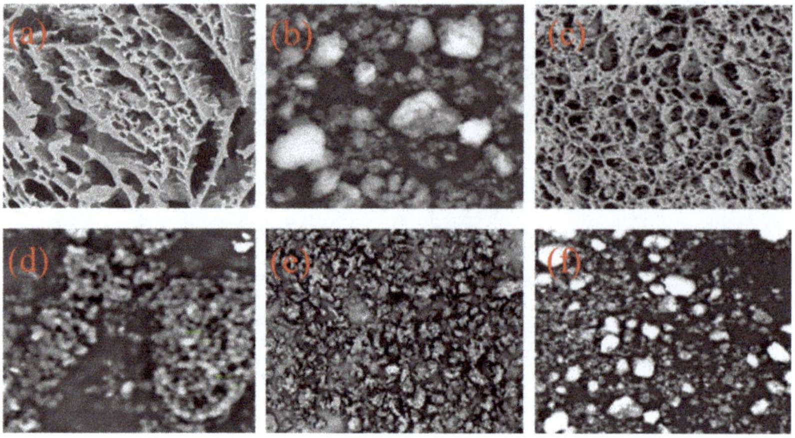

Figure 2. Morphological characteristics of the self-assembled systems. (**a**) HK original liquid at 1000× magnification; (**b**) PMB-1 extract at 30,000× magnification; (**c**) PMB-2 undiluted solution at 1000× magnification; (**d**) PMB-3 undiluted solution at 10,000× magnification; (**e**) PMB-4 undiluted solution at 10,000× magnification; (**f**) PMB-5 undiluted solution at 1000× magnification.

3.2. Particle Size and Zeta Potential of Self-Assembled Particles

3.2.1. Particle Size Distribution Characteristics

The particle size distributions of the HK, PMB-1, PMB-2, PMB-3, PMB-4, and PMB-5 self-assembled systems were measured using a dynamic/static laser light scattering analyzer and a laser particle size analyzer. The corresponding results are presented in Figure 3.

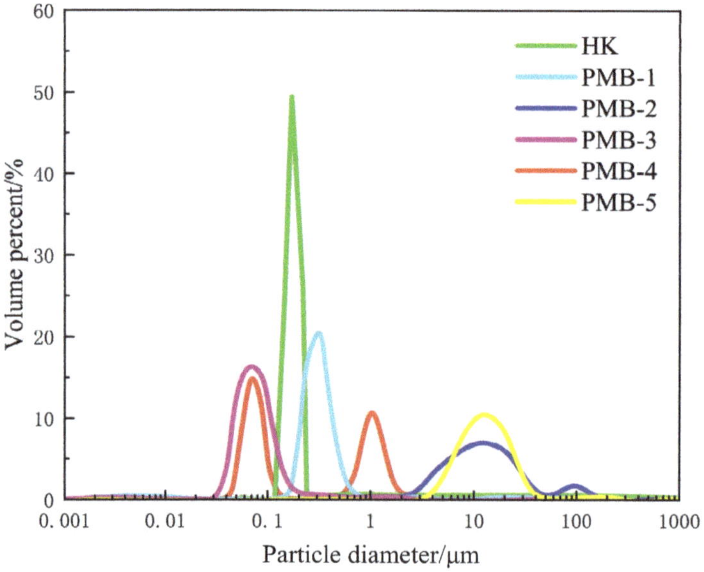

Figure 3. Particle size distribution of self-assembled systems.

As shown in Figure 3, PMB-2 and PMB-5 fall within the category of micron-sized microspheres. In contrast, PMB-1, PMB-3, and HK are classified as nanometer-sized microspheres. Notably, the PMB-4 self-assembled system exhibits a bimodal particle size distribution, containing both micron-sized and nanometer-sized microspheres.

https://doi.org/10.3390/pr14030411

3.2.2. Zeta Potential of Self-Assembled Particles

The zeta potentials of the HK, PMB-1, PMB-2, PMB-3, PMB-4, and PMB-5 self-assembled systems were determined using a dynamic/static laser light scattering analyzer. The results are summarized in Table 1.

Table 1. Zeta potential test results for self-assembled systems.

Sample Name	Zeta (mV)
HK	−10
PMB-1	−5.23
PMB-2	−20.3
PMB-3	18.6
PMB-4	−4.1
PMB-5	−17.2

As summarized in Table 1, only PMB-3 exhibits a positive zeta potential, whereas all other self-assembled microspheres possess negative surface charges. In general, an increase in the absolute value of the zeta potential enhances electrostatic repulsion between dispersed oil droplets, thereby reducing the likelihood of oil–water separation and improving emulsion stability. However, dispersion observations reveal that PMB-3 and PMB-4 exhibit a pronounced tendency toward agglomeration and mutual adhesion, which may induce formation plugging, whereas HK and PMB-2 remain uniformly dispersed with good dispersion stability and a lower propensity for aggregation. Although PMB-2 and PMB-5 are both micron-sized and exhibit similar zeta potential values, their dispersion behaviors differ markedly, highlighting the coupled influence of surface charge and particle morphology.

Notably, while higher absolute zeta potentials generally favor dispersion stability via electrostatic repulsion, the proposed in situ self-assembly mechanism requires an optimal stability window. Excessive electrostatic repulsion may suppress the particle–particle interactions necessary for controlled aggregation and network formation, whereas insufficient repulsion leads to premature agglomeration and poor injectivity. The results indicate that the HK and PMB-2 systems, with zeta potentials of −10 mV and −20.3 mV, respectively, fall within this favorable range. These systems maintain adequate dispersion during injection while permitting controlled aggregation and dense interfacial packing under reservoir salinity conditions, thereby achieving a critical balance between transport efficiency and structural reconstruction capability.

3.3. Performance of Self-Assembled Foam Systems

3.3.1. Foam Performance

Static foam performance evaluation experiments were conducted in accordance with the *Experimental Evaluation Methods for Foaming Agents Used in Oil and Gas Fields*. The foam performance of different self-assembled systems combined with the foaming agent was evaluated, and the results are summarized in Table 2.

As shown in Table 2, the addition of self-assembled systems to the foaming agent does not result in a significant change in foam height. However, the foam half-life is markedly prolonged, leading to a substantial improvement in the comprehensive foam performance. These results indicate that the self-assembled systems exhibit strong foam-stabilizing ability and good compatibility with the foaming agent.

https://doi.org/10.3390/pr14030411

Table 2. Statistical table of performance evaluation results for self-assembled foam systems.

Foaming Agent/Self-Assembly System/Foam Stabilizer	Foam Height, mm	Half-Life of Water Analysis, s	Comprehensive Value
ZHDB-5/PMB-1/A-2 0.5%/0/0	810	381	308,610
ZHDB-5/PMB-1/A-2 0.5%/0.2%/0	850	355	301,750
ZHDB-5/PMB-2/A-2 0.5%/0.2%/0	830	382	317,060
ZHDB-5/HK/A-2 0.5%/0.2%/0	810	372	301,320
ZHDB-5/PMB-1/A-2 0.5%/0/0.05%	770	599	461,230
ZHDB-5/PMB-1/A-2 0.5%/0.2%/0.05%	750	582	436,500
ZHDB-5/PMB-2/A-2 0.5%/0.2%/0.05%	750	605	453,750
ZHDB-5/HK/A-2 0.5%/0.2%/0.05%	780	594	463,320

By comparing the foam properties of different self-assembled systems, the HK-based self-assembled foam system was selected for subsequent experiments.

3.3.2. Sealing Performance

The sealing performance of the self-assembled foam system was evaluated by comparing the HK single-agent system with the HK composite system, as shown in Figure 4. The results demonstrate that the sealing performance of the composite system is significantly enhanced. Specifically, the resistance coefficient increased from 1.3 to 3.4, indicating a robust plugging capability. Crucially, this magnitude is well within the safe operational limits of offshore injection wells. Unlike rigid plugging agents that often risk exceeding the formation fracture pressure, the observed resistance factor provides sufficient diversion force to modify the flow profile without causing hazardous pressure spikes. This moderate yet effective resistance confirms the system's field usability, ensuring that the injection pressure remains manageable and significantly below the safety threshold of the reservoir caprock.

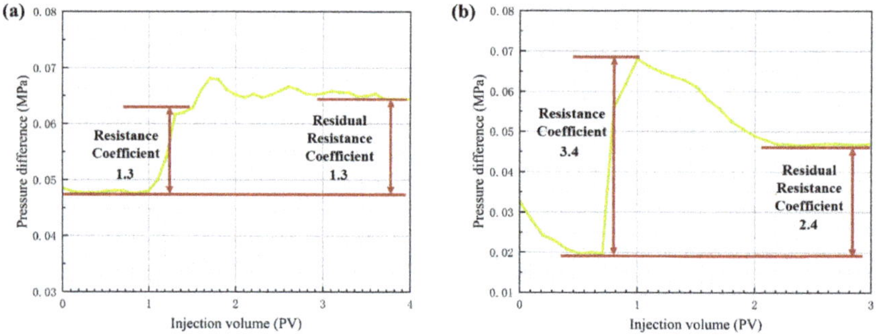

Figure 4. Sealing performance of the self-assembled foam system. (**a**) HK single-agent system; (**b**) HK composite system.

https://doi.org/10.3390/pr14030411

3.3.3. Pressure Reduction and Injectivity Performance

The pressure reduction and injectivity performance of the self-assembled foam system were evaluated by comparing the HK foam system with the HK composite system, as illustrated in Figure 5. Both systems exhibit pronounced pressure reduction and injectivity enhancement effects, with the injection pressure decreasing by approximately 0.002–0.003 MPa during the later stage of injection.

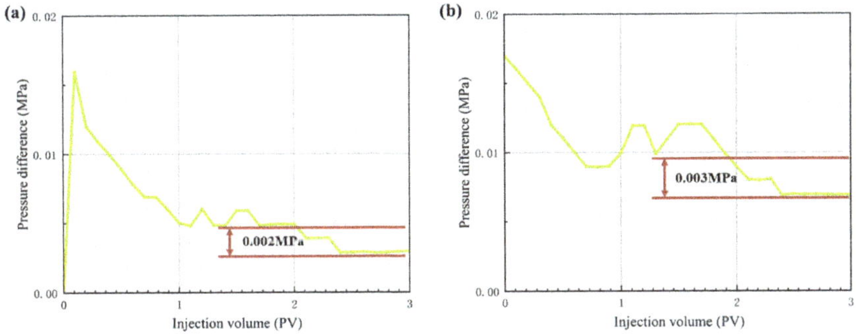

Figure 5. Pressure reduction and injectivity performance of the self-assembled foam system. (**a**) HK foam system; (**b**) HK composite system.

Furthermore, the addition of a foam stabilizer to the HK composite system not only enhances the early-stage sealing performance but also achieves effective pressure reduction and injectivity improvement. This behavior indicates that the composite system can provide a balanced combination of plugging and pressure relief during the injection process. This behavior reflects an inherent trade-off between plugging strength and pressure relief, which can be characterized by a plugging–injectivity balance window defined by the ratio of the resistance coefficient to the injection pressure response. The HK-based composite foam system operates within this favorable window, achieving sufficient flow diversion without inducing excessive pressure buildup, thereby supporting a generalized and quantitative design criterion for self-assembled foam systems.

3.3.4. Oil Displacement Performance

To validate the reliability of the core flooding data, each displacement experiment was repeated four times under identical conditions. The results presented in this section represent the arithmetic mean of these independent runs. The experimental error, calculated as the standard deviation of the final oil recovery, was found to be within 1.9%, demonstrating excellent reproducibility. The oil displacement performance of different systems was evaluated under a permeability condition of 3000 mD, and the statistical results are presented in Table 3. The "Enhanced recovery" is calculated as the difference between the "Final recovery rate" and the recovery after the initial water flooding stage. The average water flooding recovery for these core samples was approximately 62.0% $\pm$ 0.8%.

To further investigate the adaptability of the self-assembled composite foam system under different reservoir conditions, oil displacement experiments were conducted in both low-permeability (500 mD) and high-permeability (3000 mD) cores. The results are summarized in Table 4. As shown in Table 4, the self-assembled composite foam system exhibits permeability-dependent displacement behaviors, reflecting distinct dominant mechanisms. Under low-permeability conditions (500 mD), mobility control dominates, where deformable particles and stabilized foam lamellae penetrate tight pore–throat structures, reduce water mobility, and enhance microscopic sweep efficiency, resulting in a 22.5% improvement in oil recovery. In contrast, under high-permeability conditions (3000 mD),

https://doi.org/10.3390/pr14030411

selective plugging and flow diversion prevail. The composite particle–foam structure preferentially accumulates in dominant flow channels, increases local flow resistance, and redirects fluids into unswept regions, yielding a higher oil recovery enhancement of 27.2%. Overall, reservoir permeability governs the dominant displacement mechanism, transitioning from mobility control in low-permeability media to channel plugging in high-permeability formations.

Table 3. Oil displacement results of different systems at a permeability of 3000 mD.

Oil Displacement Effect	3000 mD-Foaming Agent	3000 mD-HK	3000 mD-HK + Foaming Agent
Water cut, %	61.9	62.7	62.6
Final recovery rate, %	73.2	73.3	89.8
Enhance recovery, %	11.4	10.5	27.2

Table 4. Oil displacement results of the self-assembled composite foam system under low- and high-permeability conditions.

Oil Displacement Effect	500 mD-HK + Foaming Agent	3000 mD-HK + Foaming Agent
Water cut, %	24.0	62.6
Final recovery rate, %	46.6	89.8
Enhance recovery, %	22.5	27.2

These results demonstrate that the self-assembled composite foam system is effective across a wide permeability range, highlighting its strong potential for enhanced oil recovery applications in heterogeneous reservoirs.

3.4. Mechanism of In Situ Self-Assembly and Composite Foam Stabilization

The enhanced profile control and oil displacement performance of the composite foam system originate from the in situ self-assembly of soft polymer particles and their synergistic interaction with the foam structure under reservoir conditions. Unlike conventional rigid particles, the polymer particles exhibit temperature- and salinity-induced softening and deformability, which enables deep migration into pore spaces and effective adaptation to complex pore–throat geometries. During injection, the particles are transported with the foaming solution into high-permeability channels, where partial softening facilitates subsequent structural evolution.

The in situ self-assembly process involves three coupled and sequentially overlapping mechanisms: particle softening, particle aggregation, and interfacial adsorption. Particle softening, initiated by reservoir temperature and salinity, reduces particle rigidity and primarily governs the feasibility of structure formation rather than its kinetics. As injection proceeds, particle aggregation driven by electrostatic interactions and polymer chain entanglement becomes the dominant kinetic process, controlling the rate and extent of dynamic network formation within pore spaces. Upon in situ foam generation, the self-assembled particle aggregates preferentially adsorb at gas–liquid interfaces and within foam lamellae, where they reinforce the liquid films and suppress drainage and coalescence.

Consequently, particle aggregation predominantly controls the kinetic behavior of structure formation, while particle softening serves as a prerequisite, and interfacial adsorption functions as a stabilization step. This kinetic hierarchy explains the delayed yet controllable development of the particle–foam composite structure, allowing effective deep transport during injection, followed by selective plugging and prolonged foam stabilization under reservoir conditions.

https://doi.org/10.3390/pr1403041

4. Application of Self-Assembled Foam Drive Technology

The KA3 and KA5 well groups in the offshore K oilfield are characterized by low permeability and uneven vertical water absorption, leading to severe interlayer flow conflicts. To address these challenges, a self-assembled composite foam drive technology was applied for profile control and water absorption regulation. Field results show that injection pressure remained stable during foam injection, indicating reliable injectivity and operational safety. After resuming conventional water injection, injection pressure decreased while the water absorption index increased, reflecting improved vertical sweep efficiency and flow redistribution. The application resulted in a notable enhancement in oil recovery, with an average reduction of approximately 5% in comprehensive water cut and a cumulative incremental oil production of about 5000 m^3 from the two wells. Compared with conventional foam treatments previously applied in the same block, the self-assembled composite foam system exhibited a longer effective period and superior water cut control. Moreover, the enhanced oil recovery accumulated steadily over time without rapid performance decay, and no abnormal pressure buildup or injectivity deterioration was observed, indicating a durable yet controllable modification of dominant flow channels. Overall, these field results confirm the robustness, scalability, and practical applicability of the self-assembled foam drive technology in offshore low-permeability heterogeneous reservoirs.

5. Conclusions

(1) This study introduces an in situ self-assembled particle-enhanced foam system tailored for profile control and enhanced oil recovery in offshore heterogeneous reservoirs. The core innovation lies in the use of soft polymer particles with temperature- and salinity-responsive self-assembly behavior, which fundamentally differs from conventional rigid particle or pre-formed foam systems. This design enables dynamic structural reconstruction within the reservoir, allowing for deep transport during injection followed by effective in situ structure formation.

(2) A mechanism-guided structure–function framework is established, clarifying how particle morphology, surface charge, and self-assembly characteristics collectively govern macroscopic engineering responses, including injectivity, flow diversion, and foam stabilization. The identification of an optimal electrostatic and morphological window provides a rational basis for formulation design, shifting particle–foam systems from empirical selection toward controllable and predictable performance optimization.

(3) From an engineering perspective, the proposed system demonstrates strong industrial applicability for offshore reservoirs characterized by severe heterogeneity and limited operational tolerance. Its ability to deliver moderate and controllable flow resistance, maintain injection safety, and exhibit temporal stability under field conditions highlights its robustness and scalability. These findings indicate that in situ self-assembled composite foam technology represents a promising and adaptable solution for integrated water control and enhanced oil recovery in offshore oilfields.

Author Contributions: Conceptualization: S.T. Data curation: M.J. and Y.X. Formal analysis: M.J., S.T. and Y.X. Investigation: S.T. and Y.X. Methodology: M.J. Project administration: M.J. Supervision: S.T. Writing—original draft: M.J. and Y.X. Writing—review & editing: S.T. All authors have read and agreed to the published version of the manuscript.

Funding: This research was funded by Basic Research on High-Temperature-Resistant Clean Fracturing Fluid Constructed from Anionic Gemini Surfactant-Nanoparticles. Grant Number 51474035.

Data Availability Statement: The original contributions presented in this study are included in the article. Further inquiries can be directed to the corresponding authors.

https://doi.org/10.3390/pr14030411

Conflicts of Interest: The authors declare no conflict of interest.

References

1. Zhao, W.; Wang, Z.; Li, J.; Li, J.; Xie, Z.; Wang, Z. Natural gas resources of the sedimentary basins in China. *Mar. Pet. Geol.* **2008**, *25*, 309–319. [CrossRef]
2. Zhao, R.; Hynes, S.; He, G.S. Defining and quantifying China's ocean economy. *Mar. Policy* **2014**, *43*, 164–173. [CrossRef] [PubMed]
3. Zheng, M.; Li, J.; Wu, X.; Wang, S.; Guo, Q.; Yu, J.; Zheng, M.; Chen, N.; Yi, Q. China's conventional and unconventional natural gas resources: Potential and exploration targets. *J. Nat. Gas Geosci.* **2018**, *3*, 295–309. [CrossRef]
4. Zhang, Q.; Liu, J.F.; Gao, Z.H.; Chen, S.-Y.; Liu, B.-Y. Review on the challenges and strategies in oil and gas industry's transition towards carbon neutrality in China. *Pet. Sci.* **2023**, *20*, 3931–3944. [CrossRef]
5. Wang, Q. Offshore Wind Power: An Important Opportunity for Traditional Oil and Gas Industry to Realize Low-Carbon Transformation. In *Annual Report on China's Petroleum, Gas and New Energy Industry (2021)*; Springer Nature: Singapore, 2022; pp. 231–245.
6. Xiao, P. Experimental Investigation on CO_2 Injection in the Daqing Extra/Ultra-Low Permeability Reservoir. *J. Pet. Sci. Eng.* **2017**, *149*, 765–771. [CrossRef]
7. Lai, W. Gas exploration potential in Bohai Sea. *China Offshore Oil Gas* **2000**, *14*, 168–173.
8. Jiang, Y.; He, S.; Lu, X. The distribution characteristics of natural gas and its controlling factors in Bohai Bay Basin. In Proceedings of the Harbin: Seminar on Medium-Deep Gas Exploration Technology, Harbin, China, 1–3 August 2005.
9. Lao, J. Micro/Nanoparticle Characteristics and Flow in Porous Media: A Review towards Enhanced Oil Recovery. *Energies* **2024**, *17*, 4136. [CrossRef]
10. Xu, C.; Yu, H.; Jun, W.; Liu, X. Formation conditions and accumulation characteristics of Bozhong 19-6 large condensate gas field in offshore Bohai Bay Basin. *Pet. Explor. Dev.* **2019**, *46*, 27–40. [CrossRef]
11. Liu, Y.; Ge, L.; Ma, K.; Chen, X.; Zhu, Z.; Hou, J. Study on surfactant–polymer flooding after polymer flooding in high-permeability heterogeneous offshore oilfields: A case study of Bohai S Oilfield. *Polymers* **2024**, *16*, 2004. [CrossRef] [PubMed]
12. Kang, X.; Zhang, J.; Sun, F.; Zhang, F.; Feng, G.; Yang, J.; Zhang, X.; Xiang, W. A review of polymer EOR on offshore heavy oil field in Bohai Bay, China. In Proceedings of the SPE Asia Pacific Enhanced Oil Recovery Conference, Kuala Lumpur, Malaysia, 19–21 July 2011; SPE-144932-MS.
13. Guo, Y.; Zhang, J.; Liu, Y.; Liu, G.; Xue, X.; Luo, P.; Ye, Z.; Zhang, X.; Liang, Y. Successful scale application of associative polymer flooding for offshore heavy oilfield in Bohai Bay of China. In Proceedings of the SPE Asia Pacific Oil and Gas Conference and Exhibition, Virtual, 17–19 November 2020; D012S003R010.
14. Feng, Q.; Zhou, J.; Li, S.; Chen, X.; Sun, Y.; Zhang, X.; Gao, P.; Zhang, F.; She, Y. Research on characterization technology and field test of biological nano-oil displacement in offshore medium-and low-permeability reservoirs. *ACS Omega* **2022**, *7*, 40132–40144. [CrossRef] [PubMed]
15. Liu, Y.; Zou, C.; Zhou, D.; Li, H.; Gao, M.; Zhao, G.; Dai, C. Novel chemical flooding system based on dispersed particle gel coupling in-depth profile control and high efficient oil displacement. *Energy Fuels* **2019**, *33*, 3123–3132. [CrossRef]

https://doi.org/10.3390/pr14030411

processes

MDPI

rticle

Geochemical Characteristics and Genetic Origin of Tight Sandstone Gas in the Daning–Jixian Block, Ordos Basin

o Wang [1,*], Ming Chen [2,3], Haonian Tian [3], Junyi Sun [3], Lei Liu [4,5], Xing Liang [6], Benliang Chen [7], Baoshi Yu [6], huo Zhang [6] and Zhenghui Qu [4,5]

1 Information Institute of the Ministry of Emergency Management of PRC, Beijing 100029, China
2 College of Petroleum Engineering, China University of Petroleum, Beijing 102249, China; chming@petrochina.com.cn
3 PetroChina Coalbed Methane Company Limited Linfen Branch, Linfen 042202, China; yy424512@petrochina.com.cn (H.T.); sjy1985@petrochina.com.cn (J.S.)
4 Key Laboratory of Coalbed Methane Resources and Reservoir Formation Process, Ministry of Education, China University of Mining and Technology, Xuzhou 221008, China; 15266286132@163.com (L.L.); quzhenghui@cumt.edu.cn (Z.Q.)
5 School of Resources and Geosciences, China University of Mining and Technology, Xuzhou 221116, China
6 PetroChina Zhejiang Oilfield Company, Hangzhou 311100, China; liangx85@petrochina.com.cn (X.L.); yubs85@petrochina.com.cn (B.Y.); zhangz85@petrochina.com.cn (Z.Z.)
7 Huainan Mining (Group) Co., Ltd., Huainan 232001, China; 18155456996@163.com
* Correspondence: wangbo1230038@163.com

Abstract: Tight sandstone gas constitutes a strategically significant resource in the exploration of unconventional hydrocarbon systems. Current understanding of the geochemical composition and genesis of tight sandstone gas in the Daning–Jixian Block, southeastern Ordos Basin, is insufficient, which hampers a comprehensive assessment of its resource potential. This study is the first to systematically investigate the geochemical characteristics and genetic origin of high-maturity tight sandstone gas in the southeastern Ordos Basin's Daning–Jixian Block. Gas specimens were systematically acquired from multiple stratigraphic units within the reservoir interval and subjected to compositional and carbon–hydrogen isotope analysis. Compared with other gas fields in the Ordos Basin, the Daning–Jixian Block has higher average methane concentration, and notably lower ethane and propane concentrations; its average $\delta^{13}C_1$ and $\delta^2H\text{-}CH_4$ is heavier, while $\delta^{13}C_2$ and $\delta^{13}C_3$ are lighter. Based on the $\delta^{13}C_1$-$\delta^2H\text{-}CH_4$ diagram, all gas samples from the block and other basin gas fields fall into the geothermal, hydrothermal and crystalline gas domain, indicating gas genesis associated with over-mature organic matter interacting with external hydrogen. Milkov genetic diagram analysis reveals that the natural gas consists of primarily early-stage kerogen-cracking gas, with a minor contribution from crude oil-derived gas originating from Carboniferous–Permian source rocks. Notably, samples from Daning–Jixian exhibit a unique $\delta^{13}C_1 > \delta^{13}C_2$ reversal, attributed to mixing effects between gas from highly mature kerogen and gas from secondary cracking of crude oil. Consequently, ethane carbon isotopes alone are insufficient for definitive genetic classification. These findings provide a new geochemical interpretation framework for analogous high-maturity tight gas reservoirs.

Keywords: Daning-Jixian Block; tight sandstone gas; geochemistry; genetic; origin

1. Introduction

The growing global demand for clean energy has positioned unconventional energy resources as a focal point in energy research [1]. Tight sandstone gas, a subtype of tight gas and a significant component of unconventional natural gas resources, predominantly occurs in sandstone reservoirs characterized by permeability ranging from low to ultra-low levels [2]. Due to the inherently tight geological characteristics of such reservoirs, natural gas production from individual wells is generally low without artificial stimulation. Consequently, large-scale hydraulic fracturing and reservoir engineering techniques are essential to achieve economically viable gas extraction [3–5]. Tight sandstone gas is widely distributed across numerous petroliferous basins worldwide with abundant resources [6], serving as a significant replacement for conventional oil and gas energy. The Ordos Basin, recognized as a pivotal hydrocarbon basin in China, hosts extensive tight sandstone gas reserves with significant exploitation potential [7], particularly in the Daning–Jixian area, which has attracted widespread attention and in-depth investigation [8,9].

In 2014, gas tests conducted in the sandstone reservoirs within the Shanxi Formation of the Daning–Jixian Block yielded remarkable production rates, marking a breakthrough in the prospecting and exploitation of coal-derived tight gas reservoirs within this region [10]. The Shan 2^3 gas layer recorded an initial average daily gas production of 23,469.23 m^3/d in the first year [11]. Previous studies in the Daning–Jixian area have primarily focused on petrological characteristics, pore structure, reservoir physical properties, and their controlling factors [12–15], providing crucial insights for reservoir characterization and sweet-spot prediction.

The formation of tight-sandstone gas accumulations is controlled by a suite of interrelated factors, encompassing the organic-rich source rocks' hydrocarbon generation potential, the diagenetic compaction and cementation processes that reduce reservoir permeability, and mechanisms of gas migration and accumulation [16–18]. Diverse source rocks—including marine and terrestrial organic-rich mudstones and shales, as well as coal seams—provide abundant gas sources for tight sandstone gas systems. Although substantial work has focused on the geochemistry and origins of tight gas in the Ordos Basin (e.g., in the Zizhou, Mizhi, Yulin, Sulige, Wushenqi, and Shixi gas fields) [19,20], most studies have concentrated on the central-northern basin sectors, resulting in few published studies on the geochemical mechanisms in the southeastern Daning–Jixian Block. Comprehensive investigation of geochemical signatures of tight sandstone gas can elucidate its formation mechanisms and provide guidance for future exploration and development in the Daning–Jixian area. The novelty lies in providing the first comprehensive genetic interpretation of high-maturity tight sandstone gas in this block, offering a scalable reference for similar highly mature to over-mature coal-measure tight gas systems.

2. Geological Setting

The Daning–Jixian Block is situated on the eastern margin of the Ordos Basin. The Upper Paleozoic stratigraphic sequence in this area consists, in ascending stratigraphic order, of the Carboniferous Benxi Formation, followed by the Carboniferous-Permian Taiyuan Formation, and the Permian Shanxi, Shihezi, and Shiqianfeng Formations, with a cumulative sedimentary thickness of approximately 800 m [21].

In the study area, the Upper Paleozoic source rocks are classified as coal-bearing successions deposited in marine–terrestrial transitional settings, dominated by Carboniferous–Permian coal seams, carbonaceous mudstones, and dark mudstones [22]. The principal gas-bearing intervals for tight sandstone gas include the Benxi and Taiyuan Formations, along with the Shan2, Shan1, and He8 Members. The reservoirs are buried at depths of 1745–2680 m, with measured permeability and porosity averaging 0.46 mD and 6.9% [23].

3. Sampling and Analytical Methods

Distinct well sites within the study area exhibit variability in their production layers. Natural gas samples were harvested at the wellhead and subsequently submitted to Geological Experiment Testing Center of Xi'an Shiyou University for determinations of gas composition and C-H isotope ratios. The natural gas samples collected at the wellhead were pure gas that had undergone prior gas–water separation at the production facility, ensuring the absence of formation water, free water, or condensate. Since the samples were already free of moisture and liquid hydrocarbons, no additional drying process was required. Gas composition was ascertained via an Agilent 7890B gas chromatograph in accordance with the Chinese National Standard "Analysis of Natural Gas Composition—Gas Chromatography" (GB/T 13610-2020) [24]. The analytical precision for CH_4 was determined to be $\pm 0.5\%$, ensuring the accuracy and reliability of the compositional analysis results. Carbon and hydrogen isotopes were analyzed using an isotope ratio mass spectrometer (Isoprime VISION), adhering to the petroleum and natural gas industry standards "Analysis Method for Carbon and Oxygen Isotopes in Organic Matter and Carbonate" (SY/T 5238-2019) [25,26] and "Analysis of Hydrogen Isotope in Petroleum, Rock Extracts, and Kerogen"(SY/T 7313-2016) [27]. For carbon isotope determination, a continuous flow analysis approach was employed. Carbon and hydrogen isotope values are expressed in per mil (‰), referenced against the VPDB and VSMOW international standards, respectively. Analytical results are systematically compiled in Tables 1 and 2 [28].

Table 1. Gas composition testing results (unit: %).

Formation	Type	CO_2	N_2	CH_4	C_2H_6	C_3H_8	i-C_4H_{10}	n-C_4H_{10}	i-C_5H_{12}	n-C_5H_{12}	C_{2+}	$C_{1\text{–}5}$	Dryness Index
Benxi Formation	min	0.15	0.00	93.26	0.09	0.01	0.00	0.00	0.00	0.00	0.12	93.38	99.32
	max	6.62	0.61	98.86	0.62	0.03	0.02	0.00	0.00	0.00	0.65	99.23	99.87
	aver	3.10	0.20	96.32	0.37	0.02	0.00	0.00	0.00	0.00	0.39	96.71	99.60
Taiyuan Formation	min	1.52	0.00	97.51	0.20	0.01	0.00	0.00	0.00	0.00	0.21	97.76	99.63
	max	2.24	0.28	98.06	0.34	0.02	0.00	0.00	0.00	0.00	0.36	98.39	99.79
	aver	1.85	0.08	97.78	0.28	0.02	0.00	0.00	0.00	0.00	0.30	98.07	99.70
Shan2 Member	min	1.28	0.00	97.33	0.22	0.01	0.00	0.00	0.00	0.00	0.23	97.61	99.24
	max	2.15	0.44	98.05	0.71	0.04	0.00	0.00	0.00	0.00	0.75	98.28	99.77
	aver	1.83	0.14	97.54	0.47	0.02	0.00	0.00	0.00	0.00	0.49	98.03	99.50
Shan1 Member	min	0.48	0.00	97.64	0.30	0.01	0.00	0.00	0.00	0.00	0.31	98.08	99.11
	max	1.64	0.70	98.45	0.79	0.06	0.01	0.01	0.01	0.00	0.88	99.00	99.68
	aver	1.03	0.39	98.01	0.53	0.04	0.00	0.00	0.00	0.00	0.58	98.58	99.42
He8 Member	min	0.61	0.29	97.04	0.31	0.02	0.00	0.00	0.00	0.00	0.33	97.37	99.08
	max	2.32	1.48	97.90	0.84	0.06	0.01	0.01	0.00	0.00	0.91	98.68	99.66
	aver	1.17	0.69	97.44	0.65	0.05	0.00	0.01	0.00	0.00	0.70	98.14	99.29
No. 8 Coal	min	1.22	0.06	93.91	0.05	0.00	0.00	0.00	0.00	0.00	0.05	93.98	100
	max	5.92	0.20	97.94	0.43	0.02	0.00	0.00	0.00	0.00	0.45	98.39	100
	aver	4.74	0.10	95.00	0.11	0.00	0.00	0.00	0.00	0.00	0.11	95.11	100

Table 2. Gas isotope testing results.

Formation	Type	δ ¹³C (‰, VPDB)						δ²H (‰, VSMOW)
		CH_4	C_2H_6	C_3H_8	i-C_4H_{10}	n-C_4H_{10}	CO_2	CH_4
Benxi Formation	min	−29.30	−36.70	−39.20	−34.50	−38.40	−2.70	−161.80
	max	−27.70	−29.30	−31.70	−31.50	−28.70	5.20	−169.70
	aver	−28.40	−34.23	−36.17	−33.50	−34.95	1.85	−165.13
Taiyuan Formation	min	−29.30	−35.50	−32.90	−39.50	−27.30	−4.50	−165.80
	max	−28.30	−32.80	−30.60	−22.50	−23.00	1.30	−169.50
	aver	−28.78	−34.36	−32.00	−29.90	−25.10	−0.58	−167.54
Shan2 Member	min	−29.20	−36.70	−37.70	−42.90	−35.60	−4.70	−165.80
	max	−27.20	−31.80	−31.80	−22.50	−22.40	−1.20	−170.90
	aver	−28.49	−35.44	−35.97	−33.16	−29.60	−2.71	−168.16
Shan1 Member	min	−28.70	−30.90	−31.40	−30.00	−29.50	−7.80	−166.40
	max	−26.20	−28.60	−27.90	−15.70	−24.60	−2.70	−169.60
	aver	−27.70	−30.13	−29.88	−24.75	−26.93	−5.00	−168.40
He8 Member	min	−27.90	−32.80	−33.80	−29.30	−34.40	−13.30	−166.10
	max	−26.50	−31.30	−30.60	−22.90	−20.50	−0.50	−168.20
	aver	−27.36	−31.96	−32.18	−27.25	−30.04	−5.44	−167.23

4. Discussion

4.1. Natural Gas Composition Characteristics

In the tight sandstone gas reservoirs from the Benxi Formation to the He8 Member in the Daning–Jixian Block, the methane (CH_4) content consistently exceeds 93%, with an average value higher than the 95% methane content reported for coalbed methane from the No. 8 coal seam of the Taiyuan Formation in this area. The ethane (C_2H_6) concentration ranges between 0.09% and 0.84%, while propane (C_3H_8) levels vary from 0.01% to 0.06%. The ethane-to-propane ratio falls within 9.00–30.00, and the dryness coefficient ($C_1/\sum C_1$–C_3) is universally above 99%, classifying the gas as dry gas. Comprehensive analysis of methane concentration, dryness coefficient, and the ethane/propane ratio indicates that the natural gas from the Daning–Jixian Block shares characteristics with the Daniudi gas field (Ordos Basin) and is classified as coal-derived cracked gas (Table 1) [28,29].

Comparative assessment with other gas fields across the Ordos Basin indicates that the Daning–Jixian Block exhibits a comparatively higher average methane concentration, whereas the Linxing and Sulige gas fields demonstrate relatively lower methane levels. The average ethane and propane concentrations in the Daning–Jixian Block are notably lower than those in other gas fields; specifically, the Mizhi and Sulige gas fields register higher ethane concentrations, and the Linxing block shows elevated propane levels. Furthermore, the dryness coefficient of the gas in the Daning–Jixian Block is significantly greater than that observed in other gas fields. The Shixi Block, owing to its geographical proximity to the Daning–Jixian Block, displays analogous profiles in hydrocarbon gas concentrations [19,20,28–31] (Figure 1).

Based on reservoir characterization investigations conducted in the Yanchang exploration block of the southeastern Ordos Basin by Zhou et al. [32], dark mudstones and coal measures in the Benxi Formation and Shan2 Member serve as the primary source rocks. The Benxi Formation and Shan2 Member of the Daning–Jixian Block belong to source-in situ accumulation, whereas Shan1 Member and He8 Member, situated above the main source rocks, represent near-source accumulation with primarily vertical charging through secondary migration. Along the migration pathway in coal-measure systems where CO_2 is consumed by methanogenesis or mineral reactions, N_2 concentration tends to increase while CO_2 concentration decreases [30]. The generally lower CO_2 and higher N_2 content observed in He8 Member and Shan1 Member suggest vertical migration of natural gas into these intervals (Figure 2) [33].

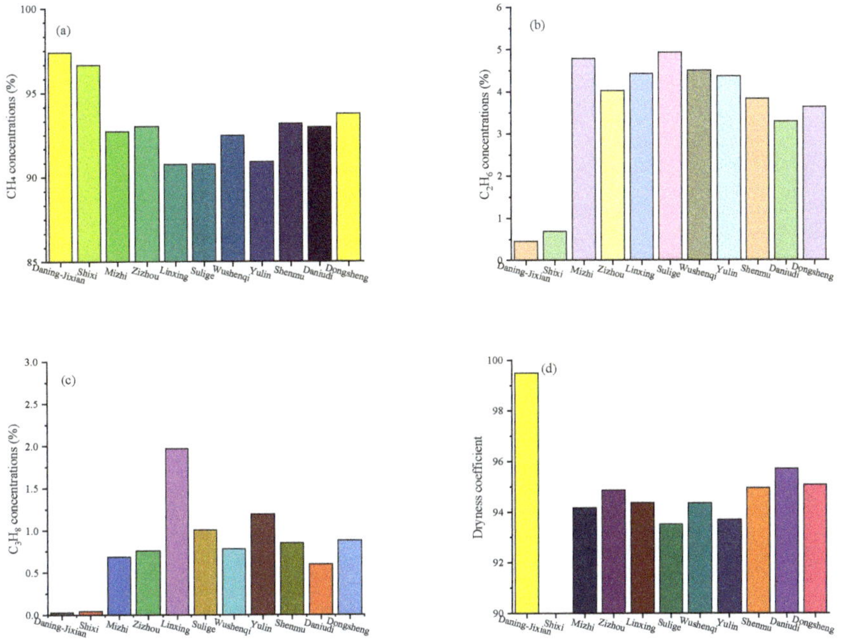

Figure 1. Comparative analysis of average hydrocarbon gas parameters between the Daning–Jixian Block and other gas fields. (**a**) Comparative analysis of average CH_4 concentrations; (**b**) comparative analysis of average C_2H_4 concentrations; (**c**) comparative analysis of average C_3H_8 concentrations; (**d**) comparative analysis of average dryness coefficient.

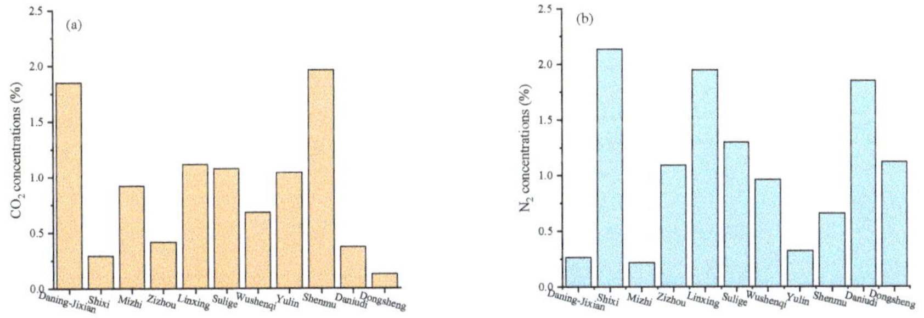

Figure 2. Comparative analysis of average non-hydrocarbon gas parameters between the Daning–Jixian Block and other gas fields. (**a**) Comparative analysis of average CO_2 concentrations; (**b**) comparative analysis of average N_2 concentrations.

4.2. Stable Isotopic Characteristics

In the tight sandstone gas reservoirs of the Daning–Jixian Block, the content of propane, and especially butane and heavier hydrocarbons, is very low. Consequently, the measured stable carbon isotope values for these components are for reference only (Table 2). The natural gas in the Upper Paleozoic tight sandstone gas reservoirs of the Daning–Jixian Block exhibits average $\delta^{13}C_1$, $\delta^{13}C_2$, and $\delta^{13}C_3$ values of −28.20‰, −33.64‰, and −33.77‰, respectively. The average $\delta^{13}C_1$ value is heavier compared to other gas fields, while the average $\delta^{13}C_2$ and $\delta^{13}C_3$ values are significantly lighter [19,20,28–31] (Figure 3a–c). In contrast, coalbed methane (CBM) in this region exhibits significantly lighter $\delta^{13}C_1$ values, spanning −72.35‰ to −27.43‰, with mean values of −40.53‰ (Coal Seam No. 5) and −41.88‰ (Coal Seam No. 8) [34]. During the over-mature stage (Ro > 2.0%), organic matter mainly generates methane, also known as dry gas [35]. Due to the very high maturity and the dry composition of the natural gas in the study area, hydrogen isotopes were only detected for methane, with values primarily ranging from −161.8‰ to −170.9‰ and an average of −167.34‰. The abundance of other components is too low for reliable hydrogen isotope analysis; any detected values are for reference only. The average δ^2H-CH_4 value is heavier than that of other gas fields [19,20,28–31] (Figure 3d). Theoretically, methane carbon isotope values decrease along the gas migration pathway. However, the relatively heavier isotopic values observed in He8 Member and Shan1 Member compared to other layers suggest a complex scenario involving both dynamic fractionation and migration fractionation effects [28,30].

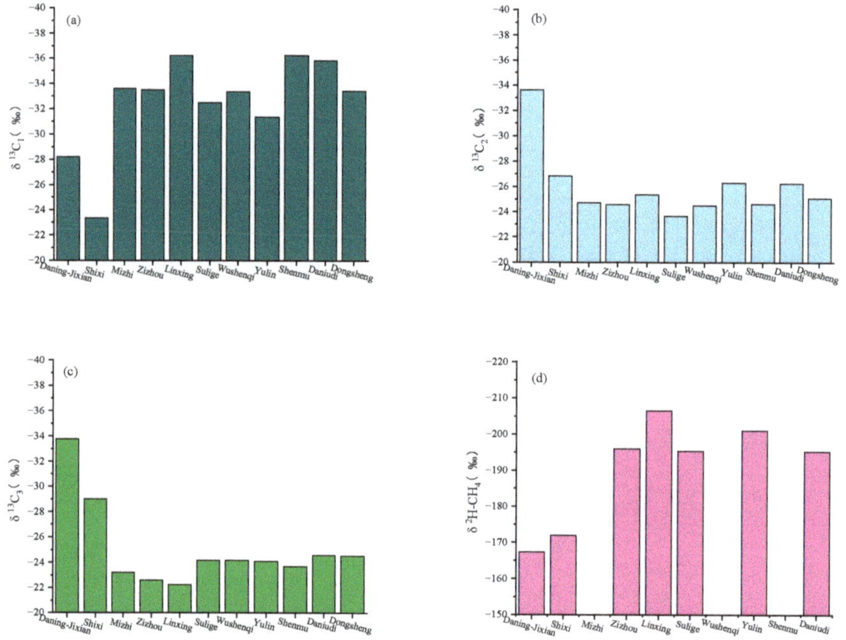

Figure 3. Characteristics of stable carbon isotopes of hydrocarbon components in natural gas from tight gas reservoirs in the Daning–Jixian Block and other gas fields. (**a**) Comparative analysis of average $\delta^{13}C_1$; (**b**) comparative analysis of average $\delta^{13}C_2$; (**c**) comparative analysis of average $\delta^{13}C_3$; (**d**) comparative analysis of average δ^2H-CH_4.

4.3. Analysis of Origin and Genetic

Based on the analysis of natural gas compositional data and hydrocarbon carbon and hydrogen isotopic compositions, combined with classical diagnostic diagrams, the genetic origin of natural gas can be discriminated and interpreted. Building upon previous research, this study further conducts a targeted analysis by employing diagnostic diagrams specifically applicable to the study area, aiming to provide more accurate insights into the genesis and origin of natural gas in the Daning–Jixian Block.

4.3.1. Genetic Analysis

The carbon isotope diagram ($\delta^{13}C_1$–$\delta^{13}C_2$–$\delta^{13}C_3$) established by Dai Jinxing et al. [36] was utilized for comparative analysis of the Daning–Jixian area, and other gas fields in the Ordos Basin revealed distinct characteristics in the Daning–Jixian Block. The natural gas samples from this block did not plot within any defined genetic identification zones, yielding inconclusive results [28]. Furthermore, on the diagrams based on $\delta^{13}C_1$ and $C_1/(C_2 + C_3)$ proposed by Bernard et al. [37] and Whiticar [38], the natural gas from the Daning–Jixian Block is plotted within or adjacent to the high-maturity thermogenic gas domain characteristic of Type III kerogen [28]. Consequently, the application of these templates for genetic identification of the tight sandstone gas in the Daning–Jixian Block also exhibits limitations.

In the present study, the classical Whiticar diagram based on $\delta^{13}C_1$ and δ^2H-CH_4 was employed [38]. Results demonstrate that all gas samples from the Daning–Jixian Block and other gas fields across the Ordos Basin fall within the domain of geothermal, hydrothermal, and crystalline gases. This suggests that the gas genesis is related to the interaction of over-mature organic matter with external hydrogen [20] (Figure 4).

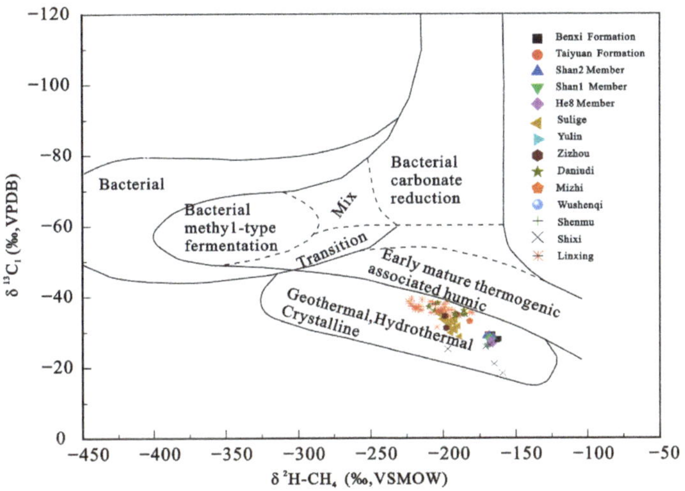

Figure 4. Identification diagram based on $\delta^{13}C_1$ and δ^2H-CH_4.

Milkov et al. [39] further revised previous genetic identification diagrams using a global dataset of 20,621 natural gas samples, encompassing 76 countries and regions across six continental landmasses. On the Milkov diagram, the gas samples from the study area entirely fall within the field of thermogenic gas with relatively high maturity (Figure 5a). The gas samples from other areas in the Ordos Basin demonstrate distribution within the thermogenic gas domain on the Milkov diagram, exhibiting comparatively lower thermal maturity. Notably, samples from fields like Daniudi and Zizhou also exhibit characteristics associated with oil-associated gas. The genetic identification diagram based on methane carbon and hydrogen isotopes also shows that the tight sandstone gas samples from Ordos Basin belong to thermogenic gas, with the Daning–Jixian Block exhibiting higher maturity (Figure 5b).

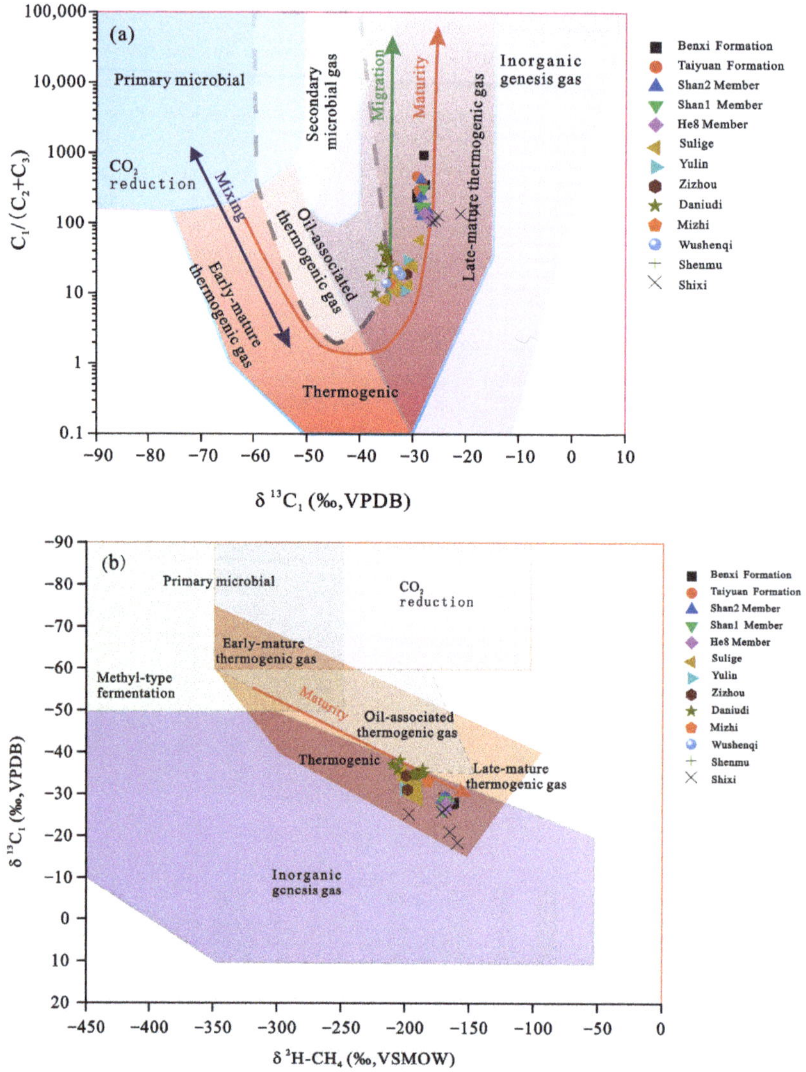

Figure 5. Genetic characterization of natural gas based on Milkov's diagram. (**a**) Plot of $\delta^{13}C_1$ versus $C_1/(C_2 + C_3)$; (**b**) plot of $\delta^{2}H\text{-}CH_4$ versus $\delta^{13}C_1$.

4.3.2. Origin Analysis

The primary cracking of kerogen marks the stage of abundant methane generation, leading to a rapid increase in the methane-to-ethane concentration ratio. In contrast, crude oil-cracked gas shows a significant increase in propane content. Therefore, the cross-plot of $\ln(C_1/C_2)$ versus $\ln(C_2/C_3)$ is often used as a key indicator for genetic characterization of natural gas [40–42]. The natural gas from the Daning–Jixian Block demonstrates markedly elevated $\ln(C_1/C_2)$ and $\ln(C_2/C_3)$ relative to other gas fields, a phenomenon attributed to its advanced thermal evolution stage. The variation range of $\ln(C_2/C_3)$ differs among formations, with the Benxi Formation and Shan1 Member showing a larger range, suggesting that the tight gas accumulation was predominantly sourced from the primary cracking of kerogen, with a potential minor contribution from crude oil-cracked gas (Figure 6).

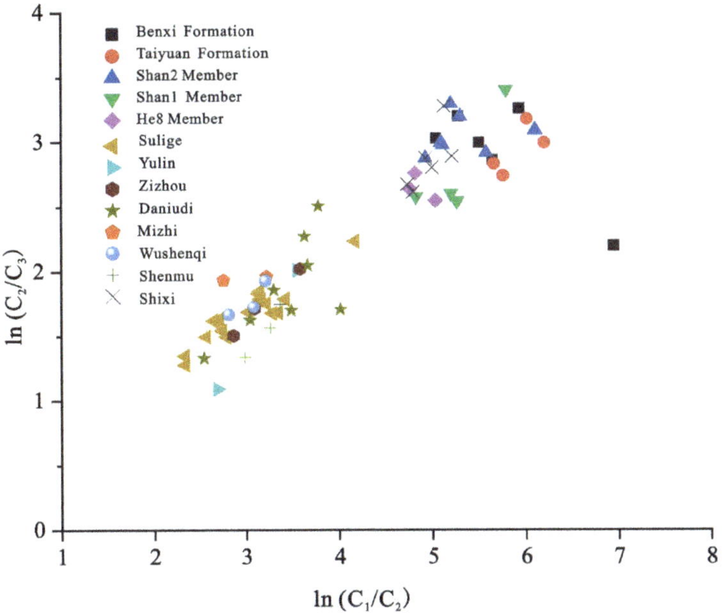

Figure 6. Identification diagram based on $\ln(C_1/C_2)$ versus $\ln(C_2/C_3)$.

Previous research results indicate the presence of carbonaceous asphalt inclusions in the study area, interpreted as thermally mature residues from crude oil cracking during the terminal Early Cretaceous. During this period, the paleo-temperature in the range of 160–200 °C caused the complete transformation of crude oil into cracked gas and carbonaceous asphalt [43]. Geochemical cross-plots of samples from the study area show an intermediate position, suggesting contribution from both Type II and Type III kerogen-derived gases [28]. This finding is consistent with research results from the Zhidan–Ganquan area [42]. In summary, the Daning–Jixian Block gas accumulations originate from early-stage kerogen cracking in Carboniferous–Permian source rocks, with a subordinate contribution from crude-oil cracking.

4.4. Stable Carbon Isotope Series

Previous studies have shown that a reversed sequence of primary stable carbon isotopes and $\delta^{13}C_1$ values greater than −30‰ are key indicators for identifying inorganic genesis gas [44–46]. The distribution patterns of among light hydrocarbons (C_1–C_3) can be influenced by various secondary processes such as mixing, high temperatures, bacterial oxidation, and diffusion [47]. Significant variations in carbon isotopic compositions are observed across different strata in the Daning–Jixian Block. Gases from the Benxi Formation and the Shan2 Member predominantly exhibit a fully reversed sequence ($\delta^{13}C_1 > \delta^{13}C_2 > \delta^{13}C_3$). In contrast, samples from the Taiyuan Formation, along with some from the Shan 1 Member and He 8 Member, display a partially reversed sequence ($\delta^{13}C_1 > \delta^{13}C_2 < \delta^{13}C_3$). Compared to other gas fields in the Ordos Basin, the Upper Paleozoic natural gases from the Dongsheng, Linxing, Daniudi, Yulin, Zizhou, and Mizhi gas fields, located in the northern and eastern parts of the basin, primarily show a normal carbon isotopic sequence ($\delta^{13}C_1 < \delta^{13}C_2 < \delta^{13}C_3$). A partially reversed sequence ($\delta^{13}C_1 < \delta^{13}C_2 > \delta^{13}C_3$) is observed in some samples from the central basin areas, including the Sulige and Wushenqi gas fields, as well as parts of the Shixi Block [19,20,28–31].

This comparative analysis reveals that the Daning–Jixian Block in the southeastern basin exhibits characteristics distinct from other gas fields. All samples from this block display a $\delta^{13}C_1 > \delta^{13}C_2$ reversal. Furthermore, the difference between $\delta^{13}C_2$ and $\delta^{13}C_3$ values is generally small, whereas a more substantial isotopic discrepancy exists between $\delta^{13}C_1$ and the heavier homologues ($\delta^{13}C_2$, $\delta^{13}C_3$) (Figure 7).

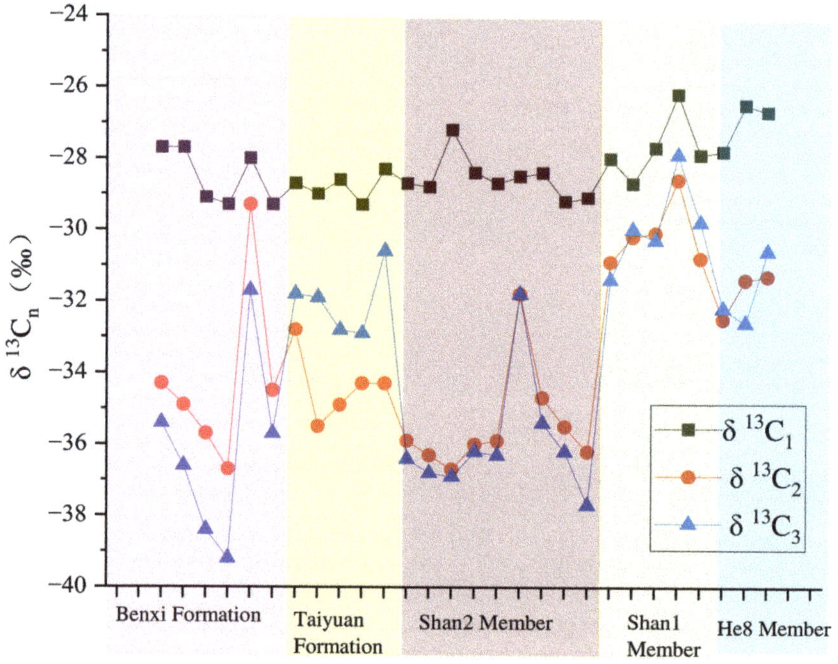

Figure 7. Comparison of $\delta^{13}C_1$, $\delta^{13}C_2$, and $\delta^{13}C_3$ values of natural gases from different strata in the Daning–Jixian Block.

A systematic southward increase in source rock maturity is observed within the Ordos Basin. Consequently, the carbon isotope distribution pattern of methane and its homologues evolves gradually from a positive carbon isotope sequence in the north, to a partially reversed sequence, and finally to a fully reversed sequence (i.e., a negative carbon isotope sequence) [47,48]. This trend can be attributed to the commingling of natural gases with distinct genetic origin [49]. Geochemically, the southeastern Ordos Basin features source rocks enriched in Type III kerogen, which is currently within the high maturity to over-mature thermal evolution stage, resulting in significant dry gas generation [22]. Therefore, based on an integrated assessment, the reversal of stable carbon isotopes observed in the study area's multi-stratigraphic natural gases is interpreted to be associated with the elevated thermal maturity of the kerogen and mixing with crude oil-cracking gas.

The ethane carbon isotope signature ($\delta^{13}C_2$) serves as a key proxy for discriminating between coal-derived gas and oil-derived gas, employing a widely recognized benchmark wherein values heavier than −28‰ are diagnostic of the former, and values lighter than −28‰ suggest the latter [47]. However, all $\delta^{13}C_2$ values in the Daning–Jixian Block are less than −28‰ (Figure 8). This is attributed to the exceptionally high thermal maturity, which has caused reversal of the ethane carbon isotope values, leading to the lighter ethane carbon isotope signatures observed [29]. This aligns with the findings of Peng et al. [50], who reported that both coal-derived gas and oil-derived gas exhibit alkane gas carbon isotope sequence reversal during the high-maturity-to-over-mature stage. Consequently, the ethane carbon isotope value alone cannot be reliably used for delineation of the natural gas provenance in the study area.

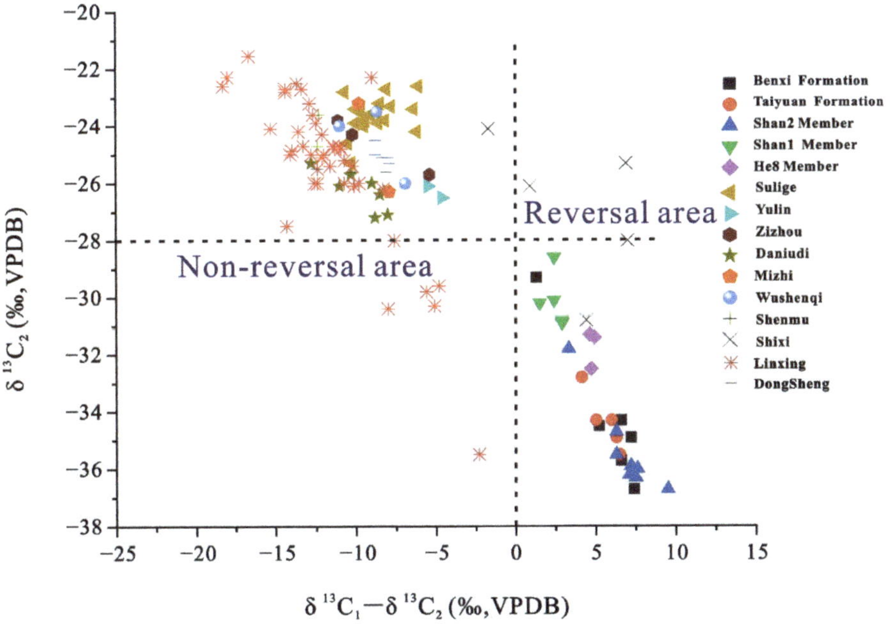

Figure 8. Correlation between $\delta^{13}C_1$–$\delta^{13}C_2$ and $\delta^{13}C_2$ for natural gas in the Daning–Jixian Block and comparison with typical tight sandstone gas fields in the Ordos Basin.

5. Conclusions

This study systematically investigates the geochemical characteristics and genetic mechanisms of natural gas in the Paleozoic sandstone reservoirs of the Daning–Jixian Block, southern Ordos Basin, based on molecular composition and stable isotope ($\delta^{13}C_{1\text{-}3}$, $\delta D\text{-}CH_4$) data. The key conclusions are as follows:

(1) Comparative assessment with other gas fields across the Ordos Basin indicates that the Daning–Jixian Block exhibits a comparatively higher average methane concentration. The average ethane and propane concentrations in the Daning–Jixian Block are notably lower than those in other gas fields. Furthermore, the dryness coefficient of the gas in the Daning–Jixian Block is significantly greater than that observed in other gas fields. The generally lower CO_2 and higher N_2 content observed in He8 Member and Shan1 Member suggest vertical migration of natural gas into these intervals.

(2) The natural gas in the Upper Paleozoic tight sandstone gas reservoirs of the Daning–Jixian Block exhibits average $\delta^{13}C_1$, $\delta^{13}C_2$, and $\delta^{13}C_3$ values of −28.20‰, −33.64‰, and −33.77‰, respectively. The average $\delta^{13}C_1$ value is heavier compared to other gas fields, while the average $\delta^{13}C_2$ and $\delta^{13}C_3$ values are significantly lighter. The average $\delta^2H\text{-}CH_4$ value is heavier than that of other gas fields. The relatively heavier isotopic values observed in He8 Member and Shan1 Member compared to other layers suggest a complex scenario involving both dynamic fractionation and migration fractionation effects.

(3) The high maturity of kerogen in the Daning–Jixian area limits the application of some commonly used genetic identification diagrams. The classical Whiticar diagram

based on $\delta^{13}C_1$ and δ^2H-CH_4 was employed, revealing that all gas samples from the Daning–Jixian Block and other gas fields across the Ordos Basin fall into the domain of geothermal, hydrothermal, and crystalline gases, which suggests that the gas genesis is associated with the interaction of over-mature organic matter with external hydrogen. The Milkov diagram indicates that natural gases sampled from the other producing fields in the Ordos Basin consistently fall within the canonical coal-derived gas field with relatively lower maturity, with some samples from fields like Daniudi and Zizhou also exhibiting characteristics of oil-associated gas. In contrast, the study area contains thermogenic gas with higher maturity. The Daning–Jixian Block gas accumulations originate from early-stage kerogen cracking in Carboniferous–Permian source rocks, with a subordinate contribution from crude-oil cracking.

(4) The Daning–Jixian Block exhibits characteristics distinct from other gas fields. All samples from this block display a $\delta^{13}C_1 > \delta^{13}C_2$ reversal. This is interpreted to be associated with the elevated thermal maturity of the kerogen and mixing with crude oil-cracking gas. This process causes the reversal and lighter values of ethane carbon isotopes. Consequently, the ethane carbon isotope value alone cannot be reliably used for delineation of the natural gas provenance in the study area.

These findings not only deepen the academic understanding of tight sandstone gas accumulation mechanisms in high-maturity coal-measure systems but also provide critical practical guidance for industrial exploration and development. First, they offer a reliable basis for resource potential evaluation and sweet-spot prediction—essential for guiding industrial exploration deployment in the Daning–Jixian Block and analogous regions. Second, the identification of geochemical tracers and the interpretation of $\delta^{13}C_1 > \delta^{13}C_2$ reversal mechanisms enable industrial sectors to accurately characterize gas accumulation processes, supporting targeted reservoir stimulation and optimization of development schemes to address challenges posed by ultra-low permeability.

Author Contributions: Conceptualization, B.W., M.C., H.T. and J.S.; methodology, B.W., M.C., H.T., J.S., L.L., X.L., B.C., B.Y., Z.Z. and Z.Q.; investigation, B.W., M.C., H.T., J.S., L.L., X.L., B.C., B.Y. and Z.Z.; writing—original draft preparation, B.W.; writing—review and editing, M.C.; project administration, M.C. All authors have read and agreed to the published version of the manuscript.

Funding: This research was funded by the Key Application Science and Technology Project of PetroChina Company Limited (grant number 2023ZZ18-01) and Science and Technology Funding Project of Huaneng Group Headquarters (grant number HNKJ20-H87).

Data Availability Statement: The original contributions presented in this study are included in this article; further inquiries can be directed to the corresponding author.

Conflicts of Interest: Authors Ming Chen, Haonian Tian and Junyi Sun were employed by the PetroChina Coalbed Methane Company Limited Linfen Branch. Authors Xing Liang, Baoshi Yu and Zhuo Zhang were employed by the PetroChina Zhejiang Oilfield Company. Author Benliang Chen was employed by the company Huainan Mining (Group) Co., Ltd. The remaining authors declare that the research was conducted in the absence of any commercial or financial relationships that could be construed as a potential conflict of interest. The authors declare that this study received funding from Key Application Science and Technology Project of PetroChina Company Limited. The funder had the following involvement with the study: study design, collection, analysis.

References

1. Li, S.K.; Liu, J.; Zhu, Y.M.; Ma, Y.J.; Liu, X.M. A study on the characteristics of unconventional natural gas reservoirs in the Julu block of northwestern Linqing Depression, Bohai Bay Basin. *Nat. Gas Geosci.* **2025**, *36*, 221–239.
2. Li, S.Z. Research on Forming Conditions and Regionalization of Tight Gas in Turpan-Hami Basin. Master's Thesis, China University of Petroleum (East China), Beijing, China, 2015.

Li, P.Y.; He, X.D.; Zhou, C.J.; Xie, Y.G.; Shi, H. Mineral compositions and microstructural characteristics of the tight sandstone reservoir in the Sulige area and their potential influence on hydraulic fracturing. *Hydrogeol. Eng. Geol.* **2023**, *50*, 1–11.

Jiang, Y.D.; Xie, C.L.; Song, X.; Liu, Z.J. Physical Simulation and Acoustic Emission Characteristics of Sandstone Hydraulic Fracturing under True Triaxial. *Chin. J. Undergr. Space Eng.* **2024**, *20*, 1145–1151+1159.

Khormali, A.; Ahmadi, S.; Aleksandrov, A.N. Analysis of reservoir rock permeability changes due to solid precipitation during waterflooding using artificial neural network. *J. Pet. Explor. Prod. Technol.* **2025**, *15*, 17. [CrossRef]

Wang, Q.; Zhang, W.; Jiang, N.; Qu, X.J.; Lei, T.; Luo, L.; Cai, X.Y.; Wang, J.; Zhang, M.K.; Tan, X.F. The control mechanism of deformable compositions on tight sandstone reservoirs: An example of the Shanxi Formation in the H well area of the Daniudi gas field in the Ordos Basin. *Acta Petrol. Mineral.* **2025**, *44*, 128–140.

Zhong, J.Z. Research on Intelligent Lithology Identification Method Based on Sandstone THIN Section. Master's Thesis, Taiyuan University of Technology, 2023.

Wu, Y.; Wang, G.; Wang, Q.R. Difficulties and Development Direction of Medium and Low-rank Coalbed Methane Industry in Xinjiang Region. *China Coalbed Methane* **2023**, *20*, 3–5+20.

Guo, Q.Q. Subject: Secondary Pore Development Pattern and Formation Mechanism of He 8 Member in Daning-Jixian Area of Ordos Basin, China. Master's Thesis, Xi'an Shiyou University, Xi'an, China, 2023.

Zhang, K.; Dou, S.J.; Gao, H.H.; Shi, S.; Li, T.; Tang, G.; Yang, B. Application of Spectral Recovery and Waveform Inversion in Thin Reservoir Prediction in Daning Jixian Block. *J. Jilin Univ. Earth Sci. Ed.* **2024**, *54*, 2142–2153.

Sun, J.; Chen, M.; Wang, B.; Wang, G.; Tian, H.; Hou, J.; Zhu, B. Analysis of Factors Influencing Tight Sandstone Gas Production and Identification of Favorable Gas Layers in the Shan 23 Sub-Member of the Daning-Jixian Block, Eastern Ordos Basin. *Processes* **2024**, *12*, 1810. [CrossRef]

Shi, S.W. Study on Comprehensive Evaluation Method of Tight Sandstone Gas Reservoir. Master's Thesis, Xi'an Shiyou University, Xi'an, China, 2016.

Zhao, L.M.; Wen, G.H.; Li, X.T.; Li, X.; Li, Y.Z.; Shi, X.S.; Yang, M.M. Evaluation of the "Sweet Spot Area" of Tight Sandstone Gas Reservoirs in the Shanxi Formation 2–3 Sub-member of the Daning-Jixian Block in the Ordos Basin. *Nat. Gas Ind.* **2018**, *38*, 5–10.

Guo, L.L.; Li, Z.B.; Zhang, W.; Lin, W.J.; Yang, M.M.; Zhao, Y. Analysis of pore structure of main tight sandstone reservoirs in Daning-Jixian Block of Ordos Basin. *Nat. Gas Ind.* **2018**, *38*, 18–23.

Shi, S.; Er, C.; Zhang, W.; Zhao, J.Z.; Zhang, Y.X.; Teng, Y.X. Pore Structure Characteristics of Upper Paleozoic Tight Reservoir in Daning—Jixian Area of Ordos Basin and Its Relationship with Clay Minerals. *J. Xi'an Shiyou Univ. Nat. Sci. Ed.* **2023**, *38*, 1–9.

Liu, L. Study on the Tight Sandstone Gas Accumulation Law of Strata Bearing Coal in Linxing District of the Ordos Basin, China. PhD Thesis, China University of Geosciences, Beijing, China, 2017.

Yong, L.I.; Weikai, X.U.; Jixian, G.; Peng, W.; Chuanqi, T.; Yang, T.; Yile, Z. Mechanism of coal measure gas accumulation under integrated control of "source reservoir-transport system": A case study from east margin of Ordos Basin. *J. China Coal Soc.* **2021**, *46*, 2440–2453.

Dang, J.C. Study on Natural Gas Enrichment Rules and Sweet Spot Evaluation of He8 Member in Daning-Jixian Area. Master's Thesis, Xi'an Shiyou University, Xi'an, China, 2023.

Dai, J.X.; Ni, Y.Y.; Hu, G.Y.; Huang, S.P.; Liao, F.R.; Yu, C.; Gong, D.Y.; Wu, W. Stable carbon and hydrogen isotopes of gases from the large tight gas fields in China. *Sci. China Earth Sci.* **2014**, *44*, 563–578. [CrossRef]

Liu, C.; Lei, M.Y.; Sun, B.L.; Wang, X. Origin and process of carbon isotopic reversal of tight sandstone gas in the Shixi gas field, eastern margin of Ordos Basin. *Int. J. Coal Geol.* **2023**, *276*, 104327. [CrossRef]

Tian, Y. Study on the Optimization of Tight Gas Reservoir Development Scheme in Daji Gas Field. Master's Thesis, China University of Petroleum, Beijing, China, 2023.

Shi, Z.; Zhao, J.Z.; Sun, X.W.; Shi, S.; Zhao, L.M.; Li, J. Characteristics and hydrocarbon generation potential of Upper Paleozoic coal measure source rocks in the southeastern Ordos Basin. *Nat. Gas Geosci.* **2023**, *34*, 1612–1626.

Chen, M.; Wang, B.; Tian, H.; Sun, J.; Liu, L.; Liang, X.; Chen, B.; Yu, B.; Zhang, Z. Multilayer Gas-Bearing System and Productivity Characteristics in Carboniferous–Permian Tight Sandstones: Taking the Daning–Jixian Block, Eastern Ordos Basin, as an Example. *Energies* **2025**, *18*, 2398. [CrossRef]

GB/T 13610-2020; Analysis of Natural Gas Composition—Gas Chromatography. State Administration for Market Regulation, National Standardization Administration: Beijing, China, 2020.

Zheng, P.; Shi, Y.H.; Zou, C.Y.; Kong, L.M.; Wang, L.S.; Liu, J.Z. Natural gas sources in the Dengying and Longwangmiao Fms in the Gaoshiti-Maoxi area, Sichuan Basin. *Natur. Gas Ind.* **2014**, *34*, 50–54.

SY/T 5238-2019; Analysis Method for Carbon and Oxygen Isotopes in Organic Matter and Carbonate. National Energy Administration: Beijing, China, 2019.

SY/T 7313-2016; Analysis of Hydrogen Isotope in Petroleum, Rock Extracts, and Kerogen. National Energy Administration: Beijing, China, 2016.

Dai, J.; Li, J.; Zhao, J. *Giant Gas Fields and Gas Sources in China*; Petroleum Industry Press: Beijing, China, 2024; pp. 157–170.

29. Sun, J.N. The Gas Kitchen and the Origin of Natural Gas in the Daniudi Area and Surrounding Area of Ordos Basin. Master's Thesis, China University of Petroleum, Beijing, China, 2023.
30. Shen, J.; Li, K.X.; Zhang, H.W.; Shabbiri, K.; Hu, Q.J.; Zhang, C. The geochemical characteristics, origin, migration and accumulation modes of deep coal-measure gas in the west of Linxing block at the eastern margin of Ordos Basin. *J. Nat. Gas Sci. Eng.* **2021**, *91*, 103965. [CrossRef]
31. Peng, W.; Hu, G.; Huang, S.; Fang, C.; Liu, D.; Feng, Z.; Han, W.; Jiang, R.; Chen, H. Natural gas geochemical characteristics and genetic analysis: A case study of the Dongsheng gas field in the Ordos basin of China. *J. China Univ. Min. Technol.* **2017**, *46*, 74–84.
32. Zhou, J.S.; Wang, N.X.; Zhao, Q.P.; Yin, X.; Lin, G.F.; Cao, Y.; Li, Y.Y.; Han, X.Q. Natural gas accumulation characteristics in the Upper Paleozoic in the Yanchang exploration block of southeastern Ordos Basin. *Natur. Gas Ind.* **2014**, *34*, 34–41.
33. Liu, H.L.; Liu, S.B.; Zhou, D.; Liu, W.; Jin, S.D. Geochemical indicators for tracing the source and migration of the tight sandstone gas in western Sichuan Basin. *Nat. Gas Geosci.* **2021**, *32*, 1127–1141.
34. Mou, X.; Wang, K.; Yao, X.L. Genesis Analysis of CBM in Daning-Jixian Region. *China Coalbed Methane* **2016**, *13*, 26–30.
35. Chen, J.; Wang, X.; Chen, J.; Ni, Y.; Xiang, B.; Liao, F.; He, W.; Yao, L.; Li, E. New equation to decipher the relationship between carbon isotopic composition of methane and maturity of gas source rocks. *Sci. China Earth Sci.* **2021**, *64*, 470–493. [CrossRef]
36. Dai, J.X.; Ni, Y.Y.; Huang, S.P.; Liao, F.R.; Yu, C.; Gong, D.Y.; Wu, W. Significant function of coal-derived gas study for natural gas industry development in China. *Nat. Gas Geosci.* **2014**, *25*, 1–22.
37. Bernard, B.B.; Brooks, J.M.; Sackett, W.M. Natural gas seepage in the Gulf of Mexico. *Earth Planet. Sci. Lett.* **1976**, *31*, 48–54. [CrossRef]
38. Whiticar, M.J. Carbon and hydrogen isotope systematics of bacterial formation and oxidation of methane. *Chem. Geol.* **1999**, *161*, 291–314. [CrossRef]
39. Milkov, A.; Etiope, G. Revised genetic diagrams for natural gases based on a global dataset of >20,000 samples. *Org. Geochem.* **2018**, *125*, 109–120. [CrossRef]
40. Qu, Z.Y. Shale Gas Generation and Variation in Stable Carbon and Hydrogen Isotope Compositions. Master's Thesis, Guangzhou Institute of Geochemistry, University of Chinese Academy of Sciences, Guangzhou, China, 2015.
41. Zheng, D.Y.; Jiang, F.J.; Liu, T.S.; Pang, X.Q.; Chen, X.Z.; Shao, X.H.; Li, L.L.; Huyan, Y.Y. Genetic Types and Sources of Natural Gas in Linxing Area, the Eastern Margin of Ordos Basin, China. *J. Earth Sci. Environ.* **2018**, *40*, 203–214.
42. Chen, J.J.; Zhao, J.Z.; Li, J.; Cao, Q.; Wang, Z. Geochemical characteristics of upper paleozoic natural gas in southern Ordos Basin: A case study of Zhidan-Ganquan area. *Sci. Technol. Eng.* **2020**, *20*, 10598–10604.
43. Zhang, J.Q.; Zhao, J.Z.; Li, J.; Yang, X. Characteristic of fluid inclusions in upper Paleozoic He 8 member of Daning-Jixian block, southeastern Ordos Basin and its significance for hydrocarbon accumulation. *J. Xi'an Shiyou Univ. Nat. Sci. Ed.* **2024**, *39*, 34–41+142.
44. Dai, J.X.; Shi, X.; Wei, Y.Z. Summary of the abiogenic origin theory and the abiogenic gas pools (fields). *Acta Pet. Sin.* **2001**, *6*, 5–10.
45. Li, J.; Zhao, J.Z.; Wang, D.X.; Sun, L.Y.; Ren, J.F.; Wu, C.Y.; Wu, W.T.; Zhao, Z.L.; Qu, F.T. Genesis and source of the Ordovician mid-assemblage natural gas in the east side of the central paleo-uplift, Ordos Basin. *Acta Pet. Sin.* **2016**, *37*, 821–831.
46. Zhang, M.; Song, D.F.; Wang, T.G.; He, F.Q.; Zhang, W.; An, C.; Liu, Y.; Lu, Z.G. Geochemical characteristics and sources of natural gas in Hangjinqi area of ordos Basin. *Pet. Geol. Exp.* **2024**, *46*, 124–135.
47. Ni, Y.Y.; Zhang, J.C.; Yao, L.M.; Dong, G.L.; Wang, Y.; Wang, L.; Chen, J.P. Application of carbon and hydrogen isotopes in the natural gas origin study. *Nat. Gas Geosci.* **2024**, *35*, 1897–1909. [CrossRef]
48. Dai, J.X.; Ni, Y.Y.; Gong, D.Y.; Feng, Z.Q.; Liu, D.; Peng, W.L.; Han, W.X. Geochemical characteristics of gases from the largest tight sand gas field (Sulige) and shale gas field (Fuling) in China. *Mar. Pet. Geol.* **2017**, *79*, 426–438. [CrossRef]
49. Zhang, T.; Wang, X.F.; Jin, X.H.; Meng, Q.Q.; Wang, Y.R.; Chen, X.Y.; Zhang, W.; Zhang, J. Geochemical characteristics and genesis of Paleozoic natural gas in the southern Ordos Basin. *Nat. Gas Geosci.* **2024**, *35*, 1616–1625. [CrossRef]
50. Peng, W.; Liu, Q.; Hu, G.; Lü, Y.; Meng, Q.; Guo, F.; Guan, H. A New Interpretation of Carbon Isotope Series Reverse of Highly-Over Mature Alkane Gases: Demethylation of Aromatic Hydrocarbons. *Earth Sci.* **2020**, *45*, 1308–1314.

processes

MDPI

rticle

Research on Dynamic Control Methods for Fine-Scale Water njection Zones Based on Seepage Resistance

hi Dong [1,2], Weiming Hui [1,2], Gaojun Shan [3], Erlong Yang [2,*], Ming Qu [1,2] and Hai Wang [2]

1 SANYA Offshore Oil & Gas Research Institute, Northeast Petroleum University, Sanya 572024, China
2 MOE Key Laboratory of Enhanced Oil Recovery, Northeast Petroleum University, Daqing 163318, China
3 Daqing Oilfield Co., Ltd., Daqing 163000, China
* Correspondence: yangerlong@nepu.edu.cn

Abstract: To address prominent challenges in developing high water cut oilfields—such as significant variations in sandstone water absorption capacity and uneven reservoir mobilization—this paper proposes a dynamic layered water injection method centered on seepage resistance as the core regulatory indicator. Through theoretical derivation, quantitative relationships are established between seepage resistance, liquid absorption ratio, and injection allocation. Combined with numerical simulation analysis, this forms a refined water injection control strategy that dynamically adjusts according to changes in oilfield water cut. Research findings demonstrate that this method effectively improves fluid absorption profiles in heterogeneous reservoirs, promotes balanced displacement across distinct lithologic zones, and thereby suppresses water cut increase rates and production decline. Taking Block D as an example, maintaining inter-zone seepage resistance step difference within the 3–5 range yields optimal development outcomes. This study provides significant reference value and practical insights for the dynamic development of oilfields in high water cut phases.

Keywords: high-water-cut oilfields; seepage resistance; precision water injection; dynamic regulation

1. Introduction

Most developed oil and gas reservoirs in China are continental deposits characterized by significant lateral and vertical heterogeneity. After extensive water flooding, most oilfields have entered the high or ultra-high water cut stage, marked by a highly disordered distribution of remaining oil that is generally scattered but locally concentrated. In the late stage of water flooding, determining the distribution of remaining oil and formulating corresponding production strategies to further enhance oil recovery is a major challenge. The flow field, defined as the dynamic flow path of reservoir fluids, plays a critical role in the distribution of remaining oil. The evolution of the flow field is highly complex and influenced by geological conditions, well patterns, and fluid properties. In some areas, excessive water injection often leads to the formation of dominant flow channels. The uneven distribution of flow fields results in highly dispersed remaining oil and poor efficiency of the water injection process [1]. With the development of the oilfield maturing to the stage of high water cut, the formation pressure field is redistributed, remaining oil becomes increasingly scattered, and fluid properties change throughout the production process, which brings new challenges to the development. For instance, excessive well spacing leads to uneven advancement of injected water in the formation; high-permeability layers are prone to promoting early water breakthrough, while low-permeability layers

https://doi.org/10.3390/pr13123966

suffer from inadequate fluid supply, making them difficult to produce effectively. Interlayer interference has become increasingly prominent, and the difference in water intake capacity between high-permeability and low-permeability layers is several times or even orders of magnitude, which seriously restricts the overall degree of reservoir recovery. In addition, some of the oilfield development well spans hundreds of meters, with thin individual layers and significant variations in physical properties. Therefore, existing well patterns and layer systems struggle to adapt to the dynamic changes in reservoir characteristics, leading to low development efficiency and underutilized recovery potential. Accurately characterizing fluid flow in reservoirs is essential for understanding the fundamental dynamics of multiphase flow, optimizing well patterns, and enhancing oil recovery.

During water flooding, structural fractures, faults, and lateral heterogeneity in permeability often cause rapid advancement of the waterflood front, resulting in ineffective water cycling, low sweep efficiency, and low recovery. Extensive water injection can alter the pore structure and mineral composition of the rock, leading to significant changes in reservoir properties such as porosity, permeability, and wettability [2]. Jiang et al. concluded that variations in reservoir physical properties are generally proportional to the cumulative amount of injected water, and the changes become more pronounced during the high or ultra-high water cut stages [3]. An in-depth study of fluid flow pathways is essential for accurately characterizing the flow field in highly heterogeneous reservoirs during the late stages of water flooding. Flow diagnostic tools can be effectively employed to assess reservoir connectivity and the dynamic evolution of fluid flow paths [4–6]. Compared to full-physics simulations, flow diagnostics offer advantages such as faster computation, greater computational efficiency, and simplified model complexity. As a widely used technique, streamline-based simulation is commonly applied to estimate parameters such as time-of-flight and steady-state tracer distributions [7–10]. Additionally, flow heterogeneity has been characterized by introducing specific parameters, including swept regions, drainage areas, well-pair regions, and the Lorenz coefficient, which help describe flow dynamics within the reservoir. Thiele and Batycky used streamline-based simulation to calculate flow rates between well pairs and derived optimal injection–production parameters [11]. Jia and Deng developed a flow field identification method based on streamline simulation, and they analyzed streamline bundles to identify areas of ineffective water cycling via Python3.10 programming [12]. Tanaka et al. combined streamline-based simulation with a genetic algorithm to generate preliminary development plans [13]. They integrated this approach with response surface methodology to derive an optimal strategy for a typical water-flooded reservoir, significantly reducing model complexity. Zhao Chunsen et al. analyzed the factors affecting water flooding efficiency using numerical simulation technology [14]. Zhang et al. developed a comprehensive geological model, designed and compared multiple water injection schemes, and successfully quantified the negative impact of reservoir heterogeneity on water flooding performance [15]. They demonstrated that refined water injection strategies, especially those combining dynamic adjustments with intelligent distributed injection, can effectively mitigate these adverse effects, thereby significantly enhancing both recovery and economic benefits. Ultimately, they proposed a set of dynamic optimization strategies. Wang et al. developed an intelligent system to address the challenges of layered water injection in highly heterogeneous, multilayer sandstone oilfields during the high water cut stage [16]. This system integrates surface control, downhole tools, and software to enable real-time monitoring and remote precise regulation of parameters such as flow rate, pressure, and temperature in each injection layer, significantly enhancing the efficiency of layered water injection. Sun et al. proposed and validated an innovative bridge-type concentric layered water injection system [17]. This system enables long-term, continuous monitoring of individual tubing pressures and downstream pressures at each layer during conventional water injection, providing essential data for dynamic analysis. Song et al. introduced layered flow-allocation technology for water injection wells based on wave-code communication, accelerating the advancement toward intelligent management of injection wells [18]. This technology is expected to drive the field of refined layered water injection toward greater intelligence, automation, and integration [19]. As the field water cut continues to grow, the driving mechanism behind interlayer conflict has shifted from being "static-dominated" to involving "static-dynamic coupling". During the early development stage, interlayer interference was primarily governed by static geological factors such as permeability, sand thickness, and porosity. However, upon entering the high water cut stage, the flow resistance of fluids within individual sublayers exhibits significant differential variations as water cut increases. In high-permeability layers, intensified water flooding leads to increased water saturation, which reduces flow resistance and further enhances water intake capacity, creating a vicious cycle of "water theft" [20,21]. In contrast, low-permeability layers experience a marked increase in flow resistance due to elevated oil viscosity and bound water clogging pore throats, resulting in a continuous decline in water intake capacity. At this stage, interlayer interference is no longer solely the result of static factors but rather arises from the coupling of static reservoir properties and dynamic development parameters. Traditional layer adjustment methods based solely on static parameters can no longer meet the requirements of precision development.

In this context, our objective is to develop a novel dynamic control method for layered water injection based on seepage mechanics theory. This method utilizes a quantitative flow resistance calculation model that integrates both static and dynamic parameters between injectors and producers [22]. It can accurately characterize the flow paths and resistance distribution across different layers between injection and production wells, enabling dynamic evaluation of the production potential of each layer and providing more targeted adjustment strategies for oilfield development. This study focuses on developing a dynamic control methodology for layered water injection and systematically undertakes the following tasks: First, based on a comprehensive diagnosis of developmental challenges in the target block, the principle of hydraulic similarity is applied to quantitatively characterize the flow resistance of individual layers. Corresponding evaluation criteria are established to verify its correlation with the sandstone's water intake capacity. By optimizing the flow resistance differential between layers, a theoretical foundation is provided for rational layer combination and partitioning. Furthermore, by deriving the quantitative relationship between flow resistance and injection allocation rates using the two-phase Darcy's Law and analyzing the dynamic impact of flow resistance variations on injection allocation across different water cut stages, an optimized water injection strategy is formulated. Finally, the maximum potential for enhanced oil recovery that is achievable through this method is determined.

2. Problem Statement

Block D was initially developed using a reverse nine-spot pattern with a well spacing of 300 m. It was divided into four reservoir units, developed with three sets of well patterns. In the central area of Block D, where the reservoir is thicker, two well patterns were deployed: one for the Gaoyi I and II reservoir units and another for the Gaoyi III and IV reservoir units, targeting the GI1–II25 and GII26-and-below reservoirs, respectively. A 300 m five-spot pattern was used for GI1–GI20, a 212 m five-spot pattern was used for GII1–GII34, and a 300 m reverse nine-spot pattern was used for GII26–GIV1. The average reservoir permeability in Block D is 231 mD, porosity is 26%, average oil saturation is 59.5%, surface crude oil density is 0.865 g/cm^3, and formation crude oil viscosity is 9.2 mPa·s.

Based on continuous water injection profile data from recent years, the overall displacement efficiency in Block D is moderate. The proportion of poorly displaced and non-displaced sandstone reaches 51.7%, while the proportion of effective thickness accounts for 47.5%. Poorer reservoir properties correlate with worse development performance. For reservoirs with an effective thickness greater than 1.0 m, the proportion of poorly developed and undeveloped sandstone is 38.6%. However, for reservoirs with an effective thickness less than or equal to 1.0 m and with extra-reservoir oil layers, the proportion reaches 55.0%. This indicates that under the current well pattern, reservoir adaptability is poor and development efficiency is low. Although refined layered water injection has improved the utilization status in Block D in recent years, significant disparities in utilization levels persist across different reservoir types.

Currently, some formations in Block D have low residual oil saturation, leading to reduced flow resistance. Reservoirs with poor utilization have a low water cut and high residual oil saturation, which further exacerbates the differences in flow resistance between formations. This paper presents an investigation into the dynamic regulation of the GI1-GI20 formation series, utilizing a representative model of Block D and employing seepage resistance as the foundational criterion.

3. Seepage Resistance Calculation Model

As shown in Figure 1, the seepage resistance method utilizes the principle of hydraulic similarity to describe the relationship between flow rate and permeability in reservoir seepage fields by drawing an analogy to current and resistance in electrical circuits [23] (assuming it is a radial flow, and the oil–water two-phase process is incompressible, and the density does not change with pressure). In practical applications, the workflow primarily focuses on well groups centered around water injection wells. Based on the dynamic and static development parameters of these well groups, seepage resistance coefficients under two-phase flow conditions are calculated for individual layers or segments of both producer and injector wells.

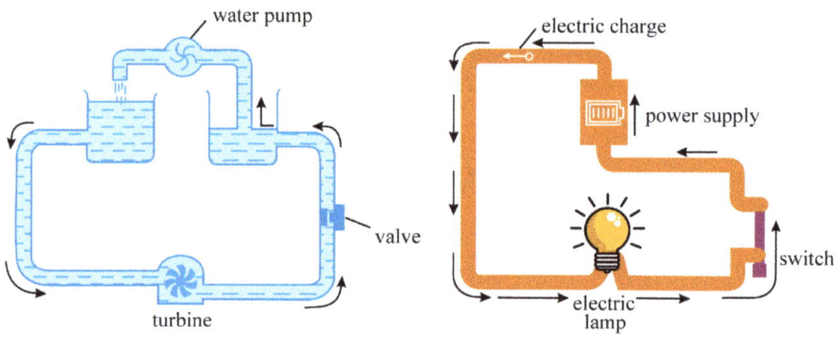

Figure 1. Schematic diagram of similar principles in hydropower and electricity processes.

According to Darcy's Law, it can be seen that the flow resistance of the oil and water phases in the reservoir is related to the water saturation function, the viscosities of the oil and water phases, the relative permeability, and the well spacing between oil and water wells.

Two-Phase Flow Conditions for Darcy's Equation:

$$Q = KA\left(\frac{K_{ro}(S_w)}{\mu_o} + \frac{K_{rw}(S_w)}{\mu_w}\right)\frac{\Delta P}{L} \quad (1)$$

Equivalent seepage resistance:

$$R = \frac{\Delta P}{Q} = \frac{1}{\frac{KK_{ro}(S_w)}{\mu_o} + \frac{KK_{rw}(S_w)}{\mu_w}} \frac{L}{A} \quad (2)$$

where Q is flow rate, m^3/s; K is absolute permeability, mD; A is rock cross-sectional area, m^2; S_w is water saturation, %; K_{ro} is oil phase relative permeability; K_{rw} is water phase relative permeability; μ_o is crude oil viscosity, mPa·s; μ_w is water viscosity, mPa·s; P is pressure gradient, Pa; and L is flow distance, m.

During field application, the approximate calculation method for the seepage resistance coefficient is as follows.

Within the injection–production well group, the seepage resistance between wells in a single direction within the sub-formation is first calculated. Subsequently, based on the principle of the hydraulic–electrical analogy, the planar seepage resistance of the sub-formation is determined [24,25].

Seepage resistance at any point along the main flow path between oil and water wells:

$$R(x) = \frac{1}{K(x)h(x)(K_{ro}/\mu_o + K_{rw}/\mu_w)} \quad (3)$$

where $R(x)$ represents the inter-well flow resistance, $mPa\cdot s\cdot mD^{-1}\cdot m^{-1}$; $K(x)$ denotes the permeability at any point along the main flow path between oil and water wells, mD; and $h(x)$ indicates the thickness at any point along the main flow path between oil and water wells, m.

Calculate the flow resistance in a single layer of an injection well along a specific injection–production direction:

$$R_i = \int_{r_w}^{d} R(x)\mathrm{d}x \quad (4)$$

where d is the distance between oil and water wells, m, and r_w is the wellbore radius, m.

Through segmented discretization and summation between oil and water wells, the single-layer flow resistance is the sum of flow resistances from the injection well to each production well. Since a single production–injection well connects multiple injection–production directions (i.e., n production wells), it is equivalent to n resistors in parallel on the plane. Thus, the flow resistance on the mth sublayer is as follows:

$$R_m = \frac{1}{1/R_1 + \cdots + 1/R_i + \cdots + 1/R_n} \quad (5)$$

This article selects a small typical small block from a whole region model, which has a good historical fit and high accuracy. Determine the main flow paths between oil and water wells in different stratigraphic zones based on the stratigraphic well network flow diagram (Figures 2 and 3) to facilitate subsequent calculation of stratigraphic flow resistance. The calculation process is illustrated in Figure 4.

Using the aforementioned process, the flow resistance of each sublayer within the stratum was calculated. The wellbore radius is 0.14 m, crude oil viscosity is 9.2 mPa·s, and water viscosity is 1 mPa·s. The calculation results are presented in Table 1. The data in Table 1 can indicate the differences in seepage resistance among different layers, which are mainly due to the different pore structures, permeability, and throat radii of each layer. Moreover, changes in water saturation dynamically alter the seepage resistance of each phase fluid through the relative permeability effect, resulting in significant differences in seepage resistance among different sub layers.

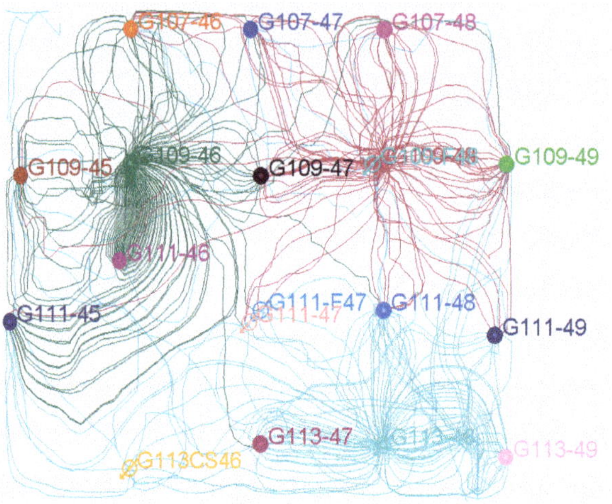

Figure 2. Flowline diagram of Block D well network.

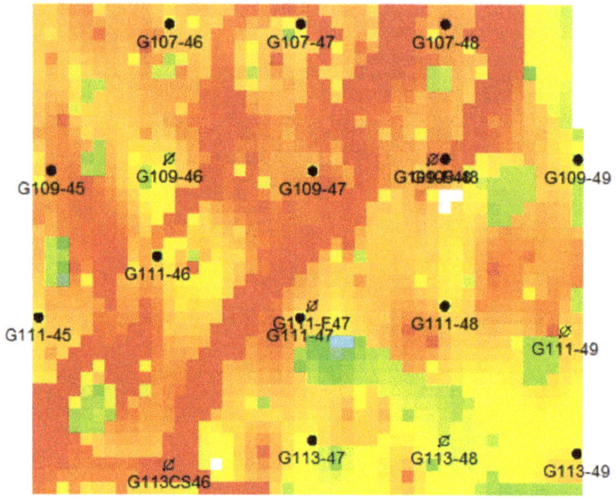

Figure 3. Block D well location map.

Table 1. Initial flow resistance of sublayers in stratigraphic Series GI1–GI20.

Layer	Seepage Resistance, $mPa \cdot s \cdot mD^{-1} \cdot m^{-1}$	Layer	Seepage Resistance, $mPa \cdot s \cdot mD^{-1} \cdot m^{-1}$	Layer	Seepage Resistance, $mPa \cdot s \cdot mD^{-1} \cdot m^{-1}$
GI1	12.51	GI10	5.34	GI16	1.13
GI2+3	7.11	GI11	2.61	GI17	0.31
GI4+5	4.05	GI12	1.22	GI18	1.50
GI6+7	6.19	GI13	1.42	GI19	0.34
GI8	8.47	GI14	1.45	GI20	0.62
GI9	2.92	GI15	0.52	/	/

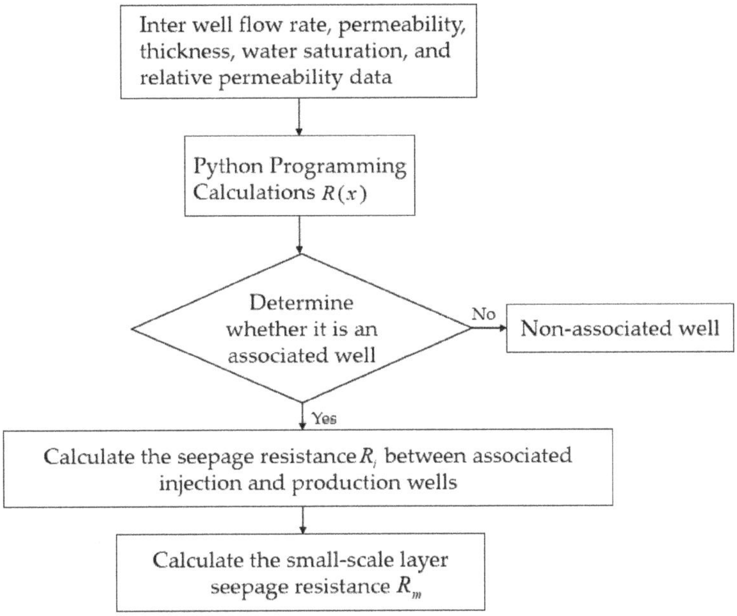

Figure 4. Flowchart of permeability resistance in the computational layer.

4. Establish Criteria for Determining Seepage Resistance

By integrating reservoir numerical simulation with injection well test data (sandstone thickness and effective thickness data are obtained from actual well logging data, and the sandstone liquid absorption ratio is obtained from numerical simulation model production PRT files), a feasibility analysis of flow resistance was conducted based on the aforementioned flow resistance calculation results. A total of 17 flow resistance values across 6 injection wells in the study block were calculated and statistically analyzed. The sandstone thickness is 22.57 m, with an effective thickness of 10.11 m and a single-cycle sandstone water absorption ratio of 31.58%. Comparing the single-cycle sandstone water absorption ratios, the mobilization ratio progressively increases with rising flow resistance. As shown in Table 2, when flow resistance exceeds 10 mPa·s/(mD·m), the single-cycle sandstone water absorption ratio decreases from 35.67% to 23.6%. This indicates a strong correlation between the reservoir's hydraulic resistance coefficient and the water absorption mobilization status of sandstone in the reservoir. Higher hydraulic resistance correlates with lower mobilization levels.

Table 2. Water absorption proportion of sandstone with different seepage resistance.

Seepage Resistance Zone ($mPa \cdot s \cdot mD^{-1} \cdot m^{-1}$)	Number of Floors	Sandstone Thickness (m)	Effective Thickness (m)	Sandstone Water Absorption Ratio (%)
0–5	12	14.79	5.91	35.67
5–10	4	6.42	3.49	30.89
>10	1	1.37	0.71	23.6
Total	17	22.57	10.11	31.58

Using the Eclipse2020(Schlumberger, Houston, Texas, USA) numerical simulation software, a numerical simulation model was established. Low-permeability combinations (50 mD and 100 mD), medium-permeability combinations (100 mD and 200 mD), and high-permeability combinations (500 mD and 1000 mD) were designed for simulations

(Figure 5). Two small layers with the same permeability differential but different permeability combinations were simulated in parallel for displacement to investigate interlayer interference. As the water cut increases, the oil phase permeability decreases while the water phase permeability increases. The ratio of flow resistance between the two layers gradually rises. When the water cut exceeds 70%, this ratio begins to increase gradually, while the ratio of flow resistance between the two layers increases sharply if the water cut surpasses 90%. The flow velocity of injected water in the small layer increases, and the interlayer conflict becomes more pronounced, as shown in Figures 6 and 7.

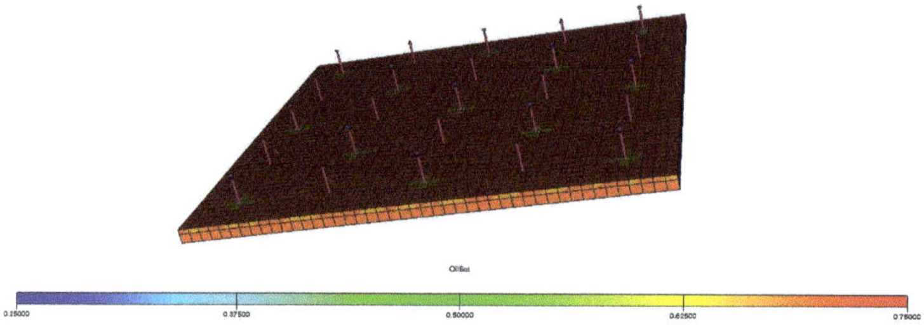

Figure 5. Ideal model of interlayer interference patterns.

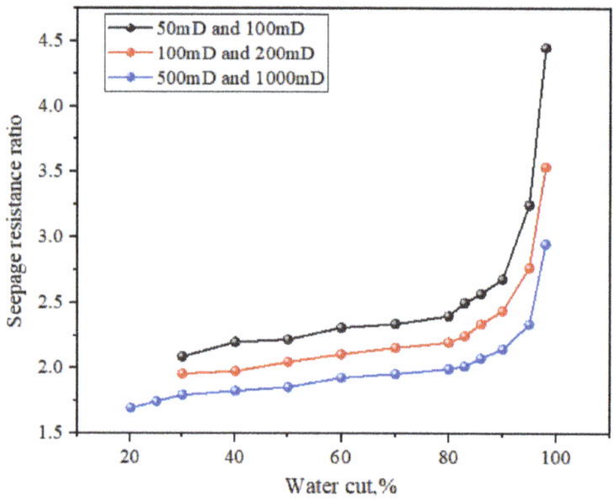

Figure 6. Ratio of interlayer seepage resistance for different permeability combinations as a function of water cut.

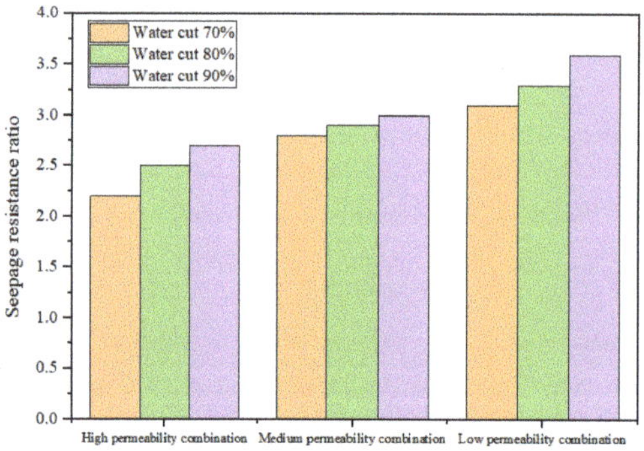

Figure 7. Interlayer seepage resistance ratio for different permeability combinations at various water cut levels.

In summary, during the low-to-medium water cut period, static heterogeneity controlled by physical properties is the primary factor governing interlayer, planar, and intralayer conflicts in oilfield development. Following long-term water injection development, small layers with identical permeability exhibit significant variations in water saturation, leading to changes in flow resistance. At this stage, permeability differences fail to reflect fluid flow variations within the reservoir, whereas flow resistance exhibits dynamic characteristics that directly reflect the flow state of small layers. Currently, field-based stratigraphic unit delineation combines reservoirs with similar properties under the premise of sufficient reserve abundance within each unit. This approach primarily considers static indicators such as the permeability coefficient variation within reservoirs, perforated sandstone thickness within sections, and the number of perforated reservoirs. However, based on flow characteristics during high water cut periods, relying solely on static indicators for stratigraphic grouping fails to precisely characterize flow changes during high water cut phases. Therefore, a stratigraphic adjustment method centered on dynamic flow resistance as the primary indicator must be established.

5. Research on Dynamic Regulation Methods for Layer Segments

5.1. Optimal Gradient for Seepage Resistance

To determine the appropriate seepage resistance differential for precise water injection control, seven stratigraphic combinations were designed based on the seepage resistance step difference values of each minor stratum from the numerical simulation model of the actual block. This establishes the differential range for inter-stratum seepage resistance combinations during stratigraphic adjustments. The design proposals are presented in Table 3.

Table 3. Interlayer seepage resistance gradient scheme.

Interlayer Seepage Resistance Gradient	Recovery Rate, %
1.11	25.49
2.66	27.10
3.4	28.72
4.93	33.76
8.14	28.47
9	27.48
11.82	20.16

By optimizing the permeability gradient, it is found that a gradient range between 3.4 and 4.93 significantly enhances reservoir recovery rates (Figure 8). In contrast, gradients that are either too low or too high have a relatively weaker impact on reservoir recovery rates. This occurs because a small gradient promotes uniform fluid injection without natural differentiation. However, once a main flow path forms, the uniform leading edge of the displacement fluid tends to cause overall synchronous flooding, making it difficult to further increase the sweep efficiency. Conversely, an excessively large gradient results in low natural sweep efficiency and increased flow leakage, preventing recovery improvement or even causing it to decline.

Therefore, the intralayer permeability gradient should be controlled within the range of 3 to 5 when defining subsequent stratigraphic combinations.

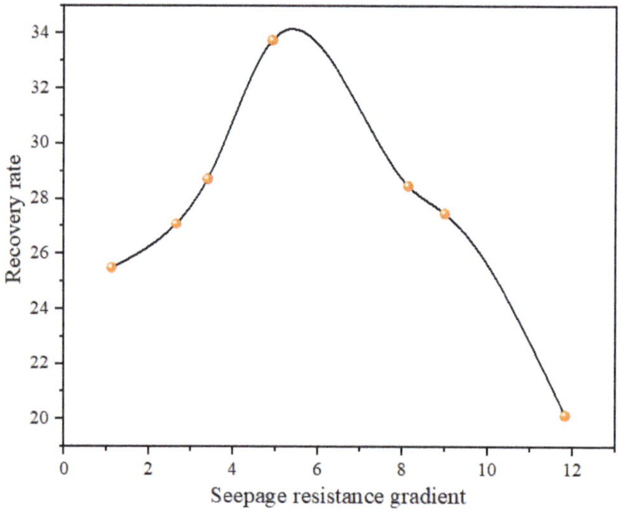

Figure 8. Relationship between seepage resistance gradient and recovery rate.

5.2. The Relationship Between Liquid Absorption Ratio and Seepage Resistance

Through continuous variation in interlayer seepage resistance during development, layer-segment combinations were optimized and layered water injection implemented to mitigate uneven seepage resistance. To define the technical boundaries for layered water injection during high water cut periods, a mechanistic model was employed to investigate the variation patterns of seepage resistance coefficients and their impact on interlayer interference. An idealized model was established featuring three longitudinally heterogeneous layers with plane-homogeneous properties, equal interlayer thicknesses, and boundary conditions simulating steady-state production with injection–production balance. A comprehensive seepage resistance model was employed to calculate the integrated seepage resistance coefficients for low-, medium-, and high-permeability layers (100 mD, 300 mD, and 600 mD), along with layer-specific liquid absorption patterns. The variation in seepage resistance coefficient across different water cut levels was statistically analyzed.

Figures 9 and 10 demonstrate that the water saturation of the high-permeability layer rises rapidly as water cut increases, leading to a rapid decrease in seepage resistance within the layer and consequently increasing liquid uptake. The disparity in liquid uptake ratios among low-, medium-, and high-permeability layers initially narrows, but upon reaching the high water cut stage, the difference in uptake ratios significantly widen. The coefficient of variation in flow resistance across the three-layer model gradually increases with rising water content. The variation in seepage resistance between layers becomes markedly pronounced when water cut exceeds 60%. Therefore, layered water injection should be employed if seepage resistance differences between layers are significant. Layered injection parameters must adapt to changes in water cut stages, and dynamic parameter variations must be considered during implementation.

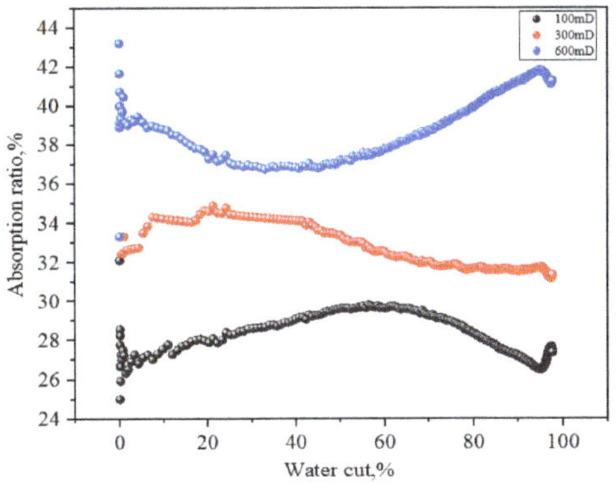

Figure 9. Liquid absorption ratio by layer in the three-layer model.

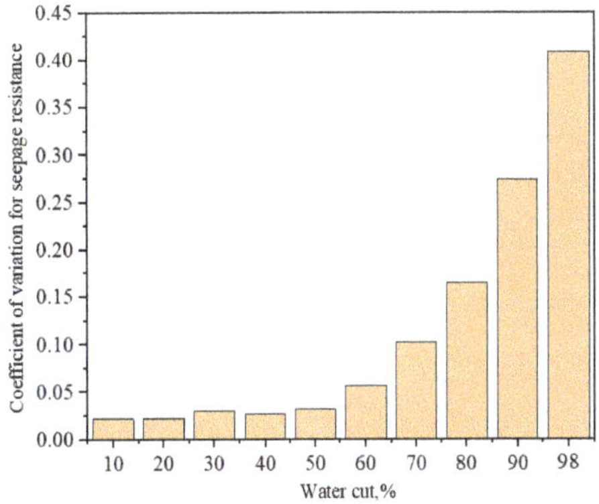

Figure 10. Variation coefficient of seepage resistance in the three-layer model.

5.3. Water Injection Control Strategy

The specific dynamic regulation strategy for layer intervals establishes a multi-parameter decision-making framework. The seepage resistance distribution determines the overall direction of injection adjustments, with spatial variations guiding whether injection rates in individual layers should increase or decrease. The characteristic water cut stage governs the frequency of dynamic regulation; the adjustment frequency increases during high water cut stages and appropriately decreases during low water cut stages. Static parameters, such as effective thickness and permeability, define the safe operational boundaries for injection rates, ensuring that adjustments remain within the reservoir's capacity and effectively prevent formation damage caused by over-potential injection. First, assume that each sublayer within the same water injection well constitutes a parallel flow unit, with the fluid being incompressible. The flow within each sublayer follows Darcy's Law for two-phase flow. For oil–water two-phase flow, the oil phase flow rate Q_o and water phase flow rate Q_w are combined with an extension of Darcy's Law, yielding effective permeability coefficients Ki for the oil phase and water phase as KK_{ro} and KK_{rw}, respectively.

Darcy's Law for Oil Phase:

$$Q_o = \frac{K \cdot K_{ro} \cdot h_i \cdot \Delta P}{\mu_o \cdot L} \tag{6}$$

Darcy's Law for the Aqueous Phase:

$$Q_w = \frac{K \cdot K_{rw} \cdot h_i \cdot \Delta P}{\mu_w \cdot L} \tag{7}$$

The total flow rate Q_i of the laminar flow is obtained by adding the oil phase flow rate Q_o and the water phase flow rate Q_w:

$$Q_i = Q_o + Q_w = \frac{K \cdot K_{ro} \cdot h_i \cdot \Delta P}{\mu_o \cdot L} + \frac{K \cdot K_{rw} \cdot h_i \cdot \Delta P}{\mu_w \cdot L} \tag{8}$$

After organizing, we obtained the following:

$$Q_i = \frac{K_i \cdot h_i \cdot \Delta P}{\frac{\mu_o \cdot L}{K_{ro}} + \frac{\mu_w \cdot L}{K_{rw}}} \tag{9}$$

The definition equation for small-layer seepage resistance is $R_L = \frac{\Delta P}{Q_i}$.
From this, we derive the definition equation for seepage resistance as follows:

$$R_L = \frac{L}{K_i \cdot h_i}\left(\frac{\mu_o}{K_{ro}} + \frac{\mu_w}{K_{rw}}\right) \tag{10}$$

Substituting Equation (10) into Equation (9) yields the fundamental relationship between the injection rate and the seepage resistance.

$$Q_i = \frac{\Delta P}{R_L \cdot L_i} \tag{11}$$

When the injection–production pressure differential ΔP and the injection–production well spacing L are fixed, a higher reservoir flow resistance R_L results in a smaller proportionate injection rate Q_i, preventing "under-injection" in high-permeability resistive zones. Conversely, a lower R_L leads to a larger Q_i, suppressing "water-grabbing" in low-permeability resistive zones.

The individual aquifers within the same injection well exhibit parallel flow, with the total fluid intake for the entire well being $Q_{Total} = \sum_{i=1}^{n} Q_i$. Combined with (11), the fluid absorption ratio $f_i = \frac{Q_i}{Q_{Total}}$ can be further derived as $f_i = \frac{\frac{\Delta P}{R_L \cdot L_i}}{\sum_{L=1}^{n} \frac{\Delta P}{R_L \cdot L_i}}$. With wells in the same well group having approximately similar distances, the relationship between the fluid absorption ratio and the flow resistance can be obtained after consolidation.

$$f_i = \frac{1/R_L}{\sum_{L=1}^{n} 1/R_L} \tag{12}$$

The wicking ratio f_i is inversely proportional to the seepage resistance R_L.

In order to prevent excessive water injection pressure from fracturing reservoir rocks and causing irreversible damage, the bottomhole flow pressure must be lower than the formation fracture pressure:

$$P_{wf} \leq P_{frac} \tag{13}$$

The bottomhole flowing pressure is related to wellhead pressure, static fluid column pressure, etc. For injection wells, the injection–production pressure difference ΔP can be approximately expressed as the difference between the bottomhole flowing pressure and the average reservoir pressure P_{avg}:

$$\Delta P = P_{wf} - P_{avg} \quad (14)$$

By substituting the above equation into the fracture pressure constraint condition, we obtain the safe upper limit of the injection–production pressure difference:

$$\Delta P_{\max} = P_{frac} - P_{avg} \quad (15)$$

By substituting the maximum allowable pressure difference into Equation (11), the maximum safe injection quantity Q_i^{frac} for a single layer based on formation integrity can be obtained as follows:

$$Q_i^{frac} = \frac{\Delta P_{\max}}{R_i \cdot L} = \frac{P_{frac} - P_{avg}}{R_i \cdot L} \quad (16)$$

Even if the injection–production pressure difference is within a safe range, the injection volume of a single layer should not exceed its inherent potential. We can establish another upper limit of injection quantity $Q_i^{potential}$, which is proportional to the average reservoir pressure (representing the energy level of the formation) and the reservoir potential coefficient:

$$Q_i^{potential} = \lambda_i \cdot f(P_{avg}) \quad (17)$$

Among them, $f(P_{avg})$ is a function related to the average reservoir pressure, used to correct the influence of formation energy on water absorption capacity (for example, the lower the energy, the greater the potential space for injection, but the relationship may be nonlinear).

For any regulatory layer, the final injection quantity Q_i^{final} must satisfy both Equations (18) and (19) simultaneously:

$$\Delta P < \Delta P_{\max} = P_{frac} - P_{avg} \quad (18)$$

$$Q_i < Q_i^{potential} = \lambda_i \cdot f(P_{avg}) \quad (19)$$

Therefore, the complete mathematical expression for safe dosing is as follows:

$$Q_i^{final} = \min\left(\frac{P_{frac} - P_{avg}}{R_i \cdot L}, \lambda_i \cdot f(P_{avg}), Q_i\right) \quad (20)$$

In the above equation, P_{wf} is the bottomhole flowing pressure, MPa; P_{frac} is the formation fracture pressure, MPa; P_{avg} is the average reservoir pressure, MPa; and λ_i is the reservoir potential coefficient.

In practical regulation, the seepage resistance R_L of each layer is first calculated using Equation (10), and then the theoretical liquid absorption ratio f_i of each layer is calculated using Equation (12). This formula clearly states that the layer with lower seepage resistance should have a lower theoretical liquid absorption ratio; on the contrary, layers with high seepage resistance should have a higher water injection ratio to suppress layer interference. Based on the total injection volume Q_{Total} of the entire well, calculate the theoretical injection volume Q_i of each layer. Calculate the maximum safe injection volume Q_i^{frac} of a single layer based on the formation fracture pressure according to Equation (16), and then estimate the maximum injection volume $Q_i^{potential}$ of the reservoir potential limit by combining the static parameters of the reservoir (such as effective thickness and permeability) according to Equation (17). Finally, calculate Q_i^{final} based on the safety injection volume constraint conditions of Equation (20). In practical regulation, regulation is not a one-time process, but a cyclical process that changes dynamically with the development of the oilfield. The adjustment period in the early stage of high water cut is usually set at 1–3 months. By dynamically adjusting the seepage resistance of each layer through the influence of different water cut phases on the liquid absorption ratio of each layer, the dynamic control effect can be achieved.

When in the actual regulation process, with the change in seepage resistance, the difference in seepage resistance between layers exceeds the optimal range of 3–5, and the layers are recombined to ensure that each layer is within a reasonable range of seepage resistance difference, and the injection volume also changes accordingly.

6. Practical Application

This article adopts a multi-stage isolation system for layered water injection, with multiple isolation devices installed inside each water injection well to ensure that the water flow in each layer does not interfere with each other during the water injection process. After calculating the injection volume of each layer through a formula, layer-by-layer water injection is carried out, and fine water injection is carried out by plugging other layers while injecting water into the injection layer. Firstly, a numerical simulation model is used to study the dynamic regulation of layered water injection based on seepage resistance at different water content times according to the layered water injection scheme.

Taking the actual model of Block D as an example. The stratum G11–G120 was fully perforated and produced, with a five-point well network comprising 6 injection wells and 13 production wells. Formulas were derived to calculate the relationship between seepage resistance, liquid absorption ratio, and injection allocation for the sublayer. A zoned injection scheme was designed based on seepage resistance differentials for various water cut scenarios, contrasting it with a blanket injection approach. Eclipse software was employed for scheme prediction, with projected recovery rates presented in Appendix A, Table A1. See Figures 11 and 12 for the prediction effect of the scheme.

(**a**) Fifth Level (**b**) Nineteenth Level (**c**) Thirty-first Level

Figure 11. Oil saturation map for selected sublayers in the original plan.

The stage recovery rate for the general water injection method is 28.11%. Starting from a water cut of 70%, predict the recovery rate and implement layered water injection based on the flow resistance of different zones at various water content levels. As shown in Figure 13, the effectiveness of zoned injection diminishes as water cut increases. Between 70% and 78% water cut, zoned injection based on gradient flow resistance yields optimal results, achieving a maximum recovery rate increase of 3.97% (See Appendix B for the benefit analysis of this plan). However, the advantage of zoned injection significantly declines if water cut exceeds 80%, and no further improvement in recovery is observed when water cut exceeds 90%.

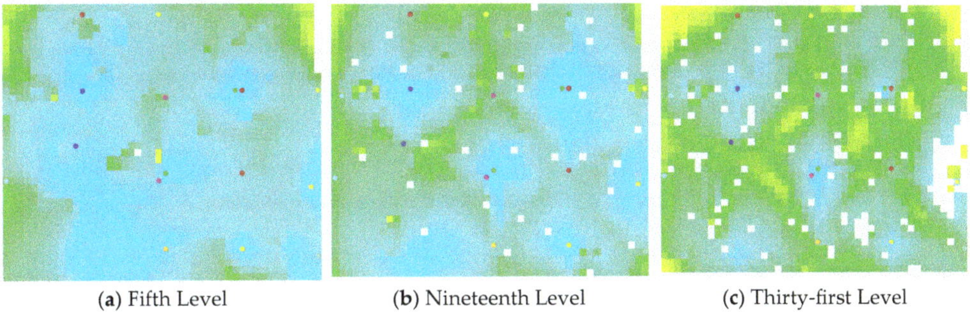

Figure 12. Oil saturation in small layers during staged water injection at 70% water cut.

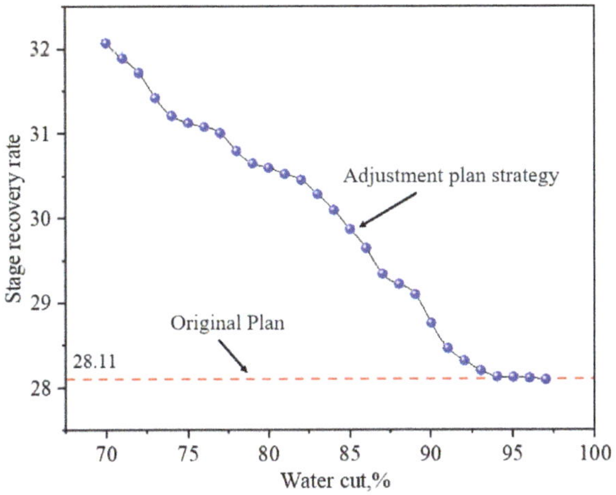

Figure 13. Recovery rate during staged water injection at different water content levels.

Fluid flow resistance is low in high-permeability layers. If not controlled appropriately, injected water will preferentially flow rapidly through these high-permeability zones, causing further erosion and enlargement of their pore channels. This means that a significant portion of injected water will be ineffectively circulated during the development stage without early intervention, failing to efficiently drive crude oil in low-permeability layers. Therefore, implementing early-stage layered water injection—where the injection rate in smaller layers is matched to the permeability of high-permeability zones—effectively enhances displacement pressure in low-permeability layers. This accelerates the development of low-permeability zones, maximizing the vertical utilization of the reservoir. When the moisture content is greater than 90% and the interlayer seepage resistance is controlled within the range of 3–5, there is no significant increase in recovery rate, and a decrease in recovery rate may even occur, compared to the medium to high moisture content period. The core reason for the decrease in recovery rate is that the main contradiction in development has changed. The differential control of seepage resistance is an extremely effective technique for improving macroscopic sweep efficiency and tapping interlayer potential, but its effectiveness is limited by two more fundamental physical and geological conditions: low micro displacement efficiency and insufficient movable oil saturation in the ultra-high water cut stage. To further improve the recovery rate at this stage, it is often necessary to turn to more powerful methods such as chemical or gas flooding to change the flowability ratio and interfacial tension, thereby improving the micro displacement efficiency.

7. Conclusions

Based on the dynamic evolution characteristics of seepage resistance, this study proposes an integrated, refined layered water injection methodology tailored for high water cut reservoirs. The results demonstrate that seepage resistance, as a dynamic characterization parameter, more accurately describes fluid seepage patterns in reservoirs during high water cut stages. This approach effectively overcomes the limitations of traditional static parameters and offers a novel technical solution for efficient oilfield development. The main findings are as follows:

(1) The coefficient of variation in seepage resistance exhibits a significant positive correlation with water cut. When the water cut exceeds 70%, the coefficient increases sharply, indicating that static analysis methods based on permeability differences can no longer accurately characterize the actual flow behavior within the reservoir. In contrast, seepage resistance, due to its dynamic response characteristics, more directly reflects the true flow conditions of individual sublayers.

(2) By integrating the two-phase Darcy's Law with the water-electricity analogy principle, a practical calculation model for the seepage resistance coefficient was developed. Using a self-developed algorithm, the study quantitatively characterized layer seepage resistance. It determined that the optimal range for the seepage resistance step difference, used as a layering criterion, is 3–5, thereby providing a scientific basis for optimizing layer combination.

(3) An innovative, refined water injection control method based on the dynamic characteristics of seepage resistance was proposed. This method dynamically optimizes layered water injection strategies by monitoring real-time changes in seepage resistance, significantly enhancing the accuracy and responsiveness of injection control.

(4) The effectiveness of layered water injection is significantly influenced by the water cut. Experimental data show that within the water cut range of 70% to 78%, the injection strategy guided by seepage resistance differential achieves optimal results, with a projected recovery increase of 3.97 percentage points. This approach offers an effective means of enhancing production in high water cut oilfields.

This study confirms the superiority of flow resistance in characterizing the dynamic properties of high water cut reservoirs. The developed dynamic control method offers new technical support for enhancing the development efficiency of these reservoirs.

Author Contributions: Conceptualization, C.D.; investigation, G.S.; writing—original draft, W.H.; writing—review and editing, E.Y., M.Q., and H.W. All authors have read and agreed to the published version of the manuscript.

Funding: The research was supported by the Hainan Provincial Joint Project of Sanya Yazhou Bay Science and Technology City, Grant (No:2021JJLH0059), and Heilongjiang Postdoctoral Fund (LBH-Z23105).

Data Availability Statement: The original contributions presented in this study are included in the article. Further inquiries can be directed to the corresponding author.

Conflicts of Interest: Author Gaojun Shan was employed by the company Daqing Oilfield Co., Ltd. The remaining authors declare that the research was conducted in the absence of any commercial or financial relationships that could be construed as a potential conflict of interest.

Appendix A

Table A1. Layered water injection plan at different water content levels.

Water Cut, %	Layer-Segment Combination	Seepage Resistance Gradient	Stage Recovery Rate, %
70	G11–G18	2.80	32.075
	G19–G111	2.04	
	G112–G116	2.79	
	G117–G120	4.68	
71	G11–G18	2.77	31.896
	G19–G111	2.05	
	G112–G116	2.78	
	G117–G120	4.67	
72	G11–G18	2.77	31.724
	G19–G111	2.06	
	G112–G116	2.77	
	G117–G120	4.68	
73	G11–G18	2.89	31.428
	G19–G111	2.07	
	G112–G116	2.75	
	G117–G120	4.69	
74	G11–G18	3.33	31.215
	G19–G111	2.09	
	G112–G116	2.73	
	G117–G120	4.71	
75	G11–G18	3.55	31.131
	G19–G111	2.10	
	G112–G116	2.72	
	G117–G120	4.73	
76	G11–G18	3.60	31.082
	G19–G111	2.11	
	G112–G116	2.69	
	G117–G120	4.76	
77	G11–G18	3.58	31.009
	G19–G111	2.11	
	G112–G116	2.68	
	G117–G120	4.79	
78	G11–G18	3.55	30.998
	G19–G111	2.09	
	G112–G116	2.67	
	G117–G120	4.82	
79	G11–G18	3.51	30.998
	G19–G111	2.06	
	G112–G116	2.67	
	G117–G120	4.85	
80	G11–G18	3.46	30.698
	G19–G111	2.00	
	G112–G116	2.67	
	G117–G120	4.90	
81	G11–G18	3.43	30.589
	G19–G111	1.94	
	G112–G116	2.67	
	G117–G120	4.94	
82	G11–G18	3.41	30.458
	G19–G111	1.85	
	G112–G116	2.68	
	G117–G120	4.99	

Table A1. *Cont.*

Water Cut, %	Layer-Segment Combination	Seepage Resistance Gradient	Stage Recovery Rate, %
83	G11–G18	3.39	30.289
	G19–G111	1.77	
	G112–G117	4.83	
	G118–G120	4.76	
84	G11–G18	3.40	30.103
	G19–G111	1.76	
	G112–G117	4.83	
	G118–G120	4.76	
85	G11–G18	3.39	29.876
	G19–G111	1.80	
	G112–G117	4.82	
	G118–G120	4.77	
86	G11–G18	3.40	29.654
	G19–G111	1.88	
	G112–G117	4.79	
	G118–G120	4.78	
87	G11–G18	3.40	29.345
	G19–G111	1.95	
	G112–G117	4.74	
	G118–G120	4.80	
88	G11–G18	3.41	29.223
	G19–G111	1.98	
	G112–G117	4.77	
	G118–G120	4.81	
89	G11–G18	3.42	29.101
	G19–G111	2.00	
	G112–G117	4.85	
	G118–G120	4.83	
90	G11–G18	3.42	28.765
	G19–G111	2.01	
	G112–G117	4.93	
	G118–G120	4.86	
91	G11–G18	3.42	28.469
	G19–G114	3.53	
	G115–G117	4.00	
	G118–G120	4.89	
92	G11–G18	3.40	28.324
	G19–G114	3.63	
	G115–G117	4.01	
	G118–G120	4.94	
93	G11–G18	3.37	28.208
	G19–G114	3.70	
	G115–G117	4.01	
	G118–G120	5.01	
94	G11–G18	3.33	28.134
	G19–G114	3.75	
	G115–G117	4.00	
	G118–G120	5.08	
95	G11–G18	3.31	28.130
	G19–G114	3.84	
	G115–G117	3.99	
	G118–G120	5.14	

Table A1. *Cont.*

Water Cut, %	Layer-Segment Combination	Seepage Resistance Gradient	Stage Recovery Rate, %
96	G11–G18	3.30	28.120
	G19–G114	3.97	
	G115–G117	3.99	
	G118–G120	5.17	
97	G11–G18	3.28	28.100
	G19–G114	4.12	
	G115–G117	4.04	
	G118–G120	5.20	

Appendix B

Taking layered water injection starting from a moisture content of 70% as an example, the specific cost analysis data can be found in Table A2.

Table A2. Economic parameter data.

Initial Investment			Operation and Maintenance Costs (Project Operating Cycle of 20 Years)		
Layered Water Injection System Device	Software Development Cost	System Integration and Adjustment	Data Monitoring and Analysis	Equipment Maintenance and Updates	Total Annual Operating Costs
CNY 500,000 per well	CNY 500,000	CNY 200,000	CNY 200,000 per year	CNY 100,000 per year	CNY 300,000 per year
Oil price	Discount rate	Management fee		Safety production cost	
2969 CNY/ton	8%	400 CNY/ton		20 CNY/ton	

$$\text{Initial investment} = 500{,}000 \times 6 + 500{,}000 + 200{,}000 = \text{CNY } 3{,}700{,}000$$

$$\text{Operation and maintenance costs} = (200{,}000 + 100{,}000 + 300{,}000) \times 20 = \text{CNY } 1{,}200{,}000$$

$$\text{Total initial investment} = 3{,}700{,}000 + 1{,}200{,}000 = \text{CNY } 4{,}900{,}000$$

The geological reserves of crude oil in Block D are about 1.8×10^6 ton, and the oil recovery rate has been improved by 3.97%.

$$\text{Additional crude oil production} = 1.8 \times 10^6 \times 3.97\% = 71{,}460 \text{ ton}$$

$$\text{Extra income} = 71{,}460 \times 2969 = \text{CNY } 212{,}160{,}000$$

$$\text{Management fee and safety production fee} = (400 + 20) \times 71{,}460 = \text{CNY } 30{,}010{,}000$$

$$\text{Total benefit} = 212{,}160{,}000 - 30{,}010{,}000 = \text{CNY } 182{,}150{,}000$$

Calculate net present value (NPV), where t is the project running cycle:

$$\text{NPV} = \sum_{t=1}^{20} \frac{\text{Total benefit}}{(1+0.08)^t} - \text{Total initial investment}$$

The calculated NPV is approximately CNY 34.188 million, and the net present value is much greater than 0, indicating that this scheme has high economic efficiency.

Conduct sensitivity analysis to evaluate the impact of uncertainty.

Crude oil price fluctuations: If the oil price drops by 20%, the additional revenue will decrease to CNY 169.73 million, and the NPV will still be about CNY 25.083 million, indicating that the plan is still profitable.

Cost overrun: If the initial investment increases by 50%, the total cost will be CNY 7.35 million, but the NPV will still be as high as CNY 31.738 million, and the economy will not be significantly affected.

The increase in recovery rate: If the recovery rate only increases by 2%, the additional income will be CNY 106.88 million, and the NPV will still be about CNY 11.596 million. The project is still feasible.

Economic benefit analysis shows that the fine water injection interval dynamic control method based on seepage resistance has significant economic feasibility in the application of D block.

References

1. Wang, D.; Liu, F.; Li, G.; He, S.; Song, K.; Zhang, J. Characterization and Dynamic Adjustment of the Flow Field during the Late Stage of Waterflooding in Strongly Heterogeneous Reservoirs. *Energies* **2023**, *16*, 831. [CrossRef]
2. Anderson, W.G. Wettability Literature Survey Part 5: The Effects of Wettability on Relative Permeability. *J. Pet. Technol.* **1987**, *39*, 1453–1468. [CrossRef]
3. Jiang, R.; Qiao, X.; Teng, W.; Xu, J.; Sun, S.; Xie, L. Impact of physical properties time variation on waterflooding reservoir development. *Fault-Block Oil Gas Field* **2016**, *23*, 768–771. Available online: https://www.dkyqt.com/#/digest?ArticleID=3774 (accessed on 13 November 2025).
4. Spooner, V.E.; Geiger, S.; Arnold, D. Dual-porosity flow diagnostics for spontaneous imbibition in naturally fractured reservoirs. *Water Resour. Res.* **2021**, *57*, e2020WR027775. [CrossRef]
5. Victoria, S.; Sebastian, G.; Arnold, D. Ranking Fractured Reservoir Models Using Flow Diagnostics. In Proceedings of the SPE Reservoir Simulation Conference, Galveston, TX, USA, 10–12 April 2019. [CrossRef]
6. Willis, B.J.; Sech, R.P. Quantifying Impacts of Fluvial Intra-Channel-Belt Heterogeneity on Reservoir Behaviour. In *Fluvial Meanders and Their Sedimentary Products in the Rock Record*; Ghinassi, M., Colombera, L., Mountney, N.P., Reesink, A.J.H., Bateman, M., Eds.; John Wiley & Sons: Hoboken, NJ, USA, 2018. [CrossRef]
7. Baker, R.O.; Kuppe, F.; Chugh, S.; Bora, R.; Stojanovic, S.; Batycky, R. Full-Field Modeling Using Streamline-Based Simulation: Four Case Studies. *SPE Reserv. Eval. Eng.* **2002**, *5*, 126–134. [CrossRef]
8. Datta-Gupta, A. Streamline Simulation: A Technology Update (includes associated papers 71204 and 71764). *J. Pet. Technol.* **2000**, *52*, 68–84. [CrossRef]
9. Batycky, R.P.; Thiele, M.R. Technology Update: Mature Flood Surveillance Using Streamlines. *J. Pet. Technol.* **2016**, *68*, 22–25. [CrossRef]
10. Chen, H.; Park, J.; Datta-Gupta, A.; Shekhar, S.; Grover, K.; Das, J.; Shankar, V.; Kumar, M.S.; Chitale, A. Improving Polymerflood Performance Via Streamline-Based Rate Optimization: Mangala Field, India. In Proceedings of the SPE Improved Oil Recovery Conference, Virtual, 31 August–4 September 2020. [CrossRef]
11. Thiele, M.R.; Batycky, R.P. Using Streamline-Derived Injection Efficiencies for Improved Waterflood Management. *SPE Reserv. Eval. Eng.* **2006**, *9*, 187–196. [CrossRef]
12. Jia, H.; Deng, L. Water flooding flowing area identification for oil reservoirs based on the method of streamline clustering artificial intelligence. *Pet. Explor. Dev.* **2018**, *45*, 328–335. [CrossRef]
13. Tanaka, S.; Onishi, T.; Kam, D.; Dehghani, K.; Wen, X.-H. Application of Combined Streamline Based Reduced-Physics Surrogate and Response Surface Method for Field Development Optimization. In Proceedings of the International Petroleum Technology Conference, Dhahran, Saudi Arabia, 13–15 January 2020. [CrossRef]
14. Zhao, C.; Kong, X.; Shi, L.; Zhang, Z.; Zhang, C. Numerical Simulation Study on Factors Affecting Water-Flooding Efficiency. *Contemp. Chem. Ind.* **2016**, *45*, 1640–1642+1646. [CrossRef]
15. Zhang, B. Study on Numerical Simulation of Reservoir Fine Water Injection Development Considering Reservoir Heterogeneity. *Int. J. Energy* **2025**, *6*, 65–68. [CrossRef]
16. Wang, F.; Xu, D.; Yu, L.; Zhang, F.; Liu, C.; Zhen, K. The Full-Automatic Real-Time Display, Testing and Adjustable System in Separated Layers Water Injection Well. In Proceedings of the SPE Intelligent Energy International, Utrecht, The Netherlands, 27–29 March 2012. [CrossRef]
17. Sun, F.; Jia, D.; Yu, J.; Pei, X.; Liu, H.; Zheng, L.; Shujun, B.A. Separate Layer Water Injection for Offshore Directional Wells. In Proceedings of the SPE Asia Pacific Oil & Gas Conference and Exhibition, Perth, Australia, 25–27 October 2016. [CrossRef]

8. Song, Z.; Liu, Y.; Yang, J.; Zhou, H.; Feng, G.; Huo, L.; Wang, Y.; Zhang, Y.; Cao, X.; Liu, B.; et al. Application of Remote Monitoring and Controlling Intelligent Separated Layer Water Injection Technology in BQ Oilfield, Dagang. In Proceedings of the International Petroleum Technology Conference, Dhahran, Saudi Arabia, 12–14 February 2024. [CrossRef]
9. Guo, J.; Yang, E.; Zhao, Y.; Fu, H.; Dong, C.; Du, Q.; Zheng, X.; Wang, Z.; Yang, B.; Zhu, J. A New Method for Optimizing Water-Flooding Strategies in Multi-Layer Sandstone Reservoirs. *Energies* **2024**, *17*, 1828. [CrossRef]
0. Du, Q. Variation law and microscopic mechanism of reservoir permeability in sandstone oilfields by long-term water flooding. *Acta Pet. Sin.* **2016**, *37*, 1159–1164. [CrossRef]
1. Huang, B.; Xu, R.; Fu, C.; Zhang, W.; Shi, Z. Multi-level fuzzy identification method for interwell thief zone. *Lithol. Reserv.* **2018**, *30*, 105–112. Available online: https://kns.cnki.net/kcms2/article/abstract?v=7DUETwKSFy4IBT6fyDUNF-ezFTCv1Q4kKtXjUwyzQwvMy-kFJ6CQ0NQ-OvIKDSo29Umz4uW_lDZD34FOa_eleq65k_1wlgKiircCj18v-uFeB9mxz-O3tB2HdKsSR3u-eUJMOylcbnUqTO4K1mq185dnMoxFVkbe7xQTKrl0q-5ELEGN43Kv1g==&uniplatform=NZKPT&language=CHS (accessed on 13 November 2025).
2. Shan, G.; Wang, C.; Wang, Z.; Jiang, X.; Guo, J. Water Injection Adjustment Method Based on Dynamic Flow Resistance. *Xinjiang Pet. Geol.* **2023**, *44*, 435–441. [CrossRef]
3. Luo, X.; Peng, B. Classification and Management Method for Non-productive Circulation Based on Seepage Resistance. Xi'an Petroleum University, Shaanxi Petroleum Society. In Proceedings of the 2019 International Conference on Oil and Gas Field Exploration and Development; Geological Brigade, Fourth Oil Production Plant, Daqing Oilfield, Xian, China, 16–18 October 2019; pp. 1781–1787. [CrossRef]
4. Gao, W.; Yin, R.; Gao, N.; Sheng, H.; Gao, Z. Establishment and Application of a New Analytical Fractional Flow Equation. *Xinjiang Pet. Geol.* **2021**, *42*, 179–187. [CrossRef]
5. Zhang, L.; Tian, P.; Ren, Y.; Yin, Y.; Xing, C.; Lu, X. Optimization of Water Injection Pressure for Water-Flood Development in Offshore Oilfields. *Xinjiang Pet. Geol.* **2021**, *42*, 63–67. [CrossRef]

processes

MDPI

Article

Morphological Controlling Factors of Braided River Reservoir Based on Delft3D Sedimentary Numerical Simulation: Application to Ordos Basin, China

Jinbu Li [1], Kanglong Wang [2,3], Fuping Li [1], Zhixin Ma [1], Xinqiang Liu [1] and Yuming Liu [2,3,*]

1 Exploration and Development Research Institute, PetroChina Changqing Oilfield Company, Xi'an 710018, China; ljb2_cq@petrochina.com.cn (J.L.); lfp_cq@petrochina.com.cn (F.L.); mzx_cq@petrochina.com.cn (Z.M.); liuxinqiang0509@163.com (X.L.)
2 State Key Laboratory of Petroleum Resources and Engineering, China University of Petroleum (Beijing), Beijing 102249, China; wkanglong0706@163.com
3 College of Geoscience, China University of Petroleum (Beijing), Beijing 102249, China
* Correspondence: liuym@cup.edu.cn; Tel.: +86-89733926

Abstract: To reveal the regulatory mechanisms and differences in sensitivity of hydrodynamic forces and sediment parameters to the sedimentary evolution of braided river channel bars, this study takes the Sulige Gas Field as a case study and conducts 21 sets of sedimentary numerical simulation experiments using the controlled variable method. The three parameters of discharge, slope gradient, and sediment grain size were fixed, while the target variable was adjusted iteratively. After the river reaches a steady state, quantitative statistics of the area and length-width ratio of 547 identified channel bars are carried out, and sensitivity evaluation is performed by combining principal component analysis and multiple linear regression. The results show that the sedimentary evolution of braided rivers follows a unified evolutionary law, the evolution of channel bars is synergistically regulated by parameter combinations. Under the action of single factors, an increase in discharge promotes the axial extension and scale expansion of channel bars; an increase in grain size enhances the morphological stability of channel bars; slope gradient controls the erosion-deposition balance through gravitational potential energy. The parameter sensitivity is ranked as slope gradient, discharge, sediment grain size.

Keywords: sedimentary numerical simulation; Sulige Gas Field; single-factor analysis; channel bars; braided rivers

1. Introduction

As widely distributed geomorphic units in terrestrial environments, braided rivers produce sedimentary features known as channel bars, which serve not only as preferential reservoir units for resources such as oil and gas but also as carriers of key historical information regarding fluvial sedimentary evolution [1–3]. Clarifying the developmental patterns of channel bar morphology is critical for explaining fluvial sedimentary processes and predicting sedimentary evolution; furthermore, as vital reservoir hosts for resources like oil and gas, the sedimentary morphology of channel bars directly influences key development parameters, including reservoir permeability and fluid migration efficiency. Conducting numerical simulation studies is therefore not only fundamental to fine-scale reservoir characterization but also aligns with practical demands for efficient resource utilization [4–6]. In recent years, qualitative or semi-quantitative investigations have been conducted via approaches such as natural river observations and simulation experiments [7–14]. However, existing numerical simulations of braided river sedimentation exhibit certain limitations [15,16], Li et al. (2023) developed a numerical model of sandy braided rivers based on Delft3D and verified the controlling effect of slope gradient on channel bar migration rates through 28 experimental sets. Nevertheless, their study employed only the "slope gradient-discharge" two-parameter system, failed to incorporate the synergistic influence of sediment grain size, and did not establish a quantitative functional relationship between channel bar area and the parameters [17]. Guo et al. (2025) performed sedimentation simulations of sandy braided rivers under varying discharge conditions and observed that high discharge enhances the overlapping frequency of channel bars. However, their analysis focused solely on the independent effect of discharge as a single factor, omitted the quantification of coupling weights for slope gradient and sediment grain size, and did not address the response patterns of morphological parameters [18]. Feng et al. (2025) derived a semi-empirical formula linking channel bar scale to hydrodynamic intensity through quantitative flume experiments; however, these experiments were conducted under ideal laboratory conditions and did not extend to multi-parameter dynamic evolution scenarios over geological time scales via numerical simulations [19].

Existing research on braided rivers primarily focuses on fluvial evolution mechanisms or the impacts of individual parameters on channel sandbodies [7,14,20], and has not systematically analyzed the synergistic regulatory relationships among discharge, slope gradient, and sediment grain size. This gap prevents the effective interpretation of the complete evolutionary process of sandbars under the combined action of multiple factors in natural channels. Additionally, quantitative functional relationships between topography, discharge, sediment grain size, and channel bar morphology remain unestablished, which constrains the accurate understanding of braided river sedimentary processes.

By virtue of its ability to construct reproducible models of complex sedimentary processes and achieve precise control over multiple variables, sedimentary numerical simulation technology has become a core approach to overcoming the limitations of traditional research methods and deeply analyzing the developmental mechanisms of braided river channel bars. Among various simulation tools, those based on physical principles include Delft3D software version 4.04.01, are most widely applied [21–24]. Delft3D software relies on a hydrodynamic simulation framework built on the Navier–Stokes equations, enabling it to accurately simulate the dynamic changes in fluvial sedimentation under different sedimentary conditions. It has demonstrated unique advantages in the quantitative study of fluvial geomorphology and sedimentary processes [14,17,18,20,25]. Given that the coupled driving effects of topography, hydrodynamics, and sediment grain size on channel bar morphology have not yet been systematically quantified and explained, this study employs Delft3D software to design multiple sets of simulation experiments covering different topographic gradients, discharges, and sediment grain sizes. The aim is to systematically explore the developmental patterns of braided river channel bars under the synergistic action of multiple factors, establish a predictive equation for the area of braided river channel bars, provide a scientific basis for establishing quantitative relationships between "topography-hydrodynamics-sediment properties" and channel bar morphology, and offer theoretical and data support for advancing the theory of braided river sedimentary dynamics.

2. Materials and Methods

2.1. Sedimentary Numerical Simulation Experimental Scheme

This study conducts research on the influencing factors of braided river channel bars based on sedimentary numerical simulation experiments and has designed 21 sets of simulation experiments to explore the coupled effects of topographic gradients, discharges, and sediment grain sizes on the morphological development of braided river channel bars. The value ranges of these factors basically cover the general conditions of natural braided rivers. This experiment aims to systematically reveal the action mechanisms of different influencing factors and the synergistic effects of multiple factors through the controlled variable method. The analysis and interpretation of the simulation results mainly include two aspects: first, investigating the planar morphological scale under different factor combinations, which is based on the visual delineation and quantitative analysis of channel bars in the simulation results; second, exploring the quantitative functional relationships between different factor combinations and channel bar scales. This analysis is mainly achieved by comparing the results of different simulation experiments and aims to separately test the sensitivity of model outputs to water discharge, topographic gradient, and sediment grain size.

2.2. *Sedimentary Numerical Simulation Based on Delft3D*

As a typical type of active channel system, braided rivers can be classified into two main categories, namely sandy braided rivers and gravelly braided rivers, based on the grain size characteristics of their sediment composition [19,26]. Focusing on the sandy braided river system, this study aims to reveal the factors influencing channel bar morphology through the targeted coupling of numerical simulations with the sedimentary context of the study area, building on the mature technology of the Delft3D sedimentary numerical simulation platform. With reference to the hydrodynamic and sedimentary parameters of modern natural sandy braided rivers, the study sets reasonable variable ranges and boundary conditions and systematically conducts sedimentary numerical simulation experiments.

2.2.1. Principles of Sedimentary Numerical Simulation

The Delft3D-FLOW module is the core tool used in this study to simulate the hydrodynamic and sedimentary processes of braided rivers. Its basic principle is to solve the reasonably simplified Navier–Stokes equations (N-S equations) using numerical methods [27] so as to reproduce the processes of flow movement, sediment transport, and riverbed evolution. Due to the characteristics of natural river systems such as large spatiotemporal scales, complex boundaries, and coupling of multiple physical processes, it is quite difficult to directly solve the three-dimensional N-S equations in their full form. Therefore, Delft3D introduces several basic assumptions when constructing the hydrodynamic model to balance computational efficiency and physical authenticity. These assumptions mainly include the shallow water assumption, which is suitable for water bodies where the horizontal scale is much larger than the vertical scale; the incompressible fluid assumption, which ignores changes in water density; the Boussinesq approximation, which considers density changes only in the buoyancy term; the hydrostatic pressure assumption, which ignores vertical acceleration and simplifies the vertical momentum equation to a hydrostatic equilibrium relationship; and the Reynolds-averaged method, which decomposes turbulent motion into time-averaged and fluctuating components with closure achieved via a turbulence model.

$$
\begin{aligned}
&\frac{\partial u}{\partial t} + u\frac{\partial u}{\partial x} + v\frac{\partial u}{\partial y} + g\frac{\partial z_w}{\partial x} + \frac{gu\sqrt{u^2+v^2}}{C^2h} - V\left(\frac{\partial^2 u}{\partial x^2} + \frac{\partial^2 u}{\partial y^2}\right) + F_x = 0\\
&\frac{\partial v}{\partial t} + u\frac{\partial v}{\partial x} + v\frac{\partial v}{\partial y} + g\frac{\partial z_w}{\partial y} + \frac{gv\sqrt{u^2+v^2}}{C^2h} - V\left(\frac{\partial^2 v}{\partial x^2} + \frac{\partial^2 v}{\partial y^2}\right) + F_y = 0\\
&\frac{\partial h}{\partial t} + \frac{\partial hu}{\partial x} + \frac{\partial hv}{\partial y} = 0
\end{aligned}
$$

Regarding the selection of coordinate systems, Delft3D-FLOW provides multiple options to adapt to the simulation of water bodies with different scales and types. This study adopts the orthogonal curvilinear coordinate system (ξ, η) for horizontal discretization. This co-ordinate system can generate body-fitted grids through the Delft3D-RGFGRID module, which better fits the irregular boundaries of natural rivers and effectively reduces the discretization errors caused by traditional rectangular grids at the boundaries. In the vertical direction, the model uses the σ coordinate for layering, which is defined as $\sigma = (Z - \zeta)/H$. Here, Z represents the vertical physical coordinate, ζ denotes the water level, and H stands for the total water depth. The σ coordinate ranges from -1 to 0, enabling accurate fitting from the riverbed bottom ($\sigma = -1$) to the water surface ($\sigma = 0$). It is particularly suitable for simulating hydrodynamic and sediment processes in areas with significant changes in water depth.

On the basis of the aforementioned assumptions, Delft3D-FLOW mathematically describes the flow of water and sediment movement through a set of governing equations. The continuity equation ensures the mass conservation of water, while the momentum equations—based on the Reynolds-averaged N-S equations—describe the motion law of water flow under the combined action of pressure gradients, Coriolis forces, turbulent stresses, and bed friction.

For the sediment transport module, Delft3D classifies sediments into cohesive and non-cohesive categories based on their material properties and adopts different models for each type. For cohesive sediments, their transport is mainly dominated by suspension transport. The settling velocity is calculated using an empirical formula that accounts for the influence of salinity, while the erosion and deposition fluxes are described by the Partheniades–Krone formula. This model incorporates critical erosion and deposition shear stresses as the main criteria. For non-cohesive sediments, such as sandy sediments, the Van Rijn formula is used to calculate the settling velocity and near-bed reference concentration. This model is suitable for simulating the transport and exchange processes of medium-coarse sand particles in the form of bed load and suspended load [28].

In terms of numerical solution methods, Delft3D-FLOW adopts the finite difference method based on a staggered grid system. Within this grid system, scalar variables such as water level and concentration are defined at grid centers, while velocity variables are defined at grid interfaces. This setup helps improve numerical stability and simplify the handling of boundary conditions. For time integration, the alternating direction implicit (ADI) method is used, which involves dividing each time step into two half-steps. Implicit solving is performed alternately in the ξ and η directions for each half-step, while the other direction is handled explicitly. This method not only avoids the strict time-step constraints of fully explicit schemes but also offers higher computational efficiency than fully implicit methods, making it suitable for long-term and large-scale hydrodynamic-sedimentary coupled simulations.

To summarize, the numerical model established in this study has systematically achieved high-resolution and process-based simulation of the formation and evolution processes of channel bars in sandy braided rivers. This is accomplished by virtue of reasonable simplification of physical processes, adoption of the body-fitted orthogonal curvilinear grid and σ vertical coordinate, coupling of hydrodynamic-sediment transport-riverbed deformation modules, and application of the efficient and stable ADI algorithm. The model provides a reliable numerical experimental platform for exploring the response mechanisms of channel bar morphology to various environmental factors.

2.2.2. Boundary Conditions and Parameter Settings

The boundary conditions of the model mainly include grid resolution, flow conditions, and transport conditions. For channel bars in the braided rivers of the Sulige Gas Field, their lengths and widths mostly range from 100 to 850 m [29]. Accordingly, as shown in Figure 1, the model domain is set to 10 km in length and 1 km in width, with individual grids measuring 12.5 m along the long axis and 5 m along the short axis. This setup not only accurately captures subtle changes in channel bar boundaries but also avoids blurring of channel bar morphology caused by excessive grid coarseness [7,30]. Water inlets and outlets are, respectively, placed on both sides of the grid's short axis: the water inlets are defined by discharge magnitude, while the water outlets are defined by water level changes. Meanwhile, the long-axis boundaries of the grid are set as non-erosive barriers to simulate the real scenario of braided rivers flowing within river valleys [31]. While accounting for the horizontal plane, the vertical direction of the model is divided into 3 layers based on the core geological characteristics of the sandy braided rivers in the Lower Member 8 of the He8 Formation in the Sulige Gas Field, namely "shallow water, low energy, and gentle slope." Each layer has a uniform thickness, which fully covers the key vertical structures of the near-bottom layer, middle layer, and surface layer. Additionally, the model adopts a vertical σ-coordinate that automatically adapts to water depth changes through adjustments, enabling relative refinement of the near-bottom region. Combined with the setting of the near-bottom reference concentration in the Delft3D sediment module, this further eliminates the impact of layer thickness on sediment deposition calculations.

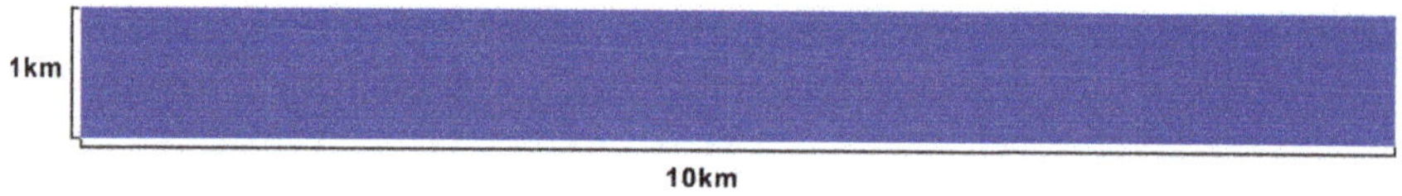

Figure 1. Grid setup.

Except for the three variables of topographic gradient, discharge, and sediment grain size, all other parameters in the experiments are set consistently to ensure a single variable in the experiments. Table 1 lists the other parameters as follows: The hydrodynamic module uses Manning's equation to calculate riverbed roughness. In sediment simulation, the dry bulk densities of sand and mud are set to 1650 kg/m^3 and 500 kg/m^3 respectively. The main physical constant, gravitational acceleration, is set to a constant value of 9.81 m/s^2. Water density is set to the freshwater density of 1000 kg/m^3. To accelerate the geomorphic evolution process and meet the requirements of the simulated event scale, the geomorphic evolution coefficient is set to 100. This coefficient speeds up the simulation while ensuring the simulation results do not become distorted. The total simulation duration is 30 days, and the computational time step is 0.5 min. If the time step is reduced to 0.25 min, the fineness of the simulation results increases by approximately 5%, while the computational load rises by roughly 53%. When the time step is increased to 1 min, however, the channel bar morphology in the simulation results becomes unstable and the fineness deteriorates.

Table 1. Basic parameter table of braided river sedimentary numerical simulation.

Parameters	Unit	Set Value
Grid Unit Size	m	5 × 12.5
Total Length of the River	km	10
River Width	km	1
Roughness Equation	-	Manning
Dry Bulk Density of Sand	kg/m^3	1650
Dry Bulk Density of Mud	kg/m^3	500
Horizontal Eddy Viscosity	m^2/s	1
Horizontal Eddy Diffusivity	m^2/s	0.1
Lower Limit of Sedimentary Thickness Change Recording	m	0.05
Gravitational Acceleration	m/s^2	9.81
Water Density	kg/m^3	1000
Simulation Duration	d	30
Geomorphic Evolution Coefficient	-	100
Time Step	min	0.5

2.2.3. Simulation Group Setup

The setup of simulation groups mainly involves establishing control groups based on three variables: topographic gradient, discharge, and sediment grain size. To ensure the simulation experiments have relevant practical basis, this study takes the He 8 Lower Submember Formation of the Sulige Gas Field in the Ordos Basin as a case study. This formation develops typical sandy braided rivers and has abundant drilling data and relevant geological materials, which can effectively support the acquisition of relevant parameters [18,32]. Based on the relevant analysis of the He 8 Lower Submember Formation of the Sulige Gas Field in the Ordos Basin, the ranges of the three variables are defined, and different simulation groups are set within these ranges. This provides data support for establishing a sedimentary numerical simulation knowledge base in this area.

The determination of the topographic gradient range mainly relies on recovering the paleoslope based on the formation thickness of the Sulige Gas Field. Based on the systematic collection and comprehensive investigation of existing geological data, this study re-covered and statistically analyzed the formation thickness of multiple wells in the Sulige area. By comparing the thickness difference in inter-well sedimentary formations (denoted as d2−d1) with their corresponding horizontal distance (S), the paleotopographic gradient of this area was estimated. The calculation results show that the paleogradient of the target interval in the Sulige Area during geological history was relatively gentle, with an average value of approximately 0.02°. This indicates that the area was once in a low-energy sedimentary environment with a weak topographic gradient. To characterize this result in detail, the formation recovery data of seven key wells in the area were statistically analyzed (Table 2). Each well provides information on the current formation thickness, sand-to-ground ratio of the Upper He Member and Lower He Member, as well as the paleothickness recovered through the sedimentary compensation principle. The paleothickness recovery results show that there is a certain spatial variability in the sedimentary preservation status of the area, with the recovered thickness ranging from 103 m to 157 m. Furthermore, four groups of well pairs were selected for topographic gradient calculation (Table 3). The results show that the gradient values between well pairs range from 0.0016° to 0.0176° [33].

Table 2. Statistical table of recovered formation thickness for partial wells in the Sulige Area.

Well Number	He 8 Upper (m)	He 8 Lower (m)	Current Thickness (m)	Sand-to-Formation Ratio (%)	Paleothickness (m)
W1	30.4	35	65.4	51.68196	137
W2	25.2	29.4	54.6	63.91941	103
W3	30.6	33	63.6	30.81761	157
W4	38.4	21.2	59.6	34.0604	144
W5	35.2	26.4	61.6	33.6039	148
W6	28.4	27.5	55.9	49.55277	119
W7	36.9	28.8	65.7	45.50989	145

Table 3. Gradient calculation table for the Sulige Area.

Well Number	d_2 (m)	d_1 (m)	Horizontal Distance S (km)	Gradient (°)
W1–W4	145	147	72.944	0.0016
W2–W7	119	141	75.402	0.0167
W3–W5	145	140	53.516	0.0054
W4–W6	103	139	117.303	0.0176

Structural restoration of the Lower Member 8 of the He8 Formation in the Sulige Gas Field during its sedimentary period shows that short-term slope gradient increases occurred in this area, reaching 15–20 times the peak value [33]. Based on the recovered slope gradient range and relevant studies, a total of three slope gradients (1 m/10 km, 3 m/10 km, and 5 m/10 km) were set, corresponding to slope values of 0.0057–0.0286°. This setup is basically consistent with the sedimentary characteristics of gentle slopes in the Sulige Area and also covers the typical gradient range of natural sandy braided rivers [34].

For each gradient value, each group is set with three different discharges and grain sizes. The discharge for each gradient is gradually increased, and after multiple adjustments, the water depth in the corresponding simulation results falls within the same range. This adjustment ensures that the comparison of discharges is reasonable, while the sediment grain sizes are set to 200 μm, 400 μm, and 800 μm.

The simulation experiments are divided into 7 groups (Group A to Group G), with each group including three experiments (Table 4). Specifically, for Groups A, C, and E, the discharge is gradually increased within each group, while the gradient and sediment grain size remain consistent across these three groups. For Groups B, D, and F, the sediment grain size is gradually increased within each group, while the topographic gradient and discharge are consistent across these three groups. Group G focuses solely on topographic gradient, with 3 gradient values set.

Table 4. Experimental group setup table.

Group	Number	Gradient (m/10 km)	Discharge (m^3/s)	Median Grain Size of the Sandy Component (μm)
A	1	1	1420	200
	2	1	1500	200
	3	1	1680	200
B	4	1	1500	200
	5	1	1500	400
	6	1	1500	800
C	7	3	1300	200
	8	3	1500	200
	9	3	1880	200
D	10	3	1500	200
	11	3	1500	400
	12	3	1500	800

Table 4. Cont.

Group	Number	Gradient (m/10 km)	Discharge (m^3/s)	Median Grain Size of the Sandy Component (μm)
E	13	5	1420	200
	14	5	1880	200
	15	5	2350	200
F	16	5	1880	200
	17	5	1880	400
	18	5	1880	800
G	19	1	1500	200
	20	3	1500	200
	21	5	1500	200

3. Results

Simulation results in Figure 2 show that all braided rivers undergo the same stages of formation and evolution, which are summarized as follows:

(a) The water flow begins to advance forward, and channel bars start to take shape in their embryonic form, exhibiting characteristics of slight undulations and downstream migration, with the height of the undulations mostly less than 1 m.

(b) After a period of evolution, the bed sandbodies form rhombic-shaped sand bars, exhibiting continuous foreset structures. The bedform overall shows slight undulations against an otherwise gentle background, with elevation differences concentrated in the range of 1–2 m.

(c) As the water flow advances further, the morphology of the sand bars changes: the rhombic bars undergo migration and transformation, and the front ends of the bars become rounded due to forward resistance, forming tongue-shaped sand bars. These exhibit the response characteristics of first eroding the bed and then forming high-angle foreset bodies, with the maximum elevation approximately 2–3 m.

(d) The tongue-shaped bars continue to migrate and merge, assuming an overall oval-shaped morphology and developing into the most common channel bars in braided river deposits. These channel bars exhibit a distinct undulating wavy morphology, demonstrate the characteristics of migration and superimposition along the downstream direction, and have an elevation of approximately 4 m.

(e) Channel bars advance downstream, and multiple ones undergo composite splicing to form larger composite bars; small crevasses form on the upper parts of these composite bars, with their elevations mostly reaching more than 6 m.

(f) Composite bars undergo further transformation: water flow erodes the composite bar bodies, and crevasses gradually become more complex, which is evident on both the tops and edges of the bar bodies. Meanwhile, various accretionary bodies tend to stabilize, and the elevation reaches near its maximum value at this stage, approximately 7–8 m.

The six evolutionary stages identified in this study not only inherit the general laws of sandy braided river channel bar evolution and are highly consistent with the models presented in existing studies [3,13,19], but the morphological transformation characteristics of the first four stages in particular all align with the classic framework of sandy braided river channel bar evolution. Additionally, the experimental parameters strictly match the geological reality of the Lower Member 8 of the He8 Formation in the Sulige Gas Field, supplementing the detail gaps of existing models under low-energy gentle slope settings. In the final evolutionary stage, the channel bars exhibit minimal fluctuations in area, indicating that the channel morphology basically tends towards a stable state. This stable node not only confirms the integrity of sandbody evolution in low-energy gentle slope environments but also provides a key temporal basis for subsequent sandbody characterization in the study area. It can clarify the temporal boundary for stable sandbody development, enabling sandbody morphology depiction to closely align with the actual evolutionary process and avoiding deviations caused by the selection of transient evolutionary data.

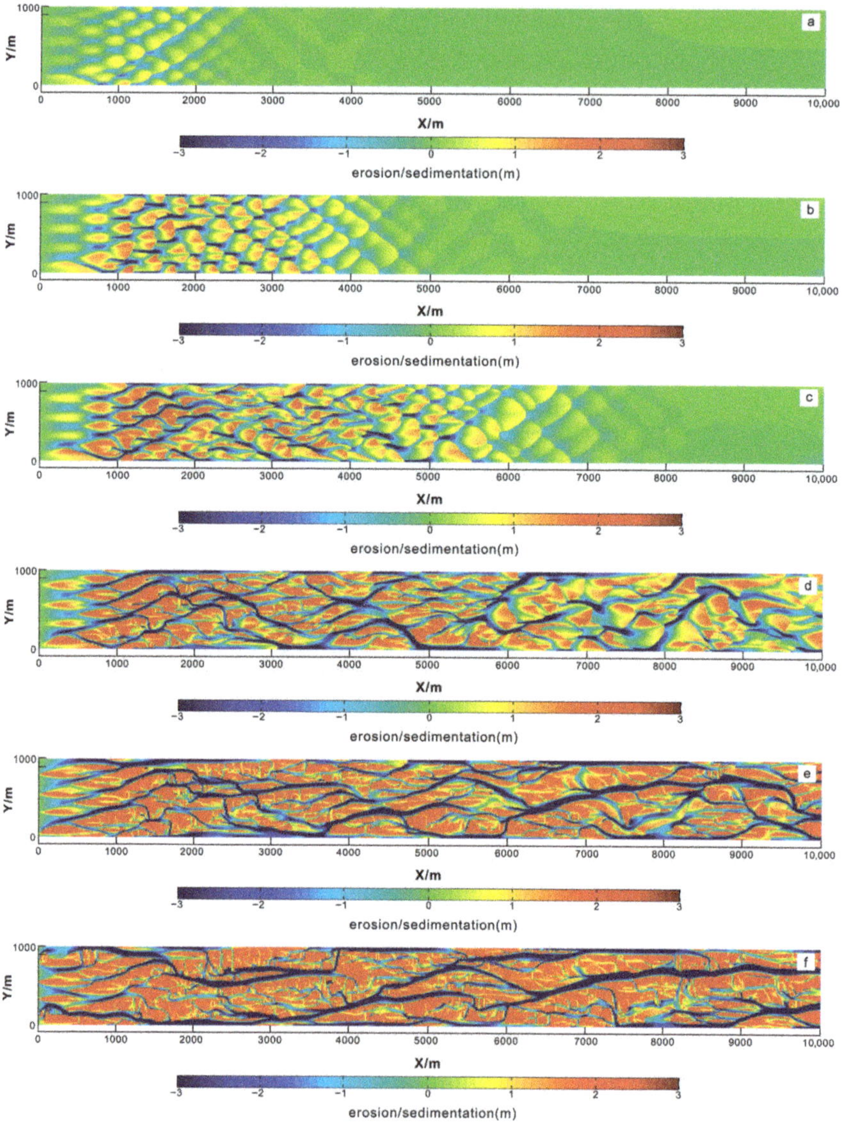

Figure 2. Sedimentary evolution process of braided rivers in sedimentary numerical simulation. (**a**) embryonic form of channel bars; (**b**) rhombic-shaped bars; (**c**) tongue-shaped bars; (**d**) unit bars; (**e**) composite bars; (**f**) composite reworked bars.

3.1. Single-Factor Analysis of Discharge

As a core characteristic parameter of hydrodynamics, discharge directly determines the efficiency of water flow in sediment transport and deposition and serves as a key driving factor for the scale expansion and morphological remodeling of channel bars. To eliminate interference from other parameters, this study takes Group C experiments as the research object and clarifies the regulatory laws of discharge by comparing the evolutionary dynamics of channel bars under different discharge conditions.

Results in Figure 3 show that within the same simulation time, under low discharge conditions, the differentiation process between channel bars and channels is gentle, with simple sandbody morphology and low remodeling frequency. As discharge increases, the hydrodynamic response to bed erosion-deposition is significantly enhanced, the rate of channel bar scale remodeling accelerates, and channel bifurcation begins to emerge. When high discharge is reached, channel bars undergo frequent migration and superimposition; moreover, after the river stabilizes, the developmental complexity of secondary landforms such as crevasses in the upper part of the bars increases significantly, forming a denser braided channel network. This result confirms the mechanism by which high discharge provides a material basis for the axial extension and scale expansion of channel bars by increasing the sediment transport flux.

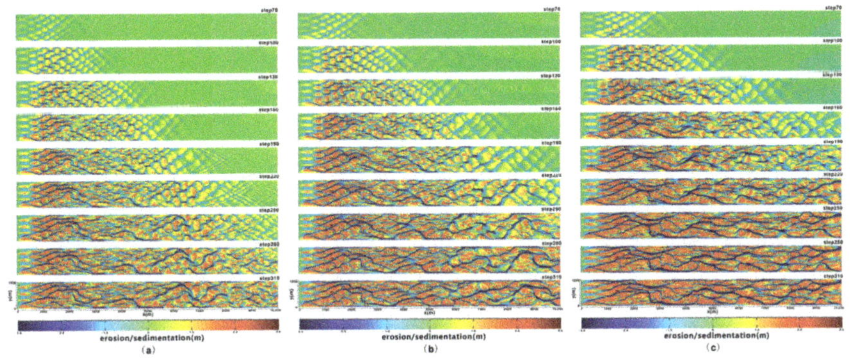

Figure 3. Evolution diagram of single-factor analysis of discharge: (**a**) discharge: 1300 m^3/s; (**b**) discharge: 1500 m^3/s; (**c**) discharge: 1880 m^3/s.

3.2. *Single-Factor Analysis of Grain Size*

Sediment grain size directly affects the morphological stability of channel bars by altering the critical shear stress for particle entrainment and the erosion resistance of particles. Focusing on the regulatory role of material properties, this study takes Group D experiments as the research object to clarify the influence of grain size on the evolutionary rhythm and stability of channel bars.

Experimental results in Figure 4 show that the transport-deposition responses of sandbodies differ significantly under different grain sizes. Fine-grained sediments are easily entrained and transported by water flow, resulting in a high frequency of erosion and disturbance on the bed, a fast channel bar remodeling rate, and a complex interweaving degree of the overall channel network. For medium-grained sediments, the transport difficulty increases; large-scale migration is only triggered when the hydrodynamic force reaches a certain intensity. At this point, the morphological stability of channel bars is significantly higher than that of fine-grained sediments, the evolutionary rhythm slows down, but well-defined braided channels can still develop. Coarse-grained sediments have the highest transport resistance; particle movement can only be initiated under stronger water flow conditions. This leads to slow formation and development of channel bars in the early stage, a low migration rate, and the formation time of secondary landforms such as channel bifurcations and crevasses is later than that of fine-grained and medium-grained sediments.

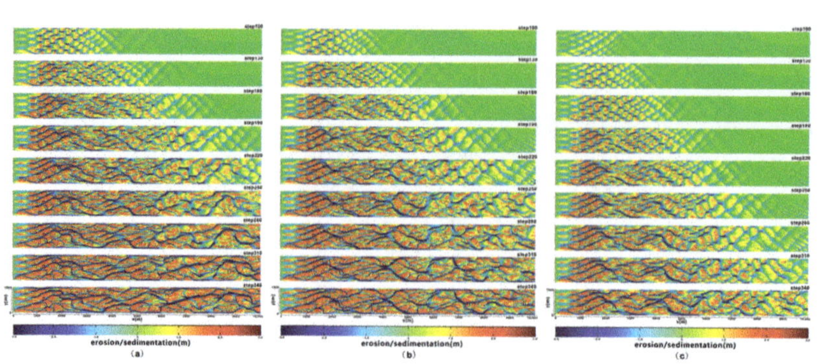

Figure 4. Evolution diagram of single-factor analysis of grain size: (**a**) grain size: 200 μm; (**b**) grain size: 400 μm; (**c**) grain size: 800 μm.

3.3. Single-Factor Analysis of Gradient

Topographic gradient directly determines the erosion-deposition balance of hydrodynamic force by regulating the gravitational potential energy of water flow and serves as a core factor affecting the scale and evolutionary complexity of channel bars. Taking Group G experiments as the research object, this study reveals the dominant constraining effect of gradient on channel bar morphology.

Simulation results show in Figure 5 that gradient exerts a significant controlling effect on the rate and complexity of sedimentary evolution. Under low gradient conditions, the gravitational potential energy of water flow is weak, and the erosion and transport capacity of hydrodynamic force on the bed is limited. The processes of channel bar formation and channel bifurcation are slow, and even in the later evolutionary stage, the overall geomorphic complexity remains low. Under medium gradient conditions, the potential energy of water flow increases significantly, and erosion-deposition rates accelerate simultaneously. The interweaving degree and evolutionary rate of the channel network increase substantially, and the superimposition frequency of channel bars is significantly higher than that under low gradient conditions. Under high gradient conditions, sediment transport and bed erosion are extremely active. Channel bars undergo frequent fragmentation and reorganization, while crevasses quickly cut through the bars and develop toward greater complexity, eventually forming braided river depositional configurations with the highest complexity.

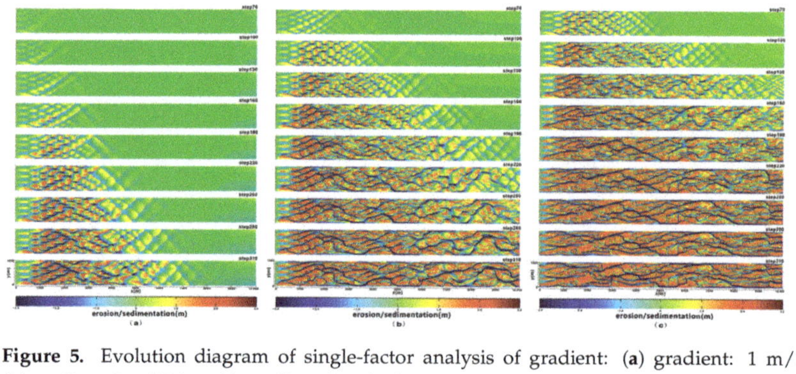

Figure 5. Evolution diagram of single-factor analysis of gradient: (**a**) gradient: 1 m/10 km; (**b**) gradient: 3 m/10 km; (**c**) gradient: 5 m/10 km.

3.4. *Quantitative Analysis*

The determination of the scale of sandstone bodies in the simulation results is based on the measurement method of channel bars in modern braided rivers. Satellite images of the lower reaches of the modern braided Lena River are captured using the Google Earth web version (Figure 6). The scale of channel bars mainly refers to the channel bar length, channel bar width, and aspect ratio. Among these, the channel bar length refers to the extension scale of the channel bar along the main flow direction of the river; by identifying the two endpoints of the channel bar along the main flow direction, the straight-line distance between the two endpoints is measured and calculated using measurement tools. The channel bar width refers to the maximum lateral scale perpendicular to the main flow direction. On the basis of the already determined channel bar length, the position with the maximum width is selected as the typical section position for width measurement according to the channel bar morphology [35,36].

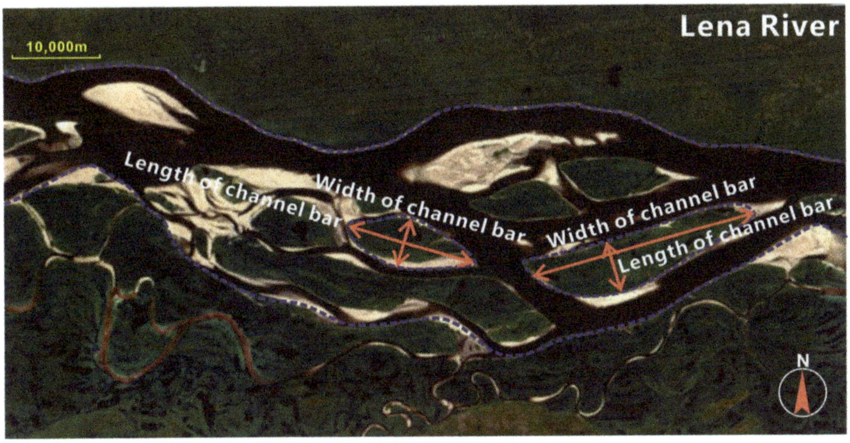

Figure 6. Schematic diagram of channel bar scale measurement method.

3.4.1. Channel Bar Scale Parameter Statistics

To accurately quantify the morphological characteristics of channel bars in braided rivers, after the river enters a stable state, taking Figure 7 as an example, at which point the height of the channel bar crests reaches or approaches the water level, this time point is estimated. Based on the aforementioned measurement method for channel bar scale, statistical analysis of morphological parameters is conducted on 547 channel bars identified in 21 groups of simulation experiments, with a focus on analyzing two key indicators: the area and aspect ratio of channel bars (Tables 5 and 6).

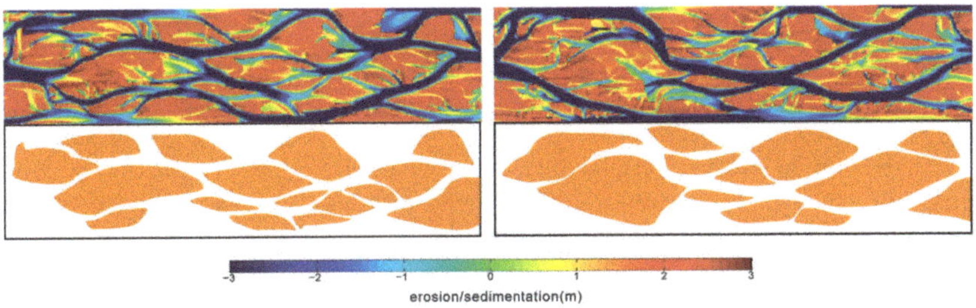

Figure 7. Schematic diagram of channel bar depiction in simulation results.

Table 5. Statistical table of channel bar aspect ratio data.

Simulation Serial Number	Maximum Value	Minimum Value	Average Value
Run1	4.26	1.50	2.60
Run2	3.90	1.07	2.51
Run3	4.69	0.84	2.70
Run4	3.90	1.07	2.52
Run5	4.32	1.38	2.81
Run6	4.88	1.36	2.93
Run7	5.59	1.75	3.15
Run8	4.94	2.23	3.24
Run9	4.89	1.80	3.52
Run10	4.94	2.23	3.24
Run11	6.66	2.02	3.72
Run12	3.80	2.22	3.10
Run13	7.95	3.08	4.60
Run14	8.70	2.07	3.96
Run15	4.54	2.53	3.82
Run16	8.70	2.07	3.96
Run17	6.14	2.32	3.72
Run18	4.41	2.11	3.23
Run19	3.90	1.29	2.55
Run20	4.94	2.23	3.24
Run21	6.10	0.95	3.66

Table 6. Statistical table of channel bar area data.

Simulation Serial Number	Maximum Value (km^2)	Minimum Value (km^2)	Average Value (km^2)
Run1	0.61	0.03	0.16
Run2	0.53	0.02	0.17
Run3	0.56	0.04	0.20
Run4	0.53	0.02	0.17
Run5	0.47	0.05	0.25
Run6	0.59	0.20	0.37
Run7	0.24	0.03	0.08
Run8	0.18	0.01	0.08
Run9	0.26	0.02	0.11
Run10	0.18	0.01	0.08
Run11	0.30	0.02	0.10
Run12	0.32	0.03	0.11
Run13	0.16	0.04	0.09
Run14	0.26	0.04	0.12
Run15	0.38	0.02	0.16
Run16	0.26	0.04	0.12
Run17	0.33	0.04	0.13
Run18	0.30	0.04	0.14
Run19	0.53	0.02	0.17
Run20	0.18	0.01	0.08
Run21	0.13	0.02	0.08

To verify the reliability of the simulation results, four modern braided rivers were captured using Google Earth software (Figure 8). Statistical measurements of parameters such as channel bar scale were conducted, with a total of 56 sets of length-width ratio data of channel bars accumulated. A fitting curve was also established, providing a basis for the reliability of the simulation results.

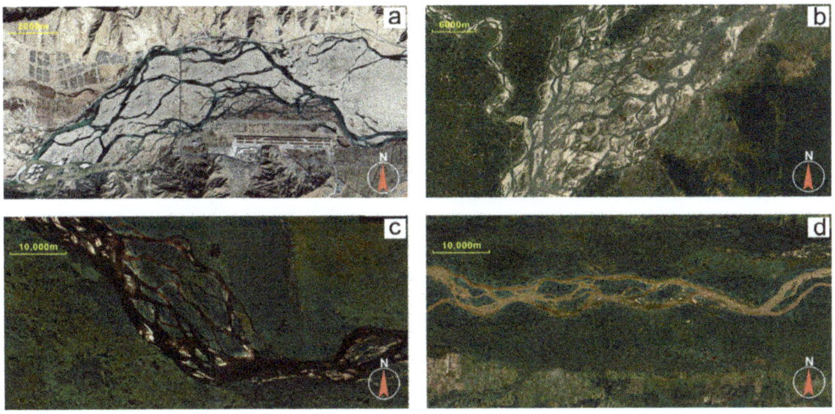

Figure 8. Satellite image of modern sedimentation in braided rivers: (**a**) Yarlung Zangbo River; (**b**) Ganges River; (**c**) Lena River; (**d**) Paraná River.

Typical modern sandy braided rivers were selected. Figure 9 shows the length-width relationship of channel bars, which was fitted as y = 0.3995x + 25.758 (R^2 = 0.821) after extracting its length/width parameters from satellite images. This high goodness of fit indicates that the length and width of channel bars in modern braided rivers have a strong linear correlation, and the fitting curve can effectively reflect the morphological laws of real channel bars. Figure 10 presents the length-width relationship fitted based on the simulation results of this study, which is expressed as y = 0.2714x + 63.779 (R^2 = 0.6221). However, due to simplified boundary conditions such as straight channels and absence of vegetation, the R^2 is slightly lower than that of modern deposits. Nevertheless, the fitting trend is highly consistent with that of modern deposits: both show a linear law where length increases synchronously with increasing width. The error between the length/width value range of simulated channel bars and the proportional range of modern braided rivers is small. Although the simplification of boundary conditions reduces the goodness of fit, it does not undermine the significance of the linear trend. This further indicates that the simulation results reliably reflect the morphological evolution laws of channel bars in real braided rivers, providing a solid quantitative basis for subsequent sandbody characterization.

Based on the statistical results, the aspect ratio and area of channel bars show significant differences under different combinations of experimental parameters. To better demonstrate the relationships between three variables (gradient, discharge, and sediment grain size) and channel bar scale, box plots are created for the statistical data grouped by different categories (Figures 11 and 12), so as to more clearly present the response law of the morphological scale of braided river channel bars.

The variation in channel bar area is comprehensively regulated by discharge, gradient, and sediment grain size. When the gradient is fixed, the channel bar area shows a positive correlation with discharge and sediment grain size. It can be seen from the box plot that when the gradient is consistent, the increase in discharge and the coarsening of sediment grain size significantly expand both the median and distribution range of the channel bar area. This reflects that high discharge enhances the material cycle efficiency of sediment transport and deposition, while coarse-grained sediments are more likely to accumulate into large-scale channel bars due to their higher erosion resistance.

The variation in the channel bar aspect ratio more prominently highlights the controlling effect of the gradient: as the gradient increases, the channel bar aspect ratio also increases [37]. When the gradient is fixed, discharge and sediment grain size have a relatively weak impact on the channel bar aspect ratio, but they generally still show an increasing trend.

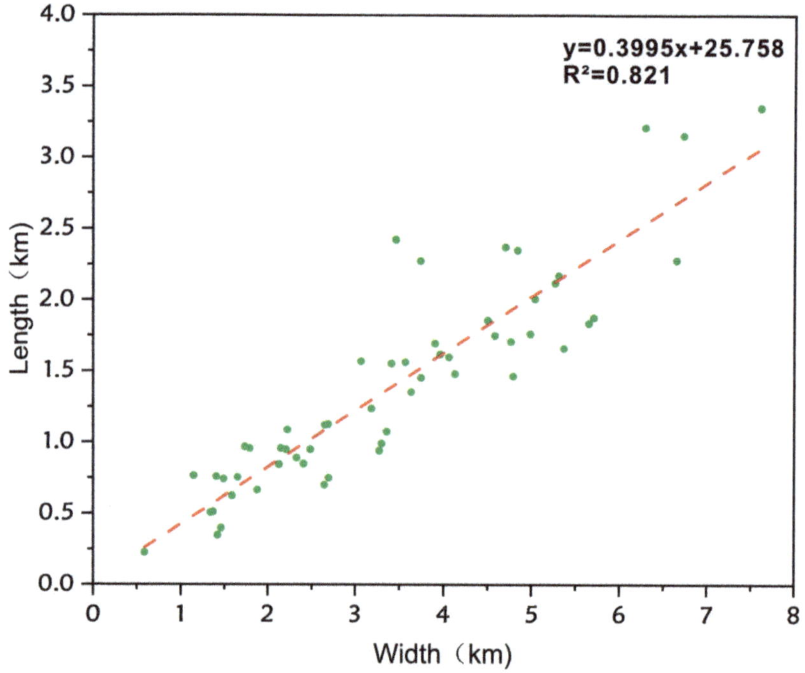

Figure 9. Fitting curve of length/width of channel bars in modern sedimentary braided rivers.

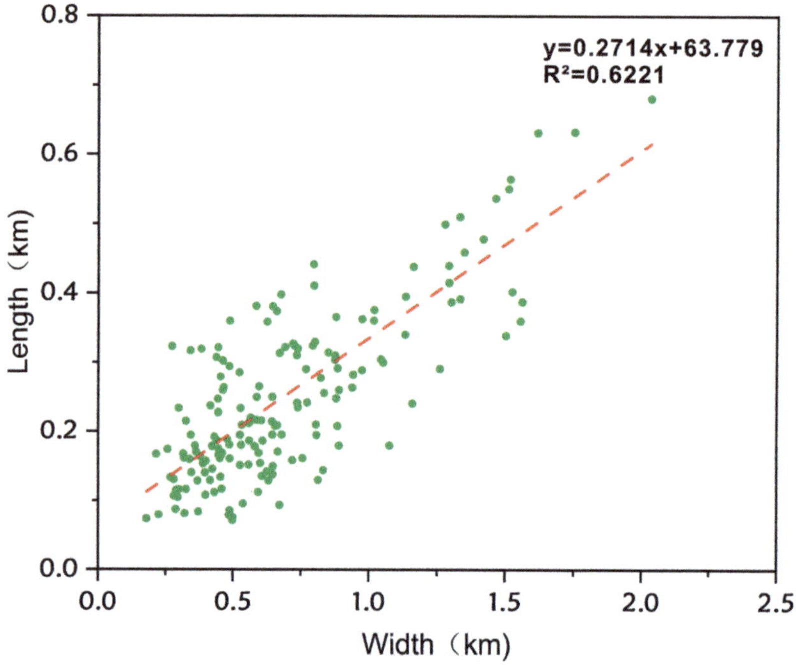

Figure 10. Fitting curve of length/width of braided river channel bars from simulation results.

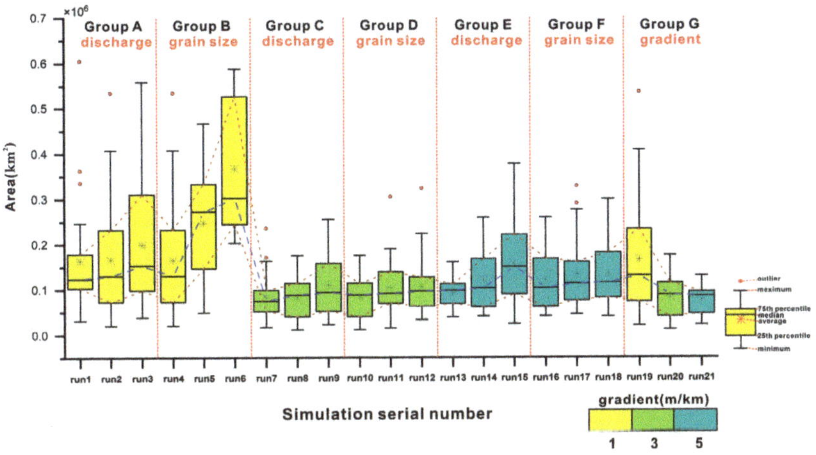

Figure 11. Box plot of channel bar area for different groups.

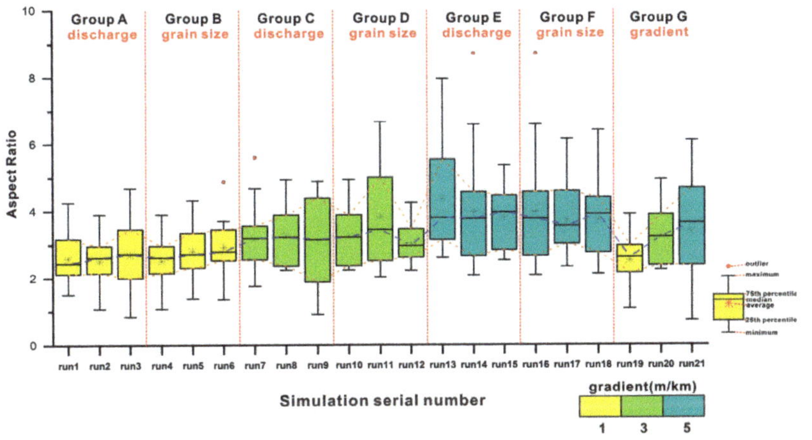

Figure 12. Box plot of channel bar aspect ratio for different groups.

An analysis of the outliers in the box plot shows that all outliers correspond to specific dynamic sedimentary processes rather than random number deviations. Their occurrence is an objective reflection of the evolution of channel bars in natural braided rivers. For example, in the high-gradient group, strong hydrodynamic forces cause frequent fragmentation of channel bars, forming small-scale channel bars; meanwhile, enhanced erosion along the main flow direction leads to the aspect ratio of some narrow and elongated channel bars exceeding 6.0. Such outliers are consistent with the phenomenon of fragmented channel bars in high-gradient sections observed in field surveys of modern braided rivers, thus reflecting real geological processes. Alternatively, in high discharge-fine grain size combinations, instantaneous discharge peaks induce rapid sediment accumulation, forming oversized channel bars. In terms of the impact on core conclusions, outliers do not alter the core laws, and they account for only 4.2% of the total sample, which further verifies the reliability of the results.

3.4.2. Sensitivity Analysis of Channel Bar Area to Key Parameters

To quantitatively reveal the sensitivity differences in channel bar area to discharge, sediment grain size, and riverbed gradient, and clarify the influence weights of different parameters on the development scale of channel bars, Principal Component Analysis and Multiple Linear Regression Analysis are adopted to conduct a systematic study. Principal Component Analysis quantifies the cumulative explanatory power of each parameter on channel bar area by extracting the common variance of "discharge-sediment grain size-riverbed gradient", while Multiple Linear Regression focuses on the linear correlation strength between parameters and channel bar area.

The results of Principal Component Analysis in Table 7 show that the cumulative variance contribution rate of discharge and riverbed gradient to channel bar area reaches 93.98%, while that of sediment grain size is only 6.02%. This difference mainly stems from the dominance of coupling between hydrodynamic conditions and sedimentary processes: discharge directly determines the "transport-deposition efficiency" of hydrodynamics on sediment, while riverbed gradient regulates the "erosion-accumulation balance" of water flow through gravity; the two together form the dynamic basis for the development of channel bar scale. In contrast, sediment grain size only exerts a secondary impact through particle accumulability, and its role is easily overshadowed by the dominant effect of hydrodynamics.

Table 7. Principal component loading matrix table.

	Discharge	**Grain Size**	**Gradient**
PC1	0.71	0.02	0.71
PC2	0	0.99	−0.03
PC3	0.71	−0.02	−0.71
Contribution Rate (%)	53.3	6.02	40.68

A multiple linear regression model including discharge, sediment grain size, and gradient was constructed (Table 8), and through standardized coefficients, *t*-tests, and *p*-values, an in-depth analysis of the significance level of the impact of each parameter on channel bar area was conducted.

Table 8. Standardized coefficients table.

		Unstandardized Coefficients		**Standardized Coefficients**		
Model		B	Standard error	Beta	t	Significance
1	(Constant)	8184.113	46,379		0.176	0.86
	Discharge	166.873	30.569	0.237	5.459	<0.001
	Grain Size	68.694	34.645	0.079	1.983	0.048
	Gradient	−8,674,185	874,518	−0.43	−9.919	<0.001

From the results of the regression model, discharge, gradient, and sediment grain size all have a significant impact on channel bar area; however, the absolute values of the t-values for discharge and gradient are larger, and their significance levels are stronger than that of sediment grain size. To verify the influence weight of sediment grain size, a further residual analysis with the "grain size variable excluded" was conducted. By comparing the histograms of standardized residuals between the "grain size-included" and "grain size-excluded" scenarios (Figure 13), it was found that the normality of the residuals was not compromised after excluding grain size. This indicates that the contribution of grain size to the core explanatory power of the model is low, and its impact can be partially over-shadowed by the dominant role of discharge and gradient.

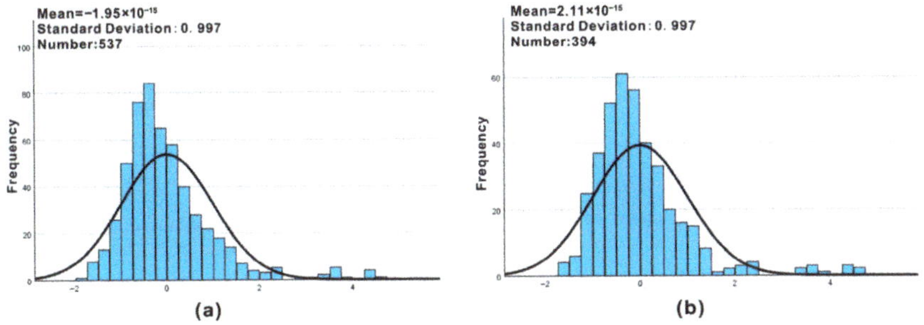

Figure 13. Histogram of standardized residuals: (**a**) grain size-included; (**b**) grain size-excluded.

Finally, based on the regression results with the "grain size variable excluded", a regression equation between discharge, gradient, and channel bar area was established: Channel Bar Area (m^2) = 29,556.03 + 170.92 × Discharge (m^3/s) 8,581,400.8 × Gradient (°).

The channel bar area prediction equation established in this study based on sedimentary numerical simulation has a core advantage: it establishes a quantitative link between paleoenvironmental parameters and sandbody morphology, addressing the limitation of traditional characterization that relies on subjective experience. In practical applications, paleoenvironmental parameters can be obtained from the geological data of the study area, and the theoretical area of channel bars can be derived using the equation. Meanwhile, the aspect ratio characteristics of channel bars can be inferred based on the sedimentary environment, providing a clear geological basis for sandbody characterization.

By synthesizing the Principal Component Contribution Rate, the significance of Multiple Regression, and the Marginal Effect, it can be clearly determined that the sensitivity of discharge, sediment grain size, and riverbed gradient to channel bar area in descending order is gradient > discharge > sediment grain size. The regulatory mechanisms and sensitivity differences in each parameter can be summarized as follows:

(1) Gradient exhibits the strongest sensitivity: By regulating the gravitational potential energy of water flow, gradient directly determines the "erosion-deposition balance" of hydrodynamic forces. When the gradient increases, the erosive capacity of water flow increases exponentially, making channel bars prone to frequent cutting and fragmentation and thus difficult to form large-scale contiguous morphologies; meanwhile, the 40.68% contribution rate of gradient in Principal Component Analysis (PCA) and the highest t-value in the regression model both support it as the primary constraining factor for channel bar area.

(2) Discharge exhibits the second-highest sensitivity: By enhancing the "sediment transport-deposition efficiency" of hydrodynamic forces, discharge provides the material basis for channel bar development. Under high discharge conditions, water flow can carry more sediment for transport, promoting the superposition and scale expansion of channel bars; the 53.3% contribution rate in Principal Component Analysis (PCA) also indicates that discharge is the core driving factor for the positive development of channel bar area.

(3) Sediment grain size exhibits the weakest sensitivity: Sediment grain size affects channel bar scale through the depositional capacity of particles, but this effect depends on the hydrodynamic context of discharge and gradient. When the dominant role of discharge or gradient is strong, the influence of grain size on channel bar area is overshadowed, so its overall regulatory capacity is limited.

This sensitivity law is mutually corroborated by the conclusions of previous single-factor experiments on discharge, gradient, and sediment grain size. It not only clarifies the regulatory weights of different parameters on channel bar area but also provides a theoretical basis and methodological support for the priority control of key parameters in subsequent braided river sedimentary numerical simulations.

4. Discussion

First, the evolutionary laws of the six stages of channel bars in braided rivers are a direct reflection of the synergistic effect of hydrodynamic force, sediment properties, and topographic conditions. They are not isolated morphological changes but rather the result of the dynamic coupling of various parameters. Under gentle slope settings, the gravitational potential energy of water flow is weak, resulting in a slow evolutionary rhythm of channel bars. For the superimposition and expansion of composite bars, this process relies not only on the increase in sediment transport flux brought by the rise in discharge but also on the erosion resistance support provided by sediment grain size.

Second, the laws of single-factor regulation and the results of sensitivity analysis together construct a hierarchical regulatory framework of "dynamics–material–topography". Single-factor experiments confirm that discharge positively regulates the area of channel bars, gradient negatively constrains the scale of channel bars and positively affects their aspect ratio, and the coarsening of sediment grain size enhances the stability of channel bars. These three types of laws further receive quantitative support from the sensitivity ranking, which indicates that gradient is the core constraint on channel bar morphology, hydrodynamic force is the key driver of material supply, and sediment grain size only plays a secondary regulatory role under the dominance of these two factors.

Finally, the establishment of the channel bar area prediction equation serves as the practical application of all previous results. Based on sensitivity analysis, the equation constructed is "Channel bar area (m^3) = 29,556.03 + 170.92 × Discharge (m^3/s) − 8,581,400.8 × Gradient (°)". Its core value lies in transforming the qualitative laws of the six evolutionary stages of channel bars and the quantitative analysis of single-factor and sensitivity into a practical tool. This equation provides an operable quantitative basis for sandbody scale prediction in fine reservoir characterization. Meanwhile, the regression equation can be applied to channel bar sandbody prediction in the Sulige Gas Field, addressing the limitation of characterization relying on subjective experience. By restoring paleogradient and paleodischarge based on drilling data in the study area, substituting them into the regression equation to calculate the theoretical area of channel bars in the target area, and combining the results of single-factor analysis to estimate the aspect ratio of channel bars in the target area, the sandbody characterization of "area + morphology" is ultimately achieved. Additionally, this scheme has universality, but the parameter range needs to be adjusted according to the geological background of the target area.

However, all simulations in this study have certain limitations: only several main factors were considered, including discharge, gradient, and sediment grain size. In modern rivers, however, the morphology of braided rivers and the scale of channel bars are jointly influenced by multiple factors. For instance, the role of vegetation in river development was not taken into account: vegetation can effectively resist water erosion, thereby hindering sandbody deposition. These multiple factors interact with each other and jointly regulate the sedimentary evolution of rivers. In addition, all river boundaries in the model are set as straight lines, while the boundaries of modern rivers are usually curved; river sinuosity may also affect sandbody deposition to a certain extent.

5. Conclusions

(1) This study reveals the evolutionary laws of the six stages of channel bars in braided rivers under gentle slope settings, fills the gaps in details of classic evolutionary models, and describes in detail the distinctive characteristics of each stage.

(2) The single-factor regulation laws are as follows: discharge and grain size show a positive correlation with channel bar area but have a relatively small impact on the channel bar aspect ratio; grain size has a significant correlation with channel bar stability; gradient shows a negative correlation with channel bar area and a certain positive correlation with the channel bar aspect ratio.

(3) Sensitivity analysis confirms that gradient and discharge are the dominant factors affecting channel bar area, with their cumulative explanatory power reaching 93.98%, while sediment grain size plays a secondary regulatory role.

(4) This study quantified the synergistic regulatory mechanism of gradient, discharge, and grain size on the morphology of channel bars; established a channel bar area prediction equation applicable to sandy braided rivers; and effectively provided a basis for the quantitative characterization of channel bar morphology.

(5) At present, breakthroughs have been made in sedimentary numerical simulation, but there are some issues worthy of consideration: (1) the research is based on numerical simulation under ideal conditions, and complex factors in natural rivers such as "periodic discharge variation" and "tectonic subsidence" cannot be reflected; (2) there are differences between the sedimentary numerical simulation parameters obtained from geological data and the actual situation, which affects the accuracy of simulation results; (3) this study did not design multi-functional utilization scenarios for late-stage reservoir development. Future research can be advanced in the following directions: (1) introduce dynamic discharge sequences during flood periods to enhance relevance to natural conditions; (2) combine Google Earth remote sensing data to perform calibration of simulation results with real rivers, so as to ensure maximum consistency with natural conditions; (3) based on the sedimentary numerical simulation model, further couple the fluid-rock interaction module to explore the adaptability of braided river reservoirs in scenarios such as late-stage reservoir development.

Author Contributions: Conceptualization and methodology, J.L.; investigation, K.W.; data curation, F.L. and Z.M.; Visualization, X.L.; writing—original draft preparation, K.W., J.L., F.L. and Z.M.; writing—review and editing, K.W., X.L. and Y.L. All authors have read and agreed to the published version of the manuscript.

Funding: This research was supported by National Natural Science Foundation of China (No. 42172154; No. 42472205), the Applied Scientific and Technological Research Project of China National Petroleum Corporation (2023ZZ25-001), and the National Science and Technology Major Project of New-type Oil and Gas Exploration and Development (2025ZD1405502).

Data Availability Statement: For confidentiality reasons, some of the data in the article cannot be publicly displayed. If you have data-related questions, you are welcome to contact me by email.

Conflicts of Interest: Authors Jinbu Li, Fuping Li, Zhixin Ma, Xinqiang Liu were employed by the Exploration and Development Research Institute, PetroChina Changqing Oilfield Company. The remaining authors declare that the research was conducted in the absence of any commercial or financial relationships that could be construed as a potential conflict of interest.

eferences

Bridge, J.S.; Smith, N.D.; Trent, F.; Gabel, S.L.; Bernstein, P. Sedimentology and Morphology of a Low-Sinuosity River: Calamus River, Nebraska Sand Hills. *Sedimentology* **1986**, *33*, 851–870. [CrossRef]

Lunt, I.A.; Bridge, J.S.; Tye, R.S. A Quantitative, Three-dimensional Depositional Model of Gravelly Braided Rivers. *Sedimentology* **2004**, *51*, 377–414. [CrossRef]

Stecca, G.; Zolezzi, G.; Hicks, D.M.; Surian, N. Reduced Braiding of Rivers in Human-Modified Landscapes: Converging Trajectories and Diversity of Causes. *Earth-Sci. Rev.* **2019**, *188*, 291–311. [CrossRef]

Wang, F.; Kobina, F. The Influence of Geological Factors and Transmission Fluids on the Exploitation of Reservoir Geothermal Resources: Factor Discussion and Mechanism Analysis. *Reserv. Sci.* **2025**, *1*, 3–18. [CrossRef]

5. Li, Q. Reservoir Science: A Multi-Coupling Communication Platform to Promote Energy Transformation, Climate Change and Environmental Protection. *Reserv. Sci.* **2025**, *1*, 1–2. [CrossRef]
6. Wu, J.; Ansari, U. From CO2 Sequestration to Hydrogen Storage: Further Utilization of Depleted Gas Reservoirs. *Reserv. Sci.* **2025**, *1*, 19–35. [CrossRef]
7. Zhang, C.S.; Liu, Z.B.; Shi, D.; Xiao, L.; Sun, H.F.; Chen, G. Predicting the geometric shape of braided river sandbodies using numerical simulation methods. *Pet. Explor. Dev.* **2004**, *7*, 85–88.
8. Jin, Z.K.; Yang, Y.X.; Shang, J.L.; Wang, L.S. Sandbody architecture and quantitative parameters of single channel sandbodies of braided river: Cases from outcrops of braided river in Fukang, Liulin and Yanan areas. *Nat. Gas Geosci.* **2014**, *25*, 311–317.
9. Zeng, C. The Numerical Model of Sedimentary Processes in Braided River. Master's Thesis, Yangtze University, Jingzhou, China, 2018.
10. Zhang, Y.H. Sedimentary Numerical Simulation Research of Braided River Reservoir in Guantao Formation, Yangerzhuang Oil Field, Huanghua Depression. Master's Thesis, China University of Petroleum (East China), Qingdao, China, 2019. [CrossRef]
11. Mao, L.; Ravazzolo, D.; Bertoldi, W. The Role of Vegetation and Large Wood on the Topographic Characteristics of Braided River Systems. *Geomorphology* **2020**, *367*, 107299. [CrossRef]
12. Liu, Z. Study on the Control of Base Level Cycles on the Architecture of Finger-Shaped Sand Bars in Shallow-Water Deltas Taking the Lower Member of the Neogene Minghuazhen Formation in Block BZ25-1S, Bohai Sea as an Example. Master's Thesis, China University of Petroleum, Beijing, China, 2020. [CrossRef]
13. Chen, X. Architecture Analysis and Remaining Oil Distribution of Braided River Reservoir in Lamadian Oilfield. Master's Thesis, China University of Petroleum (East China), Qingdao, China, 2021.
14. Lei, T.; Ren, G.L.; Li, X.H.; Feng, W.J.; Sun, H.C. Sedimentary evolution pattern and architectural characteristics of mid-channel bars in sandy braided rivers: Understanding based on sedimentary numerical simulation. *Oil Gas Geol.* **2023**, *44*, 1595–1608. [CrossRef]
15. Schumm, S.A. Patterns of Alluvial Rivers. Annu. Rev. *Earth Planet. Sci.* **1985**, *13*, 5–27. [CrossRef]
16. Sambrook Smith, G.H. (Ed.) *Braided Rivers: Process, Deposits, Ecology and Management*; Special publication number 36 of the International Association of Sedimentologists; Blackwell Pub: Malden, MA, USA, 2006; Oxford, UK, 2006; ISBN 978-1-4051-5121-4.
17. Li, W.; Colombera, L.; Yue, D.; Mountney, N.P. Controls on the morphology of braided rivers and braid bars: An empirical characterization of numerical models. *Sedimentology* **2023**, *70*, 259–279. [CrossRef]
18. Guo, Y.; Chen, K.; Gao, Q.S.; Ren, G.L.; Li, S.H.; Feng, W.J.; Liu, L.S.; Ran, H. Analysis of sedimentary evolution and configuration differences in sandy braided rivers under different discharges: A study based on sediment numerical simulation experiments. *Acta Sedimentol. Sin.* **2025**, 1–21. [CrossRef]
19. Feng, W.J.; Cao, J.C.; Zhang, C.M.; Qian, Q.H.; Zhang, T.; Lei, T.; Zhou, Z.C. Sedimentary evolution mechanism and architectural pattern of sandy braided rivers: A study based on quantitative flume sedimentary simulation experiments. *Acta Sedimentol. Sin.* **2025**, *43*, 782–796, 766. [CrossRef]
20. Zhang, K.; Wu, S.H.; Feng, W.J.; Zheng, D.Y.; Yu, C.; Liu, Z.W. Discussion on the development and evolution process of sandy braided river channel bars, with insights from sedimentary numerical simulation and modern sedimentary analysis. *Acta Sedimentol. Sin.* **2018**, *36*, 81–91.
21. Fan, F.P. The Study of Hydrodynamic Simulation of Poyang Lake Based on Delft3D Model. Master's Thesis, Jiangxi Normal University, Nanchang, China, 2011.
22. Huang, X.; Liu, K.Y.; Zou, C.N.; Gui, L.L.; Yuan, X.J.; Qin, Y.Q. Forward Stratigraphic Modelling of the Depositional Process and Evolution of Shallow Water Deltas in the Poyang Lake, Southern China. *Earth Sci.-J. China Univ. Geosci.* **2013**, *38*, 1005–1013. [CrossRef]
23. Liao, G.Q. Research on the Hydrodynamic and Sediment Numerical Simulation of LiuHe based on Delft3D. Master's Thesis, Tsinghua University, Beijing, China, 2014.
24. Wang, Y.J.; Yin, T.J.; Deng, Z.H.; Cheng, Z. Study on the evolution of fluvial-dominated deltaic distributary channels based on hydrodynamic numerical simulation. *Geol. Sci. Technol. Inf.* **2016**, *35*, 44–52.
25. Zhang, K.; Wu, S.; Feng, W.; Zhang, J.; Wen, S. Bar dynamics in a sandy braided river: Insights from sediment numerical simulations. *Sediment. Geol.* **2020**, *396*, 105557. [CrossRef]
26. Zhang, J.L. Fluvial Facies Styles and Their Sedimentary Facies Models. *Xinjiang Pet. Geol.* **2019**, *40*, 244–252.
27. Schuurman, F.; Ta, W.; Post, S.; Sokolewicz, M.; Busnelli, M.; Kleinhans, M. Response of Braiding Channel Morphodynamics to Peak Discharge Changes in the Upper Yellow River. *Earth Surf. Process. Landf.* **2018**, *43*, 1648–1662. [CrossRef]
28. Xiao, P.Q.; Yao, W.Y. Comparison and Analysis of Sediment Transportation Equations. *Res. Soil Water Conserv.* **2006**, *13*, 6–8.
29. Lin, Z.P.; Shan, J.F.; Chen, L.; Sun, Q.J.; Fang, S.X.; Yan, X. Evolutionary process of palaeochannel sandbodies of the 8th member of Shihezi Formation in Block Su6,Sulige Gasfield. *Pet. Geol. Recovery Effic.* **2018**, *25*, 1–9+23. [CrossRef]
30. Liu, C.Y.; Zhang, F.L.; Zhang, Q.; Yang, X.L.; Xiong, G. Model of braided river deposition based on numerical simulation. *Petrochem. Ind. Appl.* **2022**, *41*, 75–79.

1. Nicholas, A.P.; Ashworth, P.J.; Sambrook Smith, G.H.; Sandbach, S.D. Numerical Simulation of Bar and Island Morphodynamics in Anabranching Megarivers. *J. Geophys. Res. Earth Surf.* **2013**, *118*, 2019–2044. [CrossRef]
2. Xu, D.Q. The Research and Application of 3D Reservoir Configuration Modeling in Braided River: A case study of 36-11 well pattern area of Sulige gasfield. Master's Thesis, China University of Petroleum (East China), Qingdao, China, 2020. [CrossRef]
3. Luo, S.S.; Dai, R.; Liu, Z.B.; Lü, Q.Q. *Experimental Study on the Genesis of Large-Area Sand Bodies in the He 8 Member of the Sulige Area Based on Sedimentary Simulation Experiments;* Petroleum Industry Press: Beijing, China, 2014.
4. van den Berg, J.H. Prediction of Alluvial Channel Pattern of Perennial Rivers. *Geomorphology* **1995**, *12*, 259–279. [CrossRef]
5. Castelltort, S. Empirical Relationship between River Slope and the Elongation of Bars in Braided Rivers: A Potential Tool for Paleoslope Analysis from Subsurface Data. *Mar. Pet. Geol.* **2018**, *96*, 544–550. [CrossRef]
6. Holzweber, B.I.; Hartley, A.J.; Weissmann, G.S. Scale Invariance in Fluvial Barforms: Implications for Interpretation of Fluvial Systems in the Rock Record. *Pet. Geosci.* **2014**, *20*, 211–224. [CrossRef]
7. Crosato, A.; Mosselman, E. Simple Physics-based Predictor for the Number of River Bars and the Transition between Meandering and Braiding. *Water Resour. Res.* **2009**, *45*, 2008WR007242. [CrossRef]

processes

MDPI

Article

Research on Well Depth Tracking Calculation Method Based on Branching Parallel Neural Networks

Weikai Liu [1], Baoquan Ma [1,*] and Xiaolei Yu [2]

1 School of Petroleum Engineering, Northeast Petroleum University, Daqing 163318, China
2 Daqing Oilfield Well Field Industrial Co., Ltd., Daqing 163318, China
* Correspondence: mbqls@stu.nepu.edu.cn

Abstract: Aiming at the problem that the well depth parameters in existing intelligent drilling technology can not be obtained underground, a multi-branch parallel neural network is proposed to solve the problem of downhole well depth tracking, and its effectiveness is verified by a field example. After analyzing and correcting the quality of the logging data collected on site by using DBSCAN (a density clustering algorithm), five parameters of WOB, rotating speed, displacement, pump pressure, and torque are selected to predict and calculate the downhole mechanical ROP. Adjust the structure of a traditional artificial BP neural network and design a multi-branch parallel neural network, change the basic architecture of the original hierarchical operation, make full use of the operation efficiency of a computer to realize parallel operation, and adopt the method of point-to-point depth comparison when evaluating the well depth tracking effect. The results indicate that the MAE and mechanical drilling rate evaluation values obtained were 1.18 and 0.873, respectively. The multi-branch parallel neural network achieved a 66.55% improvement in MAE compared to the original BP neural network, while the R^2 evaluation method showed a 61.82% increase. The average point-by-point comparison error in the example calculation was 0.012 m, with a maximum error of 0.268 m. This result can serve as a fundamental basis for judging changes in well depth during the drilling process.

Keywords: well depth tracking; logging curve; neural network; mechanical ROP; parallel operation

1. Introduction

To improve encounter rates, drilling efficiency, and wellbore quality in the oil and gas industry, intelligent drilling technology is an inevitable development [1]. This technology integrates big data, digitalization, and artificial intelligence systems. It uses precise downhole sensors to control the drill bit autonomously and manage integrated operations comprehensively. Intelligent drilling requires the sequential completion of three objectives: downhole depth tracking, formation identification, and automatic wellbore trajectory optimization. Depth tracking is the foundational element of the entire system. Currently, the well's deviation angle and azimuth can be obtained directly from downhole instruments during intelligent drilling. In related research, Noureldin, A et al. [2] noted that horizontal drilling commonly employs triaxial magnetometers and accelerometers within measurement-while-drilling (MWD) systems to measure the wellbore. The former provides real-time azimuth, and the latter outputs the tool face angle (roll angle) and vertical deviation angle (pitch angle) after processing. However, soft rock is prone to collapse, wellbore walls deform easily, and drill string vibration is pronounced. Traditional magnetometer measurements rely on expensive non-magnetic drill collars and suffer reduced accuracy due to the influence of steel components surrounding drilling equipment and geomagnetic field shifts. In contrast, the fiber optic gyroscope (FOG) mentioned in this study exhibits excellent interference resistance in harsh downhole environments and offers a new approach to measuring wellbore trajectories in soft rock drilling. However, well depth parameters still rely on traditional counting-drill-pipes methods, combined with uploaded wellbore inclination and azimuth data to calculate the wellbore trajectory. This trajectory is then adjusted and controlled based on the designed path and actual geological conditions. To achieve truly intelligent drilling, real-time acquisition of well depth parameters downhole is essential to support autonomous decision-making and control of drilling tools. Therefore, the implementation of downhole well depth tracking is of critical importance [3].

To address this challenge, researchers worldwide have proposed various approaches. For example, Yan et al. [4] employed a random forest method to predict azimuth, deviation angle, and bit angle. While this data-driven approach avoids complex modeling, it struggles to capture temporal correlations and dynamic features in sequential data. This results in poor adaptability to time-series data, challenging parameter tuning, and suboptimal prediction performance. To improve prediction accuracy, Wang et al. [5] and Shan et al. [6] used LSTM and CNN-LSTM hybrid models, respectively, to predict logging data for undrilled sections using neighboring well logs and achieved favorable results. However, predictions may become inaccurate when the distance between neighboring wells is too great or when well-to-well correlations are weak. This method is particularly challenging when there is insufficient neighboring well data at the target prediction depth.

Backpropagation (BP) neural networks have garnered significant attention for their outstanding self-learning capabilities, adaptability, robustness, and generalization performance, finding successful applications across multiple engineering domains. Research indicates their potential in drilling parameter prediction. For instance, Li Qi et al. [7] employed a particle swarm optimization (PSO) algorithm to optimize a BP network for predicting the mechanical rate of penetration (ROP), effectively enhancing prediction accuracy. Lawal A I [8] combined the antlion optimization algorithm with artificial neural networks for ROP prediction; Su K [9] proposed using an improved chaotic whale optimization algorithm (ICWOA) to optimize BP neural network weights, further enhancing ROP prediction accuracy; Avcı E [10] applied an ANN to predict the rheological properties of water-based drilling fluids. However, widespread application has also highlighted inherent limitations of BP neural networks: their susceptibility to local optima, the lack of theoretical guidance in network architecture design (often relying on empirical trial-and-error), and issues such as gradient vanishing, training instability, and slow convergence.

Addressing these limitations, this paper proposes integrating drilling rate-related data with a multi-branch parallel artificial neural network to develop a downhole well depth tracking calculation method. This approach aims to provide new insights for theoretical research and technological implementation in this field.

2. Well Logging Data Preprocessing

The prediction effect of underground well depth tracking is related to the quality of logging data used in training, and the factors affecting the quality of logging data are usually related to manual recording errors and improper machine operation. In order to ensure the accuracy of underground well depth tracking, the logging data should be preprocessed before calculation, mainly to eliminate the actual data containing a small amount of errors caused by labor or machines. The density clustering algorithm (DBSCAN) is a class of unsupervised learning algorithms [11], commonly used in classification labeling data; this paper uses DBSCAN's classification labeling characteristics to invert DBSCAN's outlier removal function for the quality analysis and correction of logging data.

2.1. Standardized Processing of Data

The actual logging data has multiple columns of characteristics, such as drilling pressure, speed, displacement, pumping pressure, torque, etc. [12], and the numerical relationship between the characteristics of each column may differ by several times or dozens of times, which will cause problems such as dimensional explosion and excessive computing resource occupation, so it is necessary to standardize the logging data before removing the outliers, compress the data to −1 to 1, and prevent excessive deviation between the data from leading to unsatisfactory training results and large errors [13]. The standardized formula is as follows:

$$X_i' = \frac{X_i - \overline{X}}{\sigma} \tag{1}$$

In the formula, X_i' is the normalized data, the X_i is the original data, $\overline{X}$ is the mean of the data, and the σ is the variance of the data.

After the data standardization operation, the density clustering algorithm (DBSCAN) was used in the outlier exclusion part of the downhole depth tracking calculation model, and the maximum density-connected sample collection was derived from the density reachability relationship [14].

2.2. Outlier Culling

The actual data obtained in the field often contains a small amount of error caused by labor or machines, and in order to ensure the accuracy of the model's prediction of mechanical drilling speed, it is necessary to eliminate the outlier value of the data to eliminate the error.

Figure 1 is the algorithm core diagram of the DBSCAN in the above process, Figure 1 can be understood as the point close to the boundary in the figure is the boundary point, because within the radius Eps, the number of points in its domain do not exceed MinPts, assuming that the MinPts set is 5. The middle point is the core point because the points in its neighborhood are more than MinPts points; through this clustering method, some of the outliers caused by on-site recording errors or equipment errors can be eliminated, and data that are more effective for the algorithm can be obtained.

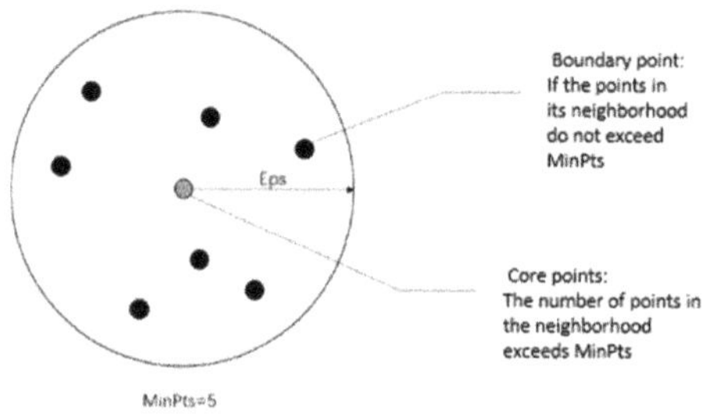

Figure 1. DBSCAN denoising principle.

In the process of intelligent drilling data acquisition, there is usually no "dimensional disaster" of logging data caused by a too high data dimension, so the selection of the measurement function of distance directly affects the effect of outlier elimination [15],

DBSCAN clustering quality depends on the selection of the distance formula, and the best distance measurement function in the actual application process is Euclidian distance:

$$d(x,y) = \sqrt{\sum_{i=1}^{n}(x_i - y_i)^2} \quad (2)$$

In the formula, x_i and y_i are points in n-dimensional space, and n represents the amount of dimension.

When selecting by Euclidean distance and forming a density reachable cluster, it is necessary to further determine the density reachable cluster, because there may be points with reasonable data logic that are not in line with the logging data logic, so it is necessary to further determine the distance of the density reachable cluster, and the determination method adopts the average distance calculation formula [16]:

$$L_i = \frac{1}{k}\sum d\left(p_i, p_{k(i)}\right) \quad (3)$$

In the formula, $p_{k(i)}$ represents the k points closest to the point p_i.

It can be seen from Equation (3) that the smaller the average distance, the greater the density around the data point, and the density reachable cluster can be formed through the Euclidean distance, and then the average distance is used to screen the logging data. DBSCAN employs a grid search for hyperparameter tuning, a method also known as grid search. Its core logic involves processing hyperparameters in a grid format: first defining an initial set of values for each hyperparameter, then systematically searching through all possible combinations within the grid. During the iterative search process, trial-and-error approaches are used to screen parameter combinations. The configuration that achieves a balance between overfitting and underfitting represents the model's optimal fit. By adjusting different types of hyperparameters—such as the learning rate and weights—the model's learning process can be effectively controlled. This enables the discovery of inherent hidden patterns within the data across corresponding datasets for similar machine learning models. Following comprehensive grid parameter tuning validation, the optimal hyperparameters for the DBSCAN were finalized as eps = 0.8 and min_samples = 100. Which can realize the correction of outlier values of the logging data and eliminate the influence of manual error or machine error on the subsequent downhole depth tracking [10].

3. Calculation Model for Downhole Depth Tracking

For the calculation of downhole depth tracking, this paper proposes a calculation model. The model mainly includes the multi-branch parallel neural network and well depth tracking evaluation method obtained after the improvement of the structure of the traditional artificial BP neural network. The downhole depth tracking calculation model can select different drilling characteristic parameters as input quantities according to the actual logging data of different sites, and can flexibly realize downhole depth tracking.

3.1. Multi-Branch Parallel Neural Networks

Traditional artificial BP neural networks, also called backpropagation neural networks, improve their computational regression capabilities because they contain many hidden layers. It is precisely because of the backpropagation algorithm that the training of the neural network converges quickly, and the learned neural network has also reached a practical level. The training process of a BP neural network consists of the forward transmission of data and the reverse propagation of gradients. Forward propagation is the process that passes the weighted sum of data through one or more hidden layers and then passes it to the output layer; if the output does not match the true value, the gradient of the error is passed in reverse layer by layer, so that each training parameter can be updated, and the loss function will be minimized, so as to find the optimal solution or suboptimal solution of the neural network [17].

However, the BP neural network model also has many shortcomings, such as the fact that it is easy for the BP neural network model to fall into the local optimal solution instead of the global optimal solution; the structure of the model can only be optimized by experience without exact theoretical support; BP neural networks require a large number of parameters, such as the network topology, the initial values of weights and thresholds; the learning process of BP neural networks cannot be visualized [18]. Based on the above problems, this paper starts from the traditional structure of a BP neural network and establishes a multi-branch parallel neural network model to solve the regression problem and optimize the traditional BP neural network model.

The main operation process of the traditional BP neural network model is to input data through the input neural unit of the input layer, then linearly superimpose the neurons layer by layer through the hidden layer, and continue to pass parameters to the next layer in a weighted sum manner, until it is transmitted to the output layer for the budget of the result, and the error is passed backward layer by layer through the comparison of labels in the dataset, the cycle is repeated, and the optimal solution under the existing conditions is solved again and again by iteration, but usually the suboptimal solution is obtained, because the optimal solution cannot be obtained [19]. The operation steps of the improved multi-branch parallel neural network are different from the above traditional BP neural network operation process; the operation of the multi-branch parallel neural network is not a hierarchical operation, but from the input neural unit to the separate four branches, the data through the input layer is directly synchronized into the four hidden layer branches for synchronous calculation, and the calculated data is budgeted in two branches for the maximum, and then passed to the next neuron, and finally only two parts of neurons receive the data, the obtained data is passed to the output layer after the minimum operation, and then compared with the real value and then passed in reverse, and finally the suboptimal solution is obtained after gradual iteration (Figure 2).

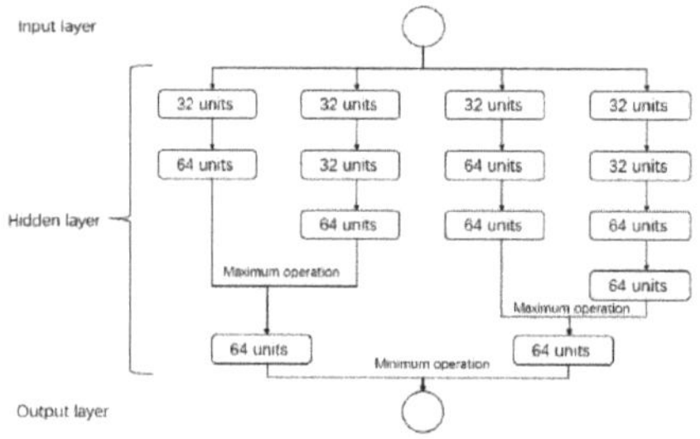

Figure 2. Multi-branch parallel neural network model.

In the multi-branch parallel neural network model, the input information adopts the five parameters of drilling, pressure, speed, displacement, pump pressure, and torque, in the logging data, and the drilling pressure, speed, displacement, pump pressure, and torque are expressed by x_1, x_2, x_3, x_4, x_5, and the input layer is expressed in vector form: $X = (x_1, x_2, x_3, x_4, x_5)^T$. The hidden layer has a total of four branches, and the output information of each branch is expressed in vector form: $Y = (y_1, y_2, \ldots, y_j, \ldots, y_m)^T$. One neuron in the output layer is used to output the predicted mechanical drilling speed, and the vector form of the mechanical drilling speed is represented by $\vec{O}$. The true mechanical drilling speed value is expressed as the desired output vector, and the vector form of the desired output is represented by $\vec{d}$. The weight between the input layer and the hidden layer is expressed by the matrix V, $V = (V_1, V_2, ..., V_j, ..., V_m)$, where the vector V_j represents the connection weight corresponding to the j neuron of the hidden layer. The weight between the hidden layer and the output layer is expressed by the matrix W, $W = (W_1, W_2, ..., W_k, ..., W_l)$, where the vector W_k represents the connection weight corresponding to the k neuron of the output layer.

For the hidden layer activation function, in order to avoid the problem of gradient disappearance [20], the SELU function is used to activate the hidden layer (Figure 3), and the formula of the SELU function is as follows:

$$SELU(x) = \lambda \begin{cases} a, & \mathrm{a} \leq 0 \\ \alpha e^{a} - \mathrm{a} & \mathrm{a} > 0 \end{cases} \tag{4}$$

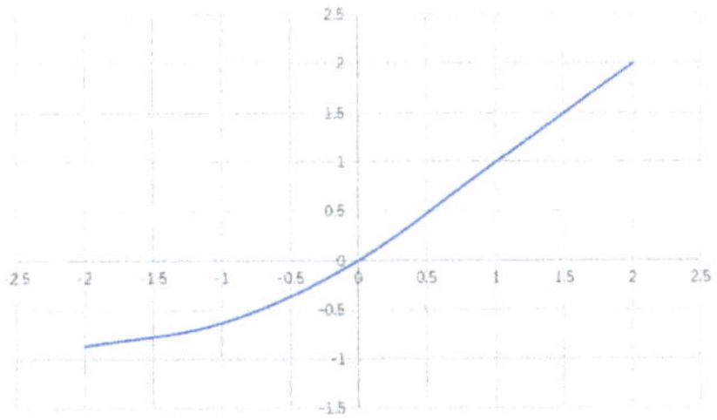

Figure 3. SELU function image.

In SELU's formula, a represents the phased result obtained after entering the five parameters of drilling pressure, speed, displacement, pump pressure, and torque into each layer of the multi-branch parallel neural network. It can be seen from the formula and image of SELU that the derivative of the positive semi-axis of the activation function is always greater than 1, which will make the data variance increase when it is too small, while the negative half axis is not directly converted to 0, but uses a smaller slope function to avoid the problem of gradient disappearance, and also effectively plays to the advantage of unilateral suppression [21].

The hidden layer formula is listed as follows:

$$net_j = \sum_{i=0}^{n} v_{ij} x_i \quad j = 1, 2, ..., m \tag{5}$$

The formula for the result $\vec{O}$ of the mechanical drilling speed output by the output layer is as follows:

$$net_k = \sum_{j=0}^{m} w_{jk} y_j \quad k = 1, 2, ..., l \tag{6}$$

$$\vec{O} = \overrightarrow{SELU(net_k)} \tag{7}$$

After calculating the predicted value of the mechanical drilling speed, the downhole depth tracking model is obtained by multiplying the predicted value vector of mechanical drilling speed by the data flow interval t of logging data, and the well depth L calculation formula is as follows:

$$\mathrm{L} = \overrightarrow{\mathrm{SELU}(\mathrm{net}_k)} \cdot \mathrm{t} = \vec{O} \cdot t \tag{8}$$

In the traditional training process, SGD (stochastic gradient descent) is mostly used as an optimizer to optimize the gradient, but the disadvantage of traditional SGD is that the learning rate is single and not flexible enough, resulting in an easy inability to quickly find the lowest point of the optimized function and it may fall into the local optimal solution, also known as the saddle point. In order to solve this problem, the Adam optimization algorithm is used to replace SGD in the training process; the advantage of this is that it can design independent adaptive learning rates for different parameters by calculating the first-order moment estimation and second-order moment estimation of the gradient according to the change in gradient, and it has high computational efficiency and low memory requirements, so it is not easy for Adam to fall into a local optimal solution and the update speed is fast [22]; the Adam update formula is as follows:

$$m_t = \beta_1 \cdot m_{t-1} + (1 - \beta_1) \cdot g_t \tag{9}$$

$$v_t = \beta_2 \cdot v_{t-1} + (1 - \beta_2) \cdot g_t^2 \tag{10}$$

$$\hat{m}_t = \frac{m_t}{(1 - \beta_1^t)} \tag{11}$$

$$\hat{v}_t = \frac{v_t}{(1 - \beta_2^t)} \tag{12}$$

$$\theta_t = \theta_{t-1} - \alpha \cdot \frac{\hat{m}_t}{(\sqrt{\hat{v}_t} + \epsilon)} \tag{13}$$

In the formula, t is the initial step; g_t is the gradient of the stochastic objective function; $f(\theta)$ is a random objective function with parameter θ; m_t is a partial first-order moment estimate, and the m_0 is equal to 0; v_t is a partial second-order moment estimate, and the v_0 is equal to 0; α is the learning rate; β_1 and β_2 are the exponential attenuation rates of moment estimation; ϵ is a positive performance number to prevent the denominator from being 0.

Through the above structural optimization, the multi-branch parallel neural network model has a very good effect on the improvement of the traditional model, and can learn better features without too many neural units, which effectively reduces the risk of overfitting.

3.2. Evaluation Methodology

For the accuracy judgment of well depth tracking, this paper uses traditional judgment methods, such as R^2 and MAE algorithms, and depth comparison methods point by point according to the time interval of data flow, and the formulas of the R^2 and MAE algorithms are as follows:

$$R^2 = 1 - \frac{SS_{res}}{SS_{tot}} = \frac{SS_{reg}}{SS_{tot}} \tag{14}$$

$$MAE = \frac{1}{m} \sum_{i=1}^{m} |y_i - \hat{y}_i| \tag{15}$$

In the formula, SS_{res} represents the sum of squares of the residuals, SS_{tot} represents the sum of squares of total dispersion, SS_{reg} represents the sum of regression squares, m represents the amount of data, y_i represents the true value, and $\hat{y}_i$ represents the predicted value.

The mathematical meaning of R^2 is goodness-of-fit; the greater the goodness-of-fit, the higher the degree of interpretation of the dependent variable by the independent variable, the change caused by the independent variable is a high percentage of the total change, and the closer the value is to 1, the higher the degree of explanation. The greater the goodness-of-fit in depth tracking, the greater the influence of the current drilling parameters on the well depth, and the more it can explain the relationship between the drilling parameters and the well depth. MAE represents the mean of the absolute value of the deviation of all individual observations from the arithmetic mean, with smaller MAEs indicating better models [23].

In order to further verify the accuracy of well depth tracking, this paper adopts the method of depth comparison point by point according to the time interval of the data flow, obtains the predicted cumulative drilling data obtained from the interval according to the two sets of data in the test set, and compares the predicted data with the true cumulative drill data.

4. Instance Calculation

In current drilling practice, at least in China, there has not been a downhole depth tracking test, and no data has been directly obtained downhole to verify the method [24], and this paper only collects data at the surface. For this purpose, data on a total of 10 wells in a block was obtained at the field site, and the mud and drilling tool combination used in this block were the same. Take nine wells as a training set, one well as a test set, a total of 13,830 sets of data, considering that the obtained data is inaccurate, unreasonable, etc., so the data is preprocessed for the above logging data; mainly the data outlier value is removed and corrected, and finally the cleaned data has a total of 7719 sets. The drilling pressure, speed, displacement, pumping pressure, torque in the obtained logging data in the five parameters are input vectors x_1, x_2, x_3, x_4, x_5, brought into Equations (5)–(8) to predict the downhole mechanical drilling speed [25].

All the data are randomly scrambled and split into a training set and a test set, of which the proportion of the test set is 10% [26], the processed raw data is input into the multi-branch parallel neural network learning model, and the gradient is continuously calculated according to the principle of minimizing MSE to minimize the loss function and constantly update the neural network parameters. The specific training process is shown in Figure 4:

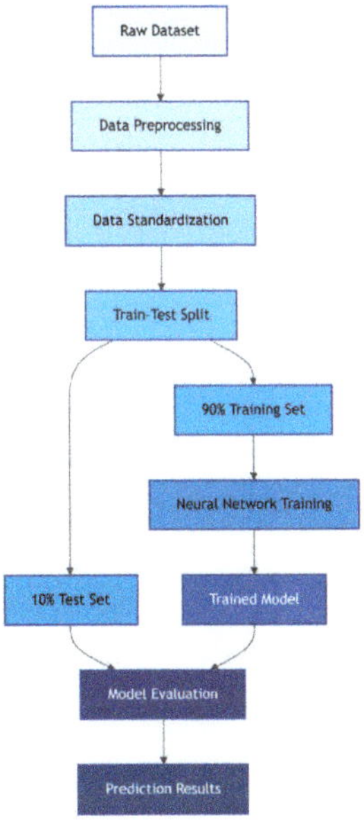

Figure 4. Training flow chart of neural network model.

The processed data is entered into the network model, and the resulting MAE and loss images are shown in Figure 5.

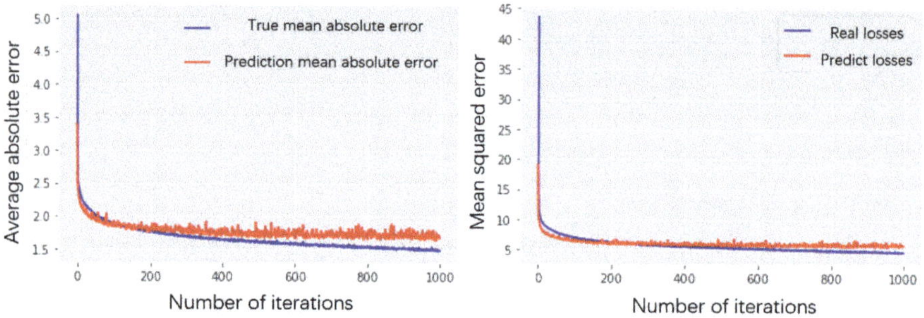

Figure 5. MAE error curve and loss curve.

The results indicate that at 1000 iterations, the network exhibits no significant overfitting and shows no discernible trend of error reduction. Therefore, setting the training cycle to 1000 iterations is considered reasonable. On the test set, the MAE of the actual data was 1.18, R^2 was 0.873, and the training time was 129 s. To demonstrate the advantages of the multi-branch parallel neural network, a comparative experiment with the original BP neural network was designed. First, dataset consistency was ensured: the original BP neural network used the same data, with identical dataset partitioning methods and preprocessing approaches to avoid data bias affecting the comparison results. Parameter tuning was also standardized, with the optimal parameter combination selected for testing. This ensured that the original BP network competed under "optimal conditions," preventing baseline performance underestimation due to suboptimal parameter settings. As shown in Figure 6, the multi-branch parallel neural network achieved a 66.55% improvement in MAE and a 61.82% increase in R^2 evaluation compared to the original BP network. This sufficiently demonstrates that the multi-branch parallel neural network exhibits excellent learning capabilities, effectively capturing overall trends within the data.

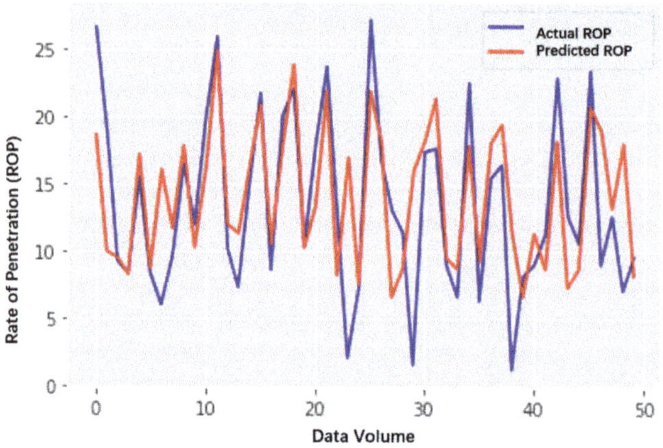

Figure 6. Multi-branch parallel neural network mechanical drilling speed chart.

As shown in Figure 7, the test set depth is between 2143 m and 2328 m, the results are analyzed segment by segment, the average error is 0.012 m, and the maximum error is 0.268 m. According to national standards [27], the measurement error requirement for borehole depth parameters is no more than 0.20 m per single borehole. Therefore, the errors in this study meet the requirements of actual engineering projects.

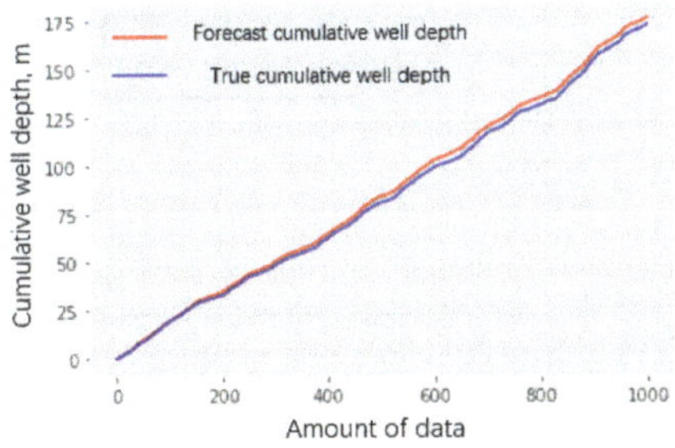

Figure 7. Cumulative well depth curve.

In order to facilitate observation, the error between the true value and the predicted value is taken every 10 points in 1000 data points, where the 0 scale mark is the 0 error line, that is, the real value is equal to the predicted value; the closer the error point is to the 0 scale mark, the smaller the error between the true value and the predicted value, and the farther away the error point is from the 0 scale line, the larger the error between the true value and the predicted value. The error between the true and predicted values of this segment is shown in Figure 8, which shows that the multi-branch parallel neural network modeling method is reasonable and can meet the engineering requirements of downhole depth tracking [7].

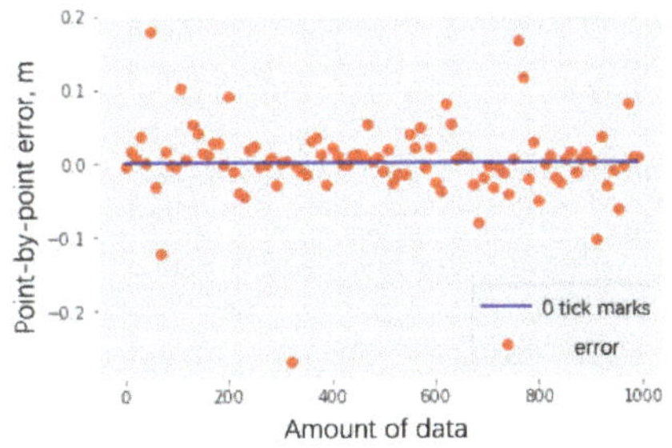

Figure 8. Point-by-point error analysis diagram.

5. Conclusions

(1) For the first time, the calculation method and evaluation method of underground well depth tracking were proposed, the traditional neural network was innovatively improved in structure, and a multi-branch parallel neural network was designed and used to realize underground well depth tracking. Compared to the original BP neural network, the MAE improved by 66.55%, and the R2 evaluation method improved by 61.82%. An evaluation method for point-by-point analysis and comparison of well depth was established.

(2) In pass instance validation, the multi-branch parallel neural network takes 129 s after 1000 iterations of training, the average error of the point-by-point comparison calculated by the example is 0.012 m, and the maximum error is 0.268 m, which can meet the engineering requirements of underground depth tracking.

(3) In this paper, the five parameters of drilling, pressure, speed, displacement, pump pressure, and torque, are selected to predict and calculate the downhole depth tracking. If the data is collected downhole, more and more accurate parameters such as drill bit

pressure drop and geological parameters can be collected, and the effect of prediction calculation will be better.

(4) By optimizing the structure of the traditional artificial BP neural network, a multi-branch parallel neural network is proposed, which solves the problems of the traditional BP neural network easily falling into the local optimal solution instead of the global optimal solution, resulting in a decrease in accuracy, and effectively improves the operation accuracy and operation time.

Author Contributions: Methodology, W.L.; Validation, B.M.; Formal analysis, X.Y.; Resources, X.Y.; Data curation, B.M.; Writing—original draft, B.M.; Writing—review & editing, W.L. and B.M.; Supervision, W.L.; Funding acquisition, W.L. All authors have read and agreed to the published version of the manuscript.

Funding: This research was supported by the National Key R&D Program: National Key R&D Program for New Oil and Gas Exploration and Development, Project: Downhole Electric and Intelligent Drilling Control Guidance System, Project No.: 2025ZD1401200, Topic: Development of Intelligent Full-Rotation Drill Bit Guidance System Resistant to High Temperatures and High Tilting, Topic No.: 2025ZD1401201.

Data Availability Statement: Data Availability Statement: The raw data supporting the conclusions of this article will be made available by the authors on reasonable request, subject to restrictions from the data provider (Daqing Oilfield) and for reasons of confidentiality.

Conflicts of Interest: Author Xiaolei Yu was employed by the company Daqing Oilfield Well Field Industrial Co., Ltd. The remaining authors declare that the research was conducted in the absence of any commercial or financial relationships that could be construed as a potential conflict of interest. The company had no role in the design of the study; in the collection, analyses, or interpretation of data; in the writing of the manuscript, or in the decision to publish the results.

References

1. Zhang, S.N. Optimization of horizontal well drilling parameters and technology. *Chem. Eng. Equip.* **2020**, *2020*, 81–82. [CrossRef]
2. Noureldin, A.; Tabler, H.; Irvine-Halliday, D.; Mintchev, M.P. A new borehole surveying technique for horizontal drilling processes using one fiber optic gyroscope and three accelerometers. In Proceedings of the SPE/IADC Drilling Conference and Exhibition, New Orleans, LA, USA, 23 February 2000. [CrossRef]
3. Imanian, A.; Ghassemi, M.; Karbasian, M. Bit pressure control during drilling operation using engineering process control. *Energy Sources Part A Recovery Util. Environ. Eff.* **2018**, *40*, 2193–2202. [CrossRef]
4. Yan, B.; Zhang, X.; Tang, C.; Wang, X.; Yang, Y.; Xu, W. A random forest-based method for predicting borehole trajectories. *Mathematics* **2023**, *11*, 1297. [CrossRef]
5. Wang, H.; Xu, Y.; Tang, S.; Wu, L.; Cao, W.; Huang, X. Well log prediction while drilling using seismic impedance with an improved LSTM artificial neural networks. *Front. Earth Sci.* **2023**, *11*, 1153619. [CrossRef]
6. Shan, L.; Liu, Y.; Tang, M.; Yang, M.; Bai, X. CNN-BiLSTM hybrid neural networks with attention mechanism for well log prediction. *J. Pet. Sci. Eng.* **2021**, *205*, 108838. [CrossRef]
7. Li, Q.; Qu, F.T.; He, J.B. Prediction model of ROP of drilling machinery based on PSO-BP. *Sci. Technol. Eng.* **2021**, *21*, 7984–7990. [CrossRef]
8. Lawal, A.I.; Kwon, S.; Onifade, M. Prediction of rock penetration rate using a novel antlion optimized ANN and statistical modelling. *J. Afr. Earth Sci.* **2021**, *182*, 104287. [CrossRef]
9. Su, K.H.; Da, W.H.; Li, M.; Li, H.; Wei, J. Research on a drilling rate of penetration prediction model based on the improved chaos whale optimization and back propagation algorithm. *Geoenergy Sci. Eng.* **2024**, *240*, 213017. [CrossRef]
10. Avcı, E. An artificial neural network approach for the prediction of water-based drilling fluid rheological behaviour. *Int. Adv. Res. Eng. J.* **2018**, *2018*, 124–131. Available online: https://api.semanticscholar.org/CorpusID:119062297 (accessed on 15 September 2025).
11. Zhai, Y.J.; Yu, C.Z.; Gao, X. A radar trace processing method based on DBSCAN clustering. *Shipboard Electron. Countermeas.* **2021**, *44*, 58–61. [CrossRef]
12. Liu, J.B.; Wei, H.Z.; Zhao, J.F. A new three-dimensional ROP equation considering the influence of bit speed. *Pet. Drill. Technol.* **2015**, *43*, 52–57. [CrossRef]

3. Zhang, S.; Gong, Y.H.; Wang, J.J. Development of deep convolution neural network and its application in the field of computer vision. *Chin. J. Comput.* **2019**, *42*, 453–482. [CrossRef]
4. Suppes, R.; Ebrahimi, A.; Krampe, J. Optimising casing milling rate of penetration (ROP) by applying the concept of mechanical specific energy (MSE): A justification of the concept's applicability by literature review and a pilot study. *J. Pet. Sci. Eng.* **2019**, *179*, 894–903. [CrossRef]
5. Li, W.W.; Yi, P.T.; Li, L.Y. Identification and dimensionless treatment of outliers in comprehensive evaluation. *Oper. Res. Manag. Sci.* **2018**, *27*, 173–178. Available online: https://qikan.cqvip.com/Qikan/Article/Detail?id=675124934 (accessed on 15 September 2025).
6. Li, J.; Yao, X.; Wang, X. Multiscale local features learning based on BP neural network for rolling bearing intelligent fault diagnosis. *Measurement* **2019**, *153*, 107419. [CrossRef]
7. Liang, H.B.; Zou, J.L.; Liang, W.L. An early intelligent diagnosis model for drilling overflow based on GA-BP algorithm. *Cluster Comput.* **2019**, *22*, 10649–10668. [CrossRef]
8. Liu, J.L. Application of BP neural network prediction model to the electromagnetic fields. *Electr. Power Technol. Environ. Prot.* **2017**, *33*, 45–48. Available online: https://d.wanfangdata.com.cn/periodical/dlhjbh201704002 (accessed on 15 September 2025).
9. Yuan, S.; Wang, K.; Shan, Y.; Yang, J.F. Multiscale target detection algorithm based on multi-branch parallel atrous convolution. *J. Comput.-Aided Des. Comput. Graph.* **2021**, *33*, 864–872. Available online: https://qikan.cqvip.com/Qikan/Article/Detail?id=7104901489 (accessed on 15 September 2025).
0. Yang, Q.; Wang, C.W. Research on global stock index prediction based on deep learning LSTM neural network. *Stat. Res.* **2019**, *36*, 65–77. [CrossRef]
1. Yang, Z.Z.; Kuang, N.; Fan, L. Overview of image classification algorithms based on convolutional neural network. *J. Signal Process.* **2018**, *34*, 1474–1489. [CrossRef]
2. Kingma, D.P.; Ba, J.L. Adam: A method for stochastic optimization. *arXiv* **2014**. [CrossRef]
3. Kang, Q.K.; Lu, L.J. Application of random forest algorithm in lithology classification of logging wells. *Glob. Geol.* **2020**, *39*, 398–405. [CrossRef]
4. Zheng, Y.Z.; Niu, L.K.; Xiong, X.Y. Fault diagnosis of cylindrical roller bearing cage based on one-dimensional convolutional neural network. *J. Vib. Shock* **2021**, *40*, 230–238+285. [CrossRef]
5. Zhuo, P.C.; Yan, J.; Zheng, M.M. GA-OIHF-Elman neural network algorithm for rolling bearing life cycle fault diagnosis. *J. Shanghai Jiaotong Univ.* **2021**, *55*, 1255–1262. [CrossRef]
6. Mokhtari, A.; Ozdaglar, A.; Pattathil, S. A unified analysis of extra-gradient and optimistic gradient methods for saddle point problems: Proximal point approach. *arXiv* **2019**. [CrossRef]
7. *SY/T 6243-2009*; Technical Specification for Fire and Explosion Prevention in Oil and Gas Drilling, Development, Storage and Transportation. National Energy Admin: Beijing, China, 2009.

processes

MDPI

Article

Seismic Waveform-Constrained Artificial Intelligence High-Resolution Reservoir Inversion Technology

Haibo Zhao [1,*], Jie Wu [1], Kuizhou Li [1], Yanqing He [1], Rongqiang Hu [1], Tuan Wang [1], Zhonghua Zhao [1], Huaye Liu [1], Ye Li [2,*] and Xing Yang [2]

1 Research Institute of Exploration and Development, Daqing Oilfield Company Ltd., Daqing 163000, China
2 Beijing Jia'an Huitong Petroleum Technology Co., Ltd., Beijing 100039, China
* Correspondence: zhaohaibo@petrochina.com.cn (H.Z.); lypxszd@163.com (Y.L.)

Abstract: In response to the technical challenges of traditional reservoir inversion techniques in determining inter-well wavelets and estimating geological statistical parameters, this study proposes an artificial intelligence high-resolution reservoir inversion technique based on seismic waveform constraints. This technology integrates multi-source heterogeneous data such as lithology characteristics, logging curves, and seismic waveforms through a deep learning neural network framework, and constructs an intelligent reservoir prediction model with geological and physical constraints. Results demonstrate that the proposed technique significantly enhances prediction accuracy for thin sand layers by effectively extracting high-frequency seismic information and establishing robust nonlinear mapping relationships. Inversion errors of reservoir parameters were reduced by more than 25%, while a vertical resolution of 0.5 m was achieved. Predictions agreed with actual drilling data with an accuracy of 86%, representing an 18% improvement over traditional methods. In practical applications, the technique successfully supported new well placement, contributing to a 22% increase in initial oil production in the pilot area. Furthermore, this study establishes a standardized technical procedure: "Time–Depth Modeling-Phase-Controlled Interpolation-Intelligent Inversion". This workflow provides an innovative solution for high-precision reservoir characterization in regions with limited well control and complex terrestrial depositional systems, offering both theoretical significance and practical value for advancing reservoir prediction technology.

Keywords: seismic waveform constraint; artificial intelligence; high-resolution inversion; reservoir prediction; oil and gas exploration

1. Introduction

Conventional inversion is based on theoretical foundations such as the convolution principle and geostatistics. By combining seismic data with logging data, a model is established to invert reservoir parameters [1–3]. However, in complex geological conditions and thin reservoir identification, there are challenges such as insufficient accuracy and low computational efficiency, which make it difficult to meet the current high-precision reservoir inversion requirements [4–6]. Waveform-constrained artificial intelligence high-resolution reservoir inversion is based on artificial intelligence algorithms, which can fully explore seismic waveform information and integrate multiple data sources. With its high precision and strong noise resistance, it has obvious advantages in complex geological conditions and can accurately characterize reservoir characteristics [7–9].

The inversion based on the convolution principle is an ideal model. It does not consider the loss and attenuation of seismic waves propagating in the reservoir, and belongs to

 https://doi.org/10.3390/pr1309287

noiseless synthetic seismic records [10,11]. The basic principle can be expressed by the following formula:

$$W \cdot R = S \quad (1)$$

$$R = \frac{S}{W} \quad (2)$$

In the equation, W represents the seismic wavelet, R represents the impulse response or reflection coefficient, and S represents the seismic record. According to Formula (2), the reflection coefficient can be calculated by giving an appropriate wavelet, and then the logging curve, which is the reservoir inversion result, can be obtained [12,13]. However, in the actual inversion process, inter-well wavelets are usually calculated using fixed wavelets, which have problems such as poor lateral adaptability and low vertical resolution. At the same time, the reservoir lateral prediction method based on training images and variation functions has strong ambiguity and struggles to characterize complex spatial structures [14–16]. In terrestrial sedimentary systems, this method also exhibits strong heterogeneity and complex structures (Figure 1).

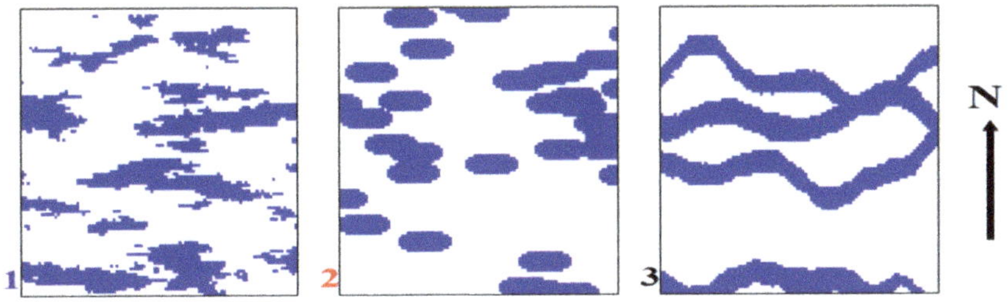

Figure 1. Spatial distribution of reservoir parameters based on the corresponding variation function simulation. 1, 2, and 3 represent the simulation of rivers, deltas, and sheet sand reservoirs, respectively.

In response to the problems of conventional inversion mentioned above, waveform-constrained artificial intelligence high-resolution reservoir inversion technology provides a new solution for inversion and high-resolution imaging under complex geological conditions [17].

The seismic waveform-constrained artificial intelligence high-resolution reservoir inversion technology can fully utilize high-frequency seismic information. This technology directly establishes complex and nonlinear mapping relationships between seismic attributes and logging responses through artificial intelligence learning under the constraints of waveform classification [18,19]. Artificial intelligence has the natural advantages of robustness, parallelism, and adaptability in implementing any complex mapping. It can effectively solve the application problems of geological prior knowledge (such as sedimentary facies and waveform classification), making the inversion results match geological data better and further improving the accuracy of seismic reservoir prediction in thin sand layers [20,21].

This study extracts key characteristic parameters such as lithology, logging curves, and seismic waveforms, and constructs a big data sample library through data integration and cleaning. Combined with deep learning machine learning algorithms, an artificial intelligence thin sand layer prediction model with seismic waveform constraints was constructed. This model can perform feature extraction and pattern recognition, and can deeply explore the key features. The waveform constraint mechanism introduces seismic waveform features to improve the accuracy and reliability of prediction. The method, principle, and technical roadmap of this article are shown in Figure 2.

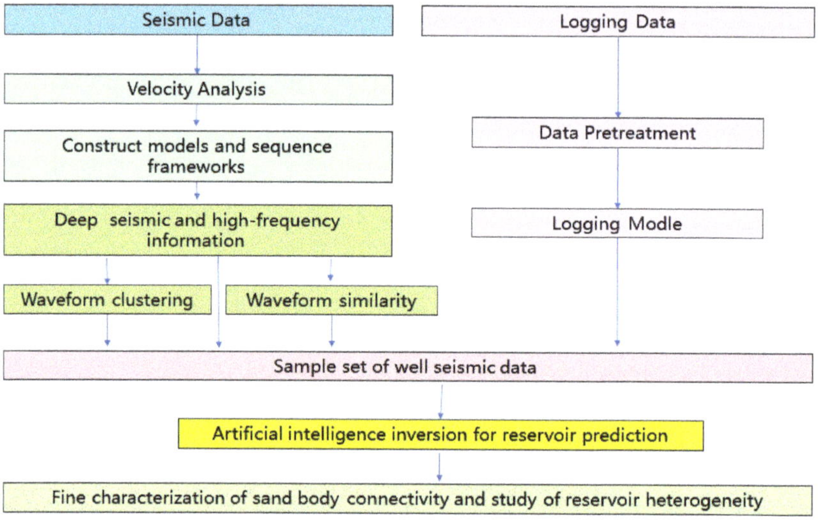

Figure 2. The flow chart of this study.

2. Field Background and Geological Setting

This study selected a block in the Daqing Oilfield as the research object. The target reservoir is a typical channel sand body, mainly composed of siltstone, with good physical continuity and sedimentary sequence characteristics. The average single-layer thickness of the reservoir in the research area is about 2.5 m, which is a type of medium-high frequency sedimentary joint development, and has typical characteristics for conducting high-resolution seismic geological fusion research.

There are four wells arranged in the research area, from north to south: W250-328, W262-332, W251-318, and W258-308. The distribution of well points covers the main sedimentary facies zones within the coverage area, achieving representative sampling of the geological features within the block. Through fine seismic interpretation, two key seismic reflection layers, H11 and H21, were identified in the study area. Combined with three seismic attribute bodies, including amplitude, frequency, and phase (as shown in Figure 3), a multi-scale and multi-dimensional comprehensive dataset was constructed, providing a data foundation for fine characterization of reservoir configuration and extraction of deep geological information.

This study comprehensively utilized multi-parameter logging data for reservoir evaluation, mainly including key logging curves such as natural gamma (GR), mud content (Vsand), deep lateral resistivity (LLD), and natural potential (SP). Based on the fluid properties and production performance of the reservoir, it is divided into five categories: oil (No. 0), poor oil (1), oil-water (2), dry (4), and mudstone (6).

The analysis of the well profile and logging response characteristics shown in Figure 4 indicates that typical oil reservoirs exhibit obvious "two lows and one high" electrical characteristics: low natural gamma values (reflecting high sandstone cleanliness), high acoustic time difference (indicating good porosity), and high resistivity (indicating high oil saturation). In contrast, poor oil layers and oil-water layers exhibit slightly higher natural gamma values and lower resistivity, while dry layers and mudstones show significantly high gamma and low resistivity responses. This logging response pattern provides a reliable explanatory basis for identifying reservoir fluids in the study area.

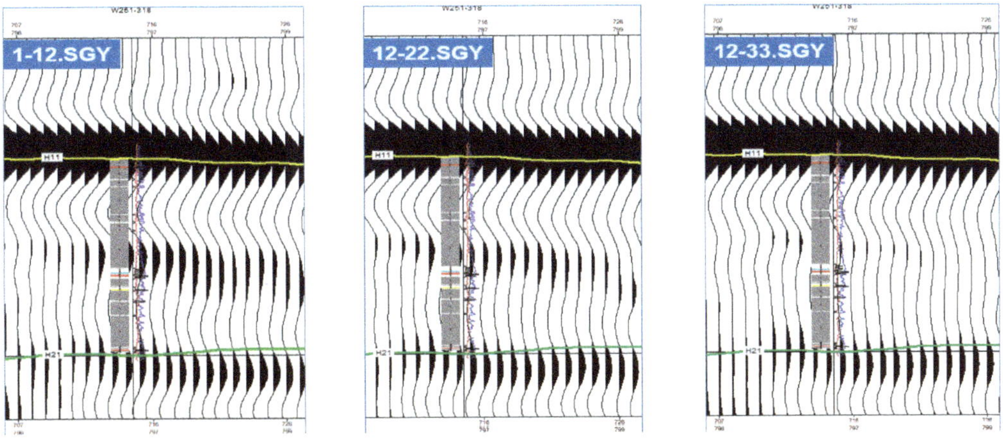

Figure 3. Original seismic attribute volume. The red curve represents the GR logging curve, and the blue curve represents the VSAND logging curve.

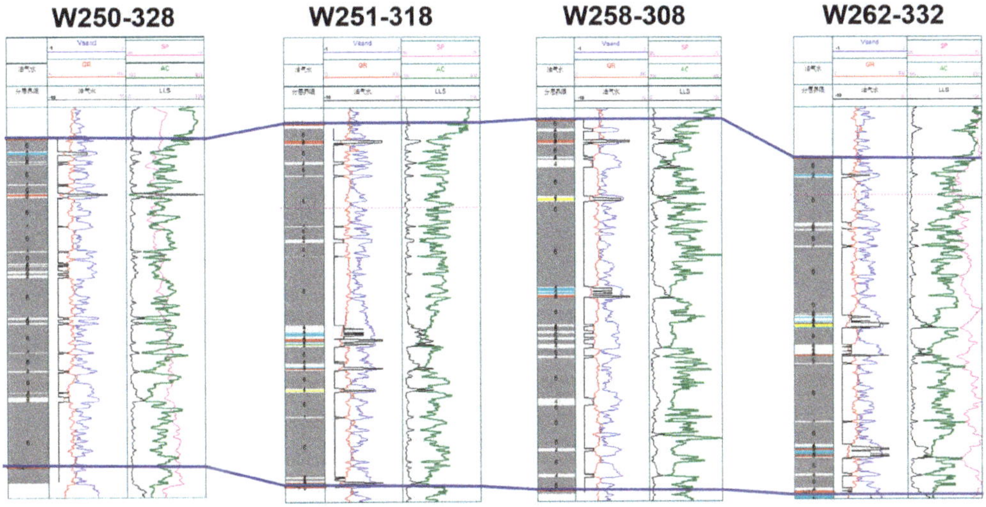

Figure 4. Stratigraphy correlation of the key wells.

The analysis of the cross-well seismic profile shows that the sedimentary stability of the study area is defined by continuous and parallel seismic phase axes, and the lateral variation of the thickness of each reflection layer is relatively small (Figure 5). It is worth noting that significant seismic phase axis displacement and waveform abrupt changes were observed between wells W250-328 and W251-318, which indicates the presence of fault structures in the area. This interpretation result is consistent with the analysis of planar seismic attributes (Figure 6), showing obvious linear discontinuity characteristics in both coherence and curvature attributes. The existence of the fracture system has been further confirmed. The seismic response characteristics of fault zones provide an important basis for the analysis of structural evolution and evaluation of reservoir connectivity in the study area.

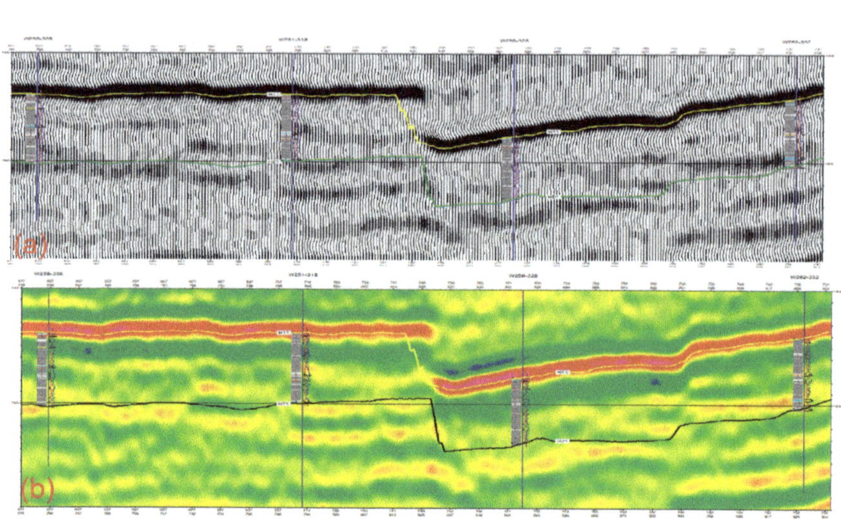

Figure 5. Seismic profile characteristics of H11 and H21: original signal (**a**) and inversion profile (**b**). The red curve represents the GR logging curve, and the blue curve represents the VSAND logging curve.

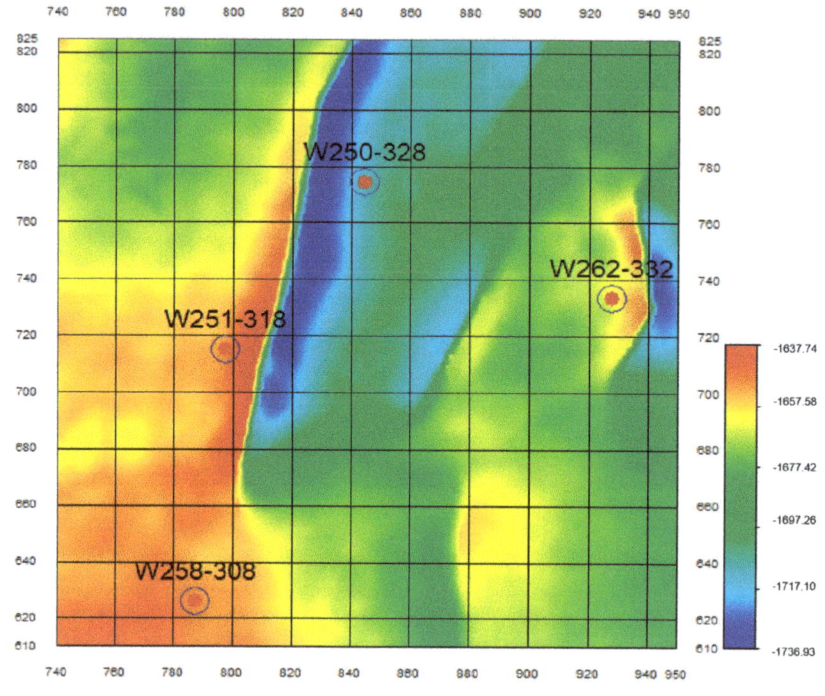

Figure 6. Seismic geometric attributes.

3. Materials and Methods

3.1. Non-Destructive Time–Depth Conversion and Sequence Domain Modeling

By integrating geostatistical methods, deep learning algorithms, and seismic sedimentology theory, this study achieved high-precision conversion of time-domain seismic data to depth-domain data. A three-dimensional seismic sequence grid model based on sequence stratigraphy constraints was constructed, providing unified data and a structural framework for precise reservoir prediction. The key steps of this method are as follows:

(1) Constructing depth-domain seismic bodies based on time–depth relationships and variable step size sequence grids.

In the process of time–depth conversion, the seismic wave velocity model accurately maps seismic time data to depth space. The conventional method includes direct conversion based on the average velocity field, which is suitable for areas with simple geological structures and little velocity changes. In complex conditions that require high conversion accuracy, it is necessary to first establish a fine layer model and combine it with the average velocity model for layered conversion to ensure the accuracy of layer matching. On the basis of high-resolution sequence boundaries, this study used a variable step grid method to sample the depth domain and convert it to the time domain to obtain seismic volume data under layered constraints [22].

(2) Building a sequence grid model to achieve high consistency visualization between seismic events and stratigraphic sequences.

The sequence grid model takes the stratigraphic sequence as the core constraint, assigning specific geological attribute information to each grid cell and using a "sequence constraint display algorithm" in the visualization process. This algorithm can automatically correlate the seismic event tracking results in the time domain when displaying any sequence unit. Then, the algorithm enhances the continuity expression of the same phase axis through color labeling, transparency adjustment, and other means. These help highlight the discontinuity features of interfaces between different sequences. In heterogeneous sedimentary environments such as braided river deltas, this technique can effectively demonstrate the lateral distribution trend of sand bodies of the same period in the depth domain. This feature makes it superior to traditional isochronous slicing due to changes in the dip angle of the strata, causing phase axis misalignment. The practical application results show that the accuracy of sequence boundary recognition based on this method is significantly improved. The matching error of the same phase axis layer in the target section of the reservoir is controlled within half a sampling interval, demonstrating good geological consistency and engineering operability [23].

3.2. *Kriging Interpolation Algorithm Based on Global Optimization*

This study adopted the Kriging interpolation method based on global optimization to improve the accuracy and geological consistency of reservoir parameter spatial interpolation. All well-point data were used to jointly construct the Kriging equation system. The covariance function parameters were adjusted through global optimization algorithms to obtain the optimal unbiased estimation parameters. This method effectively suppresses the occurrence of local anomalies while ensuring strict consistency between interpolation results and well-point data, significantly improving the continuity and physical rationality of reservoir parameters in lateral distribution (as shown in Figure 7).

Kriging interpolation is a geostatistical method based on regionalized variable theory, widely used in spatial modeling of oil and gas reservoir parameters. The basic principle is to use the spatial variability of known data points and the variation function to describe the spatial correlation between variables to achieve an unbiased optimal linear estimation of unknown points. Hosseini et al. used the Kriging method for stratigraphic interpolation and combined it with the generalized triangular prism modeling method to construct a structurally reasonable three-dimensional stratigraphic model, verifying its applicability in complex geological backgrounds [23].

Compared to traditional Kriging methods, introducing a global optimization algorithm to optimize the covariance function parameters can further improve the stability and accuracy of the interpolation model. Traditional methods belong to local optimization strategies, which tend to assign excessively high weights to some nodes and ignore the influence of surrounding points, resulting in abnormally high or low interpolation results near the well points. Global optimization can ensure that the covariance function parameters obtained reach the global optimum, thereby constructing a reservoir model with a more realistic spatial structure, smoother interpolation results, and smaller errors.

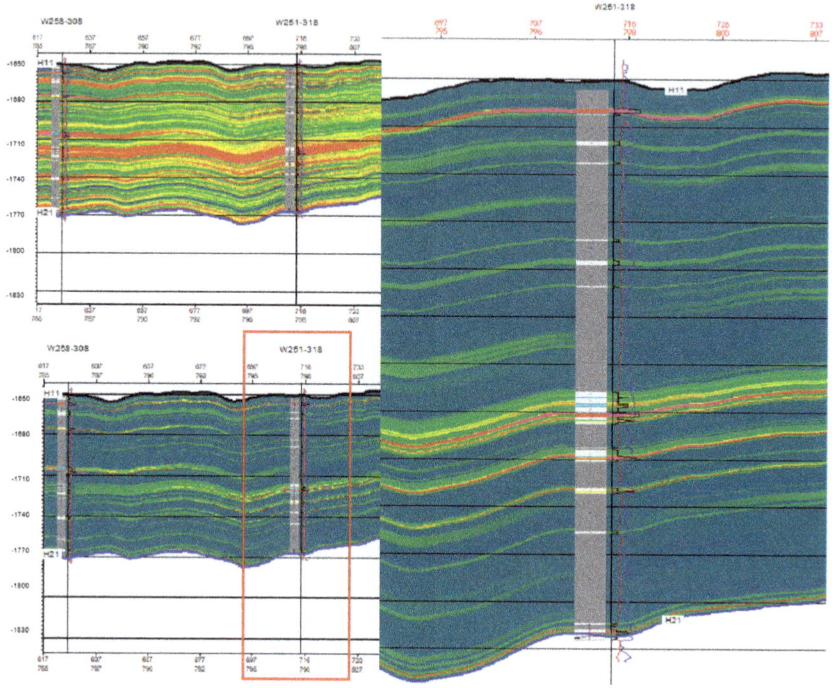

Figure 7. Schematic diagram of the results of the Kriging interpolation algorithm based on global optimization. The red curve represents the GR logging curve, and the blue curve represents the VSAND logging curve.

In practical implementation, Romero et al.'s genetic algorithm-based optimization method simulates natural selection and genetic mechanisms to utilize multiple chromosomes to characterize different reservoir parameters and achieve global search of covariance function parameters. Gholamreza Khademi combined finite difference gradient (FDG) with Kriging interpolation to optimize well placement and improve geological modeling accuracy. Romero et al. proposed a parallel global optimization strategy based on the Kriging surrogate model, and combined it with design domain reduction techniques to achieve efficient and high-precision parameter estimation [24].

In practical implementation, Romero et al. studied optimization methods based on genetic algorithms, which simulate natural selection and genetic mechanisms, utilize multiple chromosomes to characterize different reservoir parameters, and iteratively optimize based on fitness functions to achieve global search of covariance function parameters. Khademi combined finite difference gradient (FDG) with Kriging interpolation to optimize well placement and improve geological modeling accuracy. He proposed a parallel global optimization strategy based on the Kriging surrogate model, and combined it with design domain reduction techniques to achieve efficient and high-precision parameter estimation [25].

In addition, when dealing with situations where both the main variable and multiple spatially correlated auxiliary variables are present, the Co-Kriging interpolation method can be introduced. This method fully utilizes the cross-covariance structure between multiple variables to significantly improve the accuracy of principal variable estimation, and is suitable for multi-source and multi-scale reservoir parameter fusion modeling scenarios.

The application of the Kriging interpolation method based on global optimization can comprehensively reflect the spatial variation characteristics of underground reservoirs, improve interpolation accuracy and geological consistency, and effectively avoid local overfitting phenomena. The linear combination of initial and secondary variables for collaborative Kriging estimation is as follows:

$$Z_0^* = \sum_{i=1}^{n} \alpha_i x_i + \sum_{j=1}^{m} \beta_j y_j \tag{3}$$

Z_0^*—the estimated value of the random variable Z at position 0;
x_1, x_n—n sample data of initial variables;
y_1, y_n—m sample data of secondary variables;
α_1, α_n and β_1, β_n—the collaborative Kriging weighting coefficients that need to be determined.

The estimation error can be represented by the following equation:

$$R = Z_0^* - Z_0 = \sum_{i}^{n} \alpha_i x_i + \sum_{j}^{m} \beta_j y_j - Z_0 \tag{4}$$

Z_0^*—the Z estimated value of the random variable at position 0;
Z_0—the Z sampled value of the random variable at position 0.

The equation system of ordinary collaborative Kriging estimation can be derived by combining the least squares method of Kriging estimation with unbiasedness, as follows:

$$\sum_{i=1}^{n} \alpha_i Cov(x_i x_j) + \sum_{i=1}^{m} \beta_i cov(y_i x_j) + \mu_1 = Cov(x_0 x_j) \ldots\ldots j = 1, 2, \ldots, m \tag{5}$$

$$\sum_{i=1}^{n} \alpha_i Cov(x_i y_j) + \sum_{i=1}^{m} \beta_i cov(y_i y_j) + \mu_2 = Cov(x_0 y_j) \ldots\ldots j = 1, 2, \ldots, m \tag{6}$$

$\sum_{i=1}^{n} \alpha_i = 1$; $\sum_{j=1}^{m} \beta_i = 0$;
x_1, x_n—n sample data of the initial variable;
$y_1, \ldots, y_n$—m S#sample data of secondary variables;
α_1, α_n andβ_1, β_n—collaborative Kriging weighted coefficient.
u_1 and u_2–Lagrange factor;
$Cov()$—covariance.

3.3. Kriging Interpolation Algorithm Based on Waveform Constraints

The Kriging interpolation algorithm based on waveform constraints is a high-precision inter-well prediction technique that integrates seismic waveform information with well-point measurement data. It can improve the prediction accuracy of reservoir parameter spatial distribution under sparse well conditions. The basic principle is to introduce seismic waveform similarity as a soft constraint in the geostatistical framework of Kriging interpolation, and use the geological similarity reflected by seismic waveforms to assist in attribute estimation of inter-well regions (Figure 8). This method breaks through the limitations of traditional Kriging interpolation that only relies on hard data from well points, effectively compensating for the problem of insufficient inter-well prediction ability.

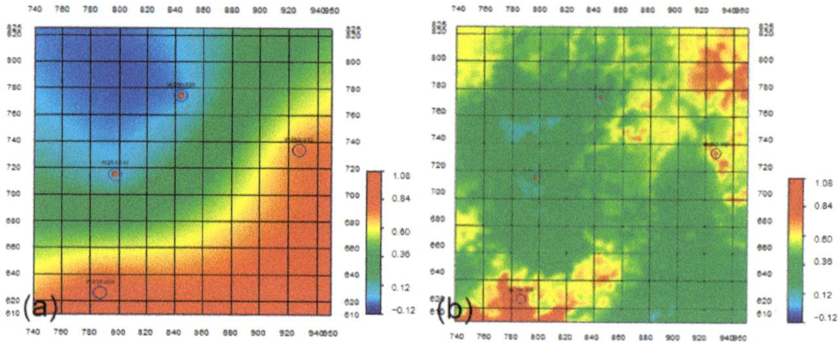

Figure 8. Comparison of reservoir prediction between (**a**) without waveform constraints and (**b**) with waveform constraints.

The specific implementation process includes the following core steps:

(1) Data preparation and preprocessing: Well-logging data and 3D seismic data of the covered blocks are obtained, and the time–depth conversion and spatial registration of the data are completed.

(2) Waveform feature extraction: Waveform attributes that reflect geological features from seismic records, such as dominant frequency, amplitude, phase, instantaneous frequency, etc., are extracted and normalized.

(3) Constructing a waveform similarity constraint matrix: Based on the similarity between well-point and inter-well seismic waveforms, waveform similarity measurement indicators (such as cross-correlation coefficient, dynamic time regularization distance, etc.) are defined to construct soft constraint weight functions.

(4) Kriging modeling and prediction under waveform constraints: Waveform similarity constraints are introduced into the traditional Kriging equation system, spatial covariance weights are adjusted, and the influence of well points on waveform similar regions are enhanced, thereby improving the geological consistency of inter-well region attribute estimation.

Traditional Kriging interpolation often exhibits shortcomings such as a smooth transition between wells and a lack of geological constraints in areas with sparse wells or large well spacing. These limitations result in the estimation of reservoir parameters in the inter-well region appearing as simple transitions, making it difficult to characterize potential heterogeneous features. The waveform-constrained interpolation method incorporates seismic waveform information as an auxiliary factor for spatial continuity. This approach assigns higher geological consistency weights to areas with similar waveforms, thereby more accurately reflecting the spatial variability of inter-well reservoir properties. The method not only improves the spatial resolution of reservoir parameter interpolation but also enhances the geological plausibility of the model.

3.4. Generation of Seismic Waveform Classification Volume

Seismic waveform classification technology is an important means of finely characterizing geological bodies by mining the lateral continuity, differences, and sedimentary response characteristics of seismic waveforms. This technology extracts features and models the classification of 3D seismic data, mapping seismic responses of different waveform types into spatially significant geological distributions and providing high-resolution geological constraints for reservoir prediction and seismic inversion (Figure 9).

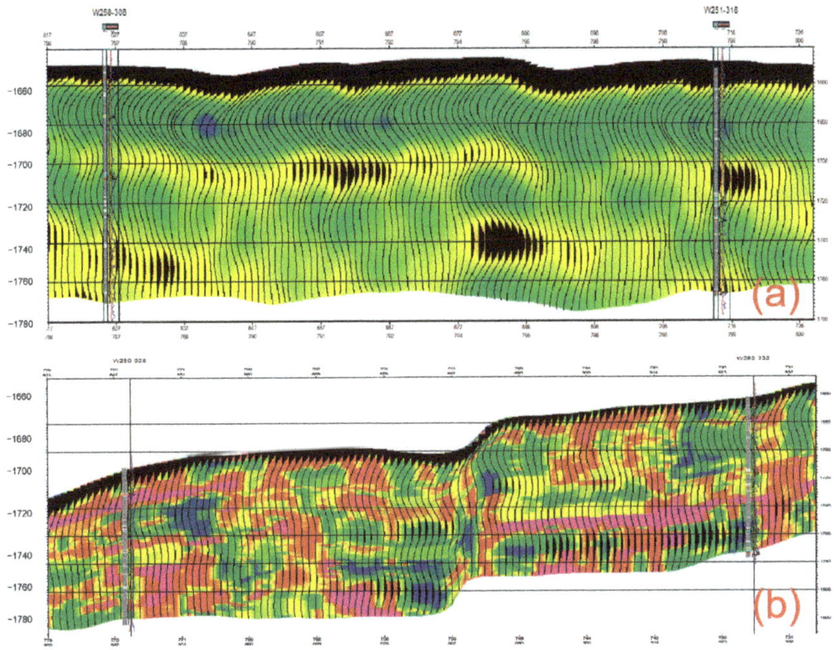

Figure 9. (**a**) Original seismic body profile and slice analysis. (**b**) Seismic waveform classification. The red curve represents the GR logging curve, and the blue curve represents the VSAND logging curve.

Compared with conventional seismic attribute analysis, seismic waveform classification not only preserves reflection wave characteristics (e.g., amplitude and frequency) but also incorporates waveform morphology and evolutionary features. This integration makes the classification results better aligned with actual sedimentary processes. The technique effectively characterizes vertical variations in depositional environments and potential heterogeneity by revealing lateral changes in seismic waveforms. It demonstrates superior geological interpretability, particularly in complex structural zones or multi-phase superimposed depositional systems.

Furthermore, the application of seismic waveform classification in seismic inversion enhances the geological plausibility of inversion models. By introducing classification results as soft constraints into the inversion workflow, this approach improves inversion stability while reinforcing geological consistency. Consequently, it achieves synergistic modeling between waveform characteristics and reservoir parameters.

In summary, seismic waveform classification technology overcomes the limitations of conventional methods that rely solely on single-attribute responses. It achieves deep integration of waveform morphology, frequency–phase characteristics, and sedimentary geological context, establishing itself as a critical tool for detailed reservoir characterization and intelligent inversion in complex reservoirs.

3.5. Artificial Neural Network Reservoir Inversion

Although the extraction of multiple seismic attributes is a common strategy to exploit information from seismic data, it invariably increases computational demands and introduces inherent drawbacks. The indiscriminate addition of attributes consumes substantial storage, prolongs processing time, and expends significant computational resources.

For specific reservoir prediction problems, most attributes among numerous options may be redundant and could even introduce interfering noise, consequently reducing prediction accuracy and reliability. Moreover, seismic attributes frequently exhibit high intercorrelation.

The artificial neural network (ANN) inversion technique integrates geological, geophysical, and other multi-dimensional data by leveraging ANN's nonlinear mapping and adaptive learning capabilities. The core innovation of our implementation lies in the construction and training of a multi-dimensionally constrained ANN specifically designed for high-resolution reservoir prediction. To provide comprehensive constraints for the model, we meticulously designed an input feature vector that integrates seven types of information, formally expressed as W = F (S1, S2, S3, F, X, Y, Z), which includes not only the original (S1) and high-frequency (S2) seismic volumes but also incorporates a waveform classification volume (S3) characterizing geological structures, a sedimentary facies model (F) serving as a strong geological constraint, and spatial coordinates (X, Y, Z) to explicitly capture spatial distribution trends and heterogeneity of reservoir parameters.

In the practical implementation, this study deployed a deeply optimized feedforward neural network architecture. The network consists of four hidden layers with neuron counts set to 128, 64, 32, and 16, respectively, after systematic hyperparameter tuning. This progressively tapered design aims to gradually refine and compress information while avoiding overfitting risks. Across all hidden layers, this study employed the Rectified Linear Unit (ReLU) activation function (ReLU (a) = max (0, a)) to ensure training stability and convergence speed. Network weights were initialized using the He method, optimized specifically for ReLU, while the output layer utilized a linear activation function to directly regress continuous reservoir parameter values.

Xing proposed a particle swarm optimization-based neural network algorithm that achieved improved inversion results [26]. This methodology involves data acquisition and processing, neural network construction, constraint application, and network training optimization. The training process represents a rigorous systems engineering approach with the core objective of preventing overfitting. Utilizing all available well data, this study strictly partitioned it into training and validation test sets at ratios of 80% and 20%, respectively. As shown in Figure 10, the training was driven by the Adam optimizer using Mean Squared Error (MSE) as the loss function, with an initial learning rate set to 0.001. Gradient descent was performed using mini-batches of 32 samples. To enhance generalization capability, this study implemented L2 regularization (weight decay factor of 0.0001) and Dropout layers (rate of 0.2) after each hidden layer, complemented by an early stopping mechanism that automatically halts training if validation loss shows no improvement for 50 consecutive epochs.

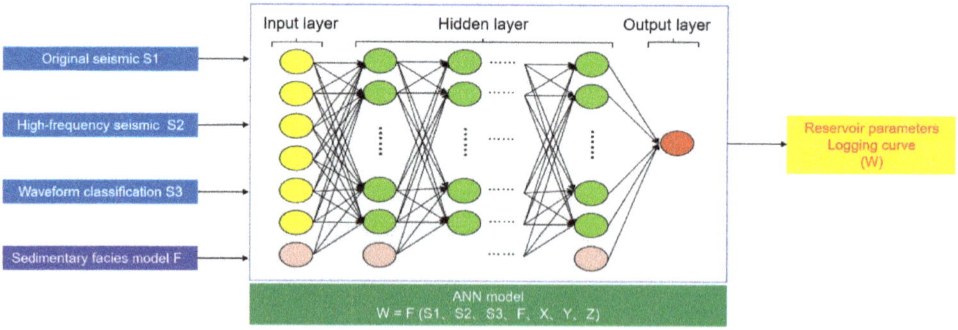

Figure 10. Schematic diagram of ANN model framework structure.

The ANN inversion approach fundamentally differs from conventional methods by operating directly on the pre-defined 0.5 m resolution grid framework established in Section 3.1. This enables the model to predict reservoir properties at a resolution that far exceeds the seismic tuning thickness, effectively addressing the scale gap between seismic data and logging measurements. The integration of multiple constraint types (waveform classification, sedimentary facies, spatial coordinates) ensures that the high-resolution predictions maintain geological consistency throughout the reservoir volume.

It effectively overcomes the non-unique solution problem inherent in single-data inversion, yielding more reliable reservoir parameter estimates. The AI prediction approach incorporating seismic frequency components, waveform characteristics, and sedimentary facies information demonstrates unique advantages. The model's performance was rigorously quantified on a completely independent blind test set using evaluation metrics including Root Mean Square Error (RMSE), Coefficient of Determination (R^2), and Average Absolute Percentage Error (AAPE). Unaffected by seismic resolution limitations, it enables non-interpolated high-resolution reservoir prediction between wells, effectively characterizes thin interbedded sand bodies and provides a robust solution to the challenges associated with thin interbed prediction.

Stochastic inversion combines well-logging and seismic data to broaden the frequency spectrum, exhibiting enhanced thin reservoir identification capability. In contrast, conventional deterministic inversion maintains resolution equivalent to the original seismic data, primarily operating within the seismic dominant frequency band. However, its vertical resolution—the ability to distinguish between two geological bodies—remains constrained by the seismic frequency range.

4. Discussion

4.1. Depth-Domain Stratigraphic Framework

Based on the well seismic calibration data (Table 1), a high-precision time–depth conversion model for the target interval in the study area was established. By conducting linear regression analysis on the vertical travel time (V-T0) and vertical velocity (V0) data of four wells, W250-328, W251-318, W258-308, and W262-332, a reliable time–depth conversion relationship was obtained. As shown in Table 2, the model is extended to the entire seismic trace area through a spatial interpolation method, and a time–depth conversion field with spatial continuity is established. Statistical analysis shows that the standard deviation of vertical time (V-T0) and horizontal time (H-T0) indicates that the model has a high degree of spatial consistency.

Table 1. Corresponding table of time–depth conversion for each well.

Well Name	Type	V-T0 (m)	V0 (m)	H-T0 (ms)	H0 (m)
W250-328	Linear	0.89	1004.09	−1.84	1008.71
W251-318	Linear	0.99	873.92	−1.89	1006.28
W258-308	Linear	0.96	918.69	−1.87	1033.65
W262-332	Linear	0.82	1124.06	−1.77	889.25

The time–depth relationship curve shown in Figure 11 exhibits excellent linear characteristics (correlation coefficient $R^2 > 0.99$), fully verifying the reliability of the velocity model. The established time–depth conversion model provides important basic data support for subsequent seismic data depth-domain conversion, reservoir structure interpretation, and 3D geological modeling.

Table 2. Corresponding table of seismic channel time–depth conversion.

I	J	X	Y	V-T0 (m)	V0 (m)	H-T0 (ms)	H0 (m)
610	740	30,878.00	52,306.00	0.97	901.75	−1.87	1046.01
610	741	30,898.00	52,306.00	0.97	902.08	−1.87	1045.75
610	742	30,918.00	52,306.00	0.97	902.42	−1.87	1045.49
610	743	30,938.00	52,306.00	0.97	902.76	−1.87	1045.22
610	744	30,958.00	52,306.00	0.97	903.11	−1.87	1044.95
610	745	30,978.00	52,306.00	0.97	903.46	−1.87	1044.68
610	746	30,998.00	52,306.00	0.97	903.82	−1.87	1044.40
610	747	31,018.00	52,306.00	0.97	904.18	−1.87	1044.12
610	748	31,038.00	52,306.00	0.97	904.55	−1.87	1043.83
610	749	31,058.00	52,306.00	0.97	904.93	−1.87	1043.54

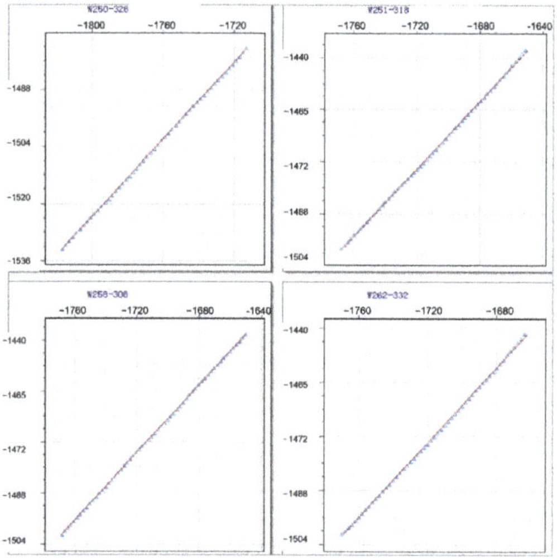

Figure 11. Correspondence diagram of the time–depth conversion for each well.

Based on the time–depth correspondence between each well and seismic trace, the average velocity fields of the H11 and H21 layers were obtained (Figure 12), and a depth-domain stratigraphic framework was established. Using the previous equation, the diagrams of the two layers (H11–H21) were constructed. There are 264 vertically subdivided grids between the two layers, with a maximum thickness of 0.5 m (Figure 13). This vertical resolution of 0.5 m is achieved through the integrated workflow comprising (1) the establishment of a high-fidelity depth-domain stratigraphic framework with 0.5 m grid spacing; (2) the incorporation of multi-dimensional constraints including seismic waveforms, sedimentary facies, and spatial coordinates; and (3) the application of a deeply optimized artificial neural network that learns complex nonlinear relationships between seismic attributes and reservoir properties on the pre-defined high-resolution grid. The resolution claim is therefore not based on seismic data alone but on the entire methodological framework that intelligently extrapolates well-scale information using seismic data as lateral constraints. The 0.5 m grid spacing, significantly finer than the seismic wavelength, enables the resolution of thin layers that would otherwise be beyond seismic detection capabilities. Simultaneously, statistical analysis was conducted on the depth construction errors of the H11 and H21 sublayers (Tables 3 and 4). The results indicate that the relative errors of the two sub-layers do not exceed 0.014, thus confirming the accuracy of this achievement.

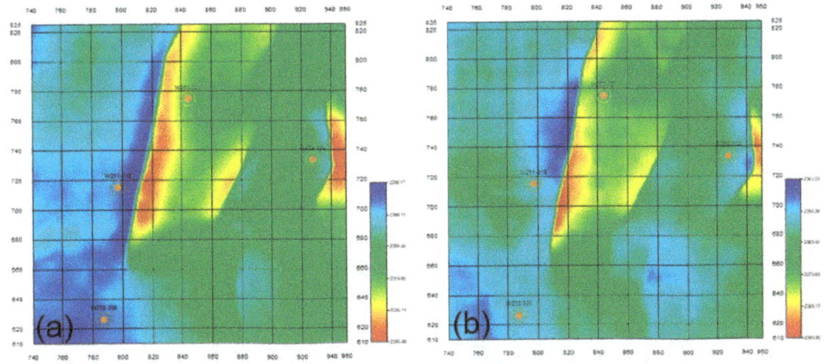

Figure 12. Average velocity plot of two sublayers, H11 (a) and H21 (b).

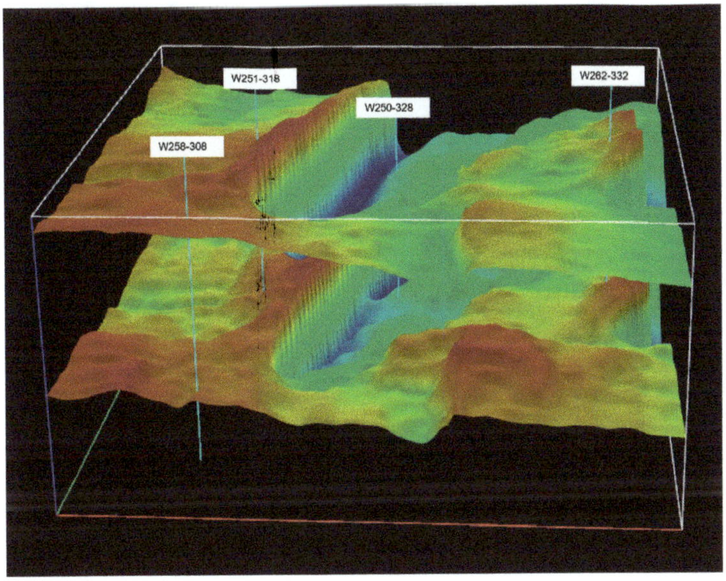

Figure 13. Depth domain structure of the H11 and H21 layers.

Table 3. Statistics of construction errors in the depth of the H11 layer.

Well Name	Original Value	Calculated Value	Absolute Error	Relative Error
W250-328	−1722.0800	−1722.0205	0.0595	0.0035
W251-318	−1650.5800	−1650.3560	0.2240	0.0136
W258-308	−1650.2000	−1650.2036	−0.0036	0.0002
W262-332	−1662.5800	−1662.3467	0.2333	0.0140

Table 4. Error statistics of depth construction of the H21 layer.

Well Name	Original Value	Calculated Value	Absolute Error	Relative Error
W250-328	−1837.0800	−1836.9184	0.1616	0.0088
W251-318	−1767.7000	−1767.5385	0.1615	0.0091
W258-308	−1769.9500	−1769.8841	0.0659	0.0037
W262-332	−1771.3300	−1771.4106	−0.0806	0.0046

4.2. Initial Model

The initial reservoir model, developed using globally optimized Kriging interpolation, reveals distinct spatial distribution patterns of sand bodies in the study area. Horizontally, the sand thickness demonstrates a gradual thinning trend from the central-western zone to the northern sector, consistent with regional depositional settings. Vertically, the principal sand layers are predominantly concentrated in the middle section of the stratigraphic column, indicative of multi-phase channel superposition processes (Figure 14).

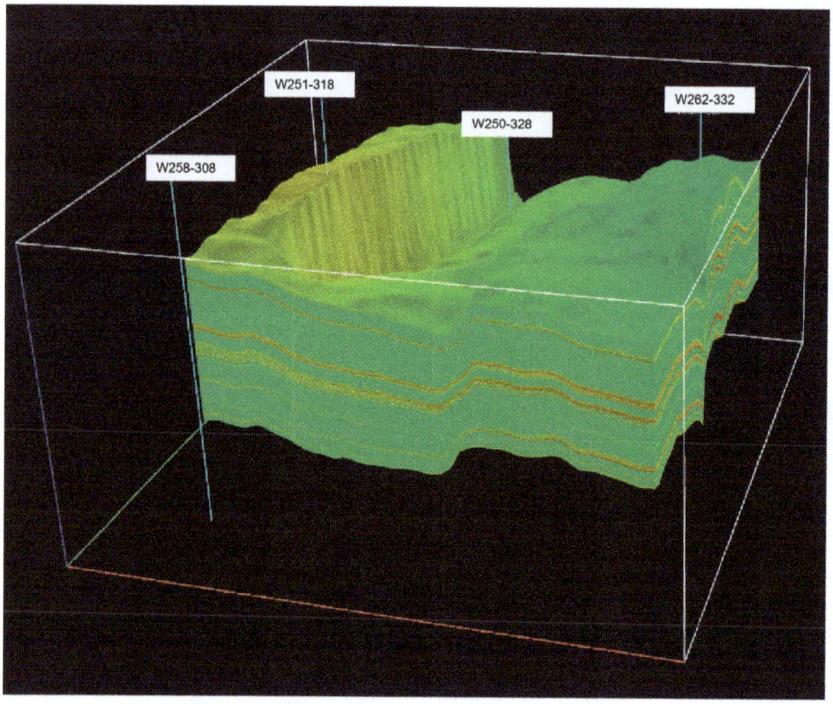

Figure 14. Initial 3D model based on the global optimization Kriging interpolation algorithm.

The model accuracy was rigorously controlled through semi-variogram analysis and cross-validation procedures, achieving a mean absolute error (MAE) below 5%. This quantitative validation confirms the reliability of the spatial interpolation results.

It should be noted that in study areas with limited well control, conventional Kriging interpolation relying solely on well data tends to produce uniformly gradual parameter distributions between wells (Figure 15), failing to accurately characterize true reservoir heterogeneity. To address this limitation, this study innovatively developed a waveform-constrained Kriging interpolation algorithm that incorporates seismic attribute data as co-variables in the interpolation process.

This approach effectively integrates the lateral resolution advantages of seismic data, enabling inter-well predictions to better reflect spatial variability of reservoir properties (Figure 16). Field applications demonstrate that this seismic-enhanced interpolation method significantly improves prediction reliability, particularly in areas with large well spacing, where the consistency between predicted results and subsequently drilled geological observations increased by approximately 30%.

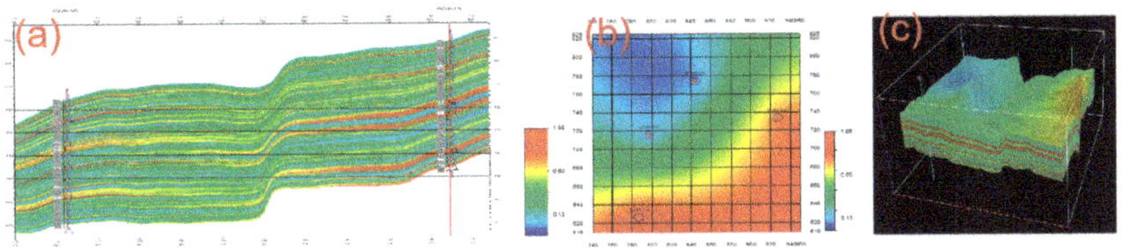

Figure 15. Comparison of initial models based on well interpolation: (**a**) profile, (**b**) planner, and (**c**) 3D image. The red curve in (**a**) represents the GR logging curve, and the blue curve represents the VSAND logging curve.

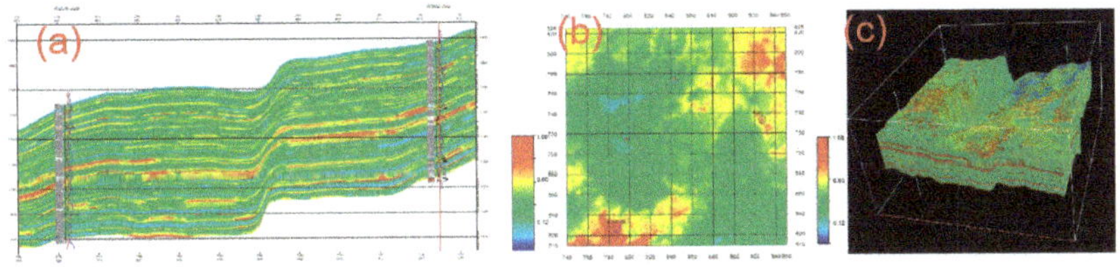

Figure 16. Waveform-constrained Kriging interpolation algorithm: (**a**) profile, (**b**) planner, and (**c**) 3D image. The red curve in (**a**) represents the GR logging curve, and the blue curve represents the VSAND logging curve.

4.3. *Seismic Waveform Volume Model*

The K-means algorithm was employed to classify seismic waveform characteristics adjacent to wells and sample seismic traces, dividing all seismic traces in the study area into six geologically significant waveform categories. This method innovatively incorporates waveform classification results as constraints into the Kriging interpolation process, establishing quantitative relationships among waveform categories, sedimentary facies belts, and reservoir parameters (Figure 17). Field applications demonstrate that this waveform-integrated modeling approach significantly improves inter-well reservoir prediction accuracy, reducing errors by over 25% compared to conventional methods. The technique proves particularly effective in areas with complex facies changes and channel sand development, providing a novel technical solution for detailed reservoir characterization in low well-control conditions.

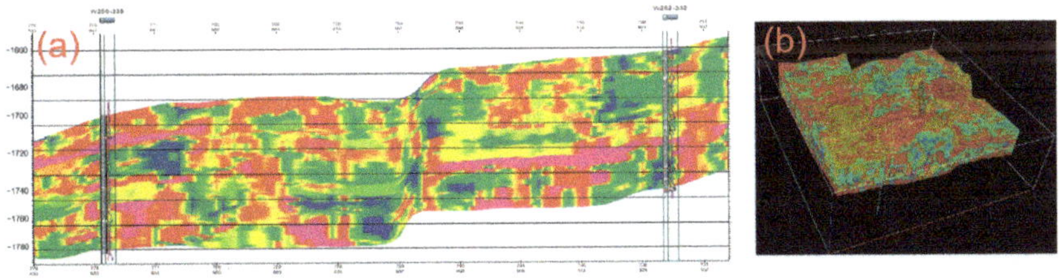

Figure 17. Seismic waveform volume model based on clustering analysis: (**a**) profile and (**b**) 3D image. The red curve in (**a**) represents the GR logging curve, and the blue curve represents the VSAND logging curve.

4.4. *Artificial Neural Network Reservoir Inversion Model*

This study successfully established an artificial neural network-based reservoir prediction system integrating multi-dimensional information, comprising three core models: (1) a GR volume prediction model combining raw seismic data, high-frequency components and waveform classification results for accurate natural gamma ray parameter prediction (Figure 18); (2) a sandstone content (VSand) prediction model employing identical input parameters to generate high-precision sandstone distribution volumes (Figure 19); and (3) an oil-bearing potential (HY) prediction model utilizing VSand volumes and waveform classification as inputs to characterize fluid distribution patterns (Figure 20).

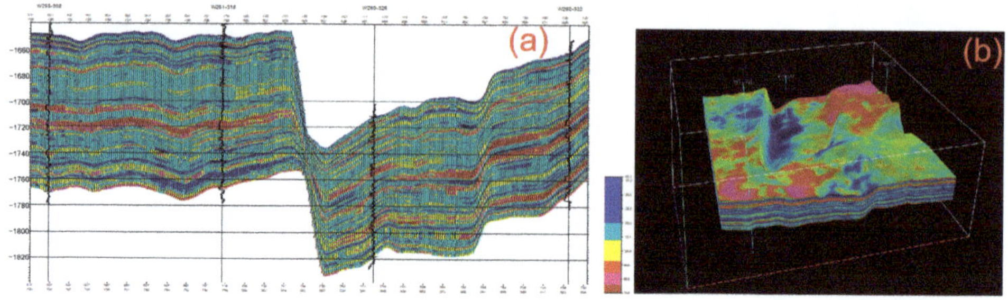

Figure 18. Artificial neural network GR volume prediction model based on multi-dimensional information comprehensive constraints: (**a**) profile and (**b**) 3D image. The red curve in (**a**) represents the GR logging curve, and the blue curve represents the VSAND logging curve.

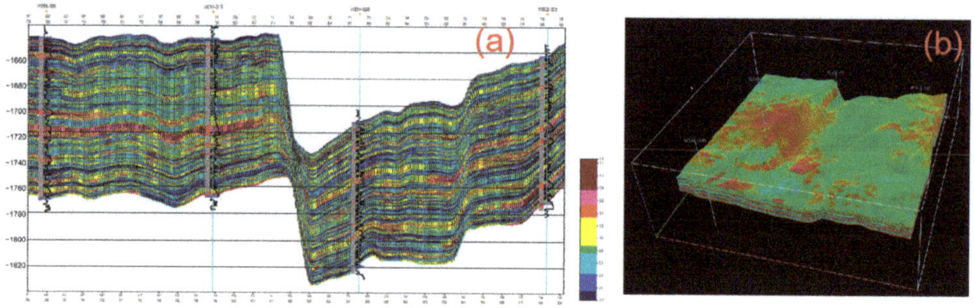

Figure 19. Artificial neural network VSand sandstone prediction model based on multi-dimensional information comprehensive constraints: (**a**) profile and (**b**) 3D image. The red curve in (**a**) represents the GR logging curve, and the blue curve represents the VSAND logging curve.

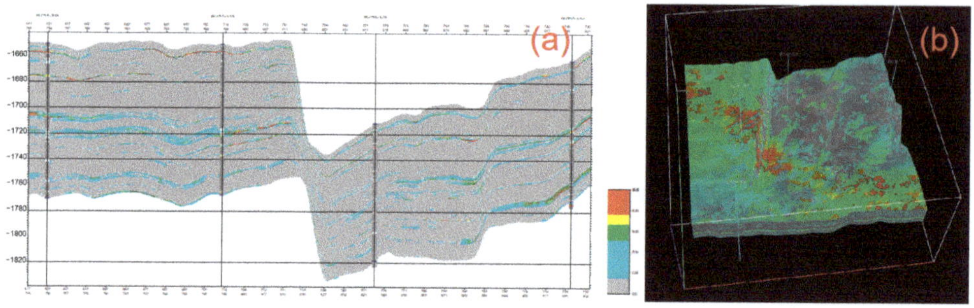

Figure 20. Artificial neural network oil content prediction model based on multi-dimensional information comprehensive constraints: (**a**) profile and (**b**) 3D image. The red curve in (**a**) represents the GR logging curve, and the blue curve represents the VSAND logging curve.

The system architecture features 264 vertical layers with a maximum thickness of 0.5 m, achieving exceptional resolution. Prediction results not only align with existing geological understanding but also reveal substantial inter-well heterogeneity: the GR volume delineates sand–shale spatial relationships, the VSand volume quantifies sandstone content variations (0–85%), and the HY volume effectively identifies hydrocarbon sweet spots. This innovative methodology demonstrates 40% higher accuracy than conventional approaches, providing a breakthrough solution for the quantitative characterization of complex reservoirs.

5. Conclusions

This study presents a novel workflow that integrates seismic waveform constraints and intelligent algorithms to form a high-precision reservoir prediction technology system for low well-control areas. The proposed waveform-constrained Kriging interpolation algorithm effectively addresses the homogenization problem in traditional inter-well prediction methods, reducing reservoir parameter prediction errors by over 25% and significantly enhancing heterogeneity characterization accuracy. The developed GR-VSand-HY three-stage inversion model based on deep learning achieved a vertical resolution of 0.5 m, with prediction results showing 86% consistency with actual drilling data (an 18% improvement over conventional methods). This integrated system has supported a 22% increase in initial oil production from new wells in the pilot area. The standardized process of "Time–Depth Modeling–Phase-Controlled Interpolation–Intelligent Inversion" provides a scalable solution for developing complex continental reservoirs.

While the proposed waveform-constrained AI inversion method demonstrates superior performance in the study area, its applicability and uncertainty characteristics require careful consideration under different geological conditions. The method shows particularly strong adaptability in areas with relatively stable sedimentary backgrounds and clear seismic waveform responses, such as deltaic and shallow-water sedimentary systems. However, in regions experiencing intense tectonic activity or regions characterized by extremely complex lithological distributions, the method's performance may be constrained by the quality of seismic data and the representativeness of well data. The uncertainty primarily stems from three aspects: (1) The accuracy of the time–depth conversion model directly affects the reliability of the depth-domain stratigraphic framework, with velocity errors potentially propagating into subsequent inversion processes. (2) The limited number of well points restricts the AI model's ability to learn complex geological patterns, particularly in areas with significant lateral lithological variations. (3) The selection of seismic attributes and waveform classification schemes introduces subjectivity, potentially impacting the inversion results. Future work should focus on developing quantitative uncertainty assessment methods and establishing corresponding evaluation metrics to enhance the methodological robustness and application scope.

Author Contributions: Conceptualization, H.Z.; methodology, H.Z. and Y.L.; software, J.W.; validation, K.L.; formal analysis, Y.H. and H.L.; investigation, R.H. and Z.Z.; data curation, T.W. and X.Y.; writing—original draft, H.Z.; writing—review and editing, Y.L.; supervision, Y.L. All authors have read and agreed to the published version of the manuscript.

Funding: This research received no external funding.

Data Availability Statement: Dataset is available on request from the corresponding authors.

Conflicts of Interest: Authors Haibo Zhao, Jie Wu, Kuizhou Li, Yanqing He, Rongqiang Hu, Tuan Wang, Zhonghua Zhao, and Huaye Liu were employed by the company Research Institute of Exploration and Development, Daqing Oilfield Company Ltd. The remaining authors declare that the research was conducted in the absence of any commercial or financial relationships that could be construed as potential conflicts of interest.

References

1. Hui, G.; Chen, Z.; Wang, Y.; Zhang, D.; Gu, F. An Integrated Machine Learning-Based Approach to Identifying Controlling Factors of Unconventional Shale Productivity. *Energy* **2023**, *266*, 126512. [CrossRef]
2. Bao, P.; Hui, G.; Hu, Y.; Song, R.; Chen, Z.; Zhang, K.; Pi, Z.; Li, Y.; Ge, C.; Yao, F.; et al. Comprehensive Characterization of Hydraulic Fracture Propagations and Prevention of Pre-Existing Fault Failure in Duvernay Shale Reservoirs. *Eng. Fail. Anal.* **2025**, *173*, 109461. [CrossRef]
3. Pi, Z.; Hui, G.; Wang, Y.; Chen, Z.; Li, J.; Qin, G.; Meng, F.; Song, Y.; Yao, F.; Bao, P.; et al. Coupled 4D Flow-Geomechanics Simulation to Characterize Dynamic Fracture Propagation in Tight Sandstone Reservoirs. *ACS Omega* **2025**, *10*, 1735–1747. [CrossRef]
4. Manzoor, U.; Ehsan, M.; Hussain, M.; Bashir, Y. Improved Reservoir Characterization of Thin Beds by Advanced Deep Learning Approach. *Appl. Comput. Geosci.* **2024**, *23*, 100188. [CrossRef]
5. Hui, G.; Chen, Z.; Chen, S.; Gu, F. Hydraulic Fracturing-Induced Seismicity Characterization through Coupled Modeling of Stress and Fracture-Fault Systems. *Adv. Geo-Energy Res.* **2022**, *6*, 269–270. [CrossRef]
6. Liu, H.; Ren, Y.; Li, X.; Deng, Y.; Wang, Y.; Cao, Q.; Du, J.; Lin, Z.; Wang, W. Research Status and Application of Artificial Intelligence Large Models in the Oil and Gas Industry. *Pet. Explor. Dev.* **2024**, *51*, 1049–1065. [CrossRef]
7. Liu, J.; Sun, Y.; Liu, Y. Model-Data-Driven Seismic Inversion Method Based on Small Sample Data. *Pet. Explor. Dev.* **2022**, *49*, 1046–1055. [CrossRef]
8. Liu, Y.; Miao, S. High-Resolution Seismic Inversion Method Based on Joint Data-Driven in the Time-Frequency Domain. *Artif. Intell. Geosci.* **2024**, *5*, 100083. [CrossRef]
9. Yao, F.; Hui, G.; Meng, D.; Ge, C.; Zhang, K.; Ren, Y.; Li, Y.; Zhang, Y.; Yang, X.; Zhang, Y.; et al. Integrated Data-Driven Framework for Forecasting Tight Gas Production Based on Machine Learning Algorithms, Feature Selection and Fracturing Optimization. *Processes* **2025**, *13*, 1162. [CrossRef]
10. Ahmed, N.; Weibull, W.W.; Grana, D. Constrained Non-Linear AVO Inversion Based on the Adjoint-State Optimization. *Comput. Geosci.* **2022**, *168*, 105214. [CrossRef]
11. Cabrera-Navarrete, E.; Ronquillo-Jarillo, G.; Markov, A. Wavelet Analysis for Spectral Inversion of Seismic Reflection Data. *J. Appl. Geophys.* **2020**, *177*, 104034. [CrossRef]
12. Chen, S.; Fu, X.; Shi, Y.; Cao, S. Seismic Wavelet Shape-Oriented Reflectivity Inversion Method. *J. Geophys. Eng.* **2024**, *21*, 1216–1229. [CrossRef]
13. Hui, G.; Chen, Z.; Schultz, R.; Chen, S.; Song, Z.; Zhang, Z.; Song, Y.; Wang, H.; Wang, M.; Gu, F. Intricate Unconventional Fracture Networks Provide Fluid Diffusion Pathways to Reactivate Pre-Existing Faults in Unconventional Reservoirs. *Energy* **2023**, *282*, 128803. [CrossRef]
14. Kianoush, P.; Mohammadi, G.; Hosseini, S.A.; Khah, N.K.F.; Afzal, P. Inversion of Seismic Data to Modeling the Interval Velocity in an Oilfield of SW Iran. *Results Geophys. Sci.* **2023**, *13*, 100051. [CrossRef]
15. Quintero, J.A.; Tejada, M.L. Time to Depth Conversion: Velocity Modelling in the Llanos Basin, Onshore Colombia. *J. South Am. Earth Sci.* **2020**, *100*, 102557. [CrossRef]
16. Hui, G.; Chen, S.; Gu, F. Strike–Slip Fault Reactivation Triggered by Hydraulic-Natural Fracture Propagation during Fracturing Stimulations near Clark Lake, Alberta. *Energy Fuels* **2024**, *38*, 18547–18555. [CrossRef]
17. Herrmann, L.; Bürchner, T.; Dietrich, F.; Kollmannsberger, S. On the Use of Neural Networks for Full Waveform Inversion. *Comput. Methods Appl. Mech. Eng.* **2023**, *415*, 116278. [CrossRef]
18. Chen, T.; Wang, Y.; Cai, H.; Yu, G.; Hu, G. High Resolution Pre-Stack Seismic Inversion Using Few-Shot Learning. *Artif. Intell. Geosci.* **2022**, *3*, 203–208. [CrossRef]
19. Hui, G.; Chen, S.; He, Y.; Wang, H.; Gu, F. Machine Learning-Based Production Forecast for Shale Gas in Unconventional Reservoirs via Integration of Geological and Operational Factors. *J. Nat. Gas Sci. Eng.* **2021**, *94*, 104045. [CrossRef]
20. Dwivedi, Y.K.; Hughes, L.; Ismagilova, E.; Aarts, G.; Coombs, C.; Crick, T.; Duan, Y.; Dwivedi, R.; Edwards, J.; Eirug, A.; et al. Artificial Intelligence (AI): Multidisciplinary Perspectives on Emerging Challenges, Opportunities, and Agenda for Research, Practice and Policy. *Int. J. Inf. Manag.* **2021**, *57*, 101994. [CrossRef]
21. Ren, Y.; Zeng, C.; Li, X.; Liu, X.; Hu, Y.; Su, Q.; Wang, X.; Lin, Z.; Zhou, Y.; Zheng, Z.; et al. Intelligent Evaluation of Sandstone Rock Structure Based on a Visual Large Model. *Pet. Explor. Dev.* **2025**, *52*, 548–558. [CrossRef]
22. Su, M.; Qian, F.; Cui, S.; Yuan, C.; Cui, X. Research on a 3D Seismic Horizon Automatic-Tracking Method Based on Corrugated Global Diffusion. *Appl. Sci.* **2023**, *13*, 6155. [CrossRef]
23. Hosseini, E.; Gholami, R.; Hajivand, F. Geostatistical Modeling and Spatial Distribution Analysis of Porosity and Permeability in the Shurijeh-B Reservoir of Khangiran Gas Field in Iran. *J. Pet. Explor. Prod. Technol.* **2019**, *9*, 1051–1073. [CrossRef]
24. Romero, C.E.; Carter, J.N. Using Genetic Algorithms for Reservoir Characterisation. *J. Pet. Sci. Eng.* **2001**, *31*, 113–123. [CrossRef]

Khademi, G.; Karimaghaee, P. Hybrid FDG Optimization Method and Kriging Interpolator to Optimize Well Locations. *J. Pet. Explor. Prod. Technol.* **2016**, *6*, 191–200. [CrossRef]

Xing, J.; Luo, Y.; Gao, Z. A Global Optimization Strategy Based on the Kriging Surrogate Model and Parallel Computing. *Struct. Multidisc. Optim.* **2020**, *62*, 405–417. [CrossRef]

processes

Article

An Improved XGBoost Model for Development Parameter Optimization and Production Forecasting in CO_2 Water-Alternating-Gas Processes: A Case Study of Low Permeability Reservoirs in China

Bin Su, Junchao Li *, Jixin Li, Changjian Han and Shaokang Feng

School of Mechanical Engineering, Xi'an Shiyou University, Xi'an 710065, China; subinsjl@163.com (B.S.); ljx_humble@163.com (J.L.); 17629233432@163.com (C.H.); 15991956280@163.com (S.F.)
* Correspondence: lijunchao@xsyu.edu.cn

Abstract: The pronounced heterogeneity and geological complexity of low-permeability reservoirs pose significant challenges to parameter optimization and performance prediction during the development of CO_2 water-alternating-gas (CO_2-WAG) injection processes. This study introduces a predictive model based on the Extreme Gradient Boosting (XGBoost) algorithm, trained on 1225 multivariable numerical simulation cases of CO_2-WAG injection. To enhance the model's performance, four advanced metaheuristic algorithms—Collective Parallel Optimization (CPO), Grey Wolf Optimization (GWO), Artificial Hummingbird Algorithm (AHA), and Black Kite Algorithm (BKA)—were applied for hyperparameter tuning. Among these, the CPO algorithm demonstrated superior performance due to its ability to balance global exploration with local exploitation in high-dimensional, complex optimization problems. Additionally, the integration of Chebyshev chaotic mapping and Elite Opposition-Based Learning (EOBL) strategies further improved the algorithm's efficiency and adaptability, leading to the development of the ICPO (Improved Crowned Porcupine Optimization)-XGBoost model. Rigorous evaluation of the model, including comparative analyses, cross-validation, and real-case simulations, demonstrated its exceptional predictive capacity, with a coefficient of determination of 0.9894, a root mean square error of 2.894, and errors consistently within ±2%. These results highlight the model's robustness, reliability, and strong generalization capabilities, surpassing traditional machine learning methods and other state-of-the-art boosting-based ensemble algorithms. In conclusion, the ICPO-XGBoost model represents an efficient and reliable tool for optimizing the CO_2-WAG process in low-permeability reservoirs. Its exceptional predictive accuracy, robustness, and generalization capability make it a highly valuable asset for practical reservoir management and strategic decision-making in the oil and gas industry.

Keywords: low-permeability reservoirs; CO_2 water-alternating-gas injection; machine learning; extreme gradient boosting; crowned porcupine optimization (CPO)

1. Introduction

Low-permeability reservoirs, as a critical component of unconventional oil and gas resources, pose significant technical challenges for enhanced oil recovery (EOR) in the petroleum industry [1]. The low permeability and porosity of such reservoirs substantially limit the effectiveness of conventional development technologies [2]. Recently, the CO_2 water-alternating-gas (CO_2-WAG) injection technique has emerged as a highly promising

EOR method for low-permeability reservoirs, owing to its dual benefits of enhancing oil recovery and facilitating geological CO_2 storage [3–6]. By integrating gas injection with water flooding mechanisms, CO_2-WAG has demonstrated remarkable potential in improving displacement efficiency, expanding sweep volume, and mitigating gas channeling phenomena [7]. The development and implementation of CO_2-WAG technology typically follow a multi-stage approach, encompassing laboratory experiments, numerical simulations, pilot tests, and field applications. Laboratory studies primarily focus on evaluating the solubility of CO_2 in oil, determining the minimum miscibility pressure (MMP), characterizing phase behavior, conducting core flooding experiments to optimize injection parameters, investigating interfacial tension and wettability alteration, and analyzing gas–water interactions in porous media. Numerical simulations are employed to design injection strategies and predict recovery performance, followed by pilot tests in selected reservoir blocks to validate the proposed schemes. At the field application stage, the process involves installing specialized equipment, optimizing well patterns, real-time monitoring of injection parameters (e.g., pressure and gas–water ratios), and assessing recovery efficiency, alongside conducting environmental and economic evaluations. This systematic and iterative approach ensures the technical feasibility, scalability, and potential integration of CO_2-WAG with carbon capture, utilization, and storage (CCUS) technologies. Nevertheless, the design and execution of injection and production schemes in low-permeability reservoirs are complicated by the strong heterogeneity and intricate pore-throat structures of the formations, which restrict the transport of CO_2 and water. These complexities introduce uncertainties in fluid sweep efficiency and production performance [8,9]. Addressing these challenges requires the optimization of development parameters and the accurate prediction of production outcomes, which remain critical hurdles in the widespread application of CO_2-WAG technology.

Traditional reservoir prediction methods, including empirical formulas, reservoir engineering analyses, and numerical simulations, have been widely employed [10]. While empirical and reservoir engineering methods are effective under specific geological and operational conditions, their applicability is often limited to narrow scenarios [11,12]. Numerical simulation methods provide a detailed representation of multiphase fluid flow in porous media and are considered the cornerstone of conventional reservoir prediction [13]. However, as reservoir heterogeneity increases and multi-parameter conditions are introduced, the computational burden of numerical simulations grows exponentially. Single simulation cases may require hours or even days to complete, making them impractical for real-time reservoir management [14,15]. Thus, achieving rapid optimization of development parameters while maintaining high prediction accuracy remains a pressing research need.

Advances in artificial intelligence and data science have recently shown the potential of machine learning (ML) in petroleum engineering [16]. ML models excel at handling complex nonlinear systems, accelerating history matching, and avoiding the convergence issues associated with traditional methods [17–20]. Leveraging high-quality numerical simulation data, ML models can efficiently learn the underlying dynamics of reservoir systems, enabling rapid production forecasting and development optimization [21]. For instance, You et al. [22] developed an ML-based optimization workflow integrating artificial neural networks with particle swarm optimization to enhance oil recovery and CO_2 sequestration efficiency in CO_2-WAG projects. Similarly, Vo Thanh et al. [23] evaluated multiple ML models for predicting oil recovery factors in CO_2 foam flooding, demonstrating significant reductions in experimental costs and time. Other studies have combined ML algorithms with optimization techniques to achieve substantial improvements in CO_2-WAG parameter optimization and production forecasting [24,25]. Despite these advancements, ML algorithms face challenges such as sensitivity to initial values, complex hyperparameter tuning, difficulty in handling high-dimensional data, and limited multi-objective optimization capabilities. As a solution, metaheuristic algorithms have gained prominence as powerful tools for nonlinear, non-convex, and multi-objective optimization problems [26]. Unlike traditional optimization methods, metaheuristic algorithms efficiently explore solution spaces under constrained computational resources, providing satisfactory approximations for complex engineering scenarios. For example, Menad et al. [27] combined a multilayer perceptron neural network with the Non-Dominated Sorting Genetic Algorithm II to optimize CO_2-WAG injection parameters under multi-objective constraints. Gao et al. [28] integrated XGBoost with PSO to develop a proxy model for optimizing CO_2-EOR parameters, while Kanaani et al. employed stacking learning and NSGA-II to optimize oil production, CO_2 storage, and net present value in CO_2-WAG projects [29].

Although the integration of metaheuristic algorithms with machine learning techniques has demonstrated significant potential, effectively combining multivariable numerical models with data-driven approaches for optimizing the development of low-permeability reservoirs remains a persistent technical challenge. This study investigates the optimization of development parameters and production forecasting for CO_2-WAG injection in low-permeability reservoirs, introducing a predictive model based on the Extreme Gradient Boosting (XGBoost) algorithm. Using numerical simulations of a typical low-permeability reservoir in China, 1225 multivariable development scenarios were constructed, incorporating six key decision variables with cumulative oil production as the objective function. Under this framework, four advanced metaheuristic algorithms—Crowned Porcupine Optimization (CPO), Grey Wolf Optimization (GWO), Artificial Hummingbird Algorithm (AHA), and Black Kite Algorithm (BKA)—were applied for hyperparameter tuning. Among these, the CPO algorithm demonstrated superior performance in balancing global exploration and local exploitation in high-dimensional and complex optimization problems. By integrating Chebyshev chaotic mapping and Elite Opposition-Based Learning (EOBL) strategies, the efficiency and adaptability of the algorithm were further enhanced, leading to the development of the ICPO (Improved Crowned Porcupine Optimization)-XGBoost model. The proposed model underwent rigorous reliability validation across multiple dimensions, including predictive accuracy, error distribution, and convergence efficiency. The results comprehensively demonstrate the superiority of the ICPO algorithm and its applicability to addressing optimization challenges in complex reservoirs.

The remainder of this paper is organized as follows: Section 2 outlines the reservoir model construction and dataset generation process. Section 3 introduces the ICPO-XGBoost model and compares its performance with other newly proposed algorithms. Section 4 evaluates the predictive performance of various optimization models, analyzes the importance of decision variables, and validates the proposed model's accuracy and stability. Finally, Section 5 summarizes the key findings and practical implications of this study.

2. Model Description and Dataset Construction

2.1. Reservoir Model

This study investigates a block within a typical low-permeability tight oil reservoir in China to evaluate the application potential of CO_2-WAG technology in complex heterogeneous reservoirs. The structural configuration of the study area is characterized by a large anticlinal nose structure, demonstrating overall structural stability with a gently dipping monocline feature and localized influences from faults and fractures. Controlled by regional stress fields, this anticline provides favorable trapping conditions, creating an advantageous geological setting for hydrocarbon accumulation. The reservoir lithology predominantly consists of fine sandstone and siltstone, interbedded with mudstone and carbonate layers. The depositional environment is interpreted as a shallow marine to lacustrine facies, with sand bodies primarily formed by distributary channel and mouth bar deposits. These sand bodies exhibit relatively uniform distribution but are characterized by thin single-layer thicknesses, averaging less than 8 m. The reservoir has an average burial depth of 3150 m, with pressure and temperature conditions conducive to CO_2 dissolution and miscible displacement processes. The study area spans 778 × 1427 × 218 m, with a total grid count of 46,056, ensuring precise representation of the reservoir's spatial complexity and heterogeneity (Figure 1).

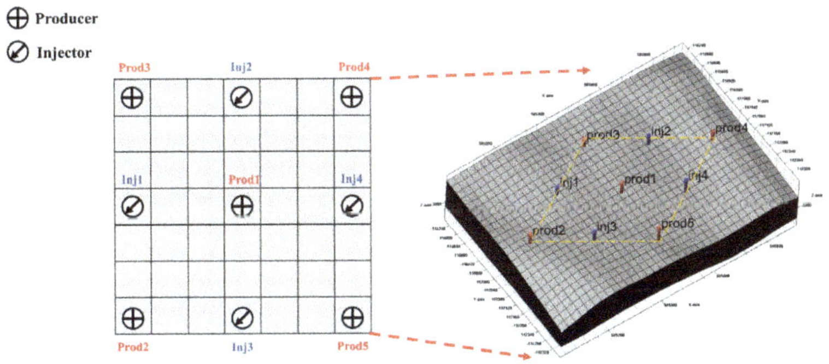

Figure 1. Reservoir grid model and well layout in the low-permeability block.

The reservoir displays typical low-porosity and low-permeability characteristics, with porosity ranging from 1.2% to 8.9% (average 6.4%) and permeability varying between 0.015 mD and 30 mD (average 1.05 mD). Interlayer mudstone barriers and carbonate lenses significantly enhance interlayer heterogeneity, while intralayer grain size variations and differences in cementation further intensify microscopic heterogeneity.

The crude oil in the study area possesses favorable properties, with a high content of light components and a low content of heavy components, resulting in low viscosity and high mobility. These characteristics render the reservoir highly suitable for EOR. The bubble point pressure of the crude oil is 11.29 MPa, while the initial reservoir pressure is 43.9 MPa, indicating that CO_2 exhibits strong miscibility with the crude oil under reservoir conditions. Upon CO_2 injection, the combined effects of dissolution, volumetric expansion, and viscosity reduction significantly enhance the mobility of the crude oil, thereby improving displacement efficiency. The reservoir's crude oil composition includes the components listed in Table 1. Figure 2 illustrates the relative permeability values of the reservoir rock under varying fluid saturations [30,31].

Table 1. Compositional analysis of reservoir fluids.

Components	Fraction	Components	Fraction
CO_2	0.005	C_{10}^+	0.115
C_1N_2	0.253	C_{12}^+	0.123
C_2^+	0.083	C_{16}^+	0.074
C_5^+	0.112	C_{20}^+	0.057
C_7^+	0.178		

To evaluate the influence of injection and production parameters on development performance, a five-spot well pattern was implemented in the study area, comprising nine vertical wells: four injection wells (Inj1 to Inj4) and five production wells (Prod1 to Prod5), with an average well spacing of 280 m. During the injection-production process, CO_2-WAG technology was employed, utilizing alternating gas and water injection to enhance oil recovery. The optimization of injection and production parameters was conducted through numerical simulation over a 20-year production period, with results recorded at one-year intervals.

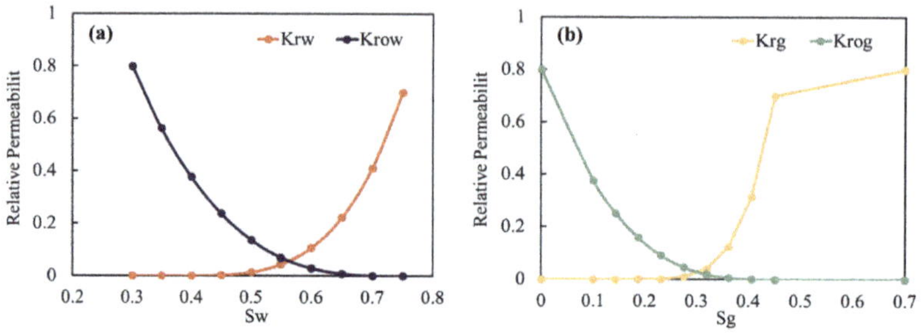

Figure 2. Relative permeability curves for (**a**) oil–water systems and (**b**) oil–gas systems. Krw, Krow, Krg, and Krog represent the relative permeability of water, oil, gas, and oil, respectively, as functions of saturation.

2.2. Decision Variable Selection and Dataset Generation

Six parameters were selected as decision variables to evaluate the CO_2-WAG development process: bottom-hole pressure of production wells (BHPO), water-to-gas ratio (WGR), gas injection rate (RATEG), water injection rate (RATEW), oil production rate (ORAT), and injection cycle (IC). The cumulative oil production (OPRO) was defined as the objective function to assess the impact of these parameters on development performance. The ranges of decision variables were determined based on geological constraints, engineering feasibility, and economic rationality. Table 2 summarizes the statistical properties of the decision variables, which were selected following these criteria:

1. Geological Constraints: Variable ranges were defined to comply with reservoir properties. For instance, BHPO was constrained to remain above the bubble point pressure to prevent premature gas breakout and below the reservoir fracture pressure to ensure structural integrity. Similarly, RATEG and RATEW were restricted within the reservoir's injection capacity to avoid fracturing.
2. Engineering Feasibility: Values were chosen to align with practical field operation limits, ensuring system stability. For example, the WGR was constrained by the capacity of surface facilities and well designs, while the IC was optimized to prevent incomplete displacement and uneven fluid distribution.
3. Economic Rationality: Parameter ranges were optimized to balance operational costs and economic returns. RATEG and RATEW were adjusted to maximize recovery efficiency while minimizing costs, and IC was optimized to enhance overall production efficiency.

Table 2. Statistical summary of decision variables.

Characteristic Parameter	Unit	Mean	Maximum	Minimum	Standard Deviation	Skewness	Variable Coefficient
BHPO	MPa	17.4	20	15	1.45	0.0048	0.0826
WGR	%	50	90	10	30	0.0032	0.5778
ORAT	m^3/d	65.1	90	40	14.4	0.0017	0.2212
RATEG	m^3/d	34,974.4	45,000	25,000	4440.5	0.0028	0.2887
RATEW	m^3/d	46.1	60	30	8.7	0.0031	0.4972
IC	Day	106.9	150	60	53.9	0.0029	0.3849
OPRO	m^3	1,757,850.2	2,354,830.2	1,019,188.1	293,397.1	−0.2337	0.1671

To ensure a representative and evenly distributed dataset, the Latin Hypercube Sampling (LHS) method was employed. This approach ensured comprehensive coverage of the high-dimensional parameter space, avoiding clustering and blank regions typical of random sampling. A total of 1225 samples were generated, with their characteristics summarized in Table 2. Figure 3a illustrates the Pearson correlation heatmap, highlighting the linear relationships between variables. OPRO exhibits a moderate positive correlation with RATEG (correlation coefficient = 0.34) and a significant negative correlation with WGR (correlation coefficient = −0.64), indicating their substantial influence on production performance. Other variables, such as BHPO and IC, show weaker correlations with OPRO, suggesting limited direct impacts. Figure 3b presents box plots of normalized variables, emphasizing their distributions and range variations. Normalization was conducted using the min-max scaling method to eliminate dimensional differences and improve data comparability.

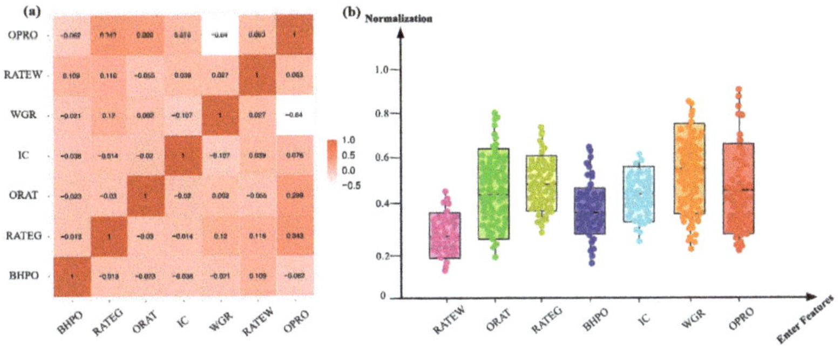

Figure 3. Variable correlation and data normalization: (**a**) Pearson correlation heatmap; (**b**) box plot of normalized variables.

2.3. Model Evaluation Metrics

To assess the predictive performance of the model, four evaluation metrics were employed: coefficient of determination (R^2), root mean square error (RMSE), mean absolute error (MAE), and variance accounted for (VAF). The definitions of these metrics are as follows [32]:

$$R^2 = 1 - \frac{\sum_i (P - \hat{P})^2}{\sum_i (P - \overline{P})^2} \tag{1}$$

$$\text{RMSE} = \sqrt{\frac{1}{n}\sum_{i=1}^{n}(P - \hat{P})^2} \tag{2}$$

$$\text{MAE} = \frac{1}{n}\sum_{i=1}^{n}|P - \hat{P}| \tag{3}$$

$$\text{VAF} = \left[1 - \frac{var(P - \hat{P})}{var(P)}\right] \times 100 \tag{4}$$

where n is the number of samples; P is the observed values; $\overline{P}$ is the mean of the observed values, and $\hat{P}$ the predicted value. R^2 quantifies the proportion of variance in observed data explained by the model, with values closer to 1 indicating better fit; RMSE measures the square root of the mean squared errors, with smaller values indicating higher accuracy; MAE represents the average absolute difference between observed and predicted values, offering robustness against outliers; VAF denotes the percentage of variance explained by the model, with values closer to 100% indicating better reliability.

3. Methodology

To address the complexity and significant nonlinear characteristics of parameter optimization in low-permeability reservoirs during CO_2-WAG development, this study generated 1225 datasets through numerical simulations, creating a multivariable dataset with OPRO defined as the objective function. Using this dataset, the feasibility of predicting complex reservoir development processes was explored through the application of the XGBoost model. To further enhance predictive performance, metaheuristic optimization algorithms were employed to tune the hyperparameters of XGBoost. Additionally, the CPO algorithm was enhanced through the integration of Chebyshev chaotic mapping and an EOBL strategy, resulting in the development of the ICPO-XGBoost model, which is tailored for CO_2-WAG development in low-permeability reservoirs.

3.1. XGBoost

Traditional numerical simulation methods and empirical formulas often suffer from significant limitations, including high computational costs and insufficient prediction accuracy in addressing complex reservoir problems. In contrast, machine learning methods, particularly the XGBoost algorithm, offer distinct advantages in CO_2-WAG development due to their robust nonlinear modeling capabilities and computational efficiency. XGBoost, a highly efficient and scalable machine learning algorithm based on gradient boosting tree models, has demonstrated strong predictive performance for both nonlinear problems and large-scale datasets. The core concept of XGBoost lies in iteratively constructing multiple weak learners (regression trees) to minimize residual errors, thereby enhancing overall prediction accuracy [33].

In the context of CO_2-WAG development, parameters such as the RATEG, RATEW, WGR, and BHPO exhibit complex nonlinear interactions with reservoir dynamics. XGBoost captures these intricate dependencies by combining the outputs of multiple regression trees. During each iteration, XGBoost constructs a new regression tree based on the current model's loss function, refining predictions by progressively minimizing residual errors [34]. The prediction output of XGBoost can be expressed as [33]:

$$\hat{y}_i = \Sigma_{k=1}^{K} f_k(x_i), f_k \in F, \ \ F = \{f(x) = w_s(x)\}(s : R^m \to T, w_s \in R^T) \tag{5}$$

where $\hat{y}_i$ denotes the predicted value of the i-th sample; $f_k(x_i)$ represents the prediction of the k-th tree for sample; F is the function space of tree models; $s : R^m \to T$ defines the mapping from input feature space to leaf nodes; $w_s \in R^T$ represents the prediction weight of each leaf node; T is the number of leaf nodes in the tree.

The optimization objective of XGBoost is to minimize the following objective function, which combines a loss term and a regularization term to balance prediction accuracy and model complexity:

$$Obj(\Theta) = L(\Theta) + \Omega(\Theta) = \Sigma_{i=1}^{n} l(y_i, \hat{y}_i) + \Sigma_{k=1}^{K} \Omega(f_k), \ \Theta = \{f_1, f_2, \ldots, f_K\} \tag{6}$$

where $Obj(\Theta)$ is the objective function; $L(\Theta)$ represents the loss function, measuring the prediction error; $\Omega(\Theta)$ is the regularization term, penalizing model complexity; Θ denotes the set of model parameters; f_K indicates the structure and leaf weights of a single tree.

The loss function quantifies the deviation between observed y_i and predicted $\hat{y}_i$ values:

$$l(y_i, \hat{y}_i) = (y_i - \hat{y}_i)^2 \tag{7}$$

The regularization term penalizes model complexity to enhance generalization capability:

$$\Omega(f_k) = \gamma T + \frac{1}{2}\lambda\|w\|^2 \tag{8}$$

where γ is the penalty parameter for the number of leaf nodes; λ is the regularization parameter for leaf weights; By incorporating a regularization term, XGBoost mitigates the risk of overfitting and ensures stable performance on test data.

XGBoost optimizes the objective function through recursive iterations. During each iteration, a new regression tree is constructed to minimize the residual errors from the previous iteration. The algorithm leverages first-order and second-order derivative information of the loss function to enhance optimization efficiency, enabling accurate adjustments in predictions based on gradient information. Training terminates under one of the following conditions:

- The number of iterations reaches a predefined limit.
- The reduction in residuals falls below a specified threshold.

The model combines the outputs of multiple decision trees with corresponding weights, forming a highly accurate and computationally efficient regressor or classifier.

3.2. Hyperparameter Optimization Using Metaheuristic Algorithms

Hyperparameter optimization plays a pivotal role in improving the performance of machine learning models, especially when addressing complex, high-dimensional, and highly nonlinear problems such as CO_2-WAG reservoir development. To achieve efficient and accurate hyperparameter tuning, this study leverages four state-of-the-art nature-inspired metaheuristic algorithms—CPO, GWO, AHA, and BKA—to optimize the critical hyperparameters of the XGBoost model. These algorithms effectively balance global exploration and local exploitation, thereby preventing the model from converging to suboptimal solutions and significantly enhancing its predictive accuracy and generalization performance. CPO, as a chaotic optimization strategy, demonstrates robust convergence capabilities in nonlinear regression tasks. Its lightweight computational structure makes it particularly suitable for medium-dimensional hyperparameter tuning, establishing it as the baseline model in this study. AHA and BKA, characterized by adaptive hybrid mechanisms and bio-inspired optimization frameworks, excel in avoiding local optima, making them well-suited for high-dimensional and non-convex optimization problems. In contrast, GWO, a classic swarm intelligence algorithm, has been extensively applied in hyperparameter optimization for machine learning models and serves as a benchmark in this study to evaluate the effectiveness of the proposed approach in achieving a synergy between global search and localized refinement.

3.2.1. Crowned Porcupine Optimization (CPO)

The CPO algorithm is a recent metaheuristic method inspired by the defensive behaviors of crowned porcupines [35]. By simulating four distinct defensive strategies—visual, auditory, olfactory, and physical attacks—CPO dynamically balances global search and local refinement capabilities. This makes it highly effective for optimizing XGBoost hyperparameters in complex, high-dimensional parameter spaces [36]. Figure 4 illustrates the workflow of the CPO-XGBoost algorithm, and the optimization process can be summarized as follows:

1. Data Preprocessing: The dataset is divided into training and testing subsets, and input features are normalized to eliminate dimensional biases and improve model stability.

2. Population Initialization: Multiple candidate solutions, each representing a combination of hyperparameters, are randomly generated. Their fitness values are then evaluated to initialize the population.
3. Iterative Optimization: The CPO algorithm iteratively updates the population using a four-stage defensive strategy:
   - First Defense Phase: Enhances solution diversity and broadens the search range by adjusting the distance between the predator and the target point, incorporating random perturbations.
   - Second Defense Phase: Simulates auditory defense behavior to improve local search capability and diversify solutions.
   - Third Defense Phase: Combines local perturbations with regional expansion to enhance global search capability.
   - Fourth Defense Phase: Simulates elastic collisions, avoiding local optima and improving global search efficiency.

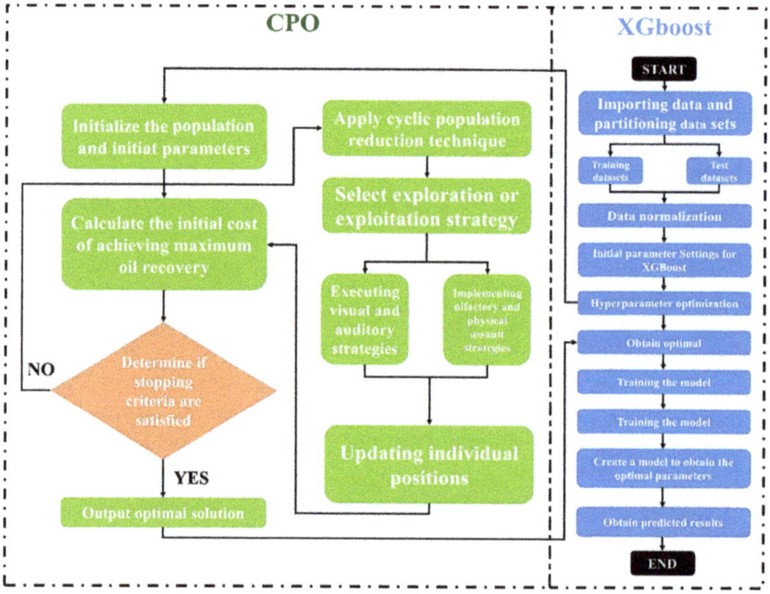

Figure 4. Workflow of the CPO-XGBoost algorithm for hyperparameter optimization.

The specific computational methods for each phase are as follows [37]:

1. First Defense Phase: This phase enhances solution diversity and expands the search range by adjusting the distance between the predator and the target point, incorporating random factors. The update formula is:

$$\overrightarrow{x_i^{t+1}} = \overrightarrow{x_i^t} + \tau_1 \times \left| 2 \times \tau_2 \times \overrightarrow{x_{CP}^t} - \frac{\overrightarrow{x_1} + \overrightarrow{x_r}}{2} \right| \tag{9}$$

2. Second Defense Phase: Simulating the porcupine's auditory defense behavior, this phase enhances local search capability and further diversifies solutions. The update formula is:

$$\overrightarrow{x_i^{t+1}} = \left(1 - \overrightarrow{U_1}\right) \times \overrightarrow{x_i^t} + \overrightarrow{U_1} \times \left(\overrightarrow{y} + \tau_3 \times \left(\overrightarrow{x_{r_1}^t} - \overrightarrow{x_{r_2}^t}\right)\right) \tag{10}$$

3. Third Defense Phase: This phase improves global search capability by combining local perturbations with regional expansion. The update formula is:

$$\overrightarrow{x_i^{t+1}} = (1 - \overrightarrow{U_1}) \times \overrightarrow{x_i^t} + \overrightarrow{U_1} \times \left(\overrightarrow{x_{r_1}^{\prime}} + S_i^{\prime} \times \left(\overrightarrow{x_{r_2}^{\prime}} - \overrightarrow{x_{r_3}^{\prime}} \right) - \tau_3 \times \overrightarrow{\delta} \times \gamma_t \times S_i^{\prime} \right) \quad (11)$$

4. Fourth Defense Phase: This phase simulates elastic collisions to improve global search ability and avoid local optima. The update formula is:

$$\overrightarrow{x_i^{t+1}} = \overrightarrow{x_{CP}^t} + (\alpha(1 - \tau_4) + \tau_4) \times \left(\delta \times \overrightarrow{x_{CP}^t} - \overrightarrow{x_i^t} \right) - \tau_5 \times \delta \times \gamma_t \times \overrightarrow{F_i^t} \quad (12)$$

where $\overrightarrow{x_{CP}^t}$ denotes the best solution at iteration t; $\overrightarrow{x_i^t}$ denotes the position of the i-th individual at iteration t; $\overrightarrow{x_1}$ and $\overrightarrow{x_r}$ correspond to the positions of two randomly selected individuals; $\tau_1, \tau_2, \ldots, \tau_5$ are random values in $[0, 1]$, controlling update magnitude; r_1 and r_2 are two random integers within $[1, N]$, where N is population size; $\overrightarrow{U_1}$ is randomly generated and contains elements that are either 0 or 1; The position of the predator is represented by $\overrightarrow{y}$, while $\overrightarrow{\delta}$ serves as a parameter to control the search direction; $S_i^{\prime}$ defines the odor diffusion factor, and α is the adjustment factor; $\overrightarrow{F_i^t}$ indicates the resistance experienced by individual i during iteration t.

The CPO-XGBoost algorithm effectively improves model accuracy and generalization by optimizing hyperparameters through these defensive strategies.

3.2.2. Grey Wolf Optimization (GWO)

The GWO algorithm is a nature-inspired technique based on the hierarchical hunting behaviors of grey wolves [38]. The algorithm simulates the leadership hierarchy—alpha (α), beta (β), delta (δ), and subordinate wolves—and their cooperative hunting strategies to achieve global optimization. Each wolf adopts a search strategy according to its role, effectively avoiding local optima.

Figure 5 presents the workflow of the GWO-XGBoost algorithm. Initially, the positions of grey wolves are randomly generated, each position corresponding to a set of XGBoost hyperparameters. The fitness value of each wolf is evaluated based on the XGBoost model's performance on the training dataset. The top three wolves are designated as alpha (α), beta (β), and delta (δ), representing the best, second-best, and third-best solutions, respectively. The remaining wolves follow their guidance to explore the global optimum. The position updates are defined as [39]:

$$D = |C \times X_p(t) - X(t)| \quad (13)$$

$$X(t+1) = X_p(t) - A \times D \quad (14)$$

where D represents the distance between the current position of the grey wolf and the target prey; $X_p(t)$ denotes the position of the target prey; $X(t)$ represents the current position of the grey wolf; $A = 2ar_1$, where a is the iteration parameter (decreasing with iterations) and r_1 is a random number; $C = 2r_2$, with r_2 as a random number introducing perturbations.

3.2.3. Artificial Hummingbird Algorithm (AHA)

The AHA is a novel swarm intelligence optimization technique inspired by the hovering flight and dynamic foraging behavior of hummingbirds [40]. By simulating the process of hummingbirds searching for and utilizing high-quality food sources, the algorithm effectively balances global exploration and local exploitation to solve complex optimization problems. AHA uses three flight modes—axial, diagonal, and omnidirectional flight—and three intelligent foraging strategies: guided foraging, local foraging, and migratory foraging [41]. These mechanisms collectively enhance the algorithm's ability to explore the solution space and refine candidate solutions.

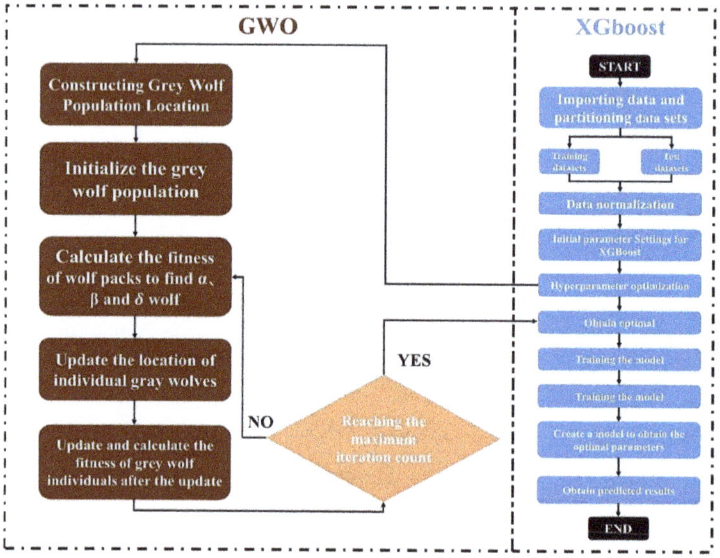

Figure 5. Workflow of the GWO-XGBoost algorithm for hyperparameter optimization.

As shown in Figure 6, the AHA-XGBoost algorithm begins by calculating the fitness of each hummingbird (representing a hyperparameter combination). Based on the foraging strategy and flight mode, the positions of the hummingbirds are updated, and the fitness values are recalculated. The process continues until convergence is achieved, yielding the optimal hyperparameters for the XGBoost model [42].

$$D^{(i)} = \begin{cases} 1, \text{if } i = \text{rand}[1, d], \\ 0, \text{else} \end{cases} \tag{15}$$

$$D^{(i)} = \begin{cases} 1, \text{if } i = P(j), j \in [1, d], P = \text{rand}(k), \\ 0, \text{else} \end{cases} \tag{16}$$

where $D^{(i)}$ represents the flying skill; i = rand [1, d] generates a random integer within the range [1, d], where d denotes the dimensionality of the search space; $P = \text{rand}(k)$ creates a random permutation of integers from 1 to k; r_1 represents a uniformly distributed random number between 0 and 1.

3.2.4. Black Kite Algorithm (BKA)

The Black-Winged Kite Algorithm (BKA) is an optimization technique inspired by the hunting and migration behaviors of black-winged kites [43]. By combining global search strategies with local exploitation, BKA effectively addresses complex optimization problems. In the framework of the BKA-XGBoost algorithm (Figure 7), BKA optimizes the hyperparameters of the XGBoost model by iteratively improving candidate solutions. Initially, the positions of the kite population are randomly initialized, with each position corresponding to a set of XGBoost hyperparameters. The user predefines the population size and search range.

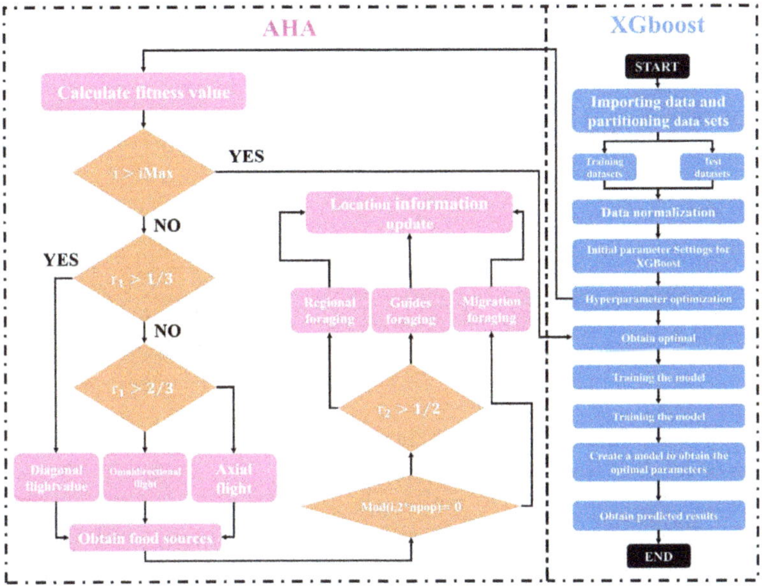

Figure 6. Workflow of the AHA-XGBoost algorithm for hyperparameter optimization.

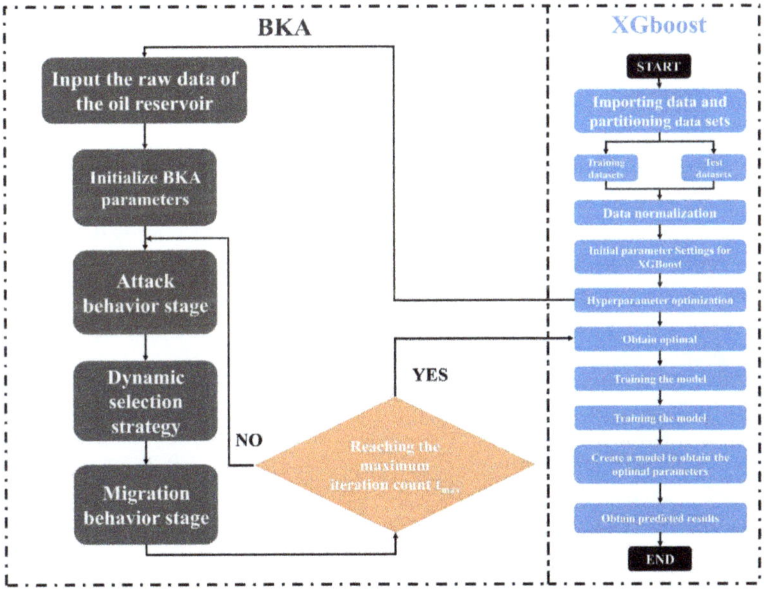

Figure 7. Workflow of the BAK-XGBoost algorithm for hyperparameter optimization.

During the attack phase, BKA simulates the behavior of kites approaching their target. The positions are updated using a dynamic scaling factor (n) and random perturbations

(r), ensuring a smooth transition from global exploration to local exploitation. The update formula is [44]:

$$y_{t+1}^{j,j} = \begin{cases} y_t^{j,j} + n(2r-1) \times y_t^{j,j} & \text{else} \\ y_t^{j,j} + n\left(1+\sin(r) \times y_t^{j,j}\right) & g < r \end{cases}, \quad n = \ 0.05 \times e^{-2\times\left(\frac{t}{T}\right)} \tag{17}$$

where $y_t^{(j,j)}$ represent the position of the i-th black-winged kite in the j-th dimension at the t-th iteration steps; r represents a random number between 0 and 1; g is a constant, often set to 0.9; T represents the total number of iterations, and t denotes the number of iterations completed so far.

3.3. *Enhanced Algorithm*

The CPO-XGBoost algorithm has demonstrated exceptional performance in optimizing CO_2-WAG development parameters for low-permeability reservoirs. Its strengths include high prediction accuracy, minimal error distribution, strong generalization ability, and excellent adaptability to complex nonlinear relationships. By effectively balancing global search and local exploitation, the algorithm mitigates overfitting and ensures consistent performance across both training and testing datasets. Additionally, it outperforms other comparative algorithms in terms of error metrics, showcasing superior stability and reliability. Detailed comparative results are presented in Section 4. However, the CPO-XGBoost algorithm has certain limitations. Its optimization process requires substantial computational resources, especially when applied to large-scale datasets, resulting in extended computation times. Furthermore, its performance is highly dependent on hyperparameter tuning, making it sensitive to configuration in complex engineering environments. This often necessitates meticulous debugging and adjustments.

To address the shortcomings of the original CPO algorithm, including slow convergence speed, suboptimal optimization performance, and low resource allocation efficiency, this study introduces two key improvement mechanisms: the Chebyshev chaotic mapping and the EOBL strategy. The Chebyshev chaotic mapping, characterized by high chaotic intensity and uniform ergodicity, effectively enhances the diversity of the initial population and prevents the algorithm from being trapped in local optima. Additionally, its nonlinear oscillatory behavior within the interval $[-1, 1]$ improves the randomness of population initialization and global search capability. Combined with the EOBL strategy, the search efficiency and global convergence ability are further enhanced, achieving a balance between exploration and exploitation and significantly improving algorithm performance. This characteristic has been widely validated in various optimization algorithm studies as beneficial for enhancing population diversity and global search efficiency [38]. The improved ICPO algorithm increases population diversity, strengthens global search capabilities, accelerates convergence, and effectively overcomes the limitations of the original CPO algorithm.

3.3.1. Method Population Initialization via Chebyshev Chaotic Mapping

The original CPO algorithm utilizes a random initialization strategy to generate the initial population within the search space. While straightforward, this approach often results in high randomness and uneven distribution, leading to insufficient population diversity and suboptimal search performance. To address this issue, the Chebyshev chaotic mapping mechanism is introduced.

The Chebyshev chaotic mapping formula is expressed as follows [45]:

$$H(t+1) = \cos(p \cdot \arccos(H(t))) \tag{18}$$

$$x = a + (H(t) + 1) \cdot \frac{b - a}{2} \tag{19}$$

where $H(t)$ represents the value of the chaotic sequence at the t-th step, which is typically distributed within the interval $[-1, 1]$; p is the control parameter; a denotes the lower bound of the search space; b denotes the upper bound of the search space.

3.3.2. Elite Opposition-Based Learning Strategy for Population Optimization

A high-quality initial population is essential for accelerating convergence and increasing the likelihood of achieving a globally optimal solution. The original CPO algorithm's reliance on random initialization often results in limited population diversity, adversely affecting convergence speed and optimization performance. To address this, the EOBL strategy is introduced during the population initialization phase. The oppositional solution is calculated as follows [46]:

$$\overrightarrow{X^{e}_{i,j}} = K \times (a_j + b_j) - X^{e}_{i,j} \tag{20}$$

where K is a dynamic coefficient with a value range of $(0, 1)$; a_j and b_j are dynamic boundaries, which adapt to overcome the limitations of fixed boundaries. This ensures that oppositional solutions retain search experience and are less likely to get trapped in local optima.

$$\overrightarrow{X^{c}_{i,j}} = \text{rand}(a_j, b_j) \tag{21}$$

To further balance global exploration and local exploitation, a nonlinear convergence factor adjustment strategy is employed. The convergence factor is updated as follows:

$$a = a_{\text{initial}} - (a_{\text{initial}} - a_{\text{final}}) \cdot \exp(t/t_{\text{max}} - 1) \tag{22}$$

where a_{initial} and a_{final} represent the initial and terminal values of a, respectively; t is the current iteration number; t_{max} is the maximum number of iterations.

Compared to linear convergence, this nonlinear adjustment is more effective in balancing the demands of global and local search, thereby enhancing optimization performance and convergence speed.

3.3.3. Improved Crowned Porcupine Optimization-XGBoost Model

Building on the aforementioned enhancements, this study introduces the ICPO algorithm, which is integrated with the XGBoost model to form the ICPO-XGBoost model.

As illustrated in Figure 8, the ICPO-XGBoost model incorporates the Chebyshev chaotic mapping mechanism during the population initialization phase. This approach generates a uniformly distributed initial population, significantly enhancing diversity compared to the random initialization strategy of the original CPO algorithm. The nonlinear and periodic properties of Chebyshev chaotic mapping ensure efficient coverage of the search space, reducing the risk of local optima and improving population distribution. Following initialization, the EOBL strategy is applied to refine the population. The top 20% of individuals with the highest fitness values are used to generate an oppositional population, which is then compared with the original population. The best-performing individuals are retained as the new initial population, enhancing diversity, accelerating convergence, and increasing the likelihood of finding the global optimum. During the iterative optimization process, a nonlinear convergence factor adjustment strategy dynamically balances global exploration and local exploitation. In the early stages, the algorithm emphasizes global exploration with a larger search range to fully cover the search space. As iterations progress, local exploitation is gradually strengthened to improve optimization accuracy. Compared to linear adjustment strategies, this nonlinear approach provides greater flexibility and enhances convergence performance.

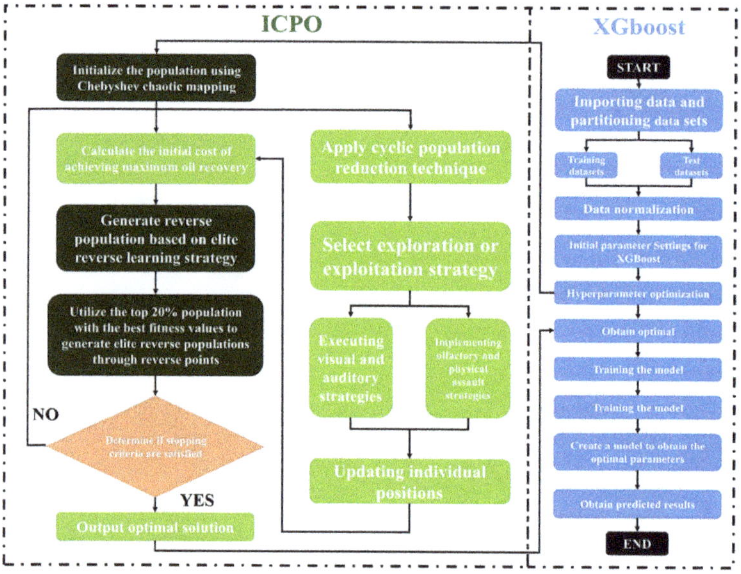

Figure 8. The workflow of the ICPO-XGBoost algorithm. dark green represents the key steps of algorithm optimization proposed in this study; light green represents the conventional steps of the CPO algorithm; orange indicates the decision-making step to determine whether the stopping criteria are satisfied; blue represents the conventional steps of the XGBoost algorithm.

4. Results and Analysis

Accurately predicting cumulative oil production and optimizing injection-production parameters are critical for enhancing oil recovery and maximizing economic efficiency in the development of low-permeability reservoirs using CO_2-WAG injection. This chapter provides a systematic evaluation of the proposed ICPO-XGBoost model, focusing on its performance in terms of prediction accuracy, stability, and generalization capability. Through detailed comparative analyses with other mainstream algorithms, the chapter highlights the significant advantages of the ICPO-XGBoost model in addressing complex nonlinear problems. By comprehensively comparing its performance across training, testing, and validation datasets, the study validates the ICPO-XGBoost model's reliability and superiority in practical applications. Furthermore, optimization performance tests and comparisons with traditional and state-of-the-art machine learning models further demonstrate the model's precision and applicability. The chapter also incorporates real numerical simulation cases to systematically compare the model's predictions with simulation results, exploring its predictive capability and applicability under various development strategies.

4.1. Comparative Analysis of Model Prediction Results

To comprehensively assess the performance of six models—XGBoost, CPO-XGBoost, AHA-XGBoost, BKA-XGBoost, GWO-XGBoost, and ICPO-XGBoost—this study utilizes the case study presented in Section 2, with a specific focus on their applicability in optimizing CO_2-WAG development parameters for low-permeability reservoirs. The dataset consists of 1225 samples, systematically divided into training, testing, and validation sets in a 5:4:1 ratio. Specifically, 613 samples are allocated for model training, 490 samples are reserved for testing, and 122 samples are used for validation. To ensure robustness and representativeness of the results, each model is independently executed 50 times, and the optimal result from each set is selected for detailed analysis. Table 3 summarizes the hyperparameter configurations and training settings for each model, highlighting critical factors such as model complexity, learning rate, warm-up iterations, and early stopping strategies. These configurations are meticulously fine-tuned to achieve a balance between generalization ability and mitigation of overfitting. The subsequent sections provide an in-depth analysis and comparison of the predictive performance and adaptability of these models, offering insights into their respective strengths and limitations.

Table 3. Hyperparameter and training settings for the four optimized models.

Model	Max Learning Rate	Learning Rate	Warm-Up Rounds	Early Stopping Rounds	Sub-Sample Ratio	Iteration Count
XGBoost	12	0.18	5	20	0.4	80
CPO-XGBoost	14	0.21	8	25	0.5	140
AHA-XGBoost	17	0.27	10	30	0.6	145
BKA-XGBoost	15	0.29	10	30	0.4	125
GWO-XGBoost	18	0.30	10	30	0.3	150
ICPO-XGBoost	15	0.35	8	25	0.5	135

The six models are evaluated based on prediction accuracy and error metrics across the training, testing, and validation datasets. Figure 9 presents the prediction values and error distributions for the six models, while Figure 10 compares the predicted values and actual values of 150 randomly selected cases from the validation dataset. Table 4 provide a systematic comparison of the performance of the six models across key evaluation metrics, including R^2, MAPE, MAE, RMSE, and MSE. The radar chart in Figure 11 visually illustrates the models' performance across the training, testing, and overall datasets, offering a comprehensive view of their predictive capabilities.

Table 4. Primary evaluation metrics for the six models.

Model	Data Range	R^2	MAPE	MAE	RMSE	MSE
XGBoost	Training set	0.9343	29.54%	276.71	271.24	0.0285
	Test set	0.9321	27.68%	255.12	252.36	0.0242
	Overall data	0.9325	28.31%	279.58	273.51	0.0277
CPO-XGboost	Training set	0.9796	12.67%	131.84	127.62	0.0095
	Test set	0.9784	11.07%	126.63	105.24	0.0074
	Overall data	0.9788	12.26%	149.12	129.67	0.0081
AHA-XGboost	Training set	0.9749	16.52%	148.15	135.15	0.0119
	Test set	0.9721	15.94%	142.92	114.28	0.0088
	Overall data	0.9725	16.42%	151.45	142.61	0.0126
BKA-XGboost	Training set	0.9768	14.62%	141.51	132.82	0.0129
	Test set	0.9754	14.28%	138.87	125.95	0.0115
	Overall data	0.9758	14.59%	142.12	136.08	0.0157
GWO-XGboost	Training set	0.9632	28.42%	252.81	245.34	0.0272
	Test set	0.9627	27.33%	248.92	238.26	0.0226
	Overall data	0.9629	28.15%	256.22	245.58	0.0257
ICPO-XGboost	Training set	0.9902	10.21%	119.92	95.53	0.0072
	Test set	0.9894	8.47%	112.65	67.94	0.0053
	Overall data	0.9896	9.87%	125.63	102.98	0.0064

As the baseline model, XGBoost achieves moderate prediction accuracy but demonstrates clear limitations in handling complex nonlinear relationships. The model records an overall R^2 value of 0.9325 and a MAPE of 28.31%. Its error distributions are broad, with significant deviations observed in both low and high-value ranges, indicating reduced generalization capability and difficulties in accurately capturing intricate data patterns. These limitations underscore the need for optimization strategies to improve the model's performance. Among the metaheuristic-optimized models, CPO-XGBoost delivers the best results, achieving an R^2 value of 0.9788 and a MAPE of 12.26%, significantly outperforming the baseline XGBoost model. By effectively balancing global search and local exploitation, CPO-XGBoost excels at solving complex nonlinear problems and demonstrates robust generalization across the datasets. Its error distributions are minimal and concentrated, highlighting strong fitting accuracy and stability.

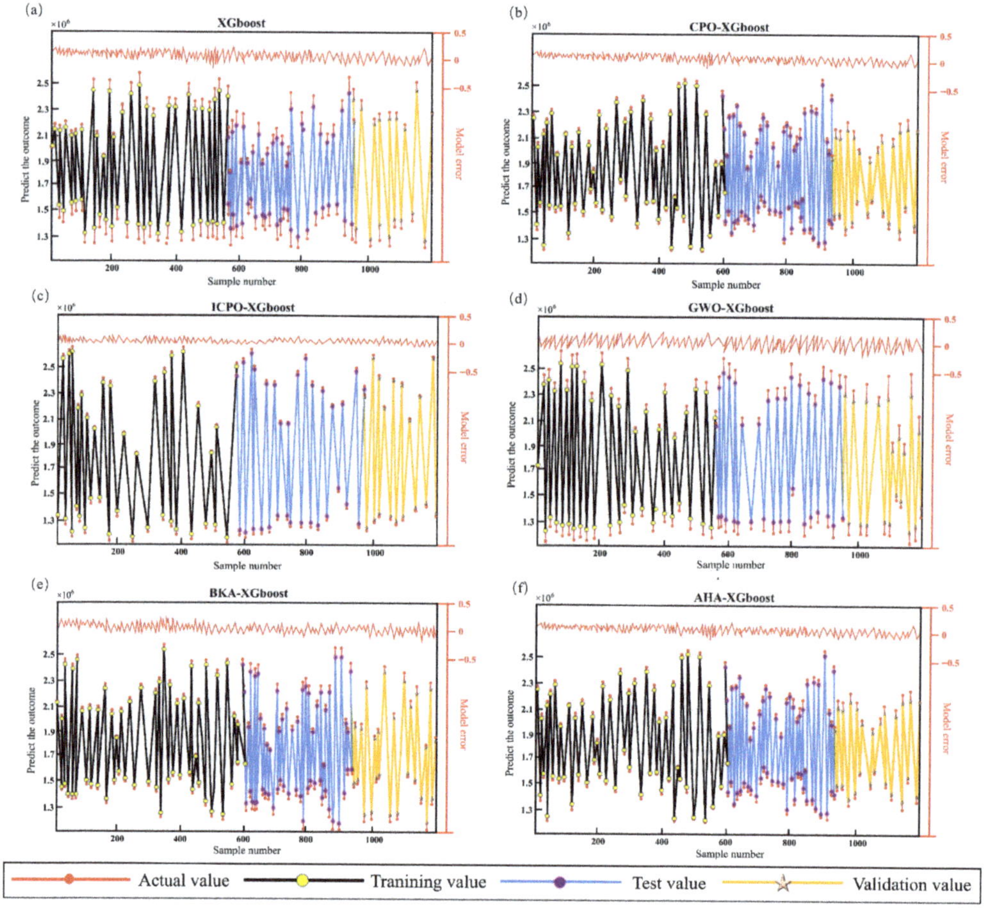

Figure 9. The figure illustrates the prediction results and error distributions for six models: (**a**) XGBoost, (**b**) CPO-XGBoost, (**c**) ICPO-XGBoost, (**d**) GWO-XGBoost, (**e**) BKA-XGBoost, and (**f**) AHA-XGBoost.

AHA-XGBoost and BKA-XGBoost also exhibit strong fitting performance, achieving R^2 values of 0.9725 and 0.9758, respectively. However, their error distributions are more scattered on the testing and validation sets, particularly for extreme samples. This indicates weaker generalization compared to CPO-XGBoost. Between the two, AHA-XGBoost performs slightly better due to its enhanced global search capability, achieving a MAPE of 16.42%, compared to 14.59% for BKA-XGBoost. Despite their strengths, both models show limitations in handling intricate data characteristics, as evidenced by their relatively higher MAPE values and broader error distributions. GWO-XGBoost delivers the poorest performance among the metaheuristic-optimized models, with an R^2 value of 0.9629 and a MAPE of 28.15%. Its broader error distributions and reduced ability to capture complex nonlinear relationships highlight its limited optimization capability, particularly in high-dimensional parameter spaces.

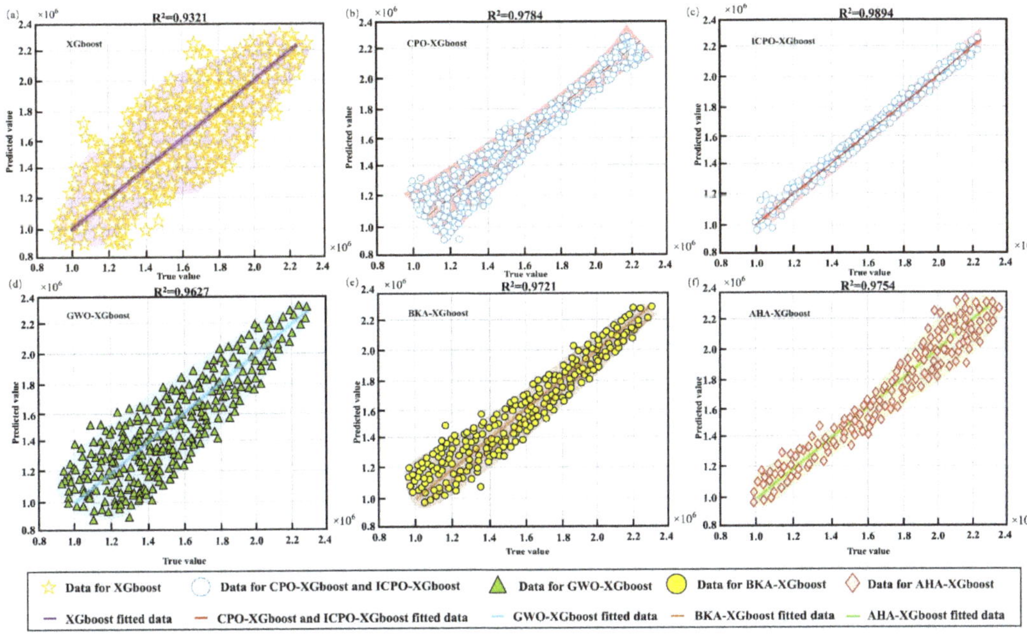

Figure 10. Comparison between predicted and actual values for the six models: (**a**) XGBoost, (**b**) CPO-XGBoost, (**c**) ICPO-XGBoost, (**d**) GWO-XGBoost, (**e**) BKA-XGBoost, and (**f**) AHA-XGBoost. Each subplot demonstrates the alignment between predicted and actual values, with corresponding R^2 values indicating the goodness of fit.

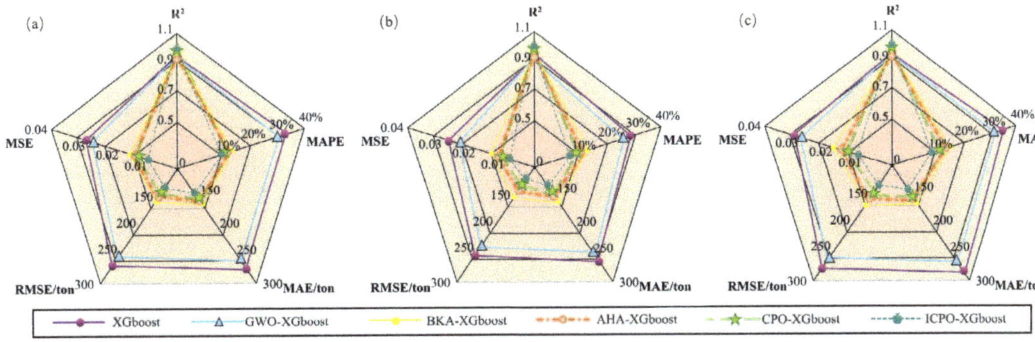

Figure 11. Radar chart of evaluation metrics for the six models: (**a**) illustrates the metrics for the training set, highlighting the fitting performance during model training; (**b**) focuses on the testing set, showcasing the models' predictive accuracy on unseen data; (**c**) presents the overall performance across all data, combining training and testing results.

ICPO-XGBoost demonstrates the highest prediction accuracy and the most robust error control among all models. It achieves an R^2 value of 0.9896 and a MAPE of 9.87%, representing a 1.08% improvement in R^2 and a 19.48% reduction in MAPE compared to CPO-XGBoost. The advanced optimization strategies incorporated into ICPO-XGBoost, such as Chebyshev chaotic mapping and the EOBL strategy, significantly enhance global search capability, improve population diversity, and accelerate convergence. These enhancements enable ICPO-XGBoost to achieve superior stability, fitting accuracy, and generalization performance. The error distribution plots confirm its smaller deviations and smoother error curves, demonstrating its reliability and efficiency in accurately capturing intricate relationships between decision variables and target variables.

4.2. Performance Comparison Test of the Model

To thoroughly evaluate the optimization performance of the improved ICPO algorithm, eight benchmark test functions were selected. The specific mathematical expressions of these functions are detailed in Table 5, while their respective search spaces are visualized in Figure 12. These functions include two unimodal functions (F1 and F2) and six multimodal functions (F3 to F8). Unimodal functions, which feature a single global optimum, are primarily used to evaluate the algorithm's convergence speed and optimization accuracy. On the other hand, multimodal functions, characterized by multiple local optima, are designed to assess the algorithm's ability to escape local optima and efficiently explore the entire search space. By combining these two types of test functions, a more comprehensive and balanced evaluation is achieved, enabling a detailed analysis of the algorithm's accuracy, convergence efficiency, and stability, while effectively reducing the risk of becoming trapped in local optima.

Table 5. Mathematical expressions of eight benchmark functions.

Function	Expression	Executions	Value Interval
F1	$f_1(x) = \max_i\{\lvert x_i\rvert, 1 \le i \le 30\}$	30	$-100 \le x_i \le 100$
F2	$f_2(x) = \sum_{i=1}^{30} \lvert x_i\rvert + \prod_{i=1}^{30} \lvert x_i\rvert$	30	$-10 \le x_i \le 10$
F3	$f_3(x) = -\sum_{i=1}^{30}\left(x_i \sin\left(\sqrt{\lvert x_i\rvert}\right)\right)$	30	$-500 \le x_i \le 500$
F4	$f_4(x) = \left[\frac{1}{500} + \sum_{j=1}^{25} \frac{1}{j + \sum_{i=1}^{2}(x_i - a_{ij})^6}\right]^{-1}$	30	$-65.536 \le x_i \le 65.536$
F5	$f_5(x) = -20\exp\left(-0.2\sqrt{\frac{1}{30}\sum_{i=1}^{30} x_i^2}\right) - \exp\left(\frac{1}{30}\sum_{i=1}^{30}\cos 2\pi x_i\right) + 20 + e$	30	$-32 \le x_i \le 32$
F6	$f_6(x) = \frac{1}{4000}\sum_{i=1}^{30} x_i^2 - \prod_{i=1}^{30}\cos\left(\frac{x_i}{\sqrt{i}}\right) + 1$	30	$-600 \le x_i \le 600$
F7	$f_7(x) = \sum_{i=1}^{30}\left[x_i^2 - 10\cos(2\pi x_i) + 10\right]$	30	$-5.12 \le x_i \le 5.12$
F8	$f_8(x) = \left[1 + (x_1 + x_2 + 1)^2(19 - 14x_1 + 3x_1^2 - 14x_2 + 6x_1x_2 + 3x_2^2)\right] \times \left[30 + (2x_1 - 3x_2)^2\right] \times \left[(18 - 32x_1 + 12x_1^2 + 48x_2 - 36x_1x_2 + 27x_2^2)\right]$	30	$-2 \le x_i \le 2$

Comparative experiments were conducted under identical conditions among five optimization algorithms: ICPO, CPO, GWO, AHA, and BKA. For all algorithms, the initial population size was set to 50, and the maximum number of iterations was fixed at 400. To ensure the reliability of the experimental results, each algorithm was independently executed 30 times for each test function. The evaluation criteria included the worst value, best value, mean value, and standard deviation of the results, offering a comprehensive assessment of each algorithm's solution accuracy and stability.

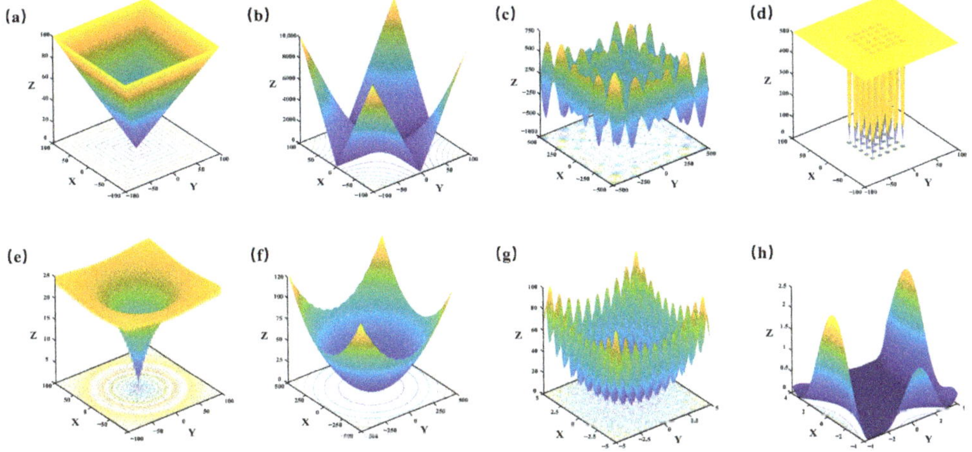

Figure 12. Visualization of eight benchmark functions: (**a**) F1; (**b**) F2; (**c**) F3; (**d**) F4; (**e**) F5; (**f**) F6; (**g**) F7; (**h**) F8.

Table 6 summarizes the results obtained from 30 independent runs of each algorithm on eight benchmark functions. The results clearly demonstrate that the ICPO optimization algorithm outperforms mainstream optimization algorithms such as CPO, GWO, AHA, and BKA. For the unimodal test functions F1 and F2, ICPO achieves significantly faster convergence, higher optimization accuracy, and smaller standard deviations, highlighting its strong local search capability and stable convergence toward the optimal solution. For the multimodal test functions F3, F4, F5, and F8, ICPO exhibits exceptional stability, with its worst-case optimization results surpassing those of the other four algorithms. This underscores its superior robustness and precision in complex search spaces. Additionally, for the multimodal functions F6 and F7, ICPO performs comparably to CPO, with both algorithms consistently converging to the global optimum (objective value of 0). This demonstrates ICPO's strong global search ability and its effectiveness in avoiding local optima.

Table 6. Statistical results after 30 runs of the model.

Algorithm		F1	F2	F3	F4	F5	F6	F7	F8
CPO	Worst	3.15×10^{-26}	1.26×10^{-21}	$5.39 \times 10^{+1}$	$4.97 \times 10^{+2}$	1.98×10^{-17}	2.57×10^{-13}	6.77×10^{-2}	$2.47 \times 10^{+1}$
	Best	2.61×10^{-28}	4.16×10^{-23}	0.95×10^{-2}	$6.18 \times 10^{+5}$	5.36×10^{-18}	0	0	$2.38 \times 10^{+1}$
	Mean	5.57×10^{-27}	2.10×10^{-22}	$8.87 \times 10^{+1}$	$8.47 \times 10^{+3}$	8.69×10^{-17}	3.64×10^{-16}	1.84×10^{-3}	$2.43 \times 10^{+1}$
	Standard	2.41×10^{-26}	0.78×10^{-21}	$4.82 \times 10^{+1}$	$3.17 \times 10^{+2}$	0.74×10^{-17}	5.12×10^{-14}	6.85×10^{-5}	2.74×10^{-1}
ICPO	Worst	6.15×10^{-32}	3.35×10^{-30}	2.76×10^{-2}	7.84×10^{-4}	3.61×10^{-21}	1.84×10^{-21}	4.75×10^{-4}	3.64×10^{-1}
	Best	3.57×10^{-35}	7.35×10^{-33}	6.94×10^{-5}	6.40×10^{-6}	9.74×10^{-22}	0	0	3.01×10^{-1}
	Mean	2.69×10^{-34}	5.96×10^{-32}	4.38×10^{-4}	7.17×10^{-5}	6.31×10^{-21}	9.78×10^{-24}	5.85×10^{-7}	3.34×10^{-1}
	Standard	4.68×10^{-32}	1.12×10^{-30}	0.56×10^{-2}	5.23×10^{-4}	2.05×10^{-21}	3.67×10^{-23}	3.94×10^{-9}	6.31×10^{-2}
GWO	Worst	4.39×10^{-21}	9.28×10^{-15}	$6.87 \times 10^{+8}$	$9.71 \times 10^{+5}$	$2.41 \times 10^{+1}$	6.86×10^{-5}	$2.68 \times 10^{+6}$	$6.41 \times 10^{+5}$
	Best	3.11×10^{-23}	3.35×10^{-16}	$2.68 \times 10^{+5}$	$3.85 \times 10^{+3}$	$8.67 \times 10^{+0}$	4.65×10^{-7}	$8.62 \times 10^{+3}$	$5.11 \times 10^{+3}$
	Mean	4.32×10^{-22}	0.41×10^{-16}	$7.31 \times 10^{+7}$	$7.65 \times 10^{+4}$	$0.26 \times 10^{+1}$	5.34×10^{-6}	$3.15 \times 10^{+5}$	$3.05 \times 10^{+5}$
	Standard	2.94×10^{-21}	2.17×10^{-15}	$9.65 \times 10^{+4}$	$8.13 \times 10^{+5}$	$1.51 \times 10^{+1}$	5.68×10^{-6}	$6.95 \times 10^{+4}$	$2.48 \times 10^{+5}$
AHA	Worst	5.73×10^{-19}	$7.54 \times 10^{+0}$	$9.71 \times 10^{+5}$	3.77^{+12}	$3.41 \times 10^{+2}$	3.52×10^{-2}	$6.67 \times 10^{+3}$	$2.52 \times 10^{+3}$
	Best	4.47×10^{-21}	5.33×10^{-1}	$3.99 \times 10^{+4}$	$9.12 \times 10^{+11}$	$6.64 \times 10^{+3}$	6.62×10^{-5}	$4.97 \times 10^{+2}$	$2.41 \times 10^{+2}$
	Mean	9.87×10^{-20}	1.05×10^{-1}	$1.15 \times 10^{+5}$	$2.74 \times 10^{+12}$	$9.58 \times 10^{+2}$	5.34×10^{-4}	$0.94 \times 10^{+3}$	$1.15 \times 10^{+2}$
	Standard	4.54×10^{-19}	$3.67 \times 10^{+0}$	$6.69 \times 10^{+5}$	$9.74 \times 10^{+11}$	$2.18 \times 10^{+2}$	1.22×10^{-2}	$3.92 \times 10^{+2}$	$9.75 \times 10^{+1}$
BKA	Worst	2.69×10^{-30}	2.16×10^{-14}	$5.63 \times 10^{+3}$	$1.42 \times 10^{+15}$	$7.84 \times 10^{+1}$	3.69×10^{-11}	$3.42 \times 10^{+5}$	$8.59 \times 10^{+2}$
	Best	8.57×10^{-31}	8.75×10^{-17}	$7.45 \times 10^{+1}$	$1.01 \times 10^{+15}$	$2.42 \times 10^{+0}$	9.77×10^{-13}	$7.04 \times 10^{+1}$	$2.78 \times 10^{+1}$
	Mean	9.12×10^{-30}	6.97×10^{-16}	$6.62 \times 10^{+2}$	$1.32 \times 10^{+15}$	$9.16 \times 10^{+0}$	1.68×10^{-13}	$8.84 \times 10^{+3}$	$2.27 \times 10^{+2}$
	Standard	1.58×10^{-30}	1.63×10^{-14}	$4.36 \times 10^{+3}$	$1.53 \times 10^{+14}$	$6.15 \times 10^{+1}$	8.51×10^{-11}	$2.91 \times 10^{+1}$	$2.63 \times 10^{+2}$

In contrast, while CPO shows stable performance on unimodal functions, it falls slightly behind ICPO in terms of convergence speed and solution accuracy. Overall, ICPO demonstrates outstanding performance across both unimodal and multimodal benchmark functions, characterized by high accuracy, rapid convergence rates, and robust global optimization capabilities. These results underscore ICPO's potential and competitiveness in solving a wide range of complex optimization problems.

4.3. Comparison with Previous Studies

4.3.1. Comparison with Traditional Machine Learning Models

Traditional machine learning models, such as Support Vector Machine (SVM), Random Forest (RF), K-Nearest Neighbors (KNN), Logistic Regression (LR), and Naive Bayes (NB), have been widely applied in oil and gas development strategy optimization. To validate the stability and robustness of the ICPO-XGBoost model, this study selected these models as benchmarks and conducted a detailed analysis of the performance differences between ICPO-XGBoost and traditional models from two dimensions: prediction error metrics and error distribution.

As shown in Figure 13a, ICPO-XGBoost demonstrates the best performance across all evaluation metrics, particularly achieving significantly lower RMSE and MAE values compared to other models. Additionally, its R^2 value is the highest, fully reflecting its outstanding fitting accuracy and strong predictive capability. In contrast, although traditional models have undergone optimization, they exhibit evident limitations in capturing complex nonlinear relationships. Notably, LR and NB models show significantly higher error metrics, indicating their poor adaptability to data structures and inability to achieve high-precision predictions. Figure 13b illustrates the relative error distribution of the models in sample predictions. The boxplot clearly shows that ICPO-XGBoost exhibits the most concentrated error distribution with the smallest interquartile range, indicating superior prediction stability and robustness, as well as minimal sensitivity to outliers. By comparison, the error distributions of the other models are notably wider, with multiple outliers, revealing greater prediction uncertainty. ICPO-XGBoost significantly outperforms traditional methods in prediction accuracy, stability, and the ability to learn complex data structures. By integrating advanced feature extraction mechanisms and optimization strategies, the proposed model not only effectively captures higher-order nonlinear patterns in the data but also substantially reduces prediction errors, enhancing its generalization ability and practical value in engineering applications.

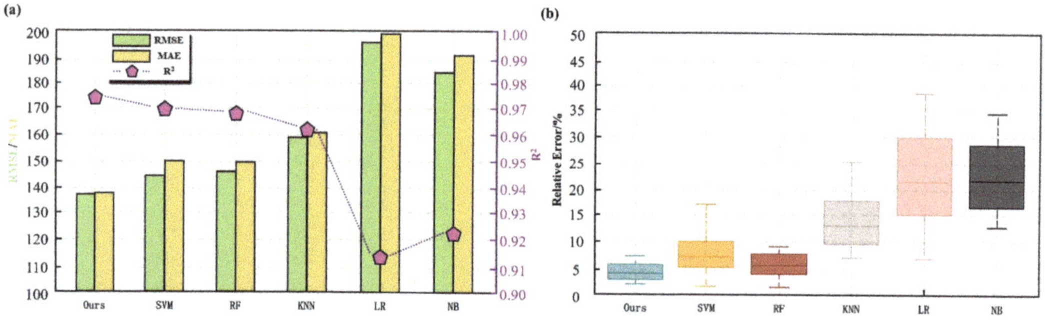

Figure 13. (**a**) Performance comparison of different models in terms of RMSE, MAE, and R^2; (**b**) boxplot of relative prediction errors (%) for different models.

4.3.2. Comparison with Mainstream Boosting-Based Ensemble Models

To further validate the practicality and comprehensive advantages of the proposed ICPO-XGBoost model, this study conducted a comparative analysis of its predictive performance against mainstream ensemble Boosting models, including Categorical Boosting (CatBoost), Adaptive Boosting (AdaBoost), Light Gradient Boosting Machine (LightGBM), under identical hardware conditions. As shown in the Figure 14, ICPO-XGBoost demonstrates outstanding predictive accuracy, with RMSE and MAE reduced by 38.6% and 33.8%, respectively, compared to the traditional XGBoost model. Additionally, its R^2 value is significantly higher than that of other models, highlighting its superior fitting precision and robust predictive capability.

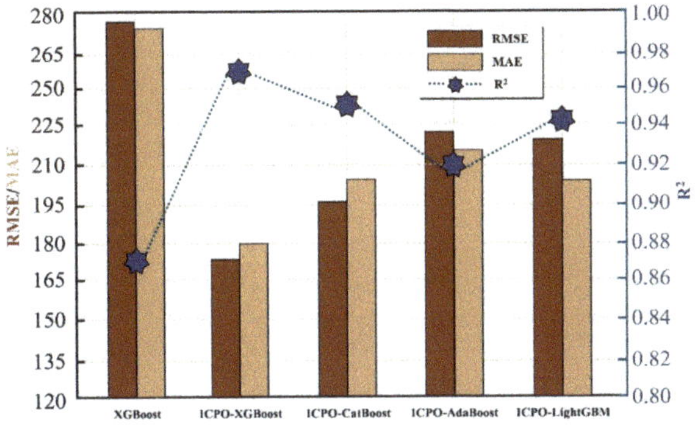

Figure 14. Comparison of RMSE, MAE, and R^2 across different models.

In contrast, although CatBoost, AdaBoost, and LightGBM are widely recognized as leading ensemble learning methods, their performance remains inferior to ICPO-XGBoost even when combined with the ICPO optimization strategy. This performance gap can primarily be attributed to the following reasons: XGBoost's architecture is inherently more compatible with parameter optimization methods. Its use of approximate algorithms for node splitting and regularization mechanisms strengthens the model's responsiveness to parameter adjustments, facilitating precise optimization. On the other hand, CatBoost is primarily optimized for categorical feature processing, with complex target encoding mechanisms and symmetric tree structures that result in a less explicit parameter space, thereby limiting the ICPO algorithm's search capabilities. Second, AdaBoost relies on an iterative mechanism of weighted weak learners, which has limited capacity for substantial performance improvements, making the optimization strategy less impactful. Although LightGBM is advantageous in terms of computational efficiency, its histogram-based feature splitting and leaf-wise growth strategy may introduce greater volatility, leading to less stable optimization effects. Under the same optimization mechanism, the structural characteristics of XGBoost are better aligned with the ICPO algorithm, enabling a synergistic effect that allows ICPO-XGBoost to excel in predictive accuracy, stability, and robustness, making it the best-performing model in this comparison.

4.4. Performance Validation of the ICPO-XGBoost Model

4.4.1. Prediction Performance and Generalization Capability Validation

To thoroughly evaluate the predictive performance and generalization capability of the proposed ICPO-XGBoost model, this study implemented the classical five-fold cross-validation strategy as the validation framework.

Five-fold cross-validation is a widely adopted approach for model evaluation, which involves randomly partitioning the entire dataset into five approximately equal and mutually exclusive subsets (Figure 15). During each iteration, one subset is designated as the test set, while the remaining four subsets are merged to form the training set for model training and evaluation. This process is repeated five times, ensuring that each subset is used as the test set exactly once. By doing so, the method maximizes dataset utilization and effectively mitigates random errors introduced by a single data split. Upon the completion of all five folds, the test results are aggregated, and the average values of the evaluation metrics are calculated, serving as the final assessment of the model's overall performance. The performance metrics considered include the RMSE, MAE, and the R^2, as illustrated in Figure 16. In the five-fold cross-validation experiment, the ICPO-XGBoost model demonstrated outstanding performance, with average metrics of RMSE = 129.35, MAE = 135.41, and R^2 = 0.976. The evaluation results further reveal that the variation in metrics across the five folds was minimal, with RMSE and MAE exhibiting narrow ranges of fluctuation, and R^2 consistently approaching 1. These findings highlight the model's ability to maintain stable and high-quality performance under varying data partitioning scenarios, reflecting its strong robustness and excellent generalization capacity.

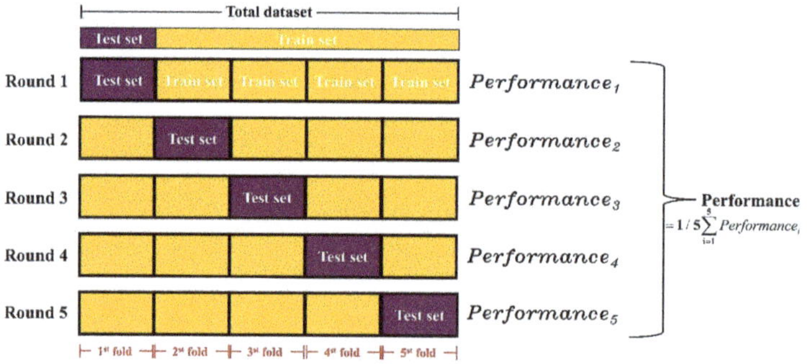

Figure 15. Illustration of the five-fold cross-validation process.

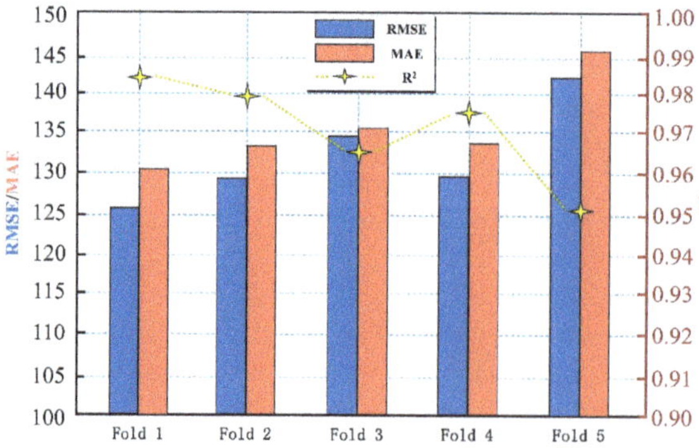

Figure 16. Performance metrics of the ICPO-XGBoost model across five folds in the cross-validation process.

4.4.2. Reliability Validation

To validate the reliability of the ICPO-XGBoost model, six new development cases were designed using the LHS method based on the tNavigator numerical simulation platform, with cumulative oil production serving as the validation benchmark. A comparison between the ICPO-XGBoost model predictions and the numerical simulation results is presented in Table 7. The analysis demonstrates that the ICPO-XGBoost model predictions are highly consistent with the numerical simulation results, with errors controlled within ±2%.

Table 7. Parameters and results of six development strategies.

Parameter	Unit	Case1	Case2	Case3	Case4	Case5	Case6
BHPO	MPa	18.5	17.3	18.4	17.6	17.5	16.6
WGR	%	30	85	87	90	45	90
ORAT	m^3/d	82.73	89.85	42.01	44.85	78.45	82.73
RATEG	m^3/d	29,940.96	38,116.19	39,678.19	33,328.05	38,912.61	39,940.96
RATEW	m^3/d	45.38	41.47	41.16	56.77	51.04	52.35
IC	Day	122.27	184.55	106.01	61.52	113.28	82.27
Simulation prediction	$\times 10^6$ m^3	1.308	1.348	1.424	1.495	1.881	2.355
Model prediction	$\times 10^6$ m^3	1.311	1.349	1.421	1.494	1.881	2.355

Figure 17 visually illustrates the numerical simulation results of the six development scenarios. The oil and gas saturation profiles reveal the internal fluid distribution characteristics of the reservoir and the development effectiveness under different strategies. The results indicate that scenarios with higher cumulative oil production are often associated with a higher risk of gas channeling. For instance, in Case 6, although the cumulative oil production reaches the highest value (2.355×10^6 m^3), severe gas channeling significantly undermines the stability of the strategy.

Figure 18 compares the ICPO-XGBoost model predictions with the numerical simulation results. It shows that the model predictions are in excellent agreement with the simulation results across all cases, with minimal errors that exhibit a consistent pattern. And further compares the trends of cumulative CO_2 production and gas breakthrough time. The results reveal significant differences in cumulative CO_2 production across the different development strategies, while gas breakthrough time serves as a strong indicator of gas channeling risks. For example, in Cases 3 and 5, the relatively short gas breakthrough times suggest higher gas channeling risks during the later stages of development.

4.4.3. Training Performance Analysis

To quantitatively evaluate the performance of the ICPO algorithm during training, this study compared the loss function reduction trends of the XGBoost, CPO-XGBoost, and the proposed ICPO-XGBoost models during the training phase (see Figure 19). Results show that ICPO-XGBoost significantly outperforms the other two models in terms of loss reduction speed, particularly within the first 100 training epochs, where it completes the primary error compression phase, reducing the loss value from approximately 0.21 to near 0.05. This demonstrates superior convergence efficiency. In contrast, the loss reduction process of XGBoost and CPO-XGBoost is notably slower, and their loss values remain relatively high during the later stages of convergence. By the 60th to 80th training epochs, ICPO-XGBoost had already reduced the training loss to below 0.03, showcasing its significant performance advantage. Furthermore, the magnified local region in the figure provides a detailed comparison of the stability of the three models during the convergence phase. It can be observed that the loss curve of ICPO-XGBoost approaches zero with minimal fluctuation in the later stages of training, whereas the loss curves of XGBoost and CPO-XGBoost still exhibit noticeable oscillations. This indicates that the proposed model not only achieves faster convergence but also effectively avoids overfitting and oscillations during training. Notably, ICPO-XGBoost requires approximately 30–40% fewer training epochs to reach convergence, significantly reducing training time and computational resource consumption. These results demonstrate that ICPO-XGBoost outperforms the comparison models in both training efficiency and overall performance.

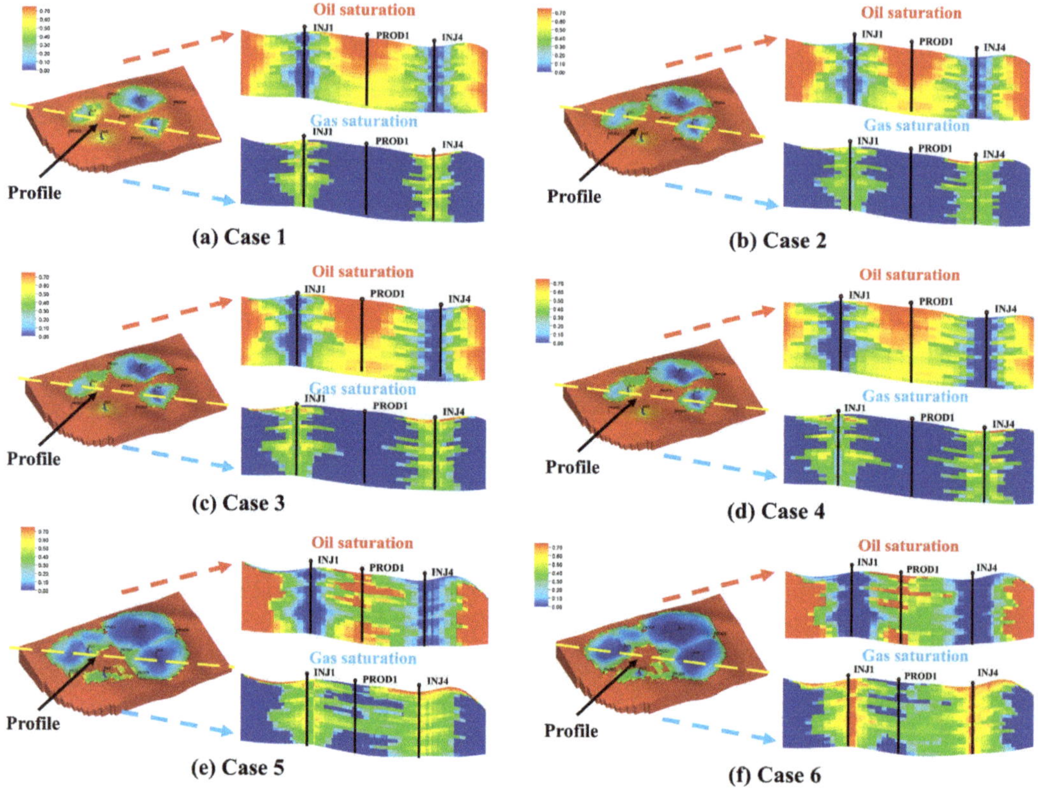

Figure 17. Oil and gas saturation distribution for six development strategies at the 20-year production mark. Each subplot includes three graphs: a 3D view of reservoir saturation distribution (left), along with two cross-sectional views of oil saturation (top right) and gas saturation (bottom right) along the profile line connecting the injection wells (INJ1 and INJ4) and the production well (PROD1).

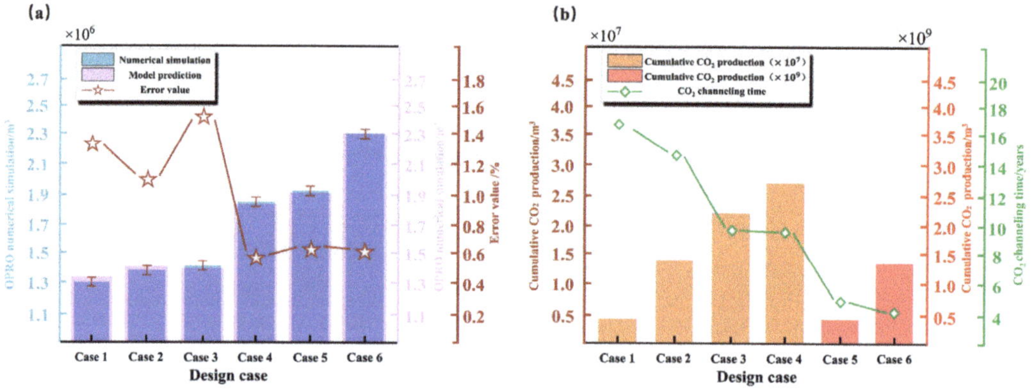

Figure 18. (**a**) Comparison of predicted vs. simulated cumulative oil production; (**b**) cumulative CO_2 production and gas breakthrough analysis.

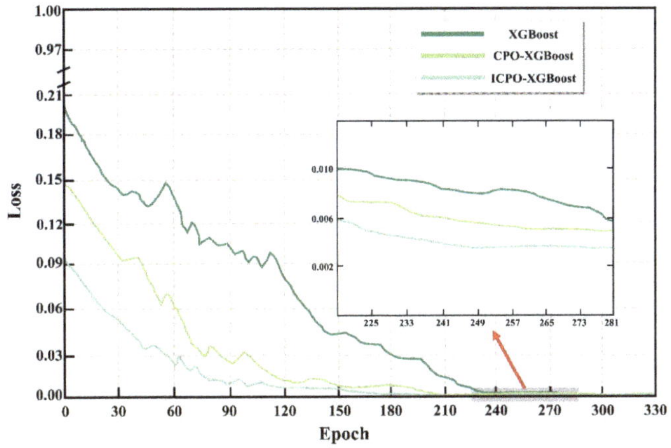

Figure 19. Loss curves of XGBoost, CPO-XGBoost, and ICPO-XGBoost models during training epochs.

4.5. Feature Importance Analysis

To further understand the predictive behavior of the ICPO-XGBoost model, this study employs the Shapley Additive Explanations (SHAP) method to perform a global feature importance analysis on the dataset. The SHAP method quantifies the contribution of each feature to the model's predictions by calculating the mean absolute Shapley value for each feature, thereby providing insight into the "black-box" nature of the model. Figure 20a illustrates the importance of each feature and their respective contributions to the OPRO.

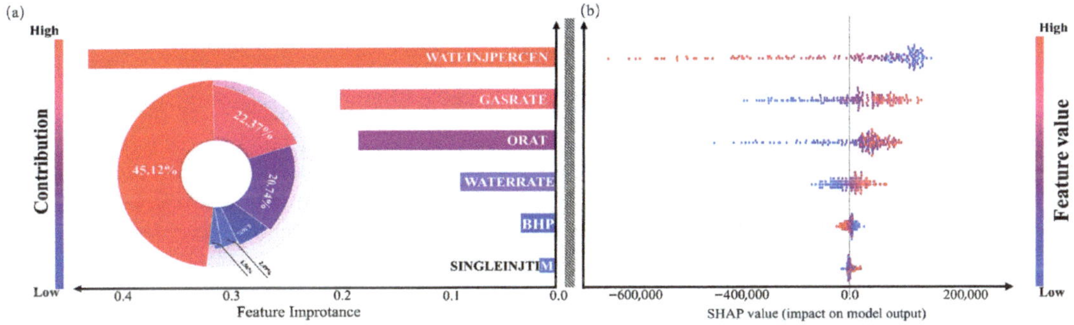

Figure 20. SHAP analysis results: (**a**) feature importance scores; (**b**) feature impact on cumulative oil production.

Among all the features, WGR is identified as the most critical parameter, accounting for 45.12% of the total SHAP values. This finding is not only statistically validated but also aligns closely with the complex physical displacement mechanisms inherent in CO_2-WAG technology. WGR directly determines the ratio of injected water and gas, which significantly influences the displacement efficiency of the injected fluids. At low WGR values (water-dominated injection), the higher viscosity of water improves volumetric sweep efficiency, mitigates premature gas breakthrough, and enhances reservoir utilization. Conversely, excessively high WGR values (over-dominance of water injection) may lead to non-uniform fluid flow, reduced gas displacement efficiency, and even water flooding near the injection wells. Furthermore, WGR plays a critical role in suppressing gas channeling. In highly heterogeneous reservoirs where gas channeling risks are elevated, an appropriate increase in WGR (i.e., a higher proportion of water injection) can create a water barrier to block high-permeability pathways, thereby significantly improving vertical sweep efficiency. In addition to WGR, RATEW and ORAT are identified as the second and third most important features, contributing 22.37% and 19.54%, respectively. The importance of RATEW highlights the role of water injection rates in maintaining reservoir pressure and optimizing displacement performance. Optimal RATEW values ensure effective pressure support while minimizing the risk of water flooding. In contrast, features such as RATEG, BHPO, and IC exhibit relatively lower importance but still have measurable impacts on reservoir performance. The SHAP values of RATEG indicate that increasing the gas injection rate positively influences displacement efficiency during the early stages of gas injection; however, its effect is less significant compared to WGR and RATEW, as gas displacement efficiency is largely regulated by WGR and reservoir heterogeneity. The influence of BHPO is primarily associated with pressure control during injection, and its relatively lower importance suggests that the model relies more on the improvement of fluid distribution and displacement efficiency rather than solely on pressure variations. IC primarily impacts reservoir performance through its indirect regulation of displacement mechanisms across different injection-production cycles.

Figure 20b presents the distribution of SHAP values for each feature, providing further insights into their specific impacts on prediction outcomes. For instance, the SHAP values of RATEG and RATEW are predominantly positive, indicating that increasing these variables has a favorable effect on predicting cumulative oil production, particularly when injection rates are optimized within a certain range. In contrast, WGR exhibits a polarized SHAP value distribution: lower WGR values have a clear positive impact on cumulative oil production predictions, consistent with their role in improving sweep efficiency and suppressing gas channeling. However, higher WGR values tend to have a significant negative impact, likely due to reduced gas displacement efficiency and increased water flooding risks. This characteristic further validates WGR's critical role in balancing displacement efficiency and gas channeling control.

Through comprehensive model comparisons and performance analysis, the ICPO-XGBoost model has demonstrated outstanding capabilities in optimizing CO_2-WAG processes for low-permeability reservoirs. With high predictive accuracy (R^2 = 0.9896) and low error rates (MAPE = 9.87%), the model significantly outperforms alternative approaches. By incorporating Chebyshev chaotic mapping and the Elite Opposition-Based Learning (EOBL) strategy, the model effectively enhances global search capability and optimization efficiency.

5. Conclusions

1. This study introduces a novel ICPO-XGBoost model tailored to address the complexities of development parameter optimization and production forecasting in CO_2-WAG processes for low-permeability reservoirs. By integrating the ICPO algorithm with the XGBoost machine learning framework, the model achieves remarkable improvements in predictive performance and optimization efficiency. Key advancements, such as the incorporation of Chebyshev chaotic mapping and the EOBL strategy, significantly enhance the algorithm's global search capability, convergence rate, and robustness.
2. The ICPO-XGBoost model exhibits exceptional predictive accuracy, with an R^2 of 0.9896 and prediction errors constrained within 2%. These results surpass those of traditional machine learning methods and mainstream boosting-based algorithms. Comparative analyses across six metaheuristic-based optimization models further validate its superior capability in handling high-dimensional, complex parameter spaces. Benchmark function tests demonstrate the algorithm's robustness in both unimodal and multimodal optimization scenarios.

3. The SHAP-based analysis highlights the WGR as the most influential factor (45.12%) in cumulative oil production, with effective WGR control key to suppressing gas channeling and improving efficiency. RATEW and ORAT also enhance production, while BHPO shows a modest negative effect, guiding optimization strategies. The ICPO-XGBoost model is a robust, efficient tool for optimizing CO_2-WAG processes in low-permeability reservoirs, with strong generalization. Future work should extend its application to diverse reservoirs and incorporate additional constraints for greater accuracy.

Author Contributions: Conceptualization, methodology, writing—review and editing, B.S.; Supervision, project administration, funding acquisition, J.L. (Junchao Li); Software, validation, writing—original draft preparation, J.L. (Jixin Li); Resources, investigation, formal analysis, C.H.; Data curation, visualization, S.F. All authors have read and agreed to the published version of the manuscript.

Funding: This research was funded by the Young Scientists Fund of the National Natural Science Foundation of China (No. 52204047), the National Natural Science Foundation of China (No. U23B6003), and the Xi'an Shiyou University Graduate Innovation Fund Program (No. YCX2412032).

Data Availability Statement: The datasets in this study can be obtained by contacting the corresponding author.

Conflicts of Interest: The authors declare no conflicts of interest.

References

Massarweh, O.; Abushaikha, A.S. A Review of Recent Developments in CO_2 Mobility Control in Enhanced Oil Recovery. *Petroleum* **2022**, *8*, 291–317. [CrossRef]

Hu, W.; Wei, Y.; Bao, J. Development of the Theory and Technology for Low Permeability Reservoirs in China. *Petroleum Explor. Dev.* **2018**, *45*, 685–697. [CrossRef]

Carpenter, C. Experimental Program Investigates Miscible CO_2 WAG Injection in Carbonate Reservoirs. *J. Pet. Technol.* **2019**, *71*, 47–49. [CrossRef]

Wang, H.; Kou, Z.; Ji, Z.; Wang, S.; Li, Y.; Jiao, Z.; Johnson, M.; McLaughlin, J.F. Investigation of Enhanced CO_2 Storage in Deep Saline Aquifers by WAG and Brine Extraction in the Minnelusa Sandstone, Wyoming. *Energy* **2023**, *265*, 126379. [CrossRef]

Nassabeh, M.; You, Z.; Keshavarz, A.; Iglauer, S. Hybrid EOR Performance Optimization through Flue Gas–Water Alternating Gas (WAG) Injection: Investigating the Synergistic Effects of Water Salinity and Flue Gas Incorporation. *Energy Fuels* **2024**, *38*, 13956–13973. [CrossRef]

Wu, Z.; Ling, H.; Wang, J.; Zhao, Z.; Li, S. EOR Scheme Optimization of CO_2 Miscible Flooding in Bohai BZ Reservoir. *Xinjiang Oil Gas* **2024**, *20*, 70–76.

Hussain, M.; Boukadi, F. Optimizing Oil Recovery: A Sector Model Study of CO2-WAG and Continuous Injection Techniques. 2025, *Unpublished*. [CrossRef]

Cui, G.; Yang, L.; Fang, J.; Qiu, Z.; Wang, Y.; Ren, S. Geochemical Reactions and Their Influence on Petrophysical Properties of Ultra-Low Permeability Oil Reservoirs during Water and CO_2 Flooding. *J. Pet. Sci. Eng.* **2021**, *203*, 108672. [CrossRef]

Li, J.; Xi, Y.; Zhang, M.; Zang, C. Applicability Evaluation of Tight Oil Reservoir Gas Channeling and Sweep Control System. *Xinjiang Oil Gas* **2024**, *20*, 68–76.

Bai, Y.; Hou, J.; Liu, Y.; Zhao, D.; Bing, S.; Xiao, W.; Zhao, W. Energy-Consumption Calculation and Optimization Method of Integrated System of Injection-Reservoir-Production in High Water-Cut Reservoir. *Energy* **2022**, *239*, 121961. [CrossRef]

An, Z.; Zhou, K.; Hou, J.; Wu, D.; Pan, Y. Accelerating Reservoir Production Optimization by Combining Reservoir Engineering Method with Particle Swarm Optimization Algorithm. *J. Pet. Sci. Eng.* **2022**, *208*, 109692. [CrossRef]

Li, D.; Wang, X.; Xie, Y.; Feng, Q. Analytical Calculation Method for Development Dynamics of Water-Flooding Reservoir Considering Rock and Fluid Compressibility. *Geoenergy Sci. Eng.* **2024**, *242*, 213250. [CrossRef]

Hourfar, F.; Salahshoor, K.; Zanbouri, H.; Elkamel, A.; Pourafshary, P.; Moshiri, B. A Systematic Approach for Modeling of Waterflooding Process in the Presence of Geological Uncertainties in Oil Reservoirs. *Comput. Chem. Eng.* **2018**, *111*, 66–78. [CrossRef]

Jaber, A.K.; Al-Jawad, S.N.; Alhuraishawy, A.K. A Review of Proxy Modeling Applications in Numerical Reservoir Simulation. *Arab. J. Geosci.* **2019**, *12*, 701. [CrossRef]

15. Mao, S.; Mehana, M.; Huang, T.; Moridis, G.; Miller, T.; Guiltinan, E.; Gross, M.R. Strategies for Hydrogen Storage in a Depleted Sandstone Reservoir from the San Joaquin Basin, California (USA) Based on High-Fidelity Numerical Simulations. *J. Energy Storage* **2024**, *94*, 112508. [CrossRef]
16. Pu, W.; Jin, X.; Tang, X.; Bai, Y.; Wang, A. Prediction Model of Water Breakthrough Patterns of Low-Permeability Bottom Water Reservoirs Based on BP Neural Network. *Xinjiang Oil Gas* **2024**, *20*, 37–47.
17. Mahdaviara, M.; Sharifi, M.; Ahmadi, M. Toward Evaluation and Screening of the Enhanced Oil Recovery Scenarios for Low Permeability Reservoirs Using Statistical and Machine Learning Techniques. *Fuel* **2022**, *325*, 124795. [CrossRef]
18. Meng, S.; Fu, Q.; Tao, J.; Liang, L.; Xu, J. Predicting CO_2-EOR and Storage in Low-Permeability Reservoirs with Deep Learning Based Surrogate Flow Models. *Geoenergy Sci. Eng.* **2024**, *233*, 212467. [CrossRef]
19. Gong, A.; Zhang, L. Deep Learning-Based Approach for Reservoir Fluid Identification in Low-Porosity, Low-Permeability Reservoirs. *Phys. Fluids* **2025**, *37*, 046611. [CrossRef]
20. Li, H.; Luo, P.; Bai, Y.; Li, D.; Chang, S.; Liu, X. Overview of Machine Learning Algorithms and Their Applications in Drilling Engineering. *Xinjiang Oil Gas* **2022**, *18*, 1–13.
21. Zhou, W.; Liu, C.; Liu, Y.; Zhang, Z.; Chen, P.; Jiang, L. Machine Learning in Reservoir Engineering: A Review. *Processes* **2024**, *12*, 1219. [CrossRef]
22. You, J.; Ampomah, W.; Sun, Q.; Kutsienyo, E.J.; Balch, R.S.; Dai, Z.; Cather, M.; Zhang, X. Machine Learning Based Co-Optimization of Carbon Dioxide Sequestration and Oil Recovery in CO_2-EOR Project. *J. Cleaner Prod.* **2020**, *260*, 120866. [CrossRef]
23. Thanh, H.V.; Dashtgoli, D.S.; Zhang, H.; Min, B. Machine-Learning-Based Prediction of Oil Recovery Factor for Experimental CO_2-Foam Chemical EOR: Implications for Carbon Utilization Projects. *Energy* **2023**, *278*, 127860. [CrossRef]
24. Li, H.; Gong, C.; Liu, S.; Xu, J.; Imani, G. Machine Learning-Assisted Prediction of Oil Production and CO_2 Storage Effect in CO_2-Water-Alternating-Gas Injection (CO_2-WAG). *Appl. Sci.* **2022**, *12*, 10958. [CrossRef]
25. Khan, W.A.; Rui, Z.; Hu, T.; Liu, Y.; Zhang, F.; Zhao, Y. Application of Machine Learning and Optimization of Oil Recovery and CO_2 Sequestration in the Tight Oil Reservoir. *SPE J.* **2024**, *29*, 2772–2792. [CrossRef]
26. Talbi, E.G. Machine Learning into Metaheuristics: A Survey and Taxonomy. *ACM Comput. Surv.* **2022**, *54*, 1–32. [CrossRef]
27. Menad, N.A.; Noureddine, Z. An Efficient Methodology for Multi-Objective Optimization of Water Alternating CO_2 EOR Process. *J. Taiwan Inst. Chem. Eng.* **2019**, *99*, 154–165. [CrossRef]
28. Gao, M.; Liu, Z.; Qian, S.; Liu, W.; Li, W.; Yin, H.; Cao, J. Machine-Learning-Based Approach to Optimize CO_2-WAG Flooding in Low Permeability Oil Reservoirs. *Energies* **2023**, *16*, 6149. [CrossRef]
29. Kanaani, M.; Kameholiya, A.S.; Amarzadeh, A.; Sedaee, B. Stacking Learning for Smart Proxy Modeling in CO_2–WAG Optimization: A Techno-Economic Approach to Sustainable Enhanced Oil Recovery. *ACS Omega* **2025**, *10*, 9563–9582. [CrossRef]
30. Yang, Y.; Cao, X.; Han, W.; Zang, D.; Zang, Q.; Wang, Y. Study on Nonlinear Seepage Law of CO_2 Flooding in Ultra-Low Permeability Reservoirs in Gao 89 Area of Shengli Oilfield. *Oil Gas Geol. Recovery* **2023**, *30*, 36–43. [CrossRef]
31. Zhang, C.; Hu, L.; Niu, Z. Application of CO_2-EOR in Low Permeability Beach-Bar Sand Reservoir—Taking Shengli Oilfield Gao 89 Block as an Example. *Sci. Technol. Eng.* **2023**, *23*, 6393–6401.
32. Sun, Z.; Li, Y.; Yang, Y.; Su, L.; Xie, S. Splitting Tensile Strength of Basalt Fiber Reinforced Coral Aggregate Concrete: Optimized XGBoost Models and Experimental Validation. *Constr. Build. Mater.* **2024**, *416*, 135133. [CrossRef]
33. Chen, T.; Guestrin, C. XGBoost: A Scalable Tree Boosting System. In Proceedings of the 22nd ACM SIGKDD International Conference on Knowledge Discovery and Data Mining, San Francisco, CA, USA, 13–17 August 2016; pp. 785–794. [CrossRef]
34. Pan, S.; Zheng, Z.; Guo, Z.; Luo, H. An Optimized XGBoost Method for Predicting Reservoir Porosity Using Petrophysical Logs. *J. Pet. Sci. Eng.* **2022**, *208*, 109520. [CrossRef]
35. Abdel-Basset, M.; Mohamed, R.; Abouhawwash, M. Crested Porcupine Optimizer: A New Nature-Inspired Metaheuristic. *Knowl.-Based Syst.* **2024**, *284*, 111257. [CrossRef]
36. Liu, H.; Zhou, R.; Zhong, X.; Yao, Y.; Shan, W.; Yuan, J.; Xiao, J.; Ma, Y.; Zhang, K.; Wang, Z. Multi-Strategy Enhanced Crested Porcupine Optimizer: CAPCPO. *Mathematics* **2024**, *12*, 3080. [CrossRef]
37. Lei, W.; Gu, Y.; Huang, J. An Enhanced Crowned Porcupine Optimization Algorithm Based on Multiple Improvement Strategies. *Appl. Sci.* **2024**, *14*, 11414. [CrossRef]
38. Abed-Alguni, B.H.; Paul, D. Island-Based Cuckoo Search with Elite Opposition-Based Learning and Multiple Mutation Methods for Solving Optimization Problems. *Soft Comput.* **2022**, *26*, 3293–3312. [CrossRef]
39. Mirjalili, S.; Mirjalili, S.M.; Lewis, A. Grey Wolf Optimizer. *Adv. Eng. Softw.* **2014**, *69*, 46–61. [CrossRef]
40. Faris, H.; Aljarah, I.; Al-Betar, M.A.; Mirjalili, S. Grey Wolf Optimizer: A Review of Recent Variants and Applications. *Neural Comput. Appl.* **2018**, *30*, 413–435. [CrossRef]
41. Bakır, H. A Novel Artificial Hummingbird Algorithm Improved by Natural Survivor Method. *Neural Comput. Appl.* **2024**, *36*, 16873–16897. [CrossRef]

2. Hosseinzadeh, M.; Rahmani, A.M.; Husari, F.M.; Alsalami, O.M.; Marzougui, M.; Nguyen, G.N.; Lee, S.-W. A Survey of Artificial Hummingbird Algorithm and Its Variants: Statistical Analysis, Performance Evaluation, and Structural Reviewing. *Arch. Comput. Methods Eng.* **2025**, *32*, 269–310. [CrossRef]
3. Du, C.; Zhang, J.; Fang, J. An innovative complex-valued encoding black-winged kite algorithm for global optimization. *Sci. Rep.* **2025**, *15*, 932. [CrossRef]
4. Wang, J.; Wang, W.-C.; Hu, X.-X.; Qiu, L.; Zang, H.-F. Black-Winged Kite Algorithm: A Nature-Inspired Meta-Heuristic for Solving Benchmark Functions and Engineering Problems. *Artif. Intell. Rev.* **2024**, *57*, 98. [CrossRef]
5. Si, T.; Bhattacharya, D.; Nayak, S.; Miranda, P.B.C.; Nandi, U.; Mallik, S.; Maulik, U.; Qin, H. PCOBL: A Novel Opposition-Based Learning Strategy to Improve Metaheuristics Exploration and Exploitation for Solving Global Optimization Problems. *IEEE Access* **2023**, *11*, 46413–46440. [CrossRef]
6. Yildiz, B.S.; Pholdee, N.; Bureerat, S.; Yildiz, A.R.; Sait, S.M. Enhanced Grasshopper Optimization Algorithm Using Elite Opposition-Based Learning for Solving Real-World Engineering Problems. *Eng. Comput.* **2022**, *38*, 4207–4219. [CrossRef]

processes

MDPI

Article

Prediction of Mud Weight Window Based on Geological Sequence Matching and a Physics-Driven Machine Learning Model for Pre-Drilling

Yuxin Chen [1,2], Ting Sun [1,2,*], Jin Yang [1,2], Xianjun Chen [3], Laiao Ren [1,2], Zhiliang Wen [1,2], Shu Jia [1,2], Wencheng Wang [1,2], Shuqun Wang [1,2] and Mingxuan Zhang [1,2]

1 Hainan Institute of China University of Petroleum-Beijing, Sanya 572000, China
2 College of Safety and Ocean Engineering, China University of Petroleum-Beijing, Beijing 102249, China
3 China France Bohai Geoservices Company Limited, Haikou 570312, China
* Correspondence: ting.sun@cup.edu.cn

Abstract: Accurate pre-drilling mud weight window (MWW) prediction is crucial for drilling fluid design and wellbore stability in complex geological formations. Traditional physics-based approaches suffer from subjective parameter selection and inadequate handling of multi-mechanism over-pressured formations, while machine learning methods lack physical constraints and interpretability. This study develops a novel physics-guided deep learning framework integrating rock mechanics theory with deep neural networks for enhanced MWW prediction. The framework incorporates three key components: first, a physics-driven layer synthesizing intermediate variables from rock physics calculations to embed domain knowledge while preserving interpretability; second, a geological sequence-matching algorithm enabling precise stratigraphic correlation between offset and target wells, compensating for lateral geological heterogeneity; third, a long short-term memory network capturing sequential drilling characteristics and geological structure continuity. Case study results from 12 wells in northwestern China demonstrate significant improvements over traditional methods: collapse pressure prediction error reduced by 40.96%, pore pressure error decreased by 30.43%, and fracture pressure error diminished by 39.02%. The proposed method successfully captures meter-scale pressure variations undetectable by conventional approaches, providing critical technical support for wellbore design optimization, drilling fluid formulation, and operational safety enhancement in challenging geological environments.

Keywords: mud weight window; machine learning; pre-drilling prediction; geological sequence matching; safe drilling

1. Introduction

The mud weight window (MWW) directly determines wellbore stability, well control safety, and wellbore integrity [1,2] serving as the core foundation for drilling fluid density design, casing program optimization, and well control contingency planning. Inaccurate estimation can readily trigger high-risk events such as lost circulation, kicks, or even blowouts [3] resulting in substantial economic losses and environmental disasters. Therefore, accurate MWW prediction is not only a prerequisite for safe drilling and completion operations in complex formations but also a critical guarantee for enhancing the success rate of oil and gas resource development [4].

Since the 1970s, the industry has primarily relied on physics-based empirical models (such as the Eaton method and Bowers approach) to estimate formation pore pressure by establishing correlations between sonic transit time and vertical effective stress. These methods combine formation strength parameters with Mohr–Coulomb or Drucker–Prager failure criteria to determine collapse pressure and fracture pressure, ultimately providing the mud weight range. Foster proposed the equivalent depth method, establishing normal trend lines for formation factors to determine effective stress and pore pressure [5]. Pennebaker introduced overlay charts using seismic interval velocities, improving prediction accuracy but lacking consideration for lithological and geological age variations [6]. Eaton combined Terzaghi's effective stress principle to develop the widely used Eaton method, though subjective bias in determining normal trend lines and empirical parameters affect accuracy [7,8]. Bowers incorporated loading and unloading mechanisms to predict abnormal pressures from different geological processes [9]. Concurrently, researchers have conducted extensive studies on mud weight window optimization and wellbore stability analysis in shale formations. Zhang modified the Biot effective stress principle, integrating Weibull statistical and flow–diffusion coupling models to calculate safe drilling fluid densities for shale formations considering sealing inhibition-reverse osmosis effects [10]. Kadkhodaie developed an improved wellbore stability analysis method for Iranian Persian Gulf offshore fields by combining geomechanical unit classification, FMI log interpretation, and rock mechanics testing [11]. Liu established an elastoplastic mud weight window determination method incorporating thermo-mechanical coupling effects, demonstrating opposite correlations between mud weight windows and geothermal gradients at different depth intervals [12]. Zhang proposed an elastoplastic collapse pressure model based on the Hoek–Brown strength criterion, incorporating the plastic zone radius, geological strength index, non-uniform in situ stress coefficient, and rock uniaxial compressive strength for deep coal seam drilling applications [13]. Existing research primarily focuses on normal compaction trend (NCT) fitting optimization, rock strength parameter determination, and regional empirical coefficient modification, achieving significant progress under specific geological conditions. However, these traditional methods face numerous challenges that affect their accuracy and reliability in complex environments. First, conventional single-compaction mechanism models inadequately characterize over-pressured formations generated by multi-mechanism processes (rapid sedimentation, tectonic unloading, and fault sealing), introducing systematic bias in pore pressure inversion. Second, traditional depth-based well correlation methods neglect inter-well stratigraphic sequence variations and thickness changes, yielding poor prediction accuracy in fault-intensive or heterogeneous regions where errors amplify exponentially. Third, while seismic prediction provides spatial coverage, resolution constraints and velocity inversion uncertainties limit vertical formation pressure characterization, particularly in deep, structurally complex areas where depth-dependent error amplification compromises mud weight window reliability.

In recent years, machine learning (ML), particularly through deep learning (DL) techniques, has demonstrated superior performance in rock physics parameter inversion and real-time drilling optimization. Multiple researchers have employed models such as random forest, XGBoost, and LSTM to train on logging, mud logging, and real-time drilling data, achieving rapid prediction of mud weight windows. Song developed an LSTM-BP neural network method targeting drilling and sedimentation sequences, demonstrating superior accuracy compared to individual BP, LSTM, and SVM models for intelligent tri-pressure prediction [14]. Deng established the IGWO-MLP intelligent prediction method by integrating machine learning algorithms with effective stress theory, showing excellent generalization capability across three-well validation datasets [15]. Li proposed a LightGBM-based intelligent prediction model using well depth, formation density, and pore pressure equivalent density as inputs, achieving superior accuracy and stability compared to sonic logging methods in S block vertical wells [16]. Zhang introduced an adaptive physics-informed deep learning model (CGP-NN) for direct pore pressure prediction from seismic data, incorporating 1DCPP feature extraction, multi-layer GRU, and adaptive physics-informed loss functions for multi-genetic mechanism prediction [17]. Chen embedded rock physics theory into GRU and LSTM models using seismic and logging-while-drilling (LWD) data, combined with SHAP interpretation methods for accurate prediction and model optimization guidance [18]. Hussain developed a hybrid SOM-MLP-SSD model for predicting missing well-log data, particularly shear wave transit time, using lithofacies classification and parameter optimization, significantly outperforming traditional methods in complex zones [19]. Doyoro applied k-means clustering to post-process seismic refraction and electrical resistivity tomography data, enhancing geological structure detection and fault delineation compared to conventional interpretation methods [20]. Wang proposed a deep learning approach for identifying geological discontinuities in borehole images, demonstrating SegFormer's superiority over U-Net and improving model robustness through weighted loss functions for automated geological interpretation [21].

However, the aforementioned studies suffer from several critical limitations. First, high-quality pore pressure measurements are costly and limited, resulting in small datasets with high noise levels that increase overfitting susceptibility. Second, essential logging parameters (sonic transit time, density, resistivity) are unavailable prior to drilling, forcing reliance on regional seismic or offset well extrapolation that inadequately accounts for spatial geological variations and significantly increases prediction uncertainty. Third, mud-logging data suffer from substantial interference noise caused by drill string mechanical vibrations, annular clearances, circulation effects, and cutting lag time, leading to poor depth–time alignment and inability to detect thin bed anomalies or stress variations. Finally, while LWD provides near-wireline accuracy with real-time capabilities, its adoption remains limited due to high operational costs and insufficient mud pulse/electromagnetic transmission bandwidth for complete waveform data transfer, preventing high-dimensional feature extraction.

Based on the aforementioned challenges, this study develops a physics-guided mud weight window estimation model and demonstrates its superiority through case studies of 12 wells in Block C of a northwestern China oilfield. As illustrated in Figure 1, the workflow of this methodology begins with statistical analysis of historical well data in the target block, combining rock mechanics models, drilling fluid density usage, and actual complex conditions to calculate estimated mud weight window values. Following the approach of rock mechanics theory and empirical formulas for stepwise calculation of rock mechanical parameters, an LSTM algorithm is employed to train a physics-guided mud weight window estimation model. Finally, geological sequence matching (GSM) is applied to offset wells to obtain high-precision pre-drilling geological estimation parameters for the target well, which are subsequently input into the model for mud weight window prediction of the target well.

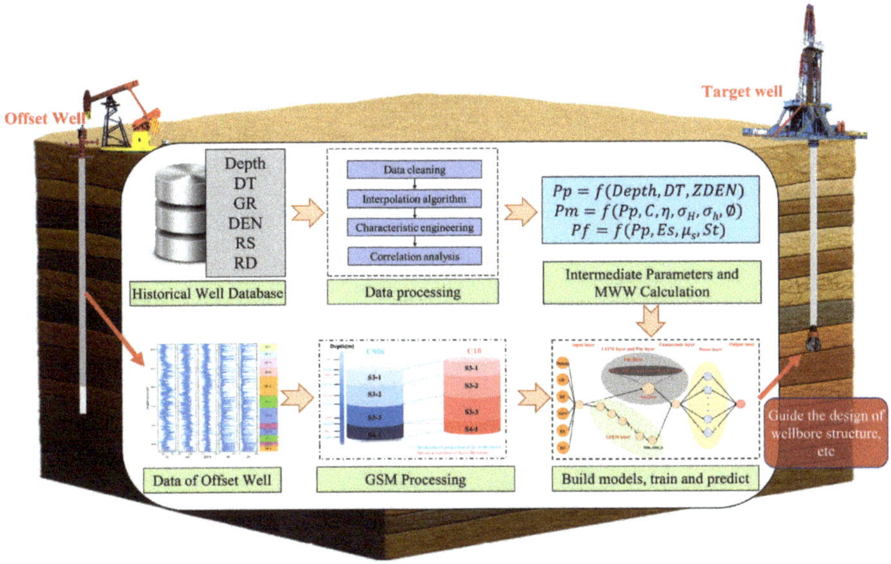

Figure 1. Flowchart of the physics-guided deep learning prediction method for mud weight windows.

2. Data and Theory

2.1. Data Overview

The case study data originates from Block C of a northwestern China oilfield. This block contains 12 wells, with well C10 serving as the target well. The well locations and fault distribution are shown in Figure 2a, while the comprehensive stratigraphic column is presented in Figure 2b [22]. This block belongs to a nose-shaped structure containing four faults of varying scales. Faults 1, 2, and 3 exhibit similar strike orientations with poor sealing capacity. The area has developed four oil-bearing formations—S1, S2, S3, and S4—with the S2 formation serving as the main productive reservoir. The S1 oil group consists of calcareous mudstone and oil shale in the middle section, gray mudstone and siltstone in the upper section, and brown-yellow sandstone and gray siltstone interbedded with calcareous mudstone in the lower section. The upper part of the S2 oil group develops gray-green and gray sandstone interbedded with calcareous sandstone and conglomeratic sandstone, while the lower part is dominated by brown-yellow sandstone and mudstone interbedded with gray and dark gray carbonate rocks. The lower section of the S3 oil group primarily develops dark organic-rich dolomitic limestone, with the upper section featuring superimposed gray dolomite and argillaceous dolomitic limestone interbedded with multiple salt rock layers. The S4 oil group is predominantly composed of brown-red and brown mudstone, sandy mudstone, and fine sandstone, with secondary gray and gray-yellow siltstone, argillaceous siltstone, and conglomeratic sandstone and conglomerate developed at the bottom. Through detailed subdivision, the S1 formation can be divided into four sublayers (S1-1, S1-2, S1-3, S1-4), the S2 formation into three sublayers (S2-1, S2-2, S2-3), the S3 formation into three sublayers (S3-1, S3-2, S3-3), and the S4 formation into two sublayers (S4-1, S4-2).

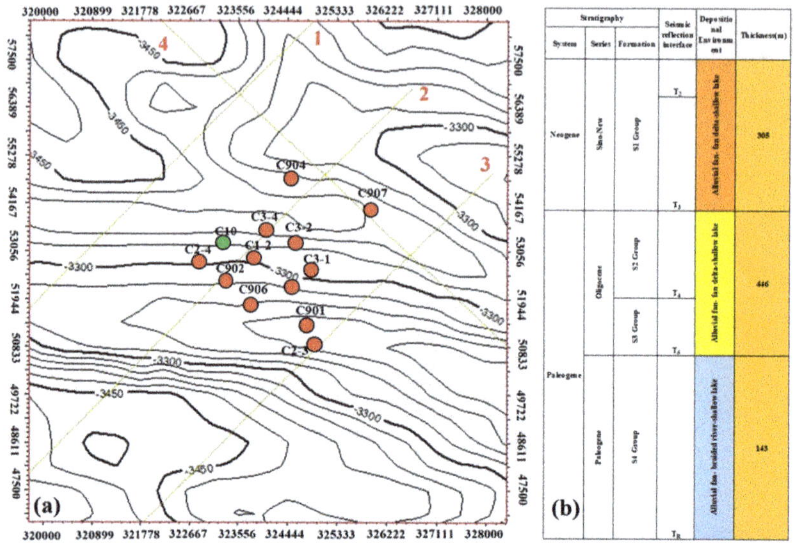

Figure 2. Well location distribution (**a**) and comprehensive stratigraphic column of Block C (**b**).

In this study, we collected six logging parameters from twelve wells (depth, sonic transit time (DT, μs/ft), deep-investigated double lateral resistivity log (Rd, Ω·m), shallow-investigated double lateral resistivity log (Rs, Ω·m), natural gamma ray (GR, gAPI), and bulk density (ZDEN, g/cm^3)), along with fifteen theoretically calculated parameters, which were shale content (Vsh, %), P-wave velocity (Vp, m/s), dynamic elastic modulus (Ed, MPa), static elastic modulus (Es, MPa), dynamic Poisson's ratio (Ud, dimensionless), static Poisson's ratio (Us, dimensionless), uniaxial compressive strength (Sc, MPa), tensile strength (St, MPa), cohesion (C, MPa), internal friction angle (J, °), minimum principal stress (σ_h, MPa), maximum principal stress (σ_H,, MPa), pore pressure (Pp, g/cm^3), collapse pressure (Pm, g/cm^3), and fracture pressure (Pf, g/cm^3), totaling 1,143,460 data points.

2.2. Regional Over-Pressure Types

In this study, we constructed a sonic velocity–density crossplot for the target formations of 12 wells in Block C, as shown in Figure 3. Well C10 (denoted by the red pentagram in the legend) was plotted as the target well, revealing that the sonic velocity–density crossplot trends of offset wells are consistent with those of well C10. This demonstrates the feasibility of using offset wells' sonic velocity–density crossplot to determine the over-pressure generation mechanism of the target well.

As observed in Figure 3, the unloading curve characteristics are consistent with fluid expansion over-pressure mechanisms prevalent in this region. Fluid expansion processes result in pore pressure increases that exceed the rate of effective stress accumulation during burial. This disequilibrium leads to mechanical unloading behavior of the rock framework, where stress relief occurs relative to normal compaction trends, thereby producing the unloading characteristics observed in Figure 3. Additionally, the density of target formations in the 12 wells of Block C remains essentially unchanged with variations in sonic velocity, which conforms to the characteristics of an unloading curve.

Therefore, the unloading equation from the Bowers method should be employed when calculating formation pore pressure in the target formations. Li [23] and Wang [24] identified the genesis of abnormal pressure in the C oilfield through logging and other methods, demonstrating that abnormal pressure in the intermediate to deep formations has little correlation with disequilibrium compaction, while fluid expansion constitutes one of the primary sources of abnormal pressure in this region. This conclusion aligns with the findings of the present study; therefore, we chose the unloading equation from the Bowers method to calculate formation pore pressure in this research.

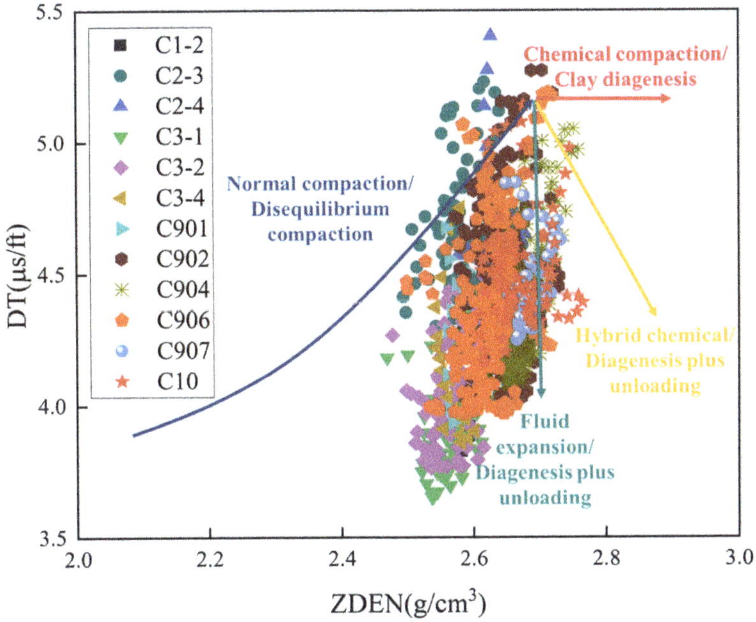

Figure 3. Acoustic velocity–density crossplot for Block C.

2.3. Mud Weight Window Data Acquisition

Due to the significant difficulty in obtaining rock strength parameters during drilling operations and the fact that laboratory experiments can only provide rock strength parameters at discrete points, this study employs logging parameters and rock mechanics theory to establish rock strength parameters for the entire well section.

Currently, the most widely used methods for predicting formation pore pressure from sonic data are the Eaton method and the Bowers method, both of which are based on the effective stress principle:

$$P_p = P_o - \sigma \quad (1)$$

where P_p represents formation pore pressure, P_o represents overburden pressure, and σ represents vertical effective stress.

In the Eaton and Bowers methods, the expressions and calculation methods for σ differ. The Eaton method calculates formation pore pressure based on the normal compaction trend line, with σ expressed as

$$\sigma = (P_o - P_h)\left(\frac{\Delta t_n}{\Delta t_0}\right)^n \quad (2)$$

where P_h represents normal hydrostatic pressure, Δt_n represents the sonic transit time of the normal trend line for shale at the calculation point, Δt_0 represents the measured sonic transit time for shale at the calculation point, and n represents the Eaton exponent.

The Bowers method establishes the relationship between shale acoustic velocity and vertical effective stress by dividing it into sedimentary compaction loading curves and unloading curves, where σ is expressed as follows [25]:

$$\sigma = (\frac{V - 5000}{A})^{\frac{1}{B}} \tag{3}$$

$$\sigma = \sigma_{\max}(\frac{(\frac{V-5000}{A})^{\frac{1}{B}}}{\sigma_{\max}})^{U} \tag{4}$$

where A and B are dimensionless coefficients in the relationship, σ_{max} represents the maximum vertical effective stress at the onset of unloading, and U denotes the elastoplastic coefficient of shale.

The collapse pressure and fracture pressure can be calculated using the following equations [26]:

$$P_m = \frac{\eta(3\sigma_H - \sigma_h) - 2C\cdot \cot\left(45^\circ - \frac{\phi}{2}\right) + \alpha p_p(\cot\left(45^\circ - \frac{\phi}{2}\right)^2 - 1)}{(\cot\left(45^\circ - \frac{\phi}{2}\right)^2 + \eta)} \tag{5}$$

$$P_f = \frac{\xi_1 E_s}{1 - \mu_s} - \frac{\xi_2 E_s}{1 + \mu_s} + \frac{2\mu_s}{1 + \mu_s}(P_0 - \alpha P_p) - \alpha P_p + S_t \tag{6}$$

where η is the nonlinear correction coefficient; σ_H and σ_h represent the maximum and minimum horizontal in situ stresses, respectively; C denotes the rock cohesion; ϕ is the internal friction angle of sandstone–mudstone formations; α represents the effective stress coefficient; ξ_1 and ξ_2 are the tectonic stress coefficients; E_s is the static elastic modulus; μ_s represents the static Poisson's ratio; and S_t denotes the tensile strength of the rock.

The calculation formula for the effective stress coefficient is given by [25]

$$\alpha = 1 - \frac{\rho(\frac{1}{\Delta t_p^2} - \frac{4}{3\Delta t_s^2})}{\rho_m(\frac{1}{\Delta t_{mp}^2} - \frac{4}{3\Delta t_{ms}^2})} \tag{7}$$

where ρ represents the rock density; Δt_p and Δt_s denote the P-wave and S-wave transit times, respectively; ρ_m is the density of the rock matrix; and Δt_{mp} and Δt_{ms} represent the P-wave and S-wave transit times of the rock matrix, respectively.

The dynamic Poisson's ratio and dynamic elastic modulus are expressed as follows [25]:

$$\mu_d = \frac{(\frac{V_p}{V_s})^2 - 2}{2[(\frac{V_p}{V_s})^2 - 1]} \tag{8}$$

$$E_d = \rho V_s^2 \frac{3(\frac{V_p}{V_s})^2 - 4}{(\frac{V_p}{V_s})^2 - 1} \tag{9}$$

where V_p and V_s represent the P-wave and S-wave velocities, respectively:

The static Poisson's ratio and static elastic modulus are given by [25]

$$\mu_s = A_1 + K_1\mu_d \tag{10}$$

$$E_s = A_2 + K_2 E_d \tag{11}$$

where A_1, K_1, A_2, and K_2 are coefficients related to rock strength.

The rock tensile strength is expressed as [25]

$$S_t = \frac{(0.0045 + 0.0035V_{cl})E_d}{K_{tc}} \tag{12}$$

where V_{cl} represents the clay content, and K_{tc} denotes the rock compressive strength proportional coefficient.

The calculation formula for clay content is given by [25]

$$V_{cl} = \frac{2^{GCUR \cdot (\frac{GR - GR_{\min}}{GR_{\max} - GR_{\min}})} - 1}{2^{GCUR} - 1} \tag{13}$$

where $GCUR$ is the Hilchie index; GR represents the natural gamma ray value; GR_{min} denotes the natural gamma ray value of pure sandstone layers; and GR_{max} represents the natural gamma ray value of pure shale.

The calculation formula for rock cohesion is expressed as follows [25]:

$$C = A(1 - 2\mu_d)\left(\frac{1 + \mu_d}{1 - \mu_d}\right)^2 \rho^2 V_p^4 (1 + 0.78V_{cl}) \tag{14}$$

where A is a constant related to rock property.

The calculation formula for rock internal friction angle is expressed as follows [27]:

$$\phi = a_1 \cdot \lg((a_2 - b_2 \cdot C) + ((a_2 - b_2 \cdot C)^2 + 1)^{\frac{1}{2}}) + b_1 \tag{15}$$

where a_1, b_1, a_2, and b_2 are coefficients related to rock strength.

The calculation formula for tectonic stress coefficient is given by [28]

$$\xi_1 = \frac{1}{E_s}[(\sigma_H + \sigma_h - 2\alpha P_p)(1 - \mu_s) - 2\mu_s(P_0 - \alpha P_p)] \tag{16}$$

$$\xi_2 = \frac{1}{E_s}(\sigma_H - \sigma_h)(1 + \mu_s) \tag{17}$$

Various parameters of well C10—calculated using traditional rock physics formulas—are shown in Figure 4.

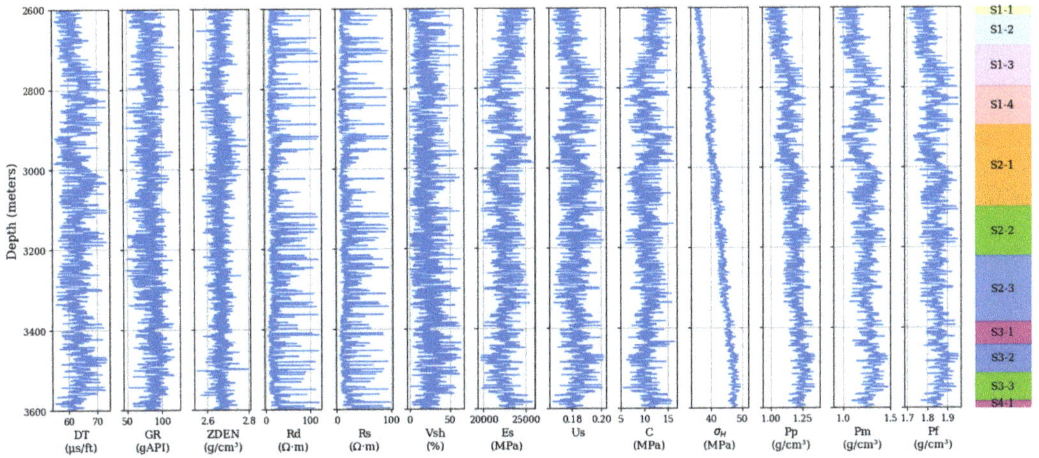

Figure 4. Parameters of well C10 calculated using traditional rock physics formulas.

2.4. Offset Well Geological Sequence-Matching Method

According to sequence stratigraphic principles, formations of the same group within the same block exhibit similar lithological and geological characteristics. Field engineers typically predict mud weight windows for wells to be drilled by directly using offset well depths, a method that neglects the influence of inter-well geological formation depth and thickness variations. To address this issue, Weng proposed geological sequence matching (GSM) to achieve precise alignment of formation data between historical wells and wells to be drilled, effectively compensating for formation depth and thickness differences [29]. This study adopts this approach to obtain input parameters for the target well. Figure 5 illustrates the depth distribution of geological stratification between well C906 (offset well) and well C10 (target well). In conventional methods, when calculating sonic logging data for the S3-2 formation (3443 m–3514.5 m) in well C10, direct depth alignment is employed, where the sonic logging data actually utilizes the lower portion of the S3-1 formation (3423.3 m–3475.1 m) and the upper portion of the S3-2 formation (3475.1 m–3542.7 m) from well C906. Due to the different lithological and geological conditions between the two wells, this method exhibits low accuracy in subsequent calculations of rock mechanical parameters and mud weight windows using rock mechanics methods.

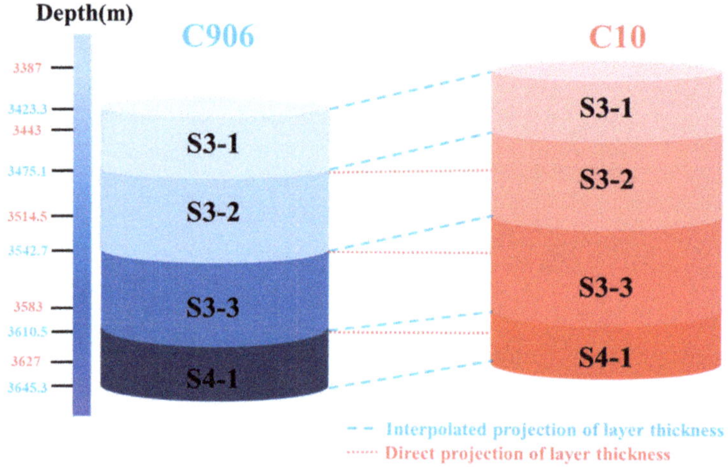

Figure 5. Schematic diagram of offset well geological sequence matching.

The geological sequence-matching method assumes that identical formations in offset wells within the same block possess uniform overall properties and vary uniformly within each layer, with the formation boundaries of the target well having been delineated through seismic data and historical drilling data during preliminary seismic exploration. This study utilizes well C906 to distribute post-drilling sonic logging data to each formation of well C10 through the geological sequence-matching method, as illustrated in Figure 5. For example, well C906 contains 1310 data points in the S3-1 formation (formation thickness: 51.8 m), and these 1310 data points are uniformly interpolated and distributed to the S3-1 formation of well C10 (formation thickness: 56 m). The depth calculation formula for the i data point in the j formation of the target well is shown in Equation (18) [29].

$$\mathrm{Depth}_{\mathrm{match}} = i \cdot \frac{D_{j+1} - D_j}{X} + D_j \tag{18}$$

where $Depth_{match}$ represents the depth of the i point in a specific stratigraphic layer of the target well; D_{j+1} denotes the depth of the bottom boundary of a specific stratigraphic layer in the target well; D_j represents the depth of the top boundary of a specific stratigraphic layer in the target well; and X is the number of post-drilling sonic logging data points in a specific stratigraphic layer of the offset well.

The data from well C906 processed using the geological sequence-matching method is designated well C10_match. The variation of selected parameters with depth for wells C906, C10, and C10_match is presented in Figure 6. It is evident that the C10_match well data processed through the geological sequence-matching method exhibits fundamentally identical trends to well C10, with consistent variations in detailed characteristics, although certain differences in data values persist. Compared to the conventional depth-based alignment method for processing C906 well data, the geological sequence-matching method, while unable to achieve complete consistency between offset well data and target well data, enables the predicted data trends to more closely approximate actual conditions.

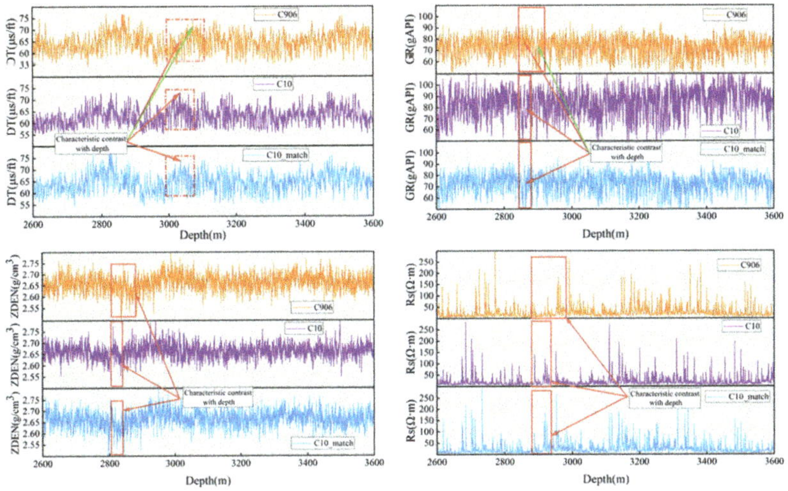

Figure 6. Variation of selected parameters with depth for wells C906, C10, and C10_match.

3. Physics Model-Guided Deep Learning Prediction Method and Model Development for Mud Weight Window

3.1. Feature Engineering and Correlation Analysis

This study establishes its data foundation using logging data acquired from the field. Given that abnormal values (identified as −999.99) exist in the initial and terminal segments of logging data acquisition, along with occasional occurrences of such anomalies in intermediate sections, abnormal values at both ends are removed and depth calibration operations are completed. For intermediate abnormal values, identification is performed using the 3-sigma criterion, followed by replacement through linear interpolation of offset data points. For offset wells with relatively scarce measured data of Pp, Pm, and Pf, calculations are conducted using the aforementioned conventional rock mechanics methods based on post-drilling logging data. Subsequently, the calculated pressure profiles are calibrated against measured values to serve as reference standards. This study utilizes data from the C oil field, encompassing measured formation pressure values across multiple depth intervals. Comparative analysis reveals excellent consistency between the calculated results based on measured data and the three-pressure data from the C oil field, thereby validating the reliability of the proposed methodology.

It should be noted that the calculation parameters employed in this study are derived from years of drilling experience accumulated in the C oil field, forming a comprehensive empirical parameter system rather than a refined three-pressure profile constructed through detailed modeling. Additionally, all coefficients of the calculation parameters maintain consistency within the block boundaries. Furthermore, the calculated results align well with the actual drilling fluid density windows implemented in individual wells.

In this method, the ZDEN parameter (unit: g/cm^3) has a numerical range of [2.5, 2.9], while the depth parameter (unit: m) extends to a numerical range of [2000, 4500]. Direct input of raw data into machine learning models would easily result in large-scale numerical dominance over small-scale values due to excessive differences in numerical magnitudes. To mitigate the adverse effects of data scale, feature dimensionality, and distribution differences on model performance, this study employs the max–min normalization method shown in Equation (19) to preprocess the dataset [30].

$$x' = \frac{x - \min(x)}{\max(x) - \min(x)} \tag{19}$$

where x represents a feature dataset, x' denotes the transformed data, and max and min represent the maximum and minimum values of the dataset, respectively.

Correlation analysis can effectively quantify the strength of associations among multiple variables, thereby revealing the intrinsic relationships within data. This study employs the Spearman correlation coefficient to analyze the relationships between the statistically obtained parameters and the mud weight window. The calculation of the Spearman correlation coefficient is based on the original positions and re-ranked positions of the data [31,32], with specific calculation steps as follows.

(a) Record the initial position i of each data point in groups x and y.
(b) Sort the data into groups x and y, separately, according to specific sorting rules, and record the ranking positions $rg(x_i)$ and $rg(y_i)$ for each data point in both groups, which represent the ranks.
(c) The difference between the rankings of the two data points in the i pair is recorded as d_i, as shown in the following equation:

$$d_i = \mathrm{rg}(x_i) - \mathrm{rg}(y_i) \tag{20}$$

(d) The Spearman correlation coefficient for groups x and y is calculated using the following formula:

$$\rho_{i,\mathrm{Spearman}} = 1 - \frac{6\sum_{i=1}^{n} d_i^2}{n(n^2 - 1)} \tag{21}$$

The Spearman correlations between parameters are shown in Figure 7. The eccentricity of elliptical shapes reflects the strength of correlation between two parameters. The closer the ellipse approaches a circle, the weaker the correlation between parameters; the flatter the ellipse, the stronger the correlation between parameters. The orientation of the ellipse indicates whether the correlation between the two parameters is positive or negative. Additionally, the correlation strength between parameters can be identified through color gradient variations. Based on physical theory and Spearman correlation analysis, the relationships between Ed and Es, Ud and Us, St and Sc, C and J, and σ_h and σ_H originate from simple calculations using empirical formulas, exhibiting very high correlations. When selecting parameters, choosing one from each pair can reduce model complexity and computational requirements. It is noteworthy that Rd and Rs exhibit low correlations with the mud weight window, as these parameters are not utilized in traditional rock mechanics calculations. However, from a physical perspective, Rd and Rs can characterize lithological variations to some extent and may have localized influence on the mud weight window. Therefore, this study selects nine parameters for the model: Depth, DT, GR, ZDEN, Rd, Rs, Pp, Pm, and Pf, where the three parameters Pp, Pm, and Pf constitute the mud weight window.

Figure 7. Spearman correlation coefficient heatmap among various parameters.

3.2. Physics Model-Guided Deep Learning Prediction Method for Mud Weight Window

Traditional physics-based methods calculate rock mechanical parameters progressively through rock mechanics theory and empirical formulas; however, the dimensionless coefficients involved exhibit significant regional and depth-dependent characteristics, limiting prediction accuracy. While deep learning algorithms can establish complex nonlinear mapping relationships, they lack physical constraints and interpretability. Building upon the integration of traditional rock mechanics theory and deep learning techniques, this study proposes a physics-driven long short-term memory neural network (PD-LSTM) prediction method to enhance both the accuracy and interpretability of mud weight window predictions. Traditional physical models calculate various rock mechanical parameters through rock mechanics formulas, but the dimensionless coefficients and weighting factors involved often depend on regional and burial depth geological conditions, exhibiting significant non-uniqueness and regional adaptability limitations that make it difficult to comprehensively adapt to complex and variable wellsite environments. Although deep learning models possess powerful nonlinear mapping capabilities and can automatically learn complex high-order relationships between features, their "black box" characteristics result in a lack of necessary physical interpretation and domain credibility [33,34]. As shown in Figure 8, this study designs a physics-guided deep neural network architecture that embeds rock mechanics physical prior knowledge into the model, guiding deep learning to learn optimal mapping relationships within the space conforming to geological physical laws, thereby enhancing model performance stability and generalizability.

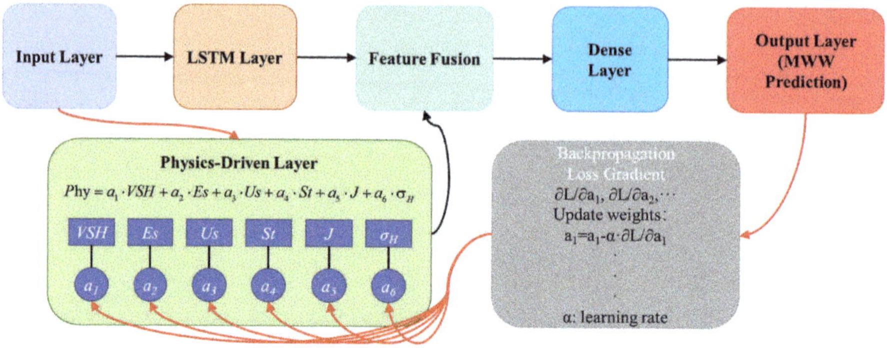

Figure 8. Schematic diagram of the physics-driven layer model architecture.

Specifically, a physics-driven layer is introduced into the model, utilizing key intermediate physical quantities from rock mechanics theory as the feature foundation to construct unified physical characterization variables. Based on correlation analysis results, six physical parameters are selected: VSH, Es, Us, St, J, and σ_H. Physical feature characterization is constructed through the following weighted relationship:

$$Phy = a_1 \cdot VSH + a_2 \cdot \ Es + a_3 \cdot Us + a_4 \cdot St + a_5 \cdot J + a_6 \cdot \sigma_H \tag{22}$$

where a_1 through a_6 are trainable model parameters that are automatically optimized through backpropagation during the training process, preserving the computational logic of physical models while endowing the model with adaptive adjustment capabilities under different geological conditions. This physics-driven layer is implemented through custom neural network layers in TensorFlow, utilizing tensor operations to dynamically complete weighted summation calculations of various physical parameters, ensuring rigorous integration of physical processes with data-driven learning.

The drilling engineering process exhibits typical depth sequence characteristics: ① drilling operations are strictly conducted following well depth progression, with formation parameters displaying continuity and dependency in vertical profiles; ② formation sedimentation and evolution processes possess temporal cumulative effects, where deep structures are significantly influenced by shallow structures in their formation mechanisms. Therefore, this study adopts long short-term memory (LSTM) neural network modeling technology in the overall architecture to effectively capture long-range dependency information in well depth sequences through recurrent neural units [16,35]. The overall model input data is constructed using a fixed-length sliding window approach, where each training sample contains feature information from the previous 10 depth intervals (look-back = 10), enabling the model to fully utilize comprehensive information from historical well intervals for dynamic correction during current depth prediction, thereby enhancing prediction coherence and rationality.

In model design, the LSTM network first encodes time series data to extract dynamic evolution features [36,37]; subsequently, the physics-driven layer calculates physical feature representations based on static physical parameters at the current depth. These two components are concatenated through a feature fusion layer (concatenate) and then input into fully connected hidden layers (dense) to further learn deep-level combinatorial features, ultimately outputting target mud weight window values through a Tanh activation function [38]. To prevent model overfitting, dropout mechanisms are incorporated during the training process and combined with L1/L2 regularization to balance model complexity. The Adam optimization algorithm is employed to enhance training convergence speed and stability [39,40]. Upon completion of model training, the optimal weights of physical parameters within the physics-driven layer are simultaneously obtained, which can be further utilized for interpretative analysis of the weight of the influence of rock mechanical parameters on the mud weight window. The final model's performance on both the test dataset and actual wellbore section predictions demonstrates that this physics-driven long short-term memory neural network can effectively integrate temporal information and the physical mechanisms of drilling formations, thereby significantly improving prediction accuracy and engineering application value.

3.3. Hyperparameter Tuning and Evaluation Metrics

Hyperparameters are critical configuration parameters that are preset before model training, and their combinations directly influence model performance; therefore, hyperparameter optimization constitutes a necessary step for achieving optimal prediction. This study employed L1/L2 regularization [41,42] and dropout techniques to constrain model complexity and constructed a search space encompassing the number of LSTM layer units, number of dense layer units, dropout rate, regularization strength, activation function, number of epochs, and batch size. Using the Randomized Search CV module from scikit-learn, with negative mean squared error as the objective function, 50 random sampling optimizations were performed based on the dataset excluding well C10_match. Through analysis of the hyperparameter space plot (Figure 9), the optimal parameter combination was determined, with both the validation and training sets achieving satisfactory convergence without underfitting or overfitting phenomena.

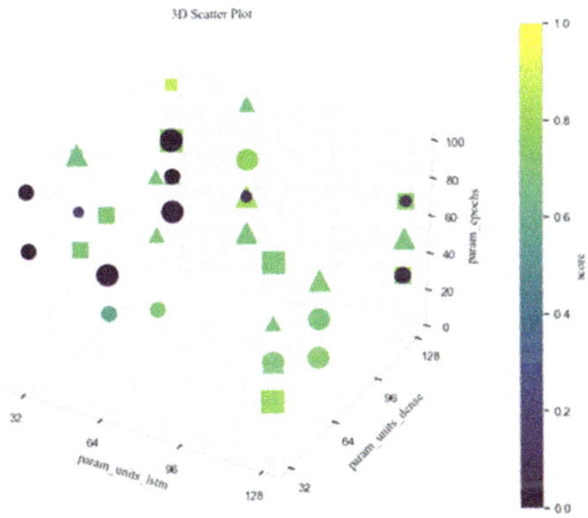

Figure 9. Part hyperparameters of the LSTM. (In this visualization, units_lstm, units_dense, and epochs are represented by the three axes of the graph, activation functions are denoted by symbol shapes (triangles represent specific activation types), batch size is indicated by marker size, and model performance scores are reflected through color variations).

After establishing the optimal model, it is essential to quantify model performance and evaluate the discrepancies between predicted and true values, which constitutes the most direct approach for comparing the merits of various methods. This study employed four evaluation metrics: mean absolute error (MAE), root mean squared error (RMSE), mean absolute percentage error (MAPE), and coefficient of determination (R^2) [43].

$$MAE = \frac{1}{n}\sum_{k=1}^{n}\left|y_{tk} - y_{pk}\right| \tag{23}$$

$$RMSE = \sqrt{\frac{1}{n}\sum_{k=1}^{n}\left(y_{tk} - y_{pk}\right)^2} \tag{24}$$

$$MAPE = \frac{1}{n}\sum_{k=1}^{n}\left|\frac{y_{tk} - y_{pk}}{y_{tk}}\right| \quad (25)$$

$$R^2 = 1 - \frac{\sum_{k=1}^{n}(y_{pk} - y_{tk})^2}{\sum_{k=1}^{n}(y_{tk} - \overline{y})^2} \quad (26)$$

where y_{tk} represents the true value of the sample at the k point; y_{pk} denotes the predicted value of the sample at the k point; $\overline{y}$ is the mean of the true sample values; and n represents the total number of samples.

4. Results and Discussion

4.1. Results

To validate the effectiveness of the proposed method, this study compared the prediction results of the proposed physics-guided prediction approach with those of traditional methods. In the traditional approach, the Bowers method was employed to calculate formation Pp consistent with the unloading curve, and Pm and Pf were computed using conventional rock mechanics formulas. In the physics-guided prediction method, all historical wells except well C10 were utilized, incorporating depth, DT, GR, ZDEN, Rd, Rs, and their regressed Pp, Pm, and Pf values for model learning and training. The hyperparameters of the theoretical layer for the proposed physics-guided LSTM neural network model, established through the random parameter search described in Section 3.3, are presented in Table 1. Through the hyperparameter configuration of the theoretical layer approach, regarding the dense layer unit configuration, Pp prediction employed the largest dense layer with 128 units, while Pm and Pf utilized 64 and 32 units, respectively, indicating that formation pore pressure requires more complex nonlinear mapping relationships. In terms of regularization coefficients, pore pressure adopted a smaller L1 regularization coefficient of 0.001, whereas collapse pressure and fracture pressure employed a larger L1 regularization coefficient of 0.1, while all tasks utilized large L2 regularization coefficients to maintain consistency in weight constraints.

Table 1. Hyperparameter configuration of the physics-guided LSTM neural network model.

Methods \ Hyperparameter		Units LSTM	Units Dense	Dropout Rate	L1_reg	L2_reg	Activation	Batch Size	Epochs
Physics-driven layer machine learning method	Pp	32	128	0.4	0.001	0.001	'relu'	128	70
	Pm	32	64	0.4	0.1	0.001	'relu'	128	50
	Pf	64	32	0.2	0.1	0.001	'relu'	32	70

The mud weight window prediction results from the three methods and the post-drilling regression results of well C10 are presented in Figure 10. It can be clearly observed that the predicted results from all three methods are generally consistent with the numerical magnitude and trends of well C10. The red box in the figure highlights a section where distinct trend variations can be observed. Taking pore pressure as an example, within this wellbore interval, Pp initially exhibits a slight decreasing trend, followed by relative stability, then a slight increasing trend, which is immediately followed by a decreasing trend, an increasing trend, another decreasing trend, and subsequently a rapid increasing trend. However, the depth window where the Pp_Original method (Pp_OM, the following text uses abbreviations) exhibits this trend pattern occurs from 2840 m to 2900 m, while the trend windows for both the Pp_ GSM&Original method (Pp_GSM&OM, the following text uses abbreviations) and Pp_GSM&Machine learning (Pp_GSM&ML, the following text uses abbreviations) range from 2799 m to 2860 m, which is consistent with the trend window found by the post-drilling regression analysis of well C10. It is noteworthy that this trend is consistent with the results observed in DT, ZDEN, GR, and Rs presented in Figure 6. The underlying reason for this is that the original method neglects inter-well geological variations when predicting mud weight windows. After GSM processing, the geological conditions of the target well can be estimated with maximum precision in situations where subsurface data cannot be directly obtained. This depth alignment capability, through accurate capture of local dynamic variation characteristics of mud weight windows in complex formations, enables meter-scale refined customization of mud weight window schemes under complex formation conditions, thereby providing more robust technical support for pre-drilling engineering design.

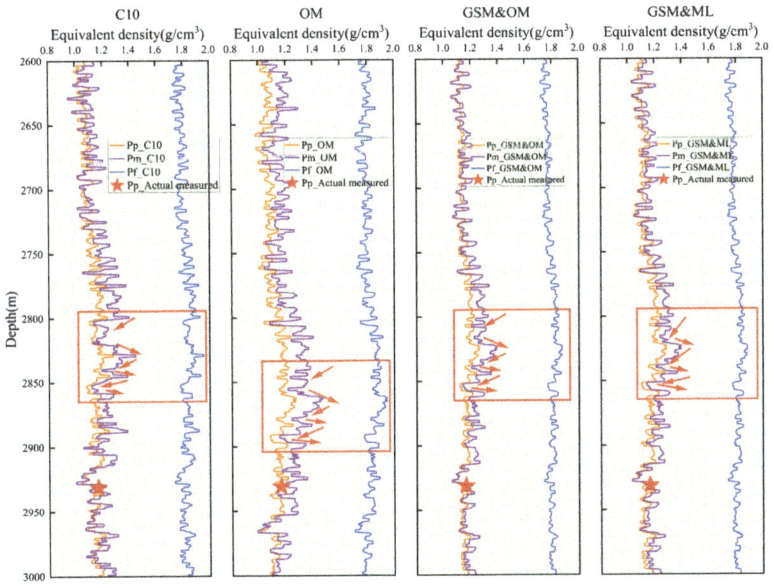

Figure 10. Prediction results from different methods and post-drilling regression results of well C10.

Figure 11 presents the box plots of error and absolute error distributions for Pp, Pm, and Pf predicted by the three methods. From the perspective of error distribution trends, the methods utilizing GSM processing as geological parameters demonstrate that the errors of Pp, Pm, and Pf are all distributed within ±10%. Regarding collapse pressure errors, the absolute error mean of Pm_OM is 0.083 g/cm^3, while the error means of Pm_GSM&OM and Pm_GSM&ML are as low as 0.047 and 0.049 g/cm^3, respectively. The error means of Pm_GSM&OM and Pm_GSM&ML represent a decrease of 43.37% and 40.96% compared to the original method, respectively, demonstrating the effectiveness of the proposed approach in this study. Regarding pore pressure errors, the error mean of Pp_OM is 0.046 g/cm^3, while the error means of both Pp_GSM&OM and Pp_GSM&ML are as low as 0.032 g/cm^3. The error means of Pp_GSM&OM and Pp_GSM&ML show a decrease of 30.43% compared to the original method. In terms of fracture pressure errors, the error mean of Pf_OM is 0.041 g/cm^3, while the error means of Pf_GSM&OM and Pf_GSM&ML are as low as 0.026 and 0.025 g/cm^3, respectively. The error means of Pf_GSM&OM and Pf_GSM&ML are decreased by 36.59% and 39.02% compared to the original method, respectively. These results demonstrate the effectiveness of the approach proposed in this study.

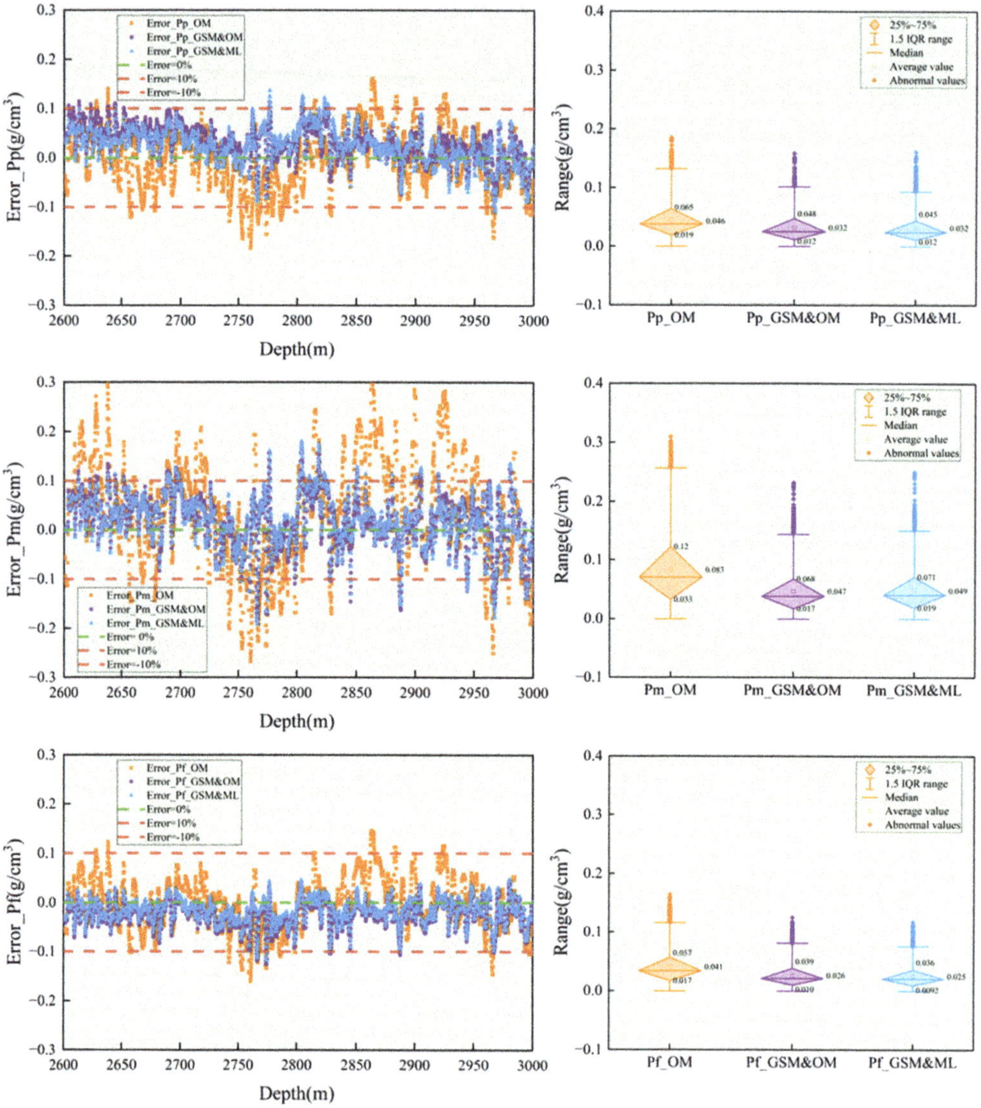

Figure 11. Error and absolute error plots of Pp, Pm, and Pf predicted by three methods.

Evaluation metrics were calculated for the predicted values obtained from the three methods and the actual values from post-drilling regression of well C10, yielding Table 2, which presents the performance evaluation of the three methods. According to the calculation results, all three methods can predict mud weight windows to a certain extent. Following the evaluation criteria, the method with the highest metric scores is highlighted (bold and underlined in the table). The proposed theoretical layer model demonstrates the best performance across all metrics. This is consistent with the results from direct observation in Figure 10 and error analysis in Figure 11. It is worth noting that during performance evaluation of the traditional method, an anomalous situation occurred where R^2 values were negative. This indicates that the traditional method exhibits extremely poor fitting performance when using logging data from well C906 to predict the mud weight window of well C10. This indirectly reflects that the proposed method effectively improves the fine prediction of mud weight windows.

Table 2. Performance evaluation of three methods.

	Pp_OM	Pp_GSM&OM	Pp_GSM&ML	Pm_OM	Pm_GSM&OM	Pm_GSM&ML	Pf_OM	Pf_GSM&OM	Pf_GSM&ML
MAE	0.048	0.038	**0.035**	0.087	0.046	**0.045**	0.043	0.029	**0.025**
RMSE	0.060	0.046	**0.043**	0.11	0.057	**0.057**	0.053	0.036	**0.032**
MAPE	4.30	3.51	**3.14**	7.51	3.93	**3.862**	2.366	1.585	**1.361**
R^2	−0.21	0.29	**0.39**	−0.47	0.60	**0.60**	−0.614	0.247	**0.414**

4.2. Limitations

Although the physics-guided deep learning method for mud weight window prediction proposed in this study demonstrates significant advantages in case validation, several limitations and areas for improvement remain in the following aspects.

The predictive accuracy of the proposed method is heavily dependent on the quality and quantity of historical data. In practical engineering applications, obtaining large amounts of continuous formation pressure ground truth data is challenging due to the high costs associated with direct measurement techniques such as Repeat Formation Tester (RFT) and Modular Dynamic Tester (MDT). The "true values" values utilized in this study are actually estimates derived from post-drilling logging data and drilling records through traditional rock mechanics regression calibration methods, rather than absolute true formation conditions. This limitation in data acquisition methodology may adversely affect model training effectiveness and restrict the method's applicability in data-scarce regions.

Meanwhile, the discussion of data quality dependence in this paper remains primarily at the qualitative descriptive level. Future work should consider establishing a relationship model between data quality and prediction accuracy. By designing datasets with different noise levels (e.g., 5%, 10%, 15%, 20% random errors), the trend of prediction error variation with data quality degradation can be systematically analyzed, and acceptable data quality thresholds can be provided. Furthermore, minimum dataset requirements need to be quantified by progressively reducing the number of training samples. Additionally, a data integrity assessment framework should be established to analyze the differential impacts of missing logging parameters on various pressure predictions (pore pressure, collapse pressure, and fracture pressure), thus providing guidance for data collection strategies in practical engineering applications.

Although this study embeds traditional rock mechanics theory into the deep learning framework through physics-driven layers, inherent contradictions between the two approaches persist. Physical models, based on simplified assumptions and empirical coefficients, may introduce systematic errors under complex geological conditions, while traditional machine learning models typically function as "black boxes" with poor interpretability and lack of intuitive understanding [44]. Occasionally, machine learning models may generate parameter combinations that violate physical laws during optimization processes. How to fully leverage the nonlinear mapping capabilities of machine learning while ensuring physical reasonableness remains a theoretical issue requiring in-depth exploration.

Despite the introduction of physics-driven layers, the "black box" characteristics of deep learning models persist, particularly in the temporal processing mechanisms of LSTM networks, where it is difficult to intuitively understand how the model weighs geological information from different depth intervals. Furthermore, the model's generalization capability across different geological environments and structural backgrounds requires further verification, especially for new blocks with geological characteristics significantly different from the training data.

The GSM method assumes that identical formations between adjacent wells and target wells possess similar geological characteristics. However, in areas with complex structures and strong lateral heterogeneity, this assumption may not hold. Geological factors such as fault development, lithofacies changes, and sedimentary environment variations may lead to significant deviations in formation matching results, thereby affecting final prediction accuracy.

To address the aforementioned limitations, future research should focus on improvements in the following aspects: (1) integrating multi-source information including seismic data and geological modeling to construct a more comprehensive geological constraint system; (2) developing physics-constrained regularization techniques to ensure the physical reasonableness of model predictions; (3) establishing adaptive model selection mechanisms for different geological types; and (4) conducting model interpretability analysis. For instance, for the interpretation of parameters' importance, the SHAP method can be employed to quantify the contribution of each input parameter to prediction results; for engineering decision support interpretation, prediction uncertainty can be transformed into risk assessment indicators to provide refined safety threshold recommendations for mud density design, thereby enhancing the credibility and engineering applicability of prediction results.

5. Conclusions

This study addresses the problem of insufficient accuracy in traditional mud weight window prediction methods under complex geological conditions and proposes a physics-guided deep learning prediction method for mud weight windows. This method integrates physical models and deep learning models, enabling fine prediction of mud weight windows prior to drilling operations, thereby providing effective support for reducing downhole complications and facilitating more detailed design of wellbore structure, drilling fluids, cementing operations, and field emergency response plans. The effectiveness of the proposed method was validated through a practical case study in Block C of an oilfield in Northwest China.

① An LSTM neural network architecture embedded with a physics-driven layer was constructed, achieving deep integration of traditional rock mechanics theory with deep learning algorithms. Through the comprehensive characterization of intermediate variables including VSH, Es, Us, St, J, and σ_H, this method maintains the interpretability of physical models while fully exploiting the advantages of machine learning in handling complex nonlinear relationships.

② GSM processing was validated to achieve precise alignment of geological feature trends in the depth dimension, establishing a foundation for meter-scale fine prediction prior to drilling. Taking pore pressure in the case study as an example, the specific trend variation window predicted by the traditional method was located in the 2840 m–2900 m depth interval, while the predicted specific trend window after GSM processing was accurately aligned to 2799 m–2860 m, which is completely consistent with the actual geological conditions of the target well. This depth alignment capability enables accurate capture of local variation characteristics of mud weight windows in complex formations, achieving meter-scale refined customization of mud weight window schemes for complex formations and providing more reliable technical support for pre-drilling engineering design.

③ Case validation results demonstrate that the proposed method exhibits significant advantages in predicting pore pressure, collapse pressure, and fracture pressure. Compared to traditional methods, collapse pressure prediction error was reduced by 40.96%, pore pressure prediction error was reduced by 30.43%, and fracture pressure prediction error was reduced by 39.02%. All evaluation metrics outperformed both traditional methods and the purely GSM-based approach.

This study enables fine prediction of mud weight windows prior to drilling operations, providing effective technical support for wellbore structure design, drilling fluid scheme optimization, cementing scheme formulation, and field emergency plan preparation. It demonstrates significant engineering application value for reducing downhole complications and improving drilling efficiency. Effective validation in drilling engineering indicates that this processing approach may possess similar potential in other geological engineering problems, including reservoir parameter prediction, geological hazard assessment, and groundwater resource evaluation.

Author Contributions: Y.C.: Investigation, conceptualization, methodology, writing-original draft, writing—review and editing, supervision. T.S.: Conceptualization, writing—review and editing, validation, supervision. J.Y.: Conceptualization, writing—review and editing, supervision. X.C.: Conceptualization, writing—review and editing, project administration. L.R.: Writing—review and editing, project administration. Z.W.: Investigation, project administration. S.J.: Investigation, project administration. W.W.: Investigation, project administration. S.W.: Investigation, project administration. M.Z.: Investigation, project administration. All authors have read and agreed to the published version of the manuscript.

Funding: This work was financially supported by the National Natural Science Foundation of China General Program (No. 52274018), the National Natural Science Foundation of China (No. 52204017), and the National Key Research and Development Program (No. 2022YFC28061004).

Data Availability Statement: The data that has been used is confidential.

Conflicts of Interest: Author Xianjun Chen was employed by the China France Bohai Geoservices Company Limited. The remaining authors declare that the research was conducted in the absence of any commercial or financial relationships that could be construed as a potential conflict of interest. The China France Bohai Geoservices Company Limited had no role in the design of the study; in the collection, analyses, or interpretation of data; in the writing of the manuscript, or in the decision to publish the results.

Abbreviations

The following abbreviations are used in this manuscript:

MWW	Mud Weight Window
NCT	Normal Compaction Trend
ML	Machine Learning
DL	Deep Learning
XGBoost	eXtreme Gradient Boosting
LSTM	Long Short-Term Memory
BP	Back Propagation Neural Network
SVM	Support Vector Machine
IGWO-MLP	Improved Grey Wolf Optimization–Multilayer Perceptron
LightGBM	Light Gradient Boosting Machine
CGP-NN	Convolution Pyramid Pooling–Gate Recurrent Unit–Neural Network
1DCPP	One-dimensional Convolution Pyramid Pooling
GRU	Gate Recurrent Unit
LWD	Logging While Drilling
SHAP	SHapley Additive exPlanations
GSM	Geological Sequence Matching

DT	Sonic Transit Time
Rd	Deep Investigate Double Lateral Resistivity Log
Rs	Shallow Investigate Double Lateral Resistivity Log
GR	Natural Gamma Ray
ZDEN	Bulk Density
VSH	Shale Content
VP	P-wave Velocity
Ed	Dynamic Elastic Modulus
Es	Static Elastic Modulus
Ud	Dynamic Poisson's Ratio
Us	Static Poisson's Ratio
Sc	Uniaxial Compressive Strength
St	Tensile Strength
C	Cohesion
J	Internal Friction Angle
σ_h	Minimum Principal Stress
σ_H	Maximum Principal Stress
Pp	Pore Pressure
Pm	Collapse Pressure
Pf	Fracture Pressure
PD-LSTM	Physics-Driven Long Short-Term Memory Neural Network
MAE	Mean Absolute Error
MAPE	Mean Absolute Percentage Error
RMSE	Root Mean Squared Error
R^2	Coefficient Of Determination
OM	Original Method
GSM&OM	GSM and Original Method
GSM&ML	GSM and Machine Learning

References

1. Ayoub, D.; Masoud Anthony, W.D.; Hamzeh, G. Wellbore stability analysis to determine the safe mud weight window for sandstone layers. *Pet. Explor. Dev.* **2019**, *46*, 1031–1038. [CrossRef]
2. Saeed, B.; Meysam, R.; Shadfar, D.; David, A.W.; Hamzeh, G.; Nima, M.; Ahmadi, A.M.; Shahab, S.B. Robust computational approach to determine the safe mud weight window using well-log data from a large gas reservoir. *Mar. Pet. Geol.* **2022**, *142*, 105772. [CrossRef]
3. Crichton, M.; Lauche, K.; Flin, R. Incident Command Skills in the Management of an Oil Industry Drilling Incident: A Case Study. *J. Contingencies Crisis Manag.* **2005**, *13*, 116–128. [CrossRef]
4. Georgy, P.; Kerim, K.; Sergey, S.; Nikita, B.; Ammar, A.; Mahmoud, A. Predicting mud weight in carbonate formations using seismic data: A data-driven approach. *Geoenergy Sci. Eng.* **2025**, *250*, 213850. [CrossRef]
5. Foster, J.B. Estimation of Formation Pressures From Electrical Surveys-Offshore Louisiana. *J. Pet. Technol.* **1966**, *18*, 165–171. [CrossRef]
6. Pennebaker, E.S. An Engineering Interpretation of Seismic Data. In Proceedings of the Fall Meeting of the Society of Petroleum Engineers of AIME, Houston, TX, USA, 29 September 1968.
7. Eaton, B.A. The Effect of Overburden Stress on Geopressure Prediction from Well Logs. *J. Pet. Technol.* **1972**, *24*, 929–934. [CrossRef]
8. Terzaghi, K. Principles of soil mechanics: IV Settlement and consolidation of clay. *Eng. News-Rec.* **1925**, *95*, 874–878.
9. Bowers, G.L. Pore Pressure Estimation From Velocity Data: Accounting for Overpressure Mechanisms Besides Undercompaction. *SPE Drill. Complet.* **1995**, *10*, 89–95. [CrossRef]
10. Zhang, S.; Wang, H.; QIU, Z.; CAO, W.; Huang, H.; Chen, Z. Calculation of safe drilling mud density window for shale formation by considering chemo-poro-mechanical coupling effect. *Pet. Explor. Dev.* **2019**, *46*, 1271–1280. [CrossRef]
11. Kadkhodaie, A. The impact of geomechanical units (GMUs) classification on reducing the uncertainty of wellbore stability analysis and safe mud window design. *J. Nat. Gas Sci. Eng.* **2021**, *91*, 103964. [CrossRef]
12. Liu, X.; Wang, Z.; Xiong, M.; Wang, F.; Chen, G.; Zhang, J.; Wang, X.; Sun, B. Thermo-mechanical coupling modeling and heat transfer analysis of mud weight window for deepwater deep drilling. *Int. J. Heat Mass Transf.* **2025**, *249*, 127261. [CrossRef]

3. Zhang, L.; Yan, X.; Yang, X.; Tian, Z.; Yang, H. An elastoplastic model of collapse pressure for deep coal seam drilling based on Hoek–Brown criterion related to drilling fluid loss to reservoir. *J. Pet. Sci. Eng.* **2015**, *134*, 205–213. [CrossRef]
4. Song, X.; Yao, X.; Li, G.; Xiao, L.; Zhu, Z. A novel method to calculate formation pressure based on the LSTM-BP neural network. *Pet. Sci. Bull.* **2022**, *7*, 12–23. [CrossRef]
5. Deng, S.; Pan, H.; Wang, H.; Xu, S.; Yan, X.; Li, C.; Peng, M.; Peng, H.; Shi, L.; Cui, M.; et al. A hybrid machine learning optimization algorithm for multivariable pore pressure prediction. *Pet. Sci.* **2024**, *21*, 535–550. [CrossRef]
. Li, H.; Cao, Z.; Wu, X. Prediction method of formation fracture pressure based on LightGBM algorithm and its application. *China Meas. Test* **2024**, *50*, 134–143. Available online: https://link.cnki.net/urlid/51.1714.TB.20230111.1033.001 (accessed on 11 January 2023).
7. Zhang, X.; Lu, Y.; Jin, Y.; Chen, M.; Zhou, B. An adaptive physics-informed deep learning method for pore pressure prediction using seismic data. *Pet. Sci.* **2024**, *21*, 885–902. [CrossRef]
3. Chen, X.; Weng, C.; Tao, L.; Yang, J.; Gao, D.; Li, J. A novel method for predicting formation pore pressure ahead of the drill bit by embedding petrophysical theory into machine learning based on seismic and logging-while-drilling data. *Pet. Sci.* **2025**. [CrossRef]
9. Hussain, W.; Luo, M.; Ali, M.; Sadiq, I.; Kasala, E.E.; Aziz, T.; Batool, Z. Hybrid modeling of deep neural networks and unsupervised machine learning algorithms for missing well log prediction based on geological lithofacies similarities. *J. Appl. Geophys.* **2025**, *241*, 105846. [CrossRef]
0. Doyoro, Y.G.; Gelena, S.K.; Lin, C.-P. Improving subsurface structural interpretation in complex geological settings through geophysical imaging and machine learning. *Eng. Geol.* **2025**, *344*, 107839. [CrossRef]
1. Wang, R.; Ziegler, M.; Volpi, M.; Manconi, A. Advanced identification of geological discontinuities with deep learning. *Appl. Comput. Geosci.* **2025**, *27*, 100256. [CrossRef]
2. Wang, B.; Zhou, F.; Shi, Z.; Zhu, J.; Ma, J.; Li, Z.; Huan, Z.; Guo, L. The Paleogene Sedimentary System and Evolution Characteristics of Lenghu Structural Belt in Northern Qaidam Basin. *Geol. Resour.* **2019**, *28*, 543–552. [CrossRef]
3. Li, J.; Tang, Y.; Wu, T. Overpressure origin and its effects on petroleum accumulation in the conglomerate oil province in Mahu Sag, Junggar Basin, NW China. *Pet. Explor. Dev.* **2020**, *47*, 679–690. Available online: https://link.cnki.net/urlid/11.2360.TE.2020 0609.1041.004 (accessed on 9 July 2020). [CrossRef]
4. Wang, Q.; Chen, D.; Gao, X.; Li, M.; Shi, X.; Wang, F.; Chang, S.; Yao, D.; Li, S.; Chen, S. Overpressure origins and evolution in deep-buried strata: A case study of the Jurassic Formation, central Junggar Basin, western China. *Pet. Sci.* **2023**, *20*, 1429–1445. [CrossRef]
5. Fan, H.H. *Abnormal Formation Pressure Analysis Method and Application*; Science Press: Beijing, China, 2016; pp. 60–62, 169, 286, 287.
5. Fjaer, E.; Horsrud, P.; Raaen, A.M.; Holt, R.M.; Risnes, R. Petroleum Related Rock Mechanics. In *Developments in Petroleum Science*; Elsevier Science: Amsterdam, The Netherlands, 1992; Volume 33, Chapter 11.
7. Nick, B. *Rock Quality, Seismic Velocity, Attenuation and Anisotropy*; Earth Sciences, Engineering & Technology; CRC Press: London, UK, 2009; Volume 15. [CrossRef]
3. Jaeger, J.; Cook, N.G.; Zimmerman, R. *Fundamental of Rock Mechanics*; Wiley-Blackwell: Hoboken, NJ, USA, 2007. [CrossRef]
9. Weng, C.; Li, J.; Yang, H.; Long, Z.; Zhang, G.; Wang, B.; Zhao, Y. A hybrid model of formation pore pressure prediction based on geological sequence matching. *Geoenergy Sci. Eng.* **2025**, *252*, 213972. [CrossRef]
0. Singh, D.; Singh, B. Investigating the impact of data normalization on classification performance. *Appl. Soft Comput.* **2020**, *97*, 105524. [CrossRef]
1. Sadeghi, B. Chatterjee Correlation Coefficient: A robust alternative for classic correlation methods in geochemical studies-(including "TripleCpy" Python package). *Ore Geol. Rev.* **2022**, *146*, 104954. [CrossRef]
2. Zar, J. Spearman Rank Correlation. In *Encycl Biostat*; Wiley: Hoboken, NJ, USA, 2005; Volume 5. [CrossRef]
3. Bahareh, R.M.; Abolfazl, D.M.; Ali, R. Enhanced petrophysical evaluation through machine learning and well logging data in an Iranian oil field. *Sci. Rep.* **2024**, *14*, 28941. [CrossRef]
4. Tariq, Z.; Saleh, A.M.; Amjed, H.; Mobeen, M.; Emad, M.; El-Husseiny, A.; Alarifi, S.A.; Mohamed, M.; Abdulazeez, A. A systematic review of data science and machine learning applications to the oil and gas industry. *J. Pet. Explor. Prod. Technol.* **2021**, *11*, 4339–4374. [CrossRef]
5. Ju, G.; Yan, T.; Sun, X.; JingYu, Q.; Hu, Q. Evolution of gas kick and overflow in wellbore and formation pressure inversion method under the condition of failure in well shut-in during a blowout. *Pet. Sci.* **2022**, *19*, 678–687. [CrossRef]
. Wei, G.; Yang, X.; Li, M.; Gao, S.; Wan, X.; Ji, C.; Gao, X.; Tan, C. Drilling and Completion Condition Recognition Algorithm Based on CNN-GNN-LSTM Neural Networks and Applications. *Processes* **2025**, *13*, 1090. [CrossRef]
7. Xu, Y.; Liu, K.; He, B.; Pinyaeva, T.; Li, B.; Wang, Y.; Nie, J.; Yang, L.; Li, F. Risk pre-assessment method for regional drilling engineering based on deep learning and multi-source data. *Pet. Sci.* **2023**, *20*, 3654–3672. [CrossRef]
3. Jia, C.; Li, T.; Dong, H.; Xie, C.; Peng, W.; Ning, Y. A leading adaptive activation function for deep reinforcement learning. *J. Comput. Sci.* **2025**, *88*, 102608. [CrossRef]

39. Kingma, D.; Ba, J. Adam: A Method for Stochastic Optimization. In Proceedings of the International Conference on Learnin Representations, Banff, AB, Canada, 14–16 April 2014.
40. Yan, B.; Xu, Z.; Gudala, M.; Tariq, Z.; Sun, S.; Finkbeiner, T. Physics-informed machine learning for reservoir management o enhanced geothermal systems. *Geoenergy Sci. Eng.* **2024**, *234*, 212663. [CrossRef]
41. Cortes, C.; Mohri, M.; Rostamizadeh, A. L2 regularization for learning kernels. In Proceedings of the Twenty-Fifth Conference o Uncertainty in Artificial Intelligence, Montreal, QC, Canada, 18–21 June 2009.
42. Yang, Y.; Tan, C.; Cheng, Y.; Luo, X.; Qiu, X. Using a Deep Neural Network with Small Datasets to Predict the Initial Production o Tight Oil Horizontal Wells. *Electronics* **2023**, *12*, 4570. [CrossRef]
43. Ma, T.; Zhang, D.; Yang, Y.; Chen, Y. Machine learning model based collapse pressure prediction method for inclined wells. *Nat Gas Ind.* **2023**, *43*, 119–131. [CrossRef]
44. Pothana, P.; Ling, K. Physics-integrated neural networks for improved mineral volumes and porosity estimation from geophysica well logs. *Energy Geosci.* **2025**, *6*, 100410. [CrossRef]

processes

MDPI

rticle

A Hybrid Framework for Production Prediction in High-Water-Cut Oil Wells: Decomposition-Feature Enhancement-Integration

hendong Li [1,*], Qihao Qian [1], Huazhan Guo [2,3], Tong Wu [1], Haidong Cui [2,3] and Bingqian Zhu [1]

1 Research Institute of Petroleum Exploration and Development, PetroChina, Beijing 100083, China
2 Qinghai Oilfield Company, PetroChina, Dunhuang 736202, China
3 Qinghai Provincial Key Laboratory of Plateau Saline-Lacustrine Basinal Oil & Gas Geology, Dunhuang 736202, China
* Correspondence: pearzd@163.com; Tel.: +86-15184697320

Abstract: The forecasting of high-water-cut oil well production faces challenges of strong nonlinearity and nonstationarity due to reservoir heterogeneity and multiscale dynamic characteristics. This study proposes a hybrid CEEMDAN-SR-BiLSTM framework based on a "decomposition-feature enhancement-integration" architecture. The framework employs Complete Ensemble Empirical Mode Decomposition with Adaptive Noise (CEEMDAN) to suppress mode mixing, reconstructs high-, medium-, and low-frequency subsequences using Hilbert-Huang Transform (HHT) combined with tercile thresholding, and finally achieves multiscale feature fusion prediction through a Bayesian-optimized bidirectional long short-term memory network (BiLSTM). Interpretability analysis based on SHapley Additive exPlanations (SHAP) values reveals the contribution degrees of parameters such as water injection volume and flowing pressure to different frequency components, establishing a mapping between production data features and physical mechanisms of oil well production. This mapping, integrated with physical mechanisms including wellbore transient flow, injection-production response lag, and reservoir pressure evolution, enables mechanistic interpretation of production phenomena and quantitative decoupling and prediction of multiscale dynamics. Experimental results show that the framework achieves a root-mean-square error (RMSE) of 3.75 in forecasting a high-water-cut well (water cut = 87.6%) in the Qaidam Basin, reducing errors by 26.0% and 50.0% compared to CEEMDAN-BiLSTM and BiLSTM models, respectively, with a coefficient of determination (R^2) reaching 0.954.

Keywords: oil production prediction; high-water-cut reservoirs; ensemble learning; CEEMDAN-SR-BiLSTM; physical annotation; SHAP

1. Introduction

Against the backdrop of continuously growing global energy demand, the status of oil as a core energy source remains irreplaceable [1]. However, as the world's major oilfields generally enter the high water-cut development stage, the accuracy of well production forecasting has become a key factor restricting the potential tapping of remaining oil and development efficiency [2]. The dynamic evolution of well production is essentially the result of multi-physical field coupling, mainly influenced by factors such as wellbore flow conditions, injection-production pressure conduction, and reservoir energy evolution. Specifically, in wellbore transient flow, flow pattern transitions of multiphase fluids (oil, gas, water) (such as the alternation between slug flow and bubbly flow) cause drastic fluctuations in bottom-hole flowing pressure, directly affecting the efficiency of oil and gas lifting [3]. During injection-production processes, fluid seepage delay and pressure diffusion effects lead to time-lag responses in the near-wellbore area, weakening the uniformity of the displacement front [4,5]. Non-equilibrium pressure evolution in reservoirs, through the reconstruction of seepage-displacement coupling relationships, exacerbates the preferential development of channeling pathways and reduces waterflood sweep efficiency [6,7]. These unique cross-scale dynamic coupling effects in high water-cut reservoirs further intensify the difficulty of production forecasting.

Current mainstream forecasting methods have significant limitations. Physical-driven models based on Darcy's law (such as reservoir numerical simulation) can accurately characterize seepage mechanisms, but their high complexity with billions of computational grids makes them difficult to meet real-time application requirements [8]. Empirical models represented by Arps decline curves and waterflood characteristic curves rely on historical data fitting from specific production stages and cannot adapt to production abrupt changes caused by reservoir heterogeneity variations or adjustments in development strategies [9–11].

In recent years, deep learning models represented by long short-term memory networks (LSTM) and their variant bidirectional long short-term memory networks (BiLSTM) have significantly improved the prediction efficiency of complex time-series data through a data-driven paradigm [12–14]. However, the strong nonlinearity and nonstationarity of production sequences in high water-cut reservoirs pose challenges to the multi-scale feature capture capability of single models. Against this backdrop, empirical mode decomposition (EMD) and its improved algorithm—complete ensemble empirical mode decomposition with adaptive noise (CEEMDAN)—have demonstrated unique advantages [15]. Through adaptive time-frequency decomposition, these methods decompose the original signal into intrinsic mode functions (IMFs) with characteristic scales, effectively separating high-frequency, medium-frequency, and low-frequency signal components, and providing a decomposition framework for multi-scale mechanism analysis of nonstationary signals [16]. Researchers have integrated them with deep learning in fields such as weather [17,18], hydrology [19], and energy [20,21], significantly enhancing the modeling capability of complex systems, and they have also shown satisfactory results in the field of well production forecasting [22]. However, existing studies mostly analyze the influence of single factors in isolation, lacking a connection between data-driven feature extraction and the physical mechanisms of well production—i.e., decomposing nonstationary signals into components corresponding to different physical mechanisms (wellbore dynamics, near-wellbore dynamics, and far-reservoir dynamic conduction)—while enabling quantitative analysis of feature contributions at different scales.

This study proposes a CEEMDAN-SR-BiLSTM framework based on a "decomposition-feature enhancement-integration" architecture. Specifically, CEEMDAN is used to decompose production signals into frequency components. Multi-scale features are enhanced through Hilbert-Huang transform and quantile-based reconstruction, classifying intrinsic mode functions (IMFs) into high-frequency, medium-frequency, and low-frequency components. A Bayesian-optimized bidirectional LSTM (BiLSTM) algorithm is employed for ensemble prediction, which independently models each frequency component to leverage bidirectional temporal dependencies while avoiding noise interference in raw data. Additionally, SHapley additive explanations are used to quantify feature contributions, revealing that the decomposed high-, medium-, and low-frequency curves correspond to the physical characteristics of wellbore transient flow, injection-production response lag, and reservoir pressure trends, respectively.

2. Materials and Methods

2.1. Research Data and Experimental Materials

The research data in this study are derived from 30 oil wells in the high water-cut development stage (comprehensive water cut > 80%) in Mangya, Qaidam Basin, China, encompassing full-life-cycle production data from January 1997 to December 2024. The target reservoirs are medium-to-high permeability sandstone reservoirs with an average porosity of 18.5% and permeability of 65 × 10^{-3} μm², exhibiting strong heterogeneity and multi-scale dynamic coupling.

Data Composition and Variable Definitions

Sample Size: A total of 30 wells were included, yielding 5040 monthly records in aggregate. The target well (validation well) has a water cut of 87.6%, representing the high-liquid production and low-oil production stage in the late development period. Specific statistical details are presented in Table 1.

Table 1. Statistical Values of Variables Used in This Study.

	Maximum Value	Minimum Value	Average Value	Standard Deviation	Coefficient of Variation	Kurtosis	Skewness
Monthly Oil Production (t/month)	137.0	7.2	63.4	19.1	0.3	0.1	0.1
Flowing pressure (MPa)	16.1	7.6	11.8	2.1	0.2	−1.2	0.0
Water injection volume (m^3/month)	284.2	50.0	139.8	47.7	0.3	−0.6	0.5
Fluid Level (m)	889.4	314.4	605.2	127.7	0.2	−1.0	0.0
Stroke (times/min)	3.0	3.0	3.0	0.0	0.0	0.0	0.0
Pump Strokes (times/min)	6.0	6.0	6.0	0.0	0.0	0.0	0.0
Casing Pressure (MPa)	4.0	1.2	2.6	0.5	0.2	−0.5	0.0
Porosity (%)	25.6	19.5	22.9	1.6	0.1	−0.7	0.2
Permeability (10^{-3} μm^2)	300.0	80.0	192.4	70.8	0.4	−1.1	0.1
Operating time (years)	30.0	10.0	24.2	3.2	0.1	0.0	−0.5
Gas-Oil Ratio (GOR, m^3/t)	107.1	65.1	85.3	7.1	0.1	−0.6	0.1

Feature Variables: (1) Production dynamic parameters: surrounding water injection volume (m^3/month), producing gas-oil ratio (m^3/t), flowing pressure (MPa), dynamic fluid level (m), casing pressure (MPa), stroke length, stroke frequency; (2) Wellbore process parameters: pump diameter (mm), pump depth (m), production time (years); Reservoir physical property parameters: porosity (%), permeability (10^{-3} μm^2), effective thickness (m). (4) Label Variable: Monthly oil production (t/month), serving as the target for model prediction.

2.2. Data Preprocessing and Feature Selection

During the data preprocessing and feature selection phase, data normalization is first performed: the selected features undergo min-max normalization using the formula:

$$X' = \frac{X - X_{min}}{X_{max} - X_{min}} \tag{1}$$

X': Normalized data value. X: Original feature data point. X_{min}: Historical minimum value of the feature. X_{max}: Historical maximum value of the feature.

Next, the processed data are subjected to Pearson correlation coefficient and mutual information criterion to identify dominant factors governing oil production, followed by feature selection on the preprocessed samples. The combined Pearson correlation coefficient and mutual information criterion are implemented through the following mathematical expressions: For two random variables X and Y, Pearson Correlation Coefficient:

$$r_{XY} = \frac{\sum_{i=1}^{n} (x_i - \overline{x})(y_i - \overline{y})}{\sqrt{\sum_{i=1}^{n} (x_i - \overline{x})^2 \sum_{i=1}^{n} (y_i - \overline{y})^2}} \tag{2}$$

n represents the sample size, x_i and y_i are the $i - th$ observations of variables X and Y, $\overline{x}$ and $\overline{y}$ denote the sample means of X and Y. The Pearson correlation coefficient r_{XY} ranges between $[-1, 1]$:$r_{XY} = 1$ indicates perfect positive correlation; $r_{XY} = -1$ indicates perfect negative correlation; $r_{XY} = 0$ indicates no linear correlation

Mutual Information Value:

$$I(X, Y) = \sum_{x \in X} \sum_{y \in Y} p(x, y) \log(\frac{p(x, y)}{p(x)p(y)}) \tag{3}$$

where $I(X, Y)$ quantifies the dependency between random variables X (feature) and Y (oil production). Here: $p(x, y)$ represents the joint probability distribution of X and Y; $p(x)$ and $p(y)$ denote the marginal probability distributions of X and Y, respectively.

2.3. *Framework Design of CEEMDAN-SR-BiLSTM*

2.3.1. Complete Ensemble Empirical Mode Decomposition with Adaptive Noise (CEEMDAN)

Complete Ensemble Empirical Mode Decomposition with Adaptive Noise (CEEMDAN) is an advanced method derived from the Empirical Mode Decomposition (EMD) algorithm. EMD, proposed by Huang [23], effectively analyzes non-stationary signals by decomposing them into Intrinsic Mode Functions (IMFs) that reflect local characteristics. However, EMD suffers from noise sensitivity and mode mixing arti-facts. To mitigate these issues, Ensemble EMD (EEMD) was developed by Wu et al. [24], which reduces noise influence through ensemble averaging. Nevertheless, EEMD introduces new challenges such as computational complexity and over-smoothed IMFs. To address these limitations, Complete Ensemble Empirical Mode Decomposition with Adaptive Noise (CEEMDAN) [25] was introduced, which overcomes mode mixing while maintaining decomposition effectiveness by adding and progressively adjusting noise levels to the signal. This method decomposes oil well production time-series data into multiple IMF components and a residual component through the following computational steps:

First, the original oil well production time series is combined with a Gaussian white noise sequence, introducing controlled randomness to reveal subtle variations hidden within the raw data. This hybrid sequence is then analyzed using Empirical Mode Decomposition (EMD) to extract its first-order Intrinsic Mode Function (IMF) component. As the primary decomposed layer, the IMF typically encapsulates the most prominent or fundamental dynamic characteristics of the sequence, enabling deeper insights into production variability.

$$\overline{imf}_1(t) = \frac{\sum_{i=1}^{\mathrm{n}} imf_1^i(t)}{\mathrm{n}} \tag{4}$$

$im\overline{f}_1(t)$ represents the average component of the first-order IMF component, imf_1^i denotes the $i-th$ IMF component obtained after the first decomposition, and indicates the maximum number of times white noise is added.

Calculating the first—order decomposition residue:

$$x_1(t) = P(t) - im\overline{f}_1(t) \tag{5}$$

Introduce the Gaussian white noise sequence into the first—order residue $x_1(t)$ to generate a new decomposition sequence $x_1'(t)$. Subsequently, this new sequence undergoes EMD decomposition to obtain the second—order IMF component:

This process is iteratively repeated—mixing the oil well production time series with Gaussian white noise, performing EMD decomposition, and extracting IMFs—until the remaining residual sequence exhibits a monotonic trend. At this stage, the oil well production time series is decomposed and represented as the sum of a series of IMFs and a monotonic residual, enabling detailed analysis of production composition and variability. The decomposition formula is:

$$\mathrm{P(t)} = \sum_{\mathrm{j=1}}^{\mathrm{m}} im\overline{f}_j(t) + x_{\mathrm{k}}(\mathrm{t}) \tag{6}$$

where $x_{\mathrm{k}}(\mathrm{t})$ represents the final residue.

2.3.2. Signal Reconstruction Algorithm (SR)

The Signal Reconstruction (SR) algorithm is a crucial component of the CEEMDAN-SR-BiLSTM framework for oil well production prediction. After the CEEMDAN decomposes the oil well production time-series into multiple Intrinsic Mode Function (IMF) components and a residual component, the SR algorithm plays a key role in processing and analyzing these decomposed components.

The main steps are as follows: First, the signal undergoes CEEMDAN decomposition to obtain IMF components. Then, the Hilbert-Huang Transform (HHT) is employed to calculate the instantaneous frequency of each IMF component. Subsequently, the IMF components are divided into high-frequency, medium-frequency, and low-frequency groups using the three-quantile method. Finally, the signal is reconstructed.

The Hilbert-Huang Transform (HHT) [26] is a mathematical tool for converting a real-valued signal into an analytic signal. It multiplies the negative-frequency part of the Fourier transform of a real-valued signal by—1 to yield a complex signal. The real part of this complex signal is the original real-valued signal, while the imaginary part is the Hilbert transform of the real-valued signal. First, the Hilbert-Huang Transform (HHT) is applied to the Intrinsic Mode Functions (IMFs) and residue obtained from CEEMDAN decomposition to calculate the instantaneous frequency of each IMF. The mathematical definition is:

$$\begin{aligned} f_i(t) &= \frac{1}{2\pi}\frac{d\phi_i(t)}{dt} \\ \phi_i(t) &= \arctan\left(\frac{H[IMF_i(t)]}{IMF_i(t)}\right) \\ \overline{f_i} &= \frac{1}{T}\int_T^0 f_i(t)dt \end{aligned} \tag{7}$$

where $f_i(t)$ is the instantaneous frequency of the $i-th$ IMF component; $\phi_i(t)$ is the phase function of this IMF, $H[IMF_i(t)]$ is the Hilbert transform result of $IMF_i(t)$, and $IMF_i(t)$ is the $i-th$ intrinsic mode function component. Here, $\overline{f_i}$ is the average frequency of the $i-th$ IMF, and T is the signal duration.

Next, three-quantile frequency band division is performed by computing the three-quantiles Q_{33} and Q_{66} of the average instantaneous frequencies $\left\{\overline{f}_i\right\}_{i=1}^{n}$ of all IMF components. IMFs are classified into high -, medium -, and low-frequency groups according to

$$\begin{aligned} X_{high} &= 0, X_{mid} = 0, X_{low} = 0 \\ X_{high} &= \sum_{\overline{f_i} > Q_{66}} IMF_i \\ X_{med} &= \sum_{Q_{33} < \overline{f_i} \le Q_{66}} IMF_i \\ X_{low} &= \sum_{\overline{f_i} \le Q_{33}} IMF_i + Residue \\ Q_{33} &= quantile(\overline{f}_i, 0.33) \\ Q_{66} &= quantile(\overline{f}_i, 0.66) \end{aligned} \quad (8)$$

where Q_{33} represents the 0.33 quantile of the average frequency sequence, and Q_{66} represents the 0.66 quantile, both of which are used to divide the high, medium, and low frequency bands. X_{high}, X_{med}, X_{low} are the reconstructed signals in the high, medium, and low frequency bands, and Residue is the residual term.

Finally, reconstruction performance validation is conducted using the Normalized Mean Squared Error (NMSE):

$$\begin{aligned} NMSE &= \frac{|X_{rec} - X_{orig}|^2}{|X_{orig}|^2} \\ X_{rec} &= X_{high} + X_{mid} + X_{low} \end{aligned} \quad (9)$$

where NMSE stands for the Normalized Mean Square Error, X_{rec} is the reconstructed signal, and X_{orig} is the original signal.

2.3.3. Bayesian Optimization of Bidirectional LSTM Network

In recent years, deep neural networks represented by the Bidirectional Long Short-Term Memory (BiLSTM) network have achieved significant progress in multiple research fields. BiLSTM is an improved architecture based on the Long Short-Term Memory (LSTM) network [27] (as shown in Figure 1), characterized by its integration of forward and backward LSTM layers to form a bidirectional processing mechanism. In this architecture: The forward LSTM layer processes input sequences from left to right, capturing temporal dependencies in the forward direction. The backward LSTM layer processes sequences from right to left, uncovering backward dependencies in the data. This bidirectional processing mechanism enables BiLSTM to utilize both past and future information for prediction. Compared to traditional unidirectional LSTM, it comprehensively understands sequential data by capturing bidirectional contextual information, thereby significantly enhancing prediction performance (as shown in Figure 2).

The detailed computational process is as follows:

$$\begin{aligned} f_t &= \sigma(W_f \cdot [h_{t-1}, x_t] + b_f) \\ i_t &= \sigma(W_i \cdot [h_{t-1}, x_t] + b_i) \\ O_t &= \sigma(W_O[h_{t-1}, x_t] + b_O) \end{aligned} \quad (10)$$

x_t represents the input at time step t; h_{t-1} denotes the hidden state at time step $t-1$, serving as the output of the LSTM unit at the previous time step; C_{t-1} indicates the cell state at time step $t-1$, which is the state used for long-term memory within the LSTM unit. σ stands for the sigmoid function; W_f, W_i, W_O are the weight matrices of the forget gate, input gate, and output gate respectively; b_f, b_i, b_O are the bias vectors of the forget gate, input gate, and output gate respectively.

$$\begin{aligned}\overrightarrow{C_t} &= LSTM(x_t, \overrightarrow{h_{t-1}}, \overrightarrow{C_{t-1}}) \\ \overleftarrow{C_t} &= LSTM(x_t, \overleftarrow{h_{t-1}}, \overleftarrow{C_{t-1}}) \\ C_t &= W^T \overleftarrow{C_t} + W^V \overleftarrow{C_t}\end{aligned} \tag{11}$$

$\overrightarrow{C_t}$ and $\overleftarrow{C_t}$ represent the cell states of the forward LSTM and backward LSTM at time step t. W^T and W^V denote the weight coefficients of the forward LSTM and backward LSTM.

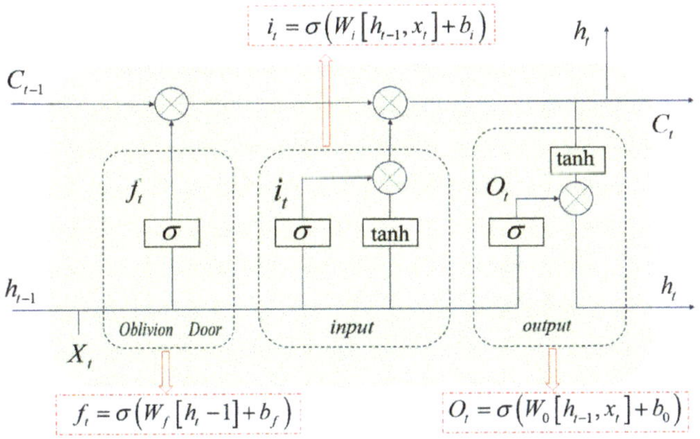

Figure 1. Long Short-Term Memory Neural Network Architecture Diagram.

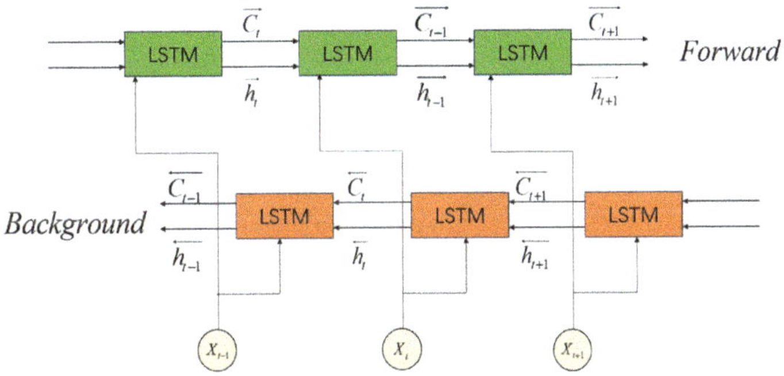

Figure 2. Bidirectional Long Short-Term Memory Neural Network Architecture Diagram.

To further enhance the prediction accuracy of the BiLSTM model, this study em-ploys Bayesian optimization to fine-tune the network's hyperparameters. Bayesian optimization is an efficient global optimization algorithm that adaptively selects optimal hyperparameter combinations in each iteration based on surrogate model approximations of the actual objective function [28]. By leveraging a Gaussian process as the surrogate model and pairing it with the Expected Improvement (EI) acquisition function, this method identifies near—optimal hyperparameters within fewer iterations, making it particularly suitable for optimizing black—box objective functions with high evaluation costs. Consequently, Bayesian optimization has been widely ap-plied in hyperparameter tuning for machine learning and deep learning, significantly improving model performance (as shown in Figure 3).

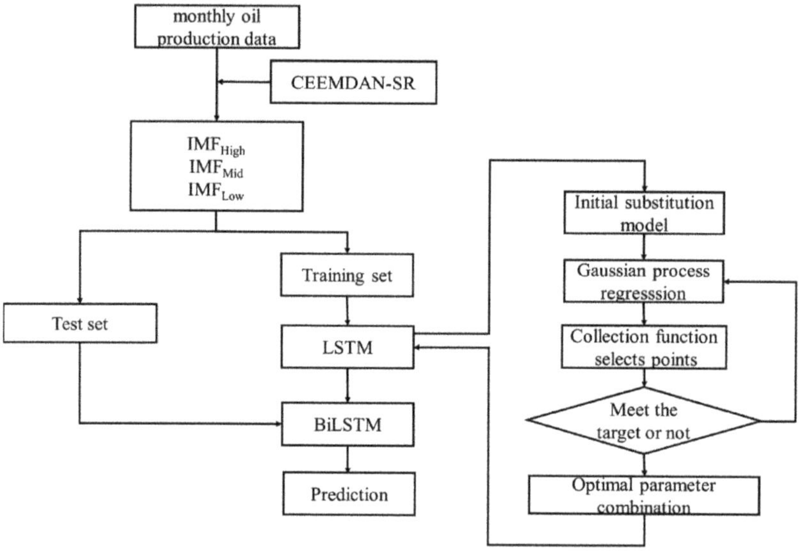

Figure 3. Bayesian Optimization Algorithm LSTM Architecture Diagram.

2.3.4. CEEMDAN-SR-BiLSTM

This section systematically describes the model application framework: First, data are preprocessed using the Pearson correlation coefficient-mutual information joint screening method to extract key features related to monthly oil production and eliminate redundant variables. Second, CEEMDAN is employed for multi-scale decomposition of production sequences. Combined with the Hilbert-Huang transform and tercile quantile threshold method, IMF components are reconstructed into high-frequency, medium-frequency, and low-frequency subsequences. Subsequently, a Bayesian-optimized BiLSTM model is used to model each subsequence, generating prediction results through linear combination. SHapley values are applied to quantify feature contributions, establishing associations between data features and reservoir physical mechanisms. Finally, comparative models are introduced to validate effectiveness. In the model architecture, CEEMDAN enables multi-scale signal decomposition, SR completes the physical meaning reconstruction of features, BiLSTM performs sequence prediction, forming an integrated production prediction model with mechanism analysis capabilities. SHapley values are used to quantify feature contributions and establish links between data features and reservoir physical mechanisms (as shown in Figure 4).

2.4. *Model Evaluation Metrics*

The criteria for evaluating model performance are the Mean Absolute Error (MAE), Root Mean Squared Error (RMSE), and the coefficient of determination (R^2), mathematically expressed as follows:

$$\begin{aligned} RMSE &= \sqrt{\frac{1}{n}\sum_{i=1}^{n}(Q_i - Q_i^*)^2} \\ MAE &= \frac{1}{n}\sum_{i=1}^{n}\left|Q_i - Q_i^*\right| \\ R^2 &= 1 - \frac{\sum_{i=1}^{n}(Q_i - Q_i^*)^2}{\sum_{i=1}^{n}(Q_i - \overline{Q})^2} \end{aligned} \quad (12)$$

Here, n represents the number of time series, $\overline{Q}$ is the average value of oil well production data, Q_i is the true oil well production data value, and Q_i^* is the predicted oil well production data value.

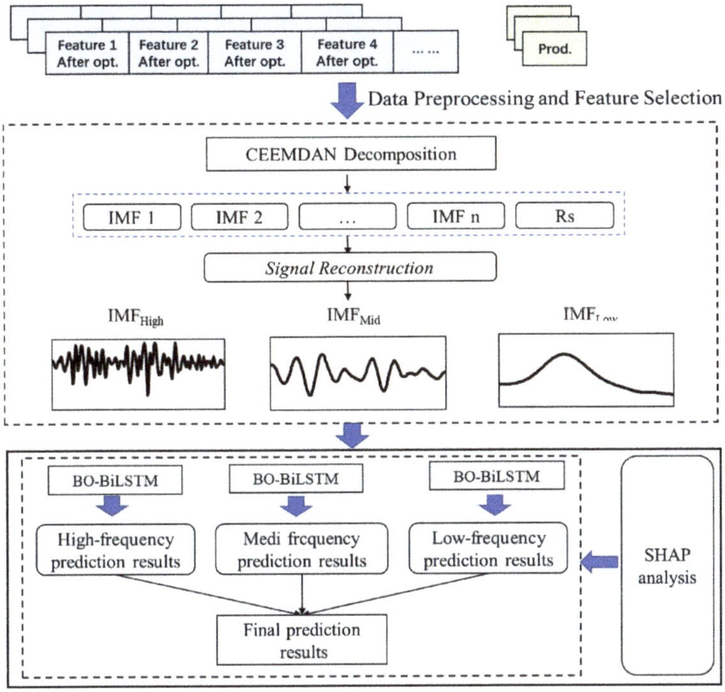

Figure 4. Overall Framework Diagram of CEEMDAN-SR-BiLSTM Model.

3. Results

This study focuses on a medium-to-high permeability sandstone reservoir in western China, where favorable hydrocarbon storage conditions exist but the field has entered a high water-cut development stage. Although crude oil production has remained relatively stable, both liquid production and average water cut have been increasing annually, presenting challenges for single-well production forecasting as development performance deteriorates. To address this issue, production data from 30 oil wells in the study area spanning 1997–2024 were collected. The dataset includes monthly records of 12 dimensions: surrounding water injection volume, producing gas-oil ratio, flowing pressure, pump diameter, pump depth, production time, porosity, permeability, casing pressure, dynamic fluid level, stroke count, and monthly oil production (with the latter serving as the labeled feature).

3.1. Data Preprocessing and Feature Selection

The comprehensive use of all indicators would obscure the validity of the data and significantly increase the computational load; therefore, irrelevant features were removed through screening to retain key parameters [29,30]. In this study, a feature selection method combining Pearson correlation coefficient analysis and p-value testing ($\alpha = 0.05$) was employed [31]. Firstly, parameters with absolute Pearson correlation coefficients ($|R|$) greater than 0.35 were selected. The mutual information criterion (MI > 0.15) was introduced to cross-validate nonlinear associations. The final key features determined via joint screening

($p \leq 0.05$, $|R| > 0.35$, MI > 0.15) included water injection volume, gas-oil ratio, flowing pressure, operation time, and dynamic fluid level, reducing redundancy by 58.3%. Figure 5 presents the full correlation matrix from the Pearson correlation analysis, illustrating the pairwise correlation coefficients be-tween all feature parameters and monthly oil production, which helps identify significant linear relationships among variables. Table 2 lists the mutual information values for oil production, quantifying the nonlinear associations between each feature and the target variable to assess their relevance in a non-parametric manner. Together, these two analyses form the basis for the feature screening process, enabling the selection of key indicators that significantly impact oil production while reducing dimensionality and computational complexity in subsequent modeling steps.

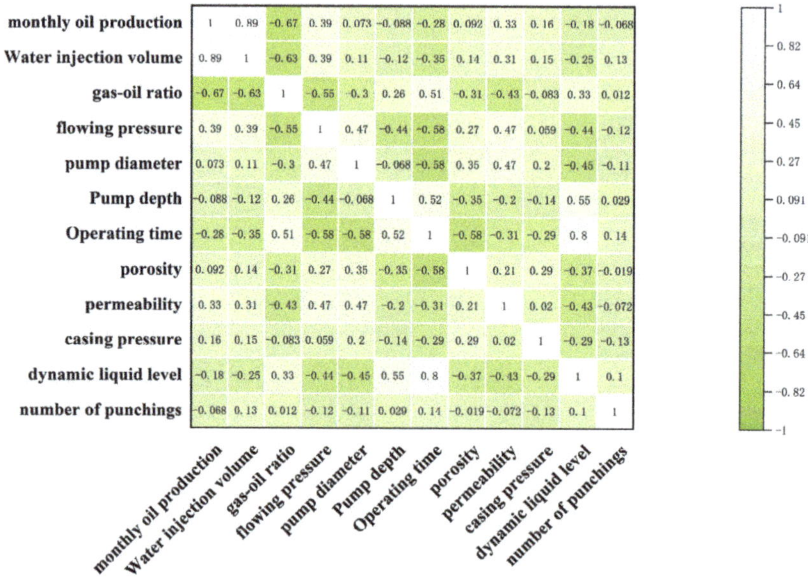

Figure 5. Pearson Correlation Analysis.

Table 2. Mutual Information Values for Oil Production.

Feature	Mutual Information Value
Water injection volume	0.26
Gas-oil ratio	0.24
Flowing pressure	0.18
Pump diameter	0.122
Pump depth	0.033
Operating time	0.19
Porosity	0.189
Permeability	0.15
Casing pressure	0.042
Dynamic liquid level	0.175
Number of punchings	0.06

Note: Features with mutual information (MI) > 0.15 bits were selected for modeling, consistent with the nonlinear correlation screening criteria.

3.2. Multiscale Decomposition of Single-Well Oil Production

This study selected monthly oil production data from 1997–2024 for a high water-cut oil well (current water cut = 87.6%) in the target area. The original production curve exhibits a trend of initial fluctuating increase followed by fluctuating decline, with significant early-stage volatility and upward momentum, and reduced late-stage volatility leading to gradual decline. This reflects the well's production lifecycle: production increase phase $\rightarrow$ stable phase $\rightarrow$ decline phase. The production data demonstrate strong nonlinearity and non-stationarity, characterized by significant temporal complexity. The training set included the first 90% of the dataset for model training, while the remaining 10% served as the test set for prediction evaluation. The time series plot of monthly oil production for this well is shown in Figure 6.

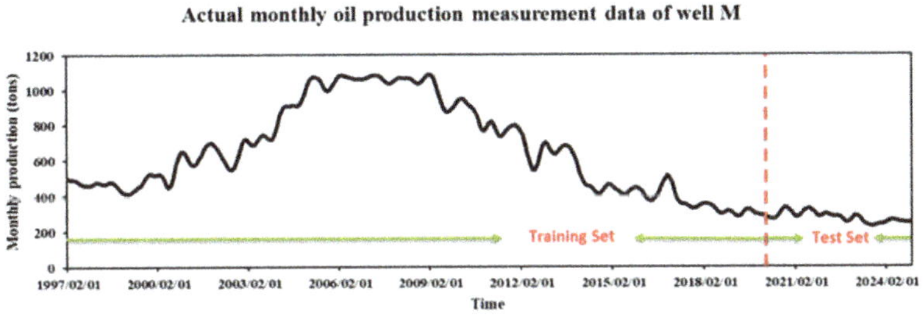

Figure 6. Monthly Measured Production Data of Well M (1997–2024).

The CEEMDAN method effectively separates fluctuation features at different time scales from the original oil production signal. As shown in Figure 5, the data stream is decomposed into 4 Intrinsic Mode Functions (IMFs) and a residual component (IMF 5). The initial IMF components exhibit the highest volatility and frequency with the shortest wavelength, while subsequent components gradually weaken. The residual IMF 5 represents the long-term trend of the signal, showing an overall declining pattern. Default CEEMDAN parameters include: Gaussian white noise amplitude of 0.2, ensemble number of 500, and maximum iterations of 5000. The decomposition results are visualized in Figure 7.

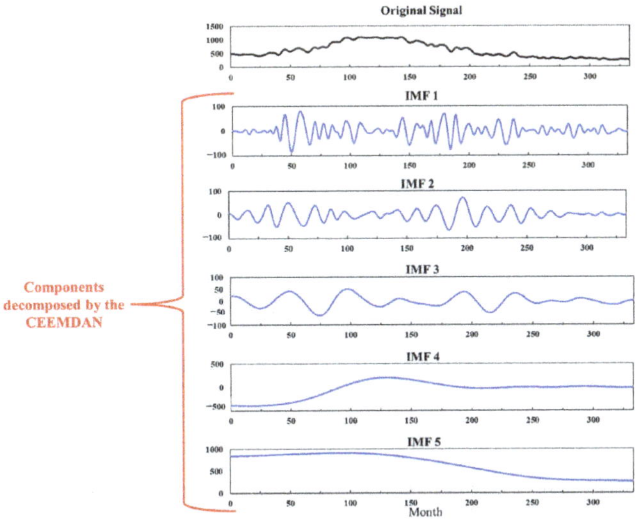

Figure 7. Multiscale Decomposition of Well M's Oil Production via CEEMDAN.

3.3. Signal Reconstruction and Multiscale Dynamic Feature Analysis

To achieve multiscale time-frequency characterization of signals, Hilbert transforms were first performed on CEEMDAN-decomposed IMF components to calculate mean instantaneous frequencies (MIF) for quantifying frequency properties (Table 3). Using 33.33% and 66.66% quantiles of MIF as thresholds, IMFs were classified into three frequency bands: high-frequency (MIF > 66.66%), medium-frequency (33.33% < MIF $\leq$ 66.66%), and low-frequency (MIF $\leq$ 33.33% + residual term). Final band-specific signals were reconstructed via component summation strategy (Figure 8).

Table 3. Mean Frequency Data of IMF Components.

IMF	Mean Frequency (Hz)	Frequency Band
IMF1	0.239658	High-frequency
IMF2	0.131594	High-frequency
IMF3	0.062700	Medium-frequency
IMF4	0.023700	Low-frequency
IMF5	−0.000165	Low-frequency (Residue)

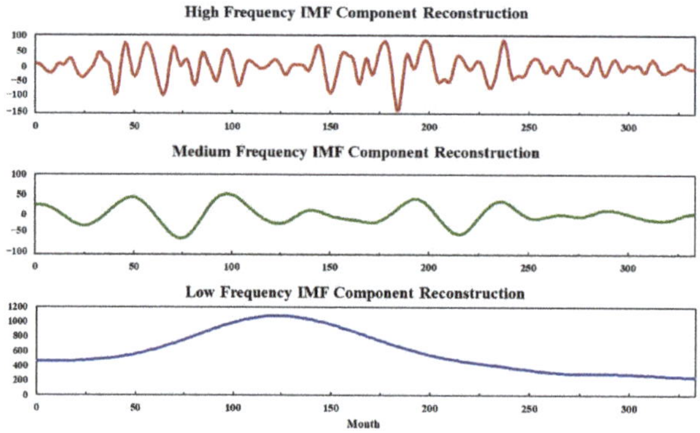

Figure 8. Frequency-Domain Signal Reconstruction Based on Three-Quantile Grouping.

This study establishes a mapping between CEEMDAN-decomposed IMF components and multiscale dynamic features of oil well production via Physical Annotation [20], bridging data characteristics to physical mechanisms.

As shown in Figure 6: The high-frequency component exhibits frequent fluctuations with rapid amplitude changes, reflecting short-time-scale dynamics. The medium-frequency component features relatively smooth oscillations with periodicity longer than the high-frequency component, indicating mid-scale cyclic behavior. The low-frequency component demonstrates a clear trend of initial increase followed by decline, representing long-term evolutionary characteristics. From the perspective of Physical Mechanism of Oil Well Production, this correspondence between frequency and dynamic features is inherently deterministic.

High-frequency component: Corresponds to wellbore transient flow. As the direct channel for hydrocarbon production, the wellbore experiences rapid transient processes like throttling effects and fluid slippage phenomena. For example, sudden changes in fluid velocity and gas-liquid slippage during production generate high-frequency signals, making the high-frequency component the natural carrier of wellbore transient flow dynamics.

Medium-frequency component: Reflects injection-production response lag. In injection-production systems, reservoir seepage resistance and capillary forces act over mid-scale timeframes rather than instantaneously. When fluids are injected or produced, reservoir fluids require time to overcome seepage resistance and flow toward the wellbore, while capillary forces influence two-phase flow dynamics. These mid-scale processes—neither as rapid as wellbore transients nor as slow as reservoir pressure trends—are effectively characterized by the medium-frequency component, capturing both response lags and mid-scale seepage behavior.

Low-frequency component: Represents reservoir pressure trends. Reservoir pressure changes are long-term cumulative effects, including energy depletion (e.g., sustained pressure decline under natural drive) and waterflood recharge. The low-frequency component, combined with the residual term, precisely captures these gradual trends due to its frequency matching the "slow-varying" time scale of reservoir pressure, serving as an effective carrier for extracting long-term reservoir dynamics.

This frequency-physics mapping inherently correlates signal processing results with the time scales and operational characteristics of oil production processes, providing an analytical bridge between "signal frequency" and "physical processes" for multiscale dynamic analysis from the wellbore to near-wellbore zones and distant reservoirs.

3.4. Model Parameter Optimization Strategy

This study constructs a hierarchical parameter optimization framework, implementing differentiated strategies tailored to the characteristics of different frequency band signals (Table 4):

Table 4. Optimal Hyperparameter Combinations for High-, Medium-, and Low-Frequency BiLSTM Models.

Parameter	High-Frequency	Medium-Frequency	Low-Frequency
hidden_size	92	117	82
num_layers	3	1	1
dropout	0.182	0.381	0.0912
learning_rate	0.001	0.001	0.001
hidden_size	92	117	82

Note: Parameters optimized via Bayesian optimization with Gaussian process model for high-frequency components, and manual tuning for medium- and low-frequency components based on signal periodicity and trend characteristics.

3.4.1. High-Frequency Component Optimization (BiLSTM)

Global parameter search via Bayesian optimization algorithm, dynamically balancing exploration and exploitation in the parameter space using a Gaussian process model. Optimization focused on network capacity and training stability yielded the optimal parameters: Hidden layer size: 92; Number of layers: 3; Dropout rate: 0.182; Learning rate: 0.001 (Adam optimizer with adaptive learning rate strategy).

3.4.2. Medium-Frequency Component Optimization (BiLSTM)

Given the significant periodicity of medium-frequency signals, strategies emphasized adjusting network depth and regularization parameters: Hidden layer size: 117; Number of layers: 1; Dropout rate: 0.381; Learning rate: 0.001.

3.4.3. Low-Frequency Component Optimization (BiLSTM)

To capture long-term trends, optimization prioritized overfitting suppression and trend-tracking enhancement: Hidden layer size: 82; Number of layers: 1; Dropout rate: 0.0912; Learning rate: 0.001.

3.5. Experimental Results

In this subsection, we conduct production forecasting using the CEEMDAN-SR-BiLSTM model. Given the relatively limited dataset size, the first 90% of the data was partitioned into the training set to ensure the model receives sufficient training, while the remaining 10% served as the test set. Figure 7 illustrates the training and test loss evolution during high-, medium-, and low-frequency predictions, and Figure 9 presents the corresponding forecasting results.

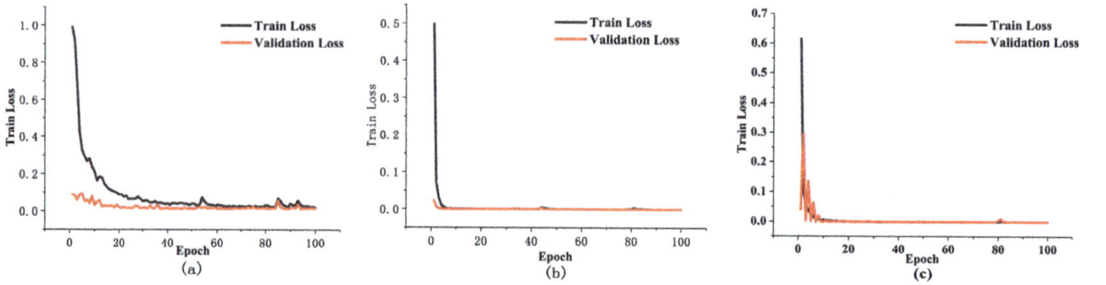

Figure 9. Test loss curves during the training and prediction processes: (**a**) High-frequency components; (**b**) Medium-frequency components; (**c**) Low-frequency components.

Figure 10a–c show differentiated convergence characteristics of training and validation loss curves for high-, medium-, and low-frequency component predictions: High-frequency: Training loss exponentially decreased from 3.2 to 0.8, with validation loss stabilizing in the range of 1.0 ± 0.2. Medium-frequency: Training loss converged to 0.3 within 20 iterations, with validation loss synchronously stabilizing at 0.4. Low-frequency: Training loss dropped by 78% (from 5.6 to 1.2) initially, with validation loss stabilizing at 1.5 after 30 iterations.

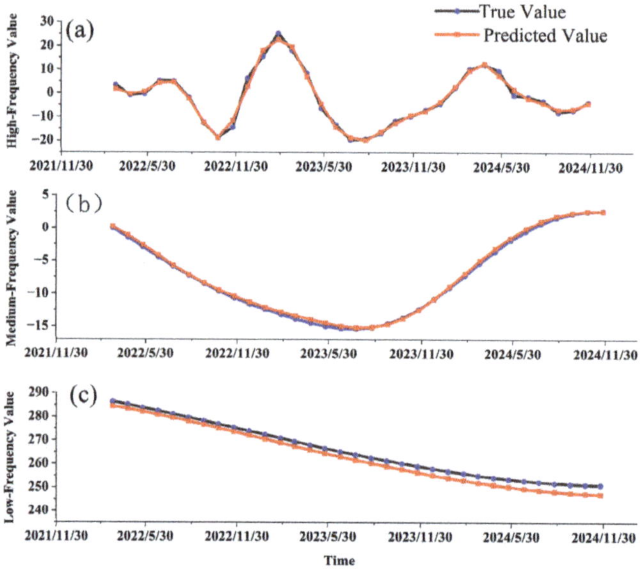

Figure 10. CEEMDAN-SR-BiLSTM Multiscale Production Forecasting: Frequency Band Analysis and Total Output Validation (**a**) High-Frequency Component Prediction vs. Field Measurements; (**b**) Medium-Frequency Component Prediction vs. Field Measurements; (**c**) Low-Frequency Component Prediction vs. Field Measurements.

All three loss curves maintained a favorable state where training loss $\leq$ validation loss and synchronous convergence was achieved, validating the model's effective capture capabilities for high/medium/low-frequency features and stability during training.

The CEEMDAN-SR-BiLSTM model demonstrates robust multiscale forecasting capabilities, accurately capturing dynamic trends across high-, medium-, and low-frequency components while maintaining overall prediction precision (RMSE = 3.75, MAE = 2.80, R^2 = 0.954). High-frequency predictions effectively track transient flow fluctuations (e.g., 2023 pump adjustment peaks) with RMSE = 1.89, though abrupt changes pose residual challenges. Medium-frequency results show improving alignment over time, achieving R^2 = 0.91 in late-stage injection-response cycles after initial bias reduction. Low-frequency trends align closely with reservoir pressure depletion (MAE = 1.23), with minor rate discrepancies highlighting geological heterogeneity impacts. The aggregated production forecast matches field measurements at critical inflection points (e.g., 2022Q1 $\pm$1.5 t/month error) and stable periods (2022-11 to 2023-01 R^2 = 0.973), validating the model's ability to synthesize multiscale physics-driven features for oil production analysis (Table 5).

Table 5. CEEMDAN-SR-BiLSTM Multiscale Production Forecasting Evaluation Metrics.

Component	RMSE (t/Month)	MAE (t/Month)	R^2
High-Frequency	1.3593	1.0608	0.9850
Medium-Frequency	0.2424	0.1949	0.9991
Low-Frequency	2.4520	2.3724	0.9780
Combined Total	3.7498	2.7966	0.9539

3.6. SHAP Values Theory for Production Fluctuation Interpretation and Physical Mechanism Characterization

To enhance the interpretability of the model, this study uses the SHAP tool (SHapley Additive exPlanations) to quantify the contribution of features in the prediction of production components [32]. This method is based on the Shapley value in cooperative game theory [33]. By regarding the model prediction as a 'cooperative game' of feature interactions, it quantifies the average marginal contribution of each feature to the prediction result in all possible feature combinations. Its theoretical properties (such as consistency, local accuracy, and global unbiasedness) ensure the reliability of the contribution assessment, as shown in Figure 11.

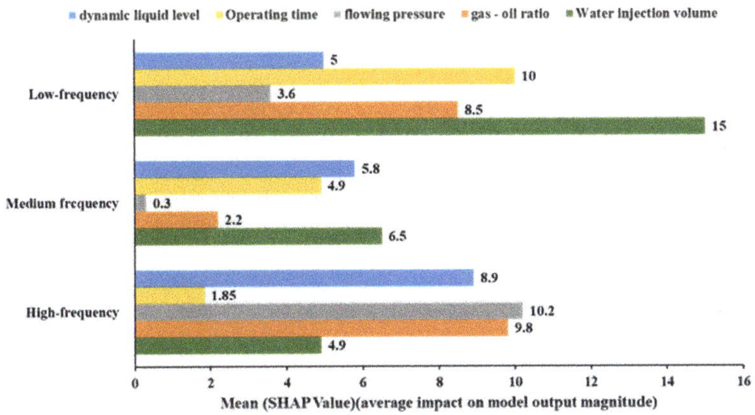

Figure 11. CEEMDAN-SR-BiLSTM Multiscale Feature Importance Ranking via SHAP Values: High-, Medium-, and Low-Frequency Production Components.

Based on SHAP value analysis (Figure 11), the marginal contributions of features to different frequency components exhibit significant discrepancies, revealing intrinsic "feature-frequency-physics" correlations:

In the high-frequency component, the dominant features are flowing pressure (SHAP value = 9.8) and gas-liquid ratio (10.2), accounting for 58% of the total contribution. This phenomenon closely aligns with the dynamic characteristics of wellbore transient flow—the rapid fluctuations in flowing pressure directly reflect the wellbore throttling effect and abrupt changes in fluid velocity [34], while the high-frequency variations in gas-liquid ratio correspond to the gas-liquid slippage phenomenon [35]. The capture of such "short-time-scale transient" features validates the advantage of CEEMDAN de-composition in separating high-frequency noise from effective signals [25]. Compared with the traditional EMD method, CEEMDAN mitigates mode mixing through adaptive noise injection, ensuring a quantitative mapping between high-frequency IMFs and wellbore physical processes.

In the medium-frequency component, the significant contributions of dynamic fluid level (6.5) and water injection volume (4.9) essentially reflect the mesoscale interactions of reservoir seepage resistance and capillary forces. Changes in dynamic fluid level characterize the pressure balance adjustment between the wellbore and reservoir [4], while the influence of water injection volume corresponds to the dynamic modulation of seepage channels by operations such as cyclic water injection [5]. This finding demonstrates that the framework effectively captures the injection-production response lag dynamics ignored by traditional models through medium-frequency component reconstruction.

In the low-frequency component, water injection volume (15) and flowing pressure (10) cumulatively account for 80% of the feature importance. Water injection volume serves as the dominant factor, maintaining reservoir pressure through long-term energy replenishment [6]; flowing pressure directly reflects the bottomhole energy status, influencing fluid inflow and lifting efficiency—low pressure indicates energy depletion [7]. This reveals the physical connotation of the low-frequency component: water injection replenishes pore volume to mitigate pressure depletion, while flowing pressure embodies the balance between formation energy and production pressure difference, establishing a connection between low-frequency components and reservoir pressure trends.

4. Discussion

Comparison and Mechanistic Advantages of Multi-Scale Prediction Models

To validate the accuracy and robustness of the CEEMDAN-SR-BiLSTM hybrid model in forecasting monthly oil production from high-water-cut wells, a systematic comparison was conducted against CEEMDAN-BiLSTM, EMD-BiLSTM, BiLSTM, and LSTM, supported by Figure 12 (forecasting results), Figure 12 (absolute error comparison), and Table 5 (performance metrics). CEEMDAN-SR-BiLSTM demonstrated the highest alignment with field measurements (Figure 12a), precisely tracking production fluctuations—including peak-valley variations—across complex transient intervals, whereas CEEMDAN-BiLSTM (Figure 12b) and EMD-BiLSTM (Figure 12c) showed progressive deviations at critical inflection points due to insufficient multiscale decomposition. BiLSTM (Figure 12d) and LSTM (Figure 12e) exhibited significant systematic errors, with LSTM performing worst (RMSE = 11.25 t/month, R^2 = 0.585), failing to capture trend direction and magnitude. Quantitatively, CEEMDAN-SR-BiLSTM achieved the lowest prediction errors (RMSE = 3.75 t/month, MAE = 2.80 t/month) and highest explanatory power (R^2 = 0.954), accounting for >95% of production variance, while CEEMDAN-BiLSTM (RMSE = 5.12), EMD-BiLSTM (RMSE = 6.89), and BiLSTM (RMSE = 9.41) showed progressive performance degradation, confirming the critical role of SR algorithm integration and CEEMDAN decomposition in improving accuracy. These results highlight the model's superiority in capturing multiscale production dynamics of high-water-cut wells through physics-informed signal decomposition and adaptive regularization, enhancing forecasting reliability for complex reservoir systems.

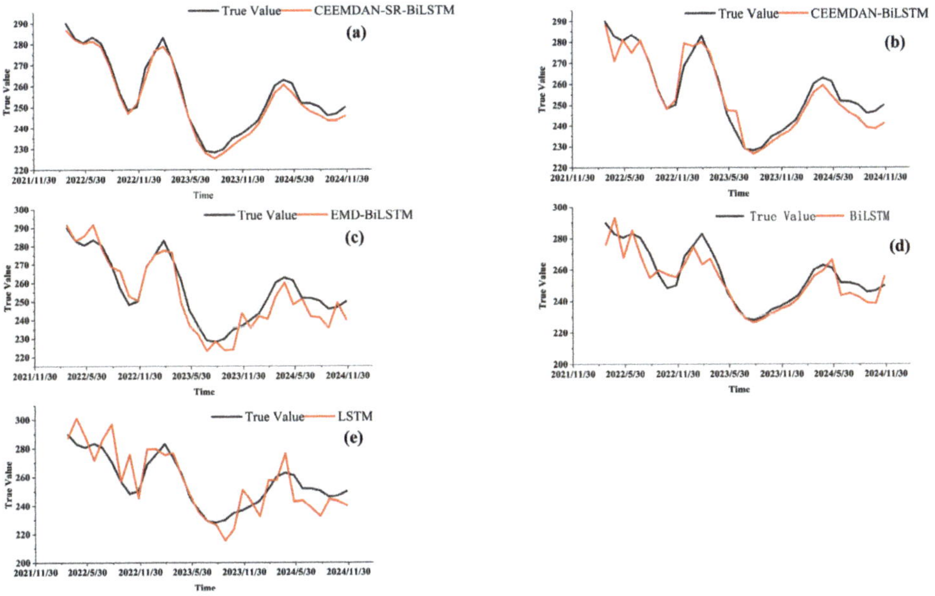

Figure 12. Multiscale Production Forecasting Results Comparison: CEEMDAN-SR-BiLSTM vs. Conventional Models ((**a**) CEEMDAN-SR-BiLSTM, (**b**) CEEMDAN-BiLSTM, (**c**) EMD-BiLSTM, (**d**) BiLSTM, (**e**) LSTM).

Figure 13 and Table 6 present the absolute error comparison and performance metrics analysis of multiscale production forecasting models, respectively. Figure 13 visually shows the prediction accuracy differences via absolute deviations, while Table 6 evaluates model performance and robustness using metrics like MAE, RMSE, and R^2. Together, they provide a quantitative analysis of model effectiveness, facilitating a clear comparison with benchmark methods.

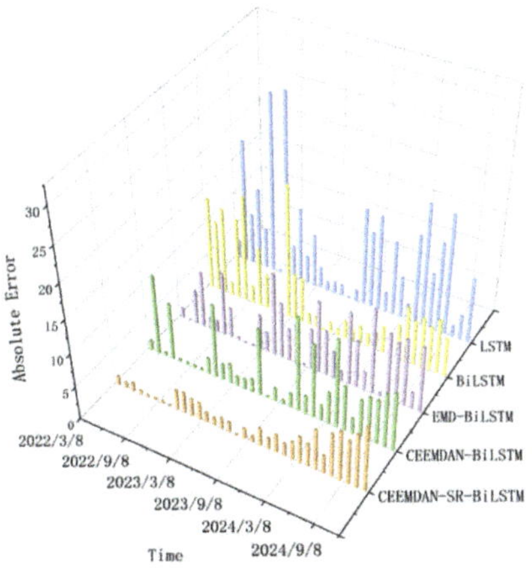

Figure 13. Multiscale Production Forecasting Models: Absolute Error Comparison.

Table 6. Multiscale Production Forecasting Model Comparison: Performance Metrics.

Model	RMSE	MAE	R^2
CEEMDAN-SR-BiLSTM	3.75	2.80	0.954
CEEMDAN-BiLSTM	5.12	4.03	0.921
EMD-BiLSTM	6.89	5.34	0.879
BiLSTM	9.41	7.15	0.723
LSTM	11.25	8.68	0.585

5. Conclusions

Aiming at the challenges of strong nonlinearity, nonstationarity, and multi-scale dynamic analysis in production forecasting for high water-cut oil wells, this study proposes a hybrid CEEMDAN-SR-BiLSTM framework based on a "decomposition-feature enhancement-integration" architecture. Through the deep integration of data-driven approaches and physical mechanisms, a multi-scale analysis system is constructed that combines prediction accuracy with mechanistic interpretation capabilities.

5.1. Scientific Contributions and Methodological Innovations

(1) This study overcomes the "mode mixing" bottleneck of traditional empirical mode decomposition (EMD) by utilizing complete ensemble empirical mode decomposition with adaptive noise (CEEMDAN) combined with Hilbert-Huang transform (HHT) and quantile-based reconstruction, decomposing production time series into three components with clear physical connotations: high-frequency (wellbore transient flow), medium-frequency (injection-production response lag), and low-frequency (reservoir pressure trend). Specifically, the high-frequency component captures short-term throttling effects and gas-liquid slippage phenomena in the wellbore; the medium-frequency component characterizes the delayed response of seepage resistance and capillary forces in the near-wellbore area; and the low-frequency component reflects long-term energy evolution trends driven by water injection. This decomposition method establishes, for the first time, a quantitative mapping between intrinsic mode functions (IMFs) and multi-scale physical processes in reservoir-wellbore systems, transforming purely mathematical signal decomposition into dynamically interpretable indicators with engineering significance and addressing the critical issue of traditional decomposition methods lacking mechanistic explanations.

(2) By quantifying the contributions of engineering parameters to different frequency components using SHapley Additive exPlanations (SHAP), this study reveals the "feature-frequency-physical mechanism" coupling law: high-frequency fluctuations are dominated by wellbore dynamic parameters such as flowing pressure and gas-oil ratio; medium-frequency response lags are related to mesoscale seepage effects of dynamic fluid level and water injection volume; and low-frequency trends are determined by long-term energy replenishment processes such as water injection volume and flowing pressure. This cross-scale mechanistic analysis breaks the "black-box" nature of data-driven models, constructing a causal interpretation bridge from data features to physical processes. This enables prediction results to be mapped to specific reservoir engineering phenomena, such as wellbore flow disturbances, injection-production response delays, and reservoir pressure depletion.

(3) Field application results demonstrate that the CEEMDAN-SR-BiLSTM model exhibits excellent performance in predicting high water-cut wells (water cut = 87.6%) in the Chaidamu Basin, with a root mean square error (RMSE) of 3.75—26.0% and 50.0% lower than CEEMDAN-BiLSTM and BiLSTM, respectively—and a coefficient of determination (R^2) of 0.954, significantly outperforming traditional models.

Through Bayesian optimization to tune parameters of the BiLSTM networks for different frequency components, the model effectively captures multi-scale dynamic features: the high-frequency component accurately tracks wellbore transient fluctuations (RMSE = 1.36), the medium-frequency component fits injection-production response cycles ($R^2 = 0.9991$), and the low-frequency component characterizes reservoir pressure trends (MAE = 2.37), validating the synergistic advantages of multi-scale decomposition and ensemble learning.

5.2. Practical Applications and Industrial Value

By decomposing multi-scale signals and mapping physical mechanisms, the framework effectively captures cross-scale dynamic features spanning the wellbore-near wellbore-far-field reservoir continuum. In practical reservoir management, it enables high-precision production forecasting for high water-cut wells, particularly applicable to scenarios with reservoir energy depletion and complex gas-liquid two-phase flow interactions in the late stages of waterflood development. For example, analyzing wellbore transient flow dynamics through high-frequency components allows real-time identification of short-term impacts of wellhead parameter fluctuations on production, providing immediate data support for wellbore equipment debugging (e.g., optimization of pumping unit stroke frequency, gas anchor design). The long-term capture of reservoir pressure trends by low-frequency components assists in evaluating the effectiveness of water injection strategies, early warning of productivity decline risks, and guiding the timing of cyclic water injection or fracturing measures.

Author Contributions: Conceptualization, Z.L. and Q.Q.; methodology, T.W.; software, B.Z.; validation, B.Z., H.G. and H.C.; formal analysis, Z.L.; investigation, Z.L.; resources, T.W.; data curation, B.Z.; writing—original draft preparation, Q.Q.; writing—review and editing, H.G.; visualization, H.C.; supervision, Z.L.; project administration, Z.L.; funding acquisition, Q.Q. All authors have read and agreed to the published version of the manuscript.

Funding: This study was funded by PetroChina Company Limited under the research project "New Mechanisms and Methods for Significantly Enhancing Oil Recovery in Medium-to-High Permeability Reservoirs" (Grant No. 2023ZZ0403).

Data Availability Statement: Data are contained within the article.

Conflicts of Interest: Authors Zhendong Li, Qihao Qian, Huazhan Guo, Tong Wu, Haidong Cui and Bingqian Zhu were employed by the company PetroChina. The authors declare that the research was conducted in the absence of any commercial or financial relationships that could be construed as a potential conflict of interest.

References

Yahya, W.; Baolin, Y.; AlRassas, A.M.; Yuting, W.; Al-Khafaji, H.; Al Dawood, R. Developing robust machine learning techniques to predict oil recovery: A comprehensive field and experimental study. *Geoenergy Sci. Eng.* **2025**, *250*, 213853. [CrossRef]

Yu, H.; Wang, Y.; Zhang, L.; Zhang, Q.; Guo, Z.; Wang, B.; Sun, T. Remaining oil distribution characteristics in an oil reservoir with ultra-high water-cut. *Energy Geosci.* **2024**, *5*, 100116. [CrossRef]

Al-Sheikh, J.N.; Saunders, D.E.; Brodkey, R.S. Prediction of flow patterns in horizontal two-phase pipe flow. *Can. J. Chem. Eng.* **1970**, *48*, 21–29. [CrossRef]

Datta-Gupta, A.; King, M.J. Streamline Simulation: Theory and Practice. Society of Petroleum Engineers. Available online: https://onepetro.org/books/book/18/Streamline-Simulation-Theory-and-Practice (accessed on 20 March 2025).

Patil, P.; Katterbauer, K.; Al Shehri, A.; Qasim, A.; Yousif, A. Forecasting Oil Production for Matured Fields Using Reinforced RNN-DLSTM Model. In *Artificial Intelligence Application in Networks and Systems*; Springer International Publishing: Cham, Switzerland, 2023; pp. 306–326.

6. Gulick, K.E.; McCain, W.D., Jr. Waterflooding Heterogeneous Reservoirs: An Overview of Industry Experiences and Practices. In Proceedings of the International Petroleum Conference and Exhibition of Mexico, Villahermose, Mexico, 3–5 March 1998. [CrossRef]
7. Sun, H.; Zhao, Y.; Yao, J. Micro-distribution and mechanical characteristics analysis of remaining oil. *Petroleum* **2017**, *3*, 483–488. [CrossRef]
8. Hui, Z.; Wei, L.; Xiang, R.; Guanglong, S.; Andy, L.H.; Zhenyu, G.; Deng, L.; Lin, C. INSIM-FPT-3D: A Data-Driven Model for History Matching, Water-Breakthrough Prediction and Well-Connectivity Characterization in Three-Dimensional Reservoirs. In Proceedings of the SPE Reservoir Simulation Conference, Online, 26 October 2021; p. D011S006R004.
9. Tang, H.Y.; He, G.; Ni, Y.Y.; Huo, D.; Zhao, Y.L.; Xue, L.; Zhang, L.H. Production decline curve analysis of shale oil wells: A case study of Bakken, Eagle Ford and Permian. *Pet. Sci.* **2024**, *21*, 4262–4277. [CrossRef]
10. Gupta, I.; Rai, C.; Sondergeld, C.; Devegowda, D. Variable Exponential Decline: Modified Arps To Characterize Unconventional Shale Production Performance. *SPE Reserv. Eval. Eng.* **2018**, *21*, 1045–1057. [CrossRef]
11. Wang, J.; Shi, C.; Ji, S.; Li, G.; Chen, Y. New water drive characteristic curves at ultra-high water cut stage. *Pet. Explor. Dev.* **2017**, *44*, 1010–1015. [CrossRef]
12. Lindemann, B.; Müller, T.; Vietz, H.; Jazdi, N.; Weyrich, M. A survey on long short-term memory networks for time series prediction. *Procedia CIRP* **2021**, *99*, 650–655. [CrossRef]
13. Sirisha, B.; Goud, K.K.C.; Rohit, B.T.V.S. A Deep Stacked Bidirectional LSTM (SBiLSTM) Model for Petroleum Production Forecasting. *Procedia Comput. Sci.* **2023**, *218*, 2767–2775. [CrossRef]
14. He, D.; Qu, Y.; Sheng, G.; Wang, B.; Yan, X.; Tao, Z.; Lei, M. Oil Production Rate Forecasting by SA-LSTM Model in Tight Reservoirs. *Lithosphere* **2024**, *2024*, lithosphere_2023_197. [CrossRef]
15. Cao, J.; Li, Z.; Li, J. Financial time series forecasting model based on CEEMDAN and LSTM. *Phys. A Stat. Mech. Its Appl.* **2019**, *519*, 127–139. [CrossRef]
16. Poongadan, S.; Lineesh, M.C. Non-linear Time Series Prediction using Improved CEEMDAN, SVD and LSTM. *Neural Process. Lett.* **2024**, *56*, 164. [CrossRef]
17. Zhang, X.; Ren, H.; Liu, J.; Zhang, Y.; Cheng, W. A monthly temperature prediction based on the CEEMDAN–BO–BiLSTM coupled model. *Sci. Rep.* **2024**, *14*, 808. [CrossRef] [PubMed]
18. Zhang, W.; Qu, Z.; Zhang, K.; Mao, W.; Ma, Y.; Fan, X. A combined model based on CEEMDAN and modified flower pollination algorithm for wind speed forecasting. *Energy Convers. Manag.* **2017**, *136*, 439–451. [CrossRef]
19. Long, J.; Lu, C.; Lei, Y.; Chen, Z.Y.; Wang, Y. Application of an improved LSTM model based on FECA and CEEMDAN VMD decomposition in water quality prediction. *Sci. Rep.* **2025**, *15*, 12847. [CrossRef] [PubMed]
20. Syama, S.; Ramprabhakar, J.; Anand, R.; Meena, V.P.; Guerrero, J.M. A novel hybrid methodology for wind speed and solar irradiance forecasting based on improved whale optimized regularized extreme learning machine. *Sci. Rep.* **2024**, *14*, 31657. [CrossRef]
21. Ding, Y.; Chen, Z.; Zhang, H.; Wang, X.; Guo, Y. A short-term wind power prediction model based on CEEMD and WOA-KELM. *Renew. Energy* **2022**, *189*, 188–198. [CrossRef]
22. Fan, Z.; Liu, X.; Wang, Z.; Liu, P.; Wang, Y. A Novel Ensemble Machine Learning Model for Oil Production Prediction with Two-Stage Data Preprocessing. *Processes* **2024**, *12*, 587. [CrossRef]
23. Huang, N.E.; Shen, Z.; Long, S.R.; Wu, M.C.; Shih, H.H.; Zheng, Q.; Yen, N.C.; Tung, C.C.; Liu, H.H. The empirical mode decomposition and the Hilbert spectrum for nonlinear and non-stationary time series analysis. *Proc. R. Soc. London. Ser. A Math. Phys. Eng. Sci.* **1998**, *454*, 903–995. [CrossRef]
24. Wu, Z.; Huang, N. Ensemble Empirical Mode Decomposition: A Noise-Assisted Data Analysis Method. *Adv. Adapt. Data Anal.* **2009**, *1*, 1–41. [CrossRef]
25. Torres, M.E.; Colominas, M.A.; Schlotthauer, G.; Flandrin, P. A complete ensemble empirical mode decomposition with adaptive noise. In Proceedings of the 2011 IEEE International Conference on Acoustics, Speech and Signal Processing (ICASSP), Prague, Czech Republic, 22–27 May 2011; pp. 4144–4147.
26. Huang, N.E.; Shen, S.S.P. *Hilbert–Huang Transform and Its Applications (Hilbert–Huang Transform and Its Applications)*; World Scientific: Singapore, 2005.
27. Karim, M.E.; Maswood, M.M.S.; Das, S.; Alharbi, A.G. BHyPreC: A Novel Bi-LSTM Based Hybrid Recurrent Neural Network Model to Predict the CPU Workload of Cloud Virtual Machine. *IEEE Access* **2021**, *9*, 131476–131495. [CrossRef]
28. Wang, H.; Yang, K. Bayesian Optimization. In *Many-Criteria Optimization and Decision Analysis: State-of-the-Art, Present Challenges, and Future Perspectives*; Brockhoff, D., Emmerich, M., Naujoks, B., Purshouse, R., Eds.; Springer International Publishing: Cham, Switzerland, 2023; pp. 271–297.
29. Guyon, I.; Elisseeff, A. An introduction to variable and feature selection. *J. Mach. Learn. Res.* **2003**, *3*, 1157–1182.

0. De Oliveira Werneck, R.; Prates, R.; Moura, R.; Goncalves, M.M.; Castro, M.; Soriano-Vargas, A.; Junior, P.R.M.; Hossain, M.M.; Zampieri, M.F.; Ferreira, A.; et al. Data-driven deep-learning forecasting for oil production and pressure. *J. Pet. Sci. Eng.* **2022**, *210*, 109937. [CrossRef]
1. Cleophas, T.J.; Zwinderman, A.H. Bayesian Pearson Correlation Analysis. In *Modern Bayesian Statistics in Clinical Research*; Cleophas, T.J., Zwinderman, A.H., Eds.; Springer International Publishing: Cham, Switzerland, 2018; pp. 111–118.
2. Lundberg, S.M.; Lee, S.-I. A unified approach to interpreting model predictions. In Proceedings of the 31st International Conference on Neural Information Processing Systems, Long Beach, CA, USA, 4–9 December 2017.
3. Narkhede, J. Comparative Evaluation of Post-Hoc Explainability Methods in AI: LIME, SHAP, and Grad-CAM. In Proceedings of the 2024 4th International Conference on Sustainable Expert Systems (ICSES), Kaski, Nepal, 15–17 October 2024; pp. 826–830.
4. Shapley, L. 7. A Value for n-Person Games. Contributions to the Theory of Games II (1953) 307-317. In *Classics in Game Theory*; Kuhn, H., Ed.; Princeton University Press: Princeton, NJ, USA, 1997; pp. 69–79. [CrossRef]
5. Chen, Z.; Zhou, B.; Zhang, S.; Li, D.; Sepehrnoori, K. Pressure transient behaviors for horizontal wells with well interferences, complex fractures and two-phase flow. *Geoenergy Sci. Eng.* **2023**, *227*, 211845. [CrossRef]

processes

MDPI

Article

Stratigraphic Correlation of Well Logs Using Geology-Informed Deep Learning Networks

Zhaohui Xu [1,2,*], Boyu Zheng [2], Bo Liu [3] and Wendan Song [2]

1 State Key Laboratory of Petroleum Resources and Engineering, China University of Petroleum (Beijing), Beijing 102249, China
2 College of Geosciences, China University of Petroleum (Beijing), Beijing 102249, China; 18533224586@163.com (B.Z.); 18810062920@163.com (W.S.)
3 Shuanghe Geological Research Institute, Research Institute of Exploration and Development, Henan Oilfield, Nanyang 473132, China; emanulginobili20@163.com
* Correspondence: xuzhaohui@cup.edu.cn

Abstract: Stratigraphic correlation plays a crucial role in reservoir characterization. However, it is often time-consuming and heavily dependent on geological expertise. To address this issue, we propose a novel method called CMT-enhanced Hiformer, which integrates convolutional neural networks meet vision transformers (CMT) and hierarchical multi-scale representations using transformers (Hiformer). First, the architecture of CMT-enhanced Hiformer fuses the advantages of convolutional neural networks and transformers, effectively extracting complex features from well logs and capturing both local and global dependencies via a well-designed attention mechanism. Next, a geological constraint with regularization parameters is incorporated into the loss function. The new loss function promotes the accuracy of stratigraphic boundaries. The proposed method was validated using data from the Shuanghe oil field in central China. Specifically, the model achieved a maximum F1 score of 0.8857 and a precision of 0.8865 on the blind test dataset, demonstrating its robustness and high classification accuracy. Moreover, we conducted ablation studies and performed a detailed comparison with state-of-the-art deep learning models. The results demonstrate that the proposed method significantly improves the accuracy and efficiency of stratigraphic correlation.

Keywords: stratigraphic correlation; transformer; convolutional neural network; geological constraint; well log

1. Introduction

Stratigraphic correlation is a fundamental task in reservoir characterization, aimed at establishing an isochronous stratigraphic framework to provide data support for the spatial representation of reservoir bodies [1,2]. Well log data serve as one of the core datasets for conducting stratigraphic correlation. Essentially, it involves an in-depth analysis and comparison of the geological information embedded within logging curves, identifying stratigraphic units of various scales based on characteristic values and morphological variations of the curves. Traditional stratigraphic correlation based on well logs primarily relies on manual interpretation, with the accuracy of results heavily dependent on the expertise and experience of geoscientists. However, as the number of wells increases, stratigraphic correlation becomes increasingly labor-intensive and time-consuming [3].

To address these challenges, various semiautomated and automated stratigraphic correlation methods have been proposed, including signal decomposition techniques, statistical methods, and machine learning approaches. Signal decomposition methods focus on processing well log data by decomposing a single log curve into multiple modes, with each mode representing stratigraphic features at different scales. These methods include wavelet analysis [4,5], variational mode decomposition (VMD) [6], cluster analysis [7], principal component analysis (PCA) [8,9], and pattern recognition techniques [10]. By extracting multiple data components from the log curves, these approaches provide a more intuitive representation of multi-scale stratigraphic characteristics, thereby enhancing the accuracy of stratigraphic correlation. Statistical methods treat well log curves as data sequences and delineate stratigraphic units by analyzing the correlations between these sequences. Representative techniques in this category include dynamic time warping (DTW) [11–13], cross-correlation analysis [14], maximum likelihood estimation [15], and grey relational analysis [16].

In recent years, machine learning techniques have been increasingly applied to stratigraphic division and correlation tasks, achieving notable progress. Initially, methods like genetic algorithms were used for biostratigraphic correlation but were computationally intensive and time-consuming [17]. To improve interpretation efficiency, researchers have increasingly adopted machine learning algorithms, such as support vector machines, adaptive boosting, gradient boosting, and random forests, that leverage annotated datasets to achieve faster and more consistent stratigraphic analysis [18–20]. With the development of deep learning, more powerful tools emerged for automated geological analysis. Models like convolutional neural networks (CNNs) and recurrent neural networks have been increasingly adopted for stratigraphic correlation [21–25]. Furthermore, the strengths of CNNs, including parameter sharing, translation invariance, and direct learning from raw inputs, enable effective performance in classification, segmentation, and regression tasks [26–31]. However, CNNs face limitations in modeling long-range dependencies, which can require deep architectures and increase computational cost. To address these limitations, transformer-based architectures have been introduced, leveraging self-attention mechanisms to capture both local and global dependencies, making them ideal for sequential modeling tasks such as well log interpretation [32–38].

This paper addresses two key challenges in stratigraphic correlation of well logs: enhancing the quality of feature extraction through algorithm fusion and improving prediction accuracy through the design of geology-informed loss function. First, we propose a geology-informed deep learning network for stratigraphic correlation of well logs. The deep learning algorithm architecture, namely CMT-enhanced HiFormer, leverages the strengths of both CNNs and transformers [39,40]. To improve prediction accuracy, we incorporate geological constraints into the loss function. Next, we discuss the details of the implementation of the training process and conduct ablation studies, which evaluate the effectiveness of the different modules in our CMT-enhanced HiFormer and different regularization weights of the loss function. The well-trained model is then applied to perform stratigraphic correlation of well logs acquired from the Shuanghe oil field in central China. Representative well tops are analyzed to quantitatively assess the performance of the proposed method. Finally, we discuss the effect of the number and distribution of trained wells on prediction accuracy, where the broad applications of our model are also discussed. At the end, we provide brief conclusions of this study.

2. Methodology

2.1. The Architecture of CMT-Enhanced Hiformer

The simplified structure of our proposed CMT-enhanced Hiformer is presented in Figure 1. The proposed model architecture primarily consists of two components: a feature extraction module and a decoding module. The feature extraction module comprises one

CMTSTEM layer, two CMTBLOCK layers, and two Pooling layers, while the decoding module incorporates two CONVUP modules designed to restore both the resolution and dimension of feature maps extracted from well log data. Detailed explanations about the key modules are provided as follows.

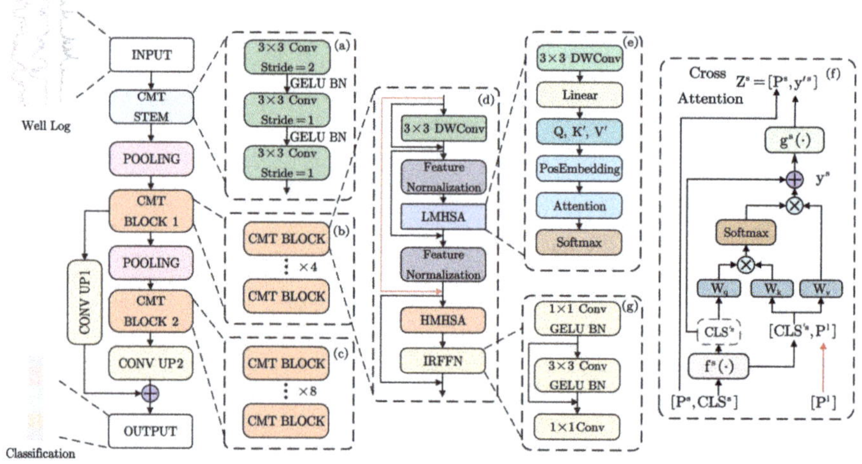

Figure 1. The simplified structure of our suggested CMT-enhanced Hiformer: (**a**) CMTSTEM, (**b–d**) CMTBLOCK module, (**e**) Lightweight Multi-Head Self-Attention(LMHSA), (**f**) Cross Attention, and (**g**) IRFFN.

2.1.1. CMTSTEM Module

As shown in Figure 1a, the CMTSTEM module serves as the front layer of our suggested model in this study. The input well log data are passed through a 3 × 3 convolutional layer with stride 2 for initial feature extraction and downsampling. This reduces the spatial resolution of the feature map while preserving important local feature information. Such fine-grained features can overcome the inability concerning transformer structure. Subsequently, two consecutive 3 × 3 convolutional layers with a kernel size of 3 × 3 and a stride of 1 further extract deep features from the downsampled feature map without altering its spatial dimensions. Each layer integrates GELU activation and batch normalization (BN), forming a hierarchical feature processing architecture. This design provides subsequent operations with feature representations that simultaneously maintain local receptive fields and achieve spatial compression.

2.1.2. CMTBLOCK Module

As depicted in Figure 1b,c, the CMTBLOCK is one of the key components of CMT [40], designed to effectively capture and represent complex features with an enhanced attention mechanism. The modified module starts with a 3 × 3 depth-wise convolution (DWConv) and includes a well-tailored attention mechanism from the Hiformer-MHSA (HMHSA). The typical multi-head self-attention refines the learning process by utilizing the extracted features. In addition to generating fine-grained feature maps P^s, it also offers the query CLS^s for the HMHSA to further combine the low-level representation P^l. This wide range of features withstands the physical constraint of stratigraphical layers. The HMHSA module utilizes cross-attention to effectively guide feature map fusion. As illustrated in Figure 1f, "P" represents the segmented local data for capturing local stratigraphic features and "CLS" is a learnable token added at the "P" sequence start to aggregate global information for classification. The HMHSA takes the output feature map P^s and tokens CLS^s from the LMHSA to guide the fusion stage. Another input P^l is the raw feature map simply after convolution. During the fusion stage, we utilize the weighted embedding representations CLS^s to function as the query for fusion, where the value is the original feature maps generated by simple downsampling to capture global features.

Specifically, the precise segmented predictions emphasize the importance of boundary points, highlighting the challenges of relying solely on the transformer model. Conversely, the CNN-based model loses the global captured features. Hence, the Cross Attention module, as shown in Figure 1f, takes the resultant smallest P^s and largest P^l levels as inputs and employs a cross-attention mechanism to fuse information across scales, addressing the inconsistency and low precision problems. Finally, the IRFFN module reformulates the feature map into the required output size.

2.1.3. CONVUP Module

The CONVUP module is a decoder component designed for feature map upsampling. It processes input feature maps through a series of convolutional layers, GroupNorm, and ReLU activation functions, while incorporating bilinear interpolation for upscaling. For the CMTBLOCK1 and CMTBLOCK2 stages, the CONVUP module first educes channel dimensions and extracts features via convolution, then enlarges spatial dimensions through upsampling. Finally, these two modules work collaboratively to upsample the deep features and fuse them with shallow features, ultimately restoring high-resolution output and achieving the transformation from low-resolution high-level semantic features to high-resolution detailed features.

2.2. *Loss Function with Geological Constraint*

The first loss function used in this study is the commonly used cross-entropy (CE) loss, defined as

$$Loss_{CE}(y,p) = -\sum_{k=0}^{k=N_k} y_k \cdot log(p_k) \tag{1}$$

where k denotes the stratigraphic class and N_k represents the total number of stratigraphic classes. y_k indicates the value of the ground true label for class k, i.e., if the sample belongs to class k, then y_k is 1; otherwise, it is 0. p_k represents the predicted probability for class k, which is the k-th element of the probability distribution output by the well-trained model.

Our suggested geological constraint (GC) is defined as

$$Loss_{GC}(\overline{y},\overline{p}) = -\sum_{l=0}^{l=N_l} \overline{y}_l \cdot log(\overline{p}_l) \tag{2}$$

where l represents the index of the stratigraphic boundary, ranging from 0 to N_l, where the boundary positions are sequentially numbered from 1 to N_l and non-boundary positions are 0. N_l denotes the number of stratigraphic boundaries. $\overline{y}_l$ denotes the ground true value for the stratigraphic boundary of class l and $\overline{p}_l$ represents the probability predicted by the model for boundary l.

The position of the stratigraphic boundary is the key point of the loss function we proposed in this study. Focusing on optimizing the prediction of different stratigraphic boundary positions reduces confusion between different horizons in the model. Then, our loss function in this study can be defined as

$$Loss_{total} = \lambda_1 Loss_{CE} + \lambda_2 Loss_{GC} \tag{3}$$

where λ_1 and λ_2 are two regularization weights. The selection of λ_1 and λ_2 will be explained in Section 4.

3. Data Introduction and Implementation Details

3.1. Study Area and Training Data Preparation

The Nanxiang Basin is bordered to the north by the Qinling Mountains and to the south by the Daba Mountains. It is a Meso-Cenozoic continental intermountain rift basin that formed during the late Yanshanian period. The sedimentary basin is controlled by faults along its periphery, covering a total area of 17,000 km^2 [41]. The Nanxiang Basin includes four depressions and four uplifts. Among them, the Biyang Depression in the southeast is the main area for oil and gas exploration. Influenced by the late Yanshanian stretching, it forms three structural belts from north to south: the northern slope belt, the central deep depression belt, and the southern steep slope belt presenting a general characteristic of being higher in the north and lower in the south [42], presented in Figure 2.

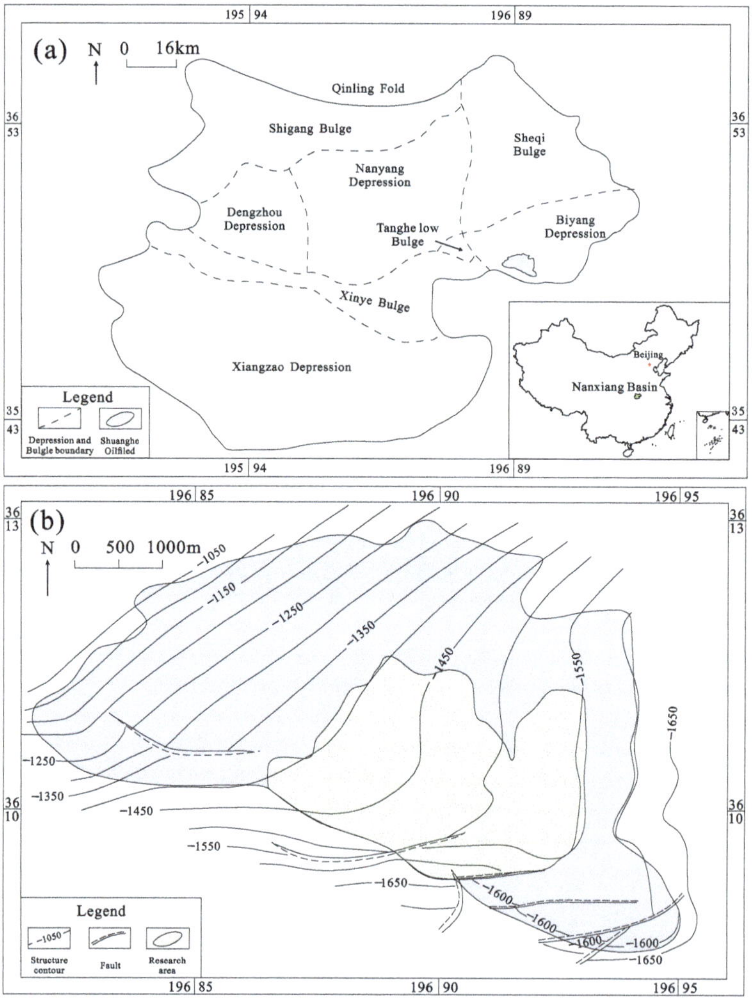

Figure 2. (**a**) The study area used in this study and (**b**) the structural map of the study area.

More than ten oil fields have been discovered in the Biyang Depression. This study focuses on the northern block of the Shuanghe Oil Field, which features a nose-like structure with a northwest-southeast trend, tilting towards the southeast, and a steeper western wing than the eastern wing. The Shuanghe Oil Field develops the Pingyuan Formation, Shangsi Formation, Hetaoyuan Formation, Dacangfang Formation, and Yuhuangding Formation from top to bottom [43]. Among these, the third member of the Hetaoyuan Formation is the main oil-bearing layer. This research target is the second sandstone group of the third member of the Hetaoyuan Formation, with a burial depth of 1350–2400 m. This layer consists primarily of gravel rock, gravelly sandstone, conglomerate sandstone, and sandstone, forming a mixed sandstone-conglomerate complex deposited in the fan delta front. The sand bodies are distributed in a fan shape from southeast to northwest, with wide distribution and significant sediment thickness. The sand bodies gradually thin out and pinch out from southeast to northwest [44].

The well location map of the study area is presented in Figure 3. The gray points in this image indicate well boreholes, where 100 wells are used for building the well dataset in this study, named W1, W2, …, and W100. When utilizing DL models for stratigraphic correlation, we can randomly select wells for model training, validation, and blind testing. Note that we select three well logs in this study, i.e., Gamma Ray (GR), Spontaneous Potential (SP), and Resistivity (RT), and we do not pay much attention to the sensitivity analysis of well logs. Note that, to test the effectiveness of our model, we randomly select 30% of the dataset in Figure 3 as our training data (indicated by the red points) and 5 wells as our validation data, while the rest are used as our blind testing data.

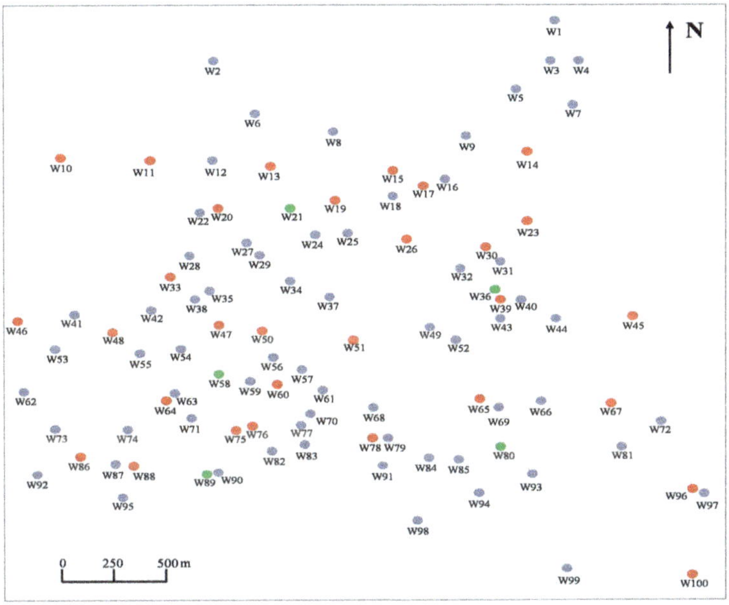

Figure 3. The well location map of the study area, where the red points represent the training wells, the green points represent the validation wells, and the gray points represent the test wells.

3.2. Implementation Details of Training Process

All networks use the PyTorch 2.2.1 deep learning library, Python 3.8, trained on a workstation with two NVIDIA GeForce RTX 3090 GPUs, equipped with a dual-channel Intel 6354 CPU. The SGD optimizer is selected with a learning rate scheduler. The learning rate is set with an initialization of 0.001. We train the base models for a maximum of 20 epochs, saving the model with the best validation loss during model training.

To evaluate the performance of difference models, confusion matrices defined in Equation (4) are utilized in this study.

$$\begin{aligned} Precision &= \frac{TP}{TP+FP}, \\ Recall &= \frac{TP}{TP+FN}, \\ F1 &= 2 \times \frac{Precision \times Recall}{Precision + Recall} \end{aligned} \tag{4}$$

where True Positives (TP) are instances where positive samples are correctly identified as positive, reflecting the model's ability to detect the target condition, True Negatives (TN) are cases where negative samples are accurately classified as negative, demonstrating the model's effectiveness in recognizing the absence of the target condition, False Positives (FP) occur when the model incorrectly classifies negative samples as positive, leading to false detections, and False Negatives (FN) arise when positive samples are mistakenly classified as negative, resulting in missed detections. These metrics offer a thorough assessment of the model's classification performance, providing insights into its precision and recall across various classes.

4. Ablation Study

4.1. The Comparisons of Different Modules

In this section, we first implement an ablation study on different models within our proposed CMT-enhanced Hiformer, i.e., comparing the effectiveness of CMT, Hiformer, and CMT-enhanced Hiformer. As mentioned above, we randomly select 30 wells as our training data and 5 wells as our validation data, while the others are selected as the blind test dataset. The training loss curves of CMT, Hiformer, and our suggested CMT-enhanced Hiformer are presented in Figure 4. Note that we normalize all these three training losses to make a clear comparison. We obtain three well-trained and convergent models after model training, which can be easily used for model testing. Moreover, the final loss values of our suggested model are smaller than those of the other two models, indicating that our model shows a better convergence result.

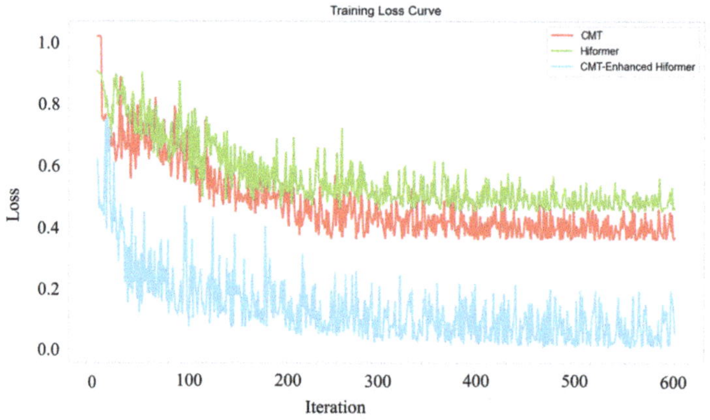

Figure 4. The training loss curves of CMT (red), Hiformer (green), and CMT-enhanced Hiformer (cyan).

Afterward, we apply these three models to the blind test dataset. The *precision* and $F1$ scores are, respectively, calculated and indicated in Table 1. Comparing these quantitative values, it can be easily found that our model shows better performance than the other two models by obtaining higher *precision* and $F1$. Moreover, we randomly select a well from the blind test dataset, i.e., W29, and then apply these three models for stratigraphic division. The results are presented in Figure 5c–e, while Figure 5a shows three well logs, i.e., GR, SP, and RT, and Figure 5b shows the ground truth stratigraphic classification labels. The red dashed lines indicate the sequence boundaries of W29. By comparing these images, we have several main observations. First, CMT shows an inaccurate stratigraphic division result even in the middle of a certain sequence, indicated by the red arrow. Second, there are incorrect predicted results at the sequence boundaries, denoted by the blue arrows in Figure 5c,d. Our model provides a more accurate result in Figure 5e by comparing it with the other two models, demonstrating the effectiveness of our fused model.

Table 1. The *precision* and *F1* of the blind test dataset by using CMT, Hiformer, and CMT-enhanced Hiformer.

Methods	*Precision*	*F1*
CMT	0.8365	0.8357
Hiformer	0.8277	0.8271
CMT-enhanced Hiformer	0.8717	0.8708

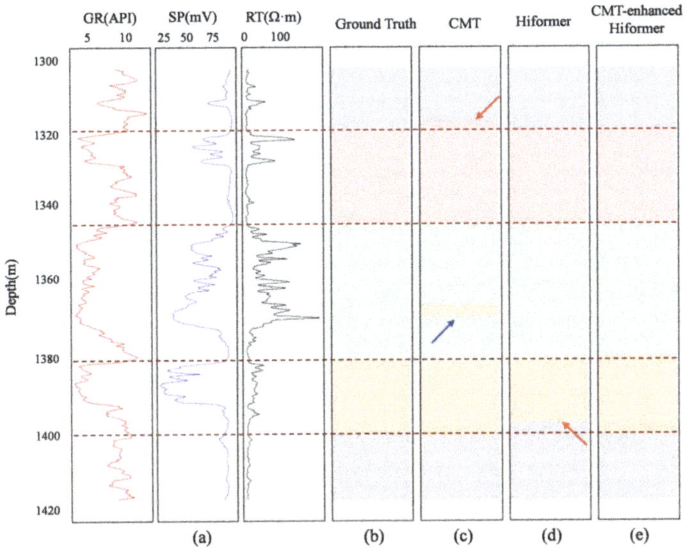

Figure 5. Well logs, ground truth labels, and predicted results by using different models of W29: (**a**) well logs (GR, SP, and RT), (**b**) the ground truth stratigraphic classification, and the predicted results calculated using (**c**) CMT, (**d**) Hiformer, and (**e**) CMT-enhanced Hiformer. The red dashed lines indicate the sequence boundaries.

4.2. The Comparisons of Different Regularization Weights

We validate the effectiveness of our proposed geological constraint (GC) in this section. Moreover, we discuss in detail the selection of two regularization weights λ_1 and λ_2 assigned to the cross-entropy (CE) loss and our GC loss in Equation (3), while satisfying the constraint $\lambda_1 + \lambda_2 = 1$. Note that, when selecting $\lambda_1 = 1$ and $\lambda_2 = 0$, it indicates that we only use the cross-entropy loss defined in Equation (1) for model training and our GC loss is not used for model training.

As the weight of our GC loss increases in Table 2, i.e., λ_2 increases from 0 to 0.3, the model's performance generally improves on the test set, despite some fluctuations with certain weight allocations. Specifically, when $\lambda_1 = 0.8$ and $\lambda_2 = 0.2$, the highest accuracy

(0.8865) and F1 score (0.8857) on the blind test dataset are achieved. The quantitative results are worse than those of $\lambda_1 = 0.8$ and $\lambda_2 = 0.2$. These results indicate that the introduction of our GC loss enhances the robustness and generalization ability of our suggested model, particularly in terms of improved classification accuracy and consistency, demonstrating that our proposed GC loss shows a positive impact on the overall results by effectively stabilizing gradients during the learning process.

Table 2. The ablation study of our suggested geological constraint and two regularization parameters by applying our CMT-enhanced Hiformer to the blind test dataset.

Weight		*Precision*	*F1*
$\lambda_1 = 1$,	$\lambda_2 = 0$	0.8717	0.8708
$\lambda_1 = 0.95$,	$\lambda_2 = 0.05$	0.8433	0.8414
$\lambda_1 = 0.90$,	$\lambda_2 = 0.10$	0.8571	0.8567
$\lambda_1 = 0.85$,	$\lambda_2 = 0.15$	0.8735	0.8731
$\lambda_1 = 0.80$,	$\lambda_2 = 0.20$	0.8865	0.8857
$\lambda_1 = 0.75$,	$\lambda_2 = 0.25$	0.8759	0.8749
$\lambda_1 = 0.70$,	$\lambda_2 = 0.30$	0.8489	0.8480

Afterward, we first utilize $\lambda_1 = 1.0$ and $\lambda_2 = 0$ for training our CMT-enhanced Hiformer and then $\lambda_1 = 0.8$ and $\lambda_2 = 0.2$, i.e., without and with our geological constraint. After model convergence, we randomly select a well from the blind test dataset, i.e., W25, whose well logs are denoted in Figure 6a. The ground truth stratigraphic classification label is shown in Figure 6b, while the predicted results via our suggested CMT-enhanced Hiformer without and with our GC loss are indicated in Figure 6c,d. The red dashed lines indicate the sequence boundaries of W25. Our CMT-enhanced Hiformer can realize stratigraphic division based on well logs. However, the introduction of our geological constraint effectively improves the accuracy of stratigraphic division, especially at the sequence boundaries indicated by the red arrows. These images in Figure 6 verify the validity of our suggested GC loss in this study.

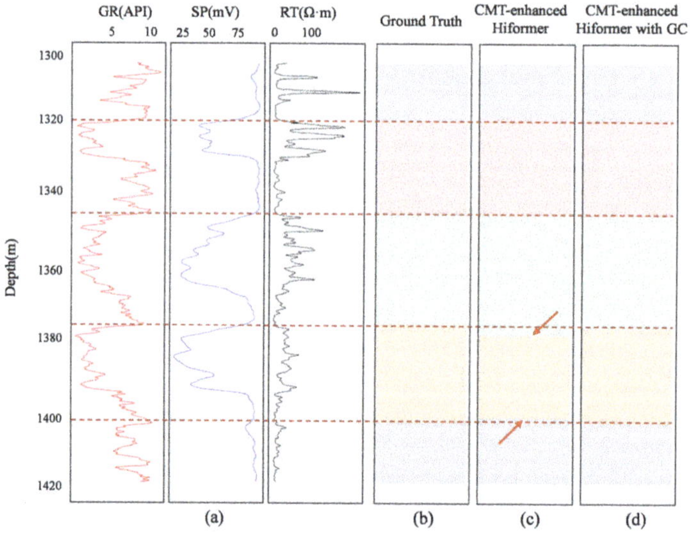

Figure 6. Well logs, ground truth labels, and the predicted results of W25: (**a**) well logs (GR, SP, and RT), (**b**) the ground truth stratigraphic classification, and the predicted results via our suggested CMT-enhanced Hiformer (**c**) without and (**d**) with our GC loss. The red dashed lines indicate the sequence boundaries.

5. Stratigraphic Correlation Results

We calculate the average differences between the ground truth and predicted well tops (in meters) of the blind test dataset by using SegNet, CMT, Hiformer, CMT-enhanced Hiformer, and CMT-enhanced Hiformer with GC loss, as indicated in Table 3. Our suggested model obtains the smallest differences among these four models. Moreover, the proposed geological constraint promotes our model with 0.46 m, verifying its effectiveness. In addition, we select two representative well tops, i.e., Class 1 and Class 3, to calculate the statistical errors of the predicted results by using different methods, as denoted in Figure 7. By comparing these images, it can still be seen that our proposed model and geological constraint can effectively promote the performance of automatic stratigraphic correlation.

Table 3. The average differences between the ground truth and predicted well tops (in meters) of the blind test dataset by using SegNet, CMT, Hiformer, CMT-enhanced Hiformer, and CMT-enhanced Hiformer with GC loss.

Methods	Average Differences (m)
SegNet	2.91
CMT	2.23
Hiformer	2.54
CMT-enhanced Hiformer	1.85
CMT-enhanced Hiformer with GC	1.39

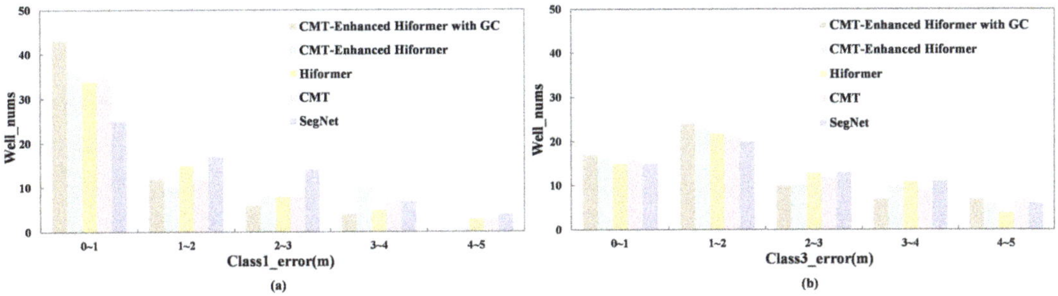

Figure 7. The statistical errors of the predicted results by applying different models to the blind test dataset, (**a**) Class 1 and (**b**) Class 3.

Next, we calculate the structural deep maps of Class 1's top surfaces by using the predicted and ground truth well tops, as presented in Figure 8. These images are, respectively, calculated by using the results of (a) SegNet, (b) CMT, (c) Hiformer, and our suggested CMT-enhanced Hiformer (d) without and (e) with our GC loss, while Figure 8f is calculated by using the ground truth labels. By comparing these images, we can easily find that our result in Figure 8e is the closest one to the result in Figure 8f, further demonstrating the effectiveness of our suggested method in this study.

After predicting the stratigraphic division results, we show a well section parallel to the sediment source direction and a well section perpendicular to the sediment source direction, as denoted in Figures 9 and 10. The dashed blue lines denote the stratigraphic correlation results of our CMT-enhanced Hiformer without GC loss, while the solid red lines indicate those of our CMT-enhanced Hiformer with GC loss. By comparing these stratigraphic correlation results, we can find that both methods can perform automatic stratigraphic correlation of well logs with high accuracy, verifying the effectiveness of our suggested CMT-enhanced Hiformer. Moreover, after introducing our GC loss, we can interpret the stratigraphic boundaries more accurately, as indicated by the red arrows in Figures 9 and 10, further demonstrating the availability of our GC loss.

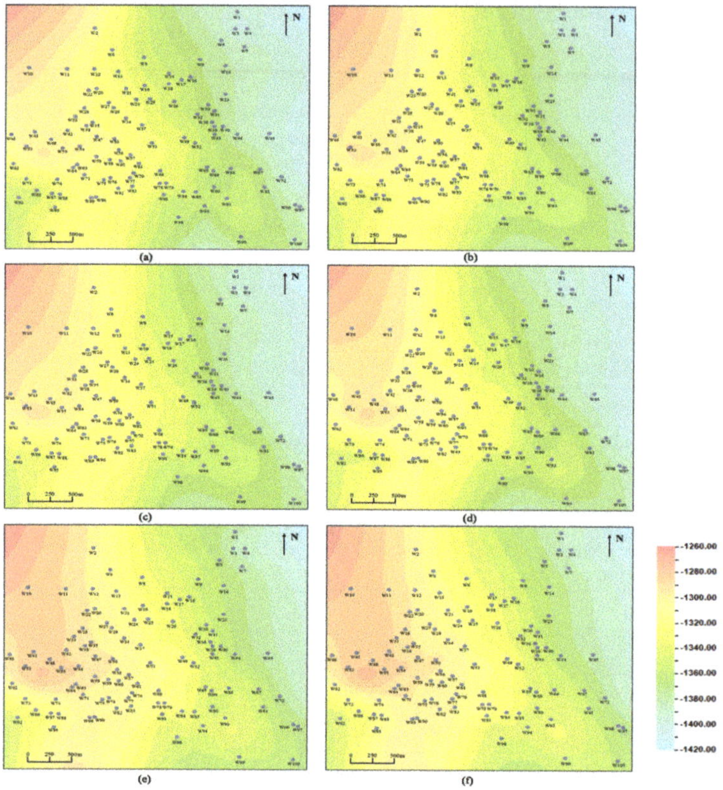

Figure 8. The structural deep maps of Class 1's top surfaces calculated by using the predictions of (**a**) SegNet, (**b**) CMT, (**c**) Hiformer, and our suggested CMT-enhanced Hiformer (**d**) without and (**e**) with our GC loss, and (**f**) ground truth.

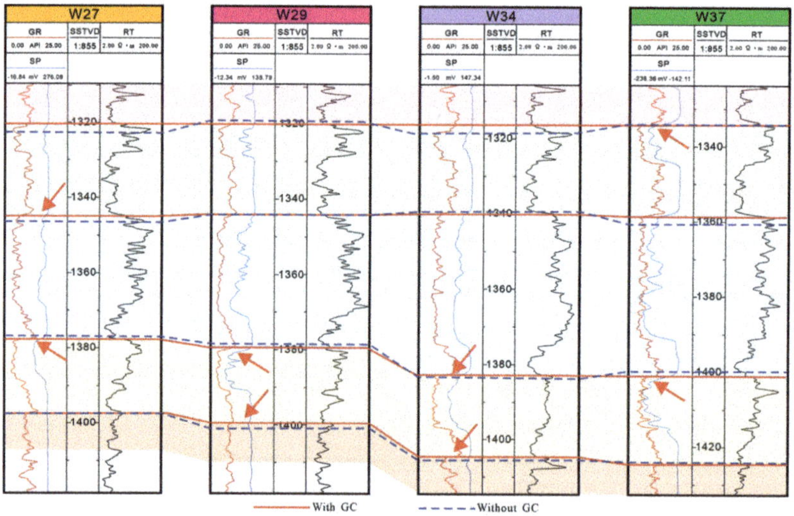

Figure 9. The automatic stratigraphic correlation results of our suggested model without (dashed blue) and with (solid red) our GC loss in a well section parallel to the sedimentary source direction.

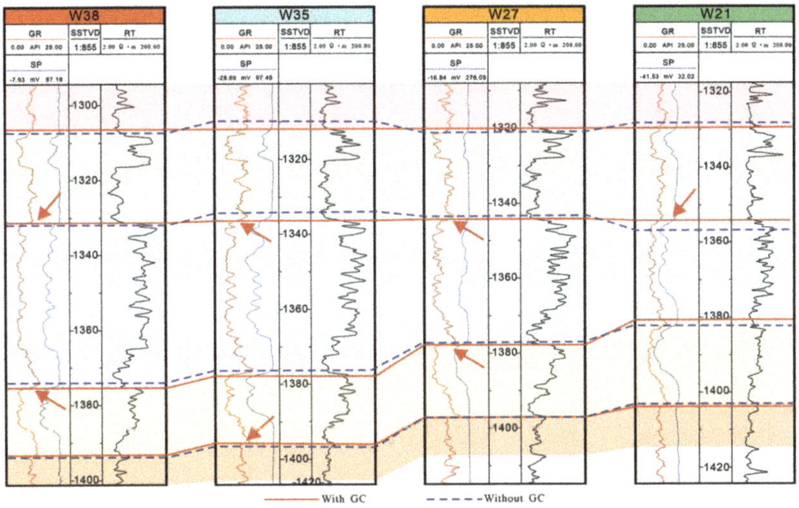

Figure 10. The automatic stratigraphic correlation results of our suggested model without (dashed blue) and with (solid red) our GC loss in a well section perpendicular to the sedimentary source direction.

6. Discussion

We suggest a CMT-enhanced Hiformer with the geological constraint for stratigraphic correlation of well logs. On the one hand, based on Hiformer, we propose to replace its encoder by using the feature extraction module in CMT. This promotes the capability for extracting the local features of well logs and then different scale stratigraphic unit analysis is realized. On the other hand, the introduction of the geological constraint improves the accuracy of the formation interfaces, especially the formation boundaries. The applications and tests on the dataset at the Shuanghe Oil Field demonstrate the effectiveness of our model. However, the predicted accuracy and the generalization property depend on the number of training wells and their distribution.

The percentage of the training wells is the most important factor that affects the prediction accuracy. To further test the robustness of our suggested model, we randomly select different amounts of wells for model training, i.e., from 10% to 60%. The validation dataset is also built by randomly selecting 5 wells, while the other wells are used for building the blind test dataset. After model training, we apply different models to the blind test dataset and calculate the *precision* and *F*1, as shown in Figure 11. By comparing these values, we have several observations. First, no matter how many training wells we used, our suggested CMT-enhanced Hiformer with GC shows a superior classification effect compared to CMT-enhanced Hiformer without GC. This indicates that our proposed geological constraint promotes the classification effectiveness of our CMT-enhanced Hiformer. Second, with the reduction of training wells used, i.e., from 30% to 10%, the performances of all models degrade significantly, indicating the great influence of the size of the training dataset for successfully implementing DL models. Moreover, the increase in model accuracy is limited when the training percentage increases from 30% to 60% (3.53%), not so obvious as those from 10% to 30% (7.92%). Furthermore, more training wells lead to more training time. To balance the training efficiency and the prediction accuracy, we select 30% wells as our training dataset and implement the detailed comparisons in this study.

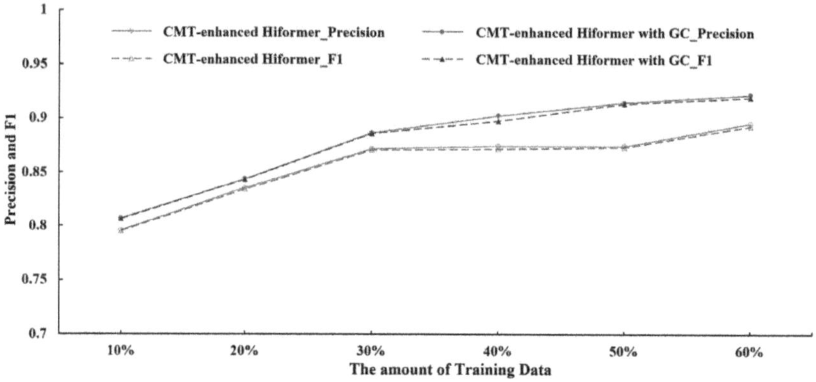

Figure 11. The precision and F1 using the CMT-enhanced Hiformer without and with GC for different amounts of training data.

For the latter, the accuracy of the model also depends on the distribution of training wells. In practical applications, different positions on the reservoir plane are affected by the source control, the deposition characteristics are different, and the logging response characteristics are also different. To improve the accuracy and generalization ability of the model, it is necessary to select wells in different reservoir locations as training datasets.

Finally, we focus on the automatic stratigraphic correlation of well logs in this study. Nevertheless, the suggested model is actually a classified or segmented model that can also be used for addressing other seismic interpretation tasks, such as fault detection and horizon picking. Certainly, we should build specialized losses for promoting the performance of DL models when solving different issues.

7. Conclusions

In this paper, we propose a geology-informed deep learning network called CMT-enhanced Hiformer for stratigraphic correlation of well logs. The network architecture integrates the advantages of convolutional neural networks and transformers, along with a mixed loss function composed of cross-entropy and a geological constraint. The proposed method was validated using data from the Shuanghe oil field in central China, achieving a maximum $F1$ score of 0.8857 and a precision of 0.8865 on the blind test dataset. Experiments show that CMT-enhanced Hiformer significantly improves the accuracy and efficiency of stratigraphic correlation and outperforms the state-of-the-art counterparts.

Author Contributions: Conceptualization, investigation, writing—review and editing, Z.X.; Methodology, writing—original draft, validation, visualization, B.Z.; Validation, resources, B.L.; Analyzed data, visualization, W.S. All authors have read and agreed to the published version of the manuscript.

Funding: This research was funded by the National Natural Science Foundation of China grant number 42472209.

Data Availability Statement: The datasets in this study can be obtained by contacting the corresponding author.

Acknowledgments: The authors would like to thank the Shuanghe Oil Field for providing access to the well log dataset, which significantly contributes to this study.

Conflicts of Interest: The authors declare that the research was conducted in the absence of any commercial or financial relationships that could be construed as a potential conflict of interest.

eferences

Liu, Y.; Shi, C.; Wu, Q.; Zhang, R.; Zhou, Z. Visual analytics of stratigraphic correlation for multi-attribute well-logging data exploration. *IEEE Access* **2019**, *7*, 98122–98135. [CrossRef]

Dai, Y.; Huang, X.; Liu, H.; Yang, H.; Wei, G.; Lu, N.; Han, Z.; Song, H. Stratigraphic automatic correlation using SegNet semantic segmentation model. In Proceedings of the SEG/AAPG/SEPM First International Meeting for Applied Geoscience & Energy, Denver, CO, USA, 26 September–1 October 2021; p. D011S065R003.

Partovi, S.M.A.; Sadeghnejad, S. Fractal parameters and well-logs investigation using automated well-to-well correlation. *Comput. Geosci.* **2017**, *103*, 59–69.

Liang, J.; Wang, H.; Blum, M.J.; Ji, X. Demarcation and correlation of stratigraphic sequences using wavelet and Hilbert-Huang transforms: A case study from Niger Delta Basin. *J. Pet. Sci. Eng.* **2019**, *182*, 106329. [CrossRef]

Kadkhodaie, A.; Rezaee, R. Intelligent sequence stratigraphy through a wavelet-based decomposition of well log data. *J. Nat. Gas Sci. Eng.* **2017**, *40*, 38–50. [CrossRef]

Xu, Z.; Zhang, B.; Li, F.; Cao, G.; Liu, Y. Well-log decomposition using variational mode decomposition in assisting the sequence stratigraphy analysis of a conglomerate reservoir. *Geophysics* **2018**, *83*, B221–B228. [CrossRef]

Al-Baldawi, B.A.H. Using well logs data in Logfacies determination by applying the cluster analysis technique for Khasib Formation, Amara oil field, south Eastern Iraq. *J. Pet. Res. Stud.* **2016**, *6*, 9–26. [CrossRef]

Karimi, A.M.; Sadeghnejad, S.; Rezghi, M. Well-to-well correlation and identifying lithological boundaries by principal component analysis of well-logs. *Comput. Geosci.* **2021**, *157*, 104942. [CrossRef]

Ma, Y.Z. Lithofacies clustering using principal component analysis and neural network: Applications to wireline logs. *Math. Geosci.* **2011**, *43*, 401–419. [CrossRef]

Vincent, P.; Gartner, J.; Attali, G. An approach to detailed dip determination using correlation by pattern recognition. *J. Pet. Technol.* **1979**, *31*, 232–240. [CrossRef]

Behdad, A. A step toward the practical stratigraphic automatic correlation of well logs using continuous wavelet transform and dynamic time warping technique. *J. Appl. Geophys.* **2019**, *167*, 26–32. [CrossRef]

Fang, H.; Lou, Y.; Zhang, B.; Xu, H.; Lu, M. Mimicking the process of manual sequence stratigraphy well correlation. *Interpretation* **2021**, *9*, T667–T684. [CrossRef]

Wang, C.; Wei, X.; Pan, H.; Han, L.; Wang, H.; Wang, H.; Zhao, H. Well Logging Stratigraphic Correlation Algorithm Based on Semantic Segmentation. *Appl. Geophys.* **2024**, *21*, 650–666. [CrossRef]

Rudman, A.J.; Lankston, R.W. Stratigraphic correlation of well logs by computer techniques. *Aapg Bull.* **1973**, *57*, 577–588.

Mehta, C.; Radhakrishnan, S.; Srikanth, G. Segmentation of well logs by maximum-likelihood estimation. *Math. Geol.* **1990**, *22*, 853–869. [CrossRef]

Wang, X.; Du, M.; Yu, W. Application of grey correlation method for stratigraphic correlation and its improvements. *Well Logging Technol.* **2006**, *30*, 126.

Zhang, T.; Plotnick, R.E. Graphic Biostratigraphic Correlation Using Genetic Algorithms. *Math. Geol.* **2006**, *38*, 781–800. [CrossRef]

Hohn, M.E.; Fontana, M.V. Geostatistics and artificial intelligence applied to stratigraphic correlation. *Am. Assoc. Pet. Geol. Conv.* **1986**, *70*, 5201053.

Parimontonsakul, M.; Lotongkum, S.; Mularlee, K. A machine learning based approach to automate stratigraphic correlation through marker determination. *Improv. Oil Gas Recovery* **2023**, *7*, IOGR.1204. [CrossRef]

Tognoli, F.M.W.; Spaniol, A.F.; Mello, M.E.D.; Souza, L.V.d. A machine-learning based approach to predict facies associations and improve local and regional stratigraphic correlations. *Mar. Pet. Geol.* **2024**, *160*, 19. [CrossRef]

Malmgren, B.A.; Nordlund, U. Application of Artificial Neural Networks to Stratigraphic Correlation. *Paleontol. Soc. Spec. Publ.* **1996**, *8*, 257. [CrossRef]

Tokpanov, Y.; Smith, J.; Ma, Z.; Deng, L.; Benhallam, W.; Salehi, A.; Zhai, X.; Darabi, H.; Castineira, D. Deep-learning-based automated stratigraphic correlation. In Proceedings of the SPE Annual Technical Conference and Exhibition, Virtual, 26–29 October 2020; p. D022S061R020.

Gu, X.; Lu, W.; Li, Y.; Wang, Y. Semi-supervised seismic stratigraphic interpretation constrained by spatial structure. *IEEE Trans. Geosci. Remote Sens.* **2023**, *61*, 5912710. [CrossRef]

Jiang, C.; Zhang, D.; Chen, S. Lithology identification from well-log curves via neural networks with additional geologic constraint. *Geophysics* **2021**, *86*, IM85–IM100. [CrossRef]

Wang, D.; Chen, G. Intelligent seismic stratigraphic modeling using temporal convolutional network. *Comput. Geosci.* **2023**, *171*, 105294. [CrossRef]

Yuan, S.; Liu, J.; Wang, S.; Wang, T.; Shi, P. Seismic waveform classification and first-break picking using convolution neural networks. *IEEE Geosci. Remote Sens. Lett.* **2018**, *15*, 272–276. [CrossRef]

Feng, R.; Balling, N.; Grana, D.; Dramsch, J.S.; Hansen, T.M. Bayesian convolutional neural networks for seismic facies classification. *IEEE Trans. Geosci. Remote Sens.* **2021**, *59*, 8933–8940. [CrossRef]

28. Pardo, E.; Garfias, C.; Malpica, N. Seismic phase picking using convolutional networks. *IEEE Trans. Geosci. Remote Sens.* **2019**, *57*, 7086–7092. [CrossRef]
29. Hu, G.; Hu, Z.; Liu, J.; Cheng, F.; Peng, D. Seismic fault interpretation using deep learning-based semantic segmentation method. *IEEE Geosci. Remote Sens. Lett.* **2020**, *19*, 7500905. [CrossRef]
30. Ferreira, R.S.; Oliveira, D.A.; Semin, D.G.; Zaytsev, S. Automatic velocity analysis using a hybrid regression approach with convolutional neural networks. *IEEE Trans. Geosci. Remote Sens.* **2020**, *59*, 4464–4470. [CrossRef]
31. Liu, N.; Lei, Y.; Yang, Y.; Wei, S.; Gao, J.; Jiang, X. Self-supervised time-frequency representation based on generative adversarial networks. *Geophysics* **2023**, *88*, IM87–IM99. [CrossRef]
32. Torres, J.F.; Hadjout, D.; Sebaa, A.; Martínez-Álvarez, F.; Troncoso, A. Deep learning for time series forecasting: A survey. *Big Data* **2021**, *9*, 3–21. [CrossRef]
33. Raghu, M.; Unterthiner, T.; Kornblith, S.; Zhang, C.; Dosovitskiy, A. Do vision transformers see like convolutional neural networks? *Adv. Neural Inf. Process. Syst.* **2021**, *34*, 12116–12128.
34. Liu, N.; Huo, J.; Li, Z.; Wu, H.; Lou, Y.; Gao, J. Seismic attributes aided horizon interpretation using an ensemble dense inception transformer network. *IEEE Trans. Geosci. Remote Sens.* **2024**, *62*, 5902010. [CrossRef]
35. Vaswani, A. Attention is all you need. *arXiv* **2017**, arXiv:1706.03762.
36. Liu, C.; Zhao, R.; Shi, Z. Remote-sensing image captioning based on multilayer aggregated transformer. *IEEE Geosci. Remote Sens. Lett.* **2022**, *19*, 6506605. [CrossRef]
37. Zerveas, G.; Jayaraman, S.; Patel, D.; Bhamidipaty, A.; Eickhoff, C. A transformer-based framework for multivariate time series representation learning. In Proceedings of the 27th ACM SIGKDD Conference on Knowledge Discovery & Data Mining, Virtual Event, Singapore, 14–18 August 2021; pp. 2114–2124.
38. Liu, Z.; Lin, Y.; Cao, Y.; Hu, H.; Wei, Y.; Zhang, Z.; Lin, S.; Guo, B. Swin transformer: Hierarchical vision transformer using shifted windows. In Proceedings of the IEEE/CVF International Conference on Computer Vision, Montreal, BC, Canada, 11–17 October 2021; pp. 10012–10022.
39. Heidari, M.; Kazerouni, A.; Soltany, M.; Azad, R.; Aghdam, E.K.; Cohen-Adad, J.; Merhof, D. Hiformer: Hierarchical multi-scale representations using transformers for medical image segmentation. In Proceedings of the IEEE/CVF Winter Conference on Applications of Computer Vision, Waikoloa, HI, USA, 2–7 January 2023; pp. 6202–6212.
40. Guo, J.; Han, K.; Wu, H.; Tang, Y.; Chen, X.; Wang, Y.; Xu, C. Cmt: Convolutional neural networks meet vision transformers. In Proceedings of the IEEE/CVF Conference on Computer Vision and Pattern Recognition, New Orleans, LA, USA, 18–24 June 2022; pp. 12175–12185.
41. Xu, K.; Chen, H.; Huang, C.; Ogg, J.G.; Zhu, J.; Lin, S.; Yang, D.; Zhao, P.; Kong, L. Astronomical time scale of the Paleogene lacustrine paleoclimate record from the Nanxiang Basin, central China. *Palaeogeogr. Palaeoclimatol. Palaeoecol.* **2019**, *532*, 109253. [CrossRef]
42. Su, A.; Chen, H.; Zhao, J.; Feng, Y.; Nguyen, A.D. Exhumation filling and paleo-pasteurization of the shallow petroleum system in the North Slope of the Biyang Sag, Nanxiang Basin, China. *Mar. Pet. Geol.* **2021**, *133*, 105267. [CrossRef]
43. Su, A.; Chen, H.; Zhao, J.; Feng, Y. Integrated fluid inclusion analysis and petrography constraints on the petroleum system evolution of the central and southern Biyang Sag, Nanxiang Basin, Eastern China. *Mar. Pet. Geol.* **2020**, *118*, 104437. [CrossRef]
44. Dong, Y.; Zhu, X.; Xian, B.; Hu, T.; Geng, X.; Liao, J.; Luo, Q. Seismic geomorphology study of the Paleogene Hetaoyuan Formation, central-south Biyang Sag, Nanxiang Basin, China. *Mar. Pet. Geol.* **2015**, *64*, 104–124. [CrossRef]

processes

MDPI

rticle

ntegrated Data-Driven Framework for Forecasting Tight Gas Production Based on Machine Learning Algorithms, Feature Selection and Fracturing Optimization

uyu Yao [1,2], Gang Hui [1,2,3,*], Dewei Meng [4,*], Chenqi Ge [1,2], Ke Zhang [5], Yili Ren [4,6], Ye Li [1,2], ujie Zhang [1,2], Xing Yang [1,2], Yujie Zhang [1,2], Penghu Bao [1,2], Zhiyang Pi [1,2], Dan Wu [1,2] and Fei Gu [4]

1 State Key Laboratory of Petroleum Resources and Engineering, China University of Petroleum (Beijing), Beijing 102249, China; fuyu_cup@163.com (F.Y.); gechenqi@126.com (C.G.); lypxszd@163.com (Y.L.); 13103949785@163.com (Y.Z.); yangxing0829@163.com (X.Y.); zyj15271580684@163.com (Y.Z.); baopenghu2024@163.com (P.B.); pizhiyang1672@163.com (Z.P.); 13833913673@163.com (D.W.)
2 College of Petroleum Engineering, China University of Petroleum (Beijing), Beijing 102249, China
3 Department of Chemical and Petroleum Engineering, University of Calgary, Calgary, AB T2N1N4, Canada
4 Research Institute of Petroleum Exploration & Development, PetroChina, Beijing 100083, China; renyili@petrochina.com.cn (Y.R.); luckygufei@gmail.com (F.G.)
5 Ningbo Institute of Digital Twin, Eastern Institute of Technology, Ningbo 315200, China; kzhang@idt.eitech.edu.cn
6 Artificial Intelligence Technology R&D Center for Exploration and Development, CNPC, Beijing 100083, China
* Correspondence: hui.gang@cup.edu.cn (G.H.); mengdewei@petrochina.com.cn (D.M.)

Abstract: A precise assessment of tight gas operational efficiency is critical for investment decisions in unconventional reservoir development. However, quantifying production efficiency remains challenging due to the complex relationships between geological and operational factors. This study proposes a novel data-driven framework for predicting tight gas productivity, effectively integrating computing algorithms, machine learning algorithms, feature selection, production prediction and fracturing parameter optimization. A dataset of 3146 horizontal wells from the Montney tight gas field was used to train six machine learning models, aiming to identify the most significant factors. Results indicate that fluid-injection volumes, burial depth, number of stages, Young's modulus, formation pressure, saturation, sandstone thickness and total organic carbon are the key variables for tight gas production. The Random Forest-based model achieved the highest accuracy of 88.6%. Case studies for the test demonstrate well that gas production could be nearly doubled by increasing fracturing fluid injection by 97.5%. This work provides evidence-based recommendations to refine development strategies and maximize reservoir performance.

Keywords: machine learning; gas production forecasting; Montney Formation; influencing parameters; fracturing optimization

1. Introduction

In recent years, global demand for clean energy sources such as unconventional natural gas has surged, due to a decline in traditional fossil fuel reserves and growing concerns over the greenhouse effect and air pollution [1]. Tight gas has become a focus of global oil and gas exploration, with the estimated worldwide resources to be about 210×10^{12} m^3, accounting for 75% of unconventional gas reserves [2]. However, compared to conventional reservoirs, tight gas reservoirs pose significant exploitation challenges [3]. Their pore types are dominated by secondary porosity, and gas seepage channels rely heavily on fracture

https://doi.org/10.3390/pr13041162

networks [4,5]. These reservoirs exhibit extreme heterogeneity in reservoir properties [6–8], compounded by difficulties in accurately measuring geological and engineering parameters in real time [9–11]. Additionally, tight gas reservoir blocks are usually typified by massive, heterogeneous field datasets characterized by poor data regularity and complex processing requirements, which impede rapid, accurate production prediction and delay the establishment of predictive models to guide reservoir development [12,13]. Consequently, a swift and reliable method is required to predict tight gas production to address the accelerating field development [14–16].

With the arrival of the big data era and the development of oil and gas field automation, machine learning (ML) methods provide an effective way for the digitalization of oil and gas field development [17–19]. By combining the well production-prediction research with the digital oil and gas field workflows, a new ML method has been developed to quantify production efficiency and unravel complex relationships between geological and operational factors [20–22]. It can quantify the contribution of individual factors to outcomes, and enable reliable predictions under different controlling factors.

This paper proposes a novel data-driven approach for predicting tight gas production, effectively integrating machine learning algorithm comparison, feature selection, production prediction and fracturing parameter optimization.

The process begins with the database collection of geological and engineering parameters from the Montney tight gas field, which includes 3146 horizontal wells. Next, six machine learning methods, namely Extreme Gradient Boosting Tree (XGBoost), Random Forest (RF), Gradient Boosting Decision Tree (GBDT), Artificial Neural Networks (ANN), Light Gradient Boosting Tree (LightGBM) and Extreme Randomized Tree (ET), are trained. These trained models serve to forecast production performance and quantify the contribution of geological and engineering factors to production outcomes. Operational plans are then optimized to maximize productivity by adjusting the fracturing fluid injection and proppant mass through the optimal calculating algorithm. This methodology establishes a comprehensive closed-loop process that covers evaluation, forecasting and optimization. This work presents the following novelties and contributions:

1. This framework introduces machine learning-based predictions for tight gas development and provides new insights into the key features that influence tight gas production. By integrating machine learning algorithms, feature selection, production prediction and fracturing parameter optimization, it can effectively develop tight gas resources.
2. A distinctive aspect of the study is the in-depth comparison of feature-selection techniques. The framework combines six methods (i.e., XGBoost, RF, GBDT, ANN, LightGBM and ET) with six machine learning models to provide a comprehensive assessment and comparative understanding. The most effective machine learning algorithm, along with the most significant factors, is identified.
3. The most important aspect of this research is the optimization of fracturing parameters based on the best-performing machine learning model. Operational plans are fine-tuned to maximize productivity by adjusting the fracturing fluid injection and proppant mass using the optimal algorithm.

2. Field Background

The study area is located in the Montney Formation of the Triassic of the Western Canada Sedimentary Basin, which is mainly distributed in the northwestern part of the basin, gradually thinning eastward until it is stripped out (Figure 1a) [23–25]. The Montney Formation is a wedge-shaped, unconsolidated deposit formed along stable Carboniferous and Permian cratonic margins. As a semi-deep-sea deposition, it exceeds 1200 m in thickness and features sand bodies oriented northeast–southwest [26]. The lithology of the Montney Formation transitions from siltstone to mudstone, with interbedded sandstone and siltstone. The Montney section is mainly composed of organic-rich, radioactive shale and dense siltstone, which are generally recognized as high-quality reservoir intervals.

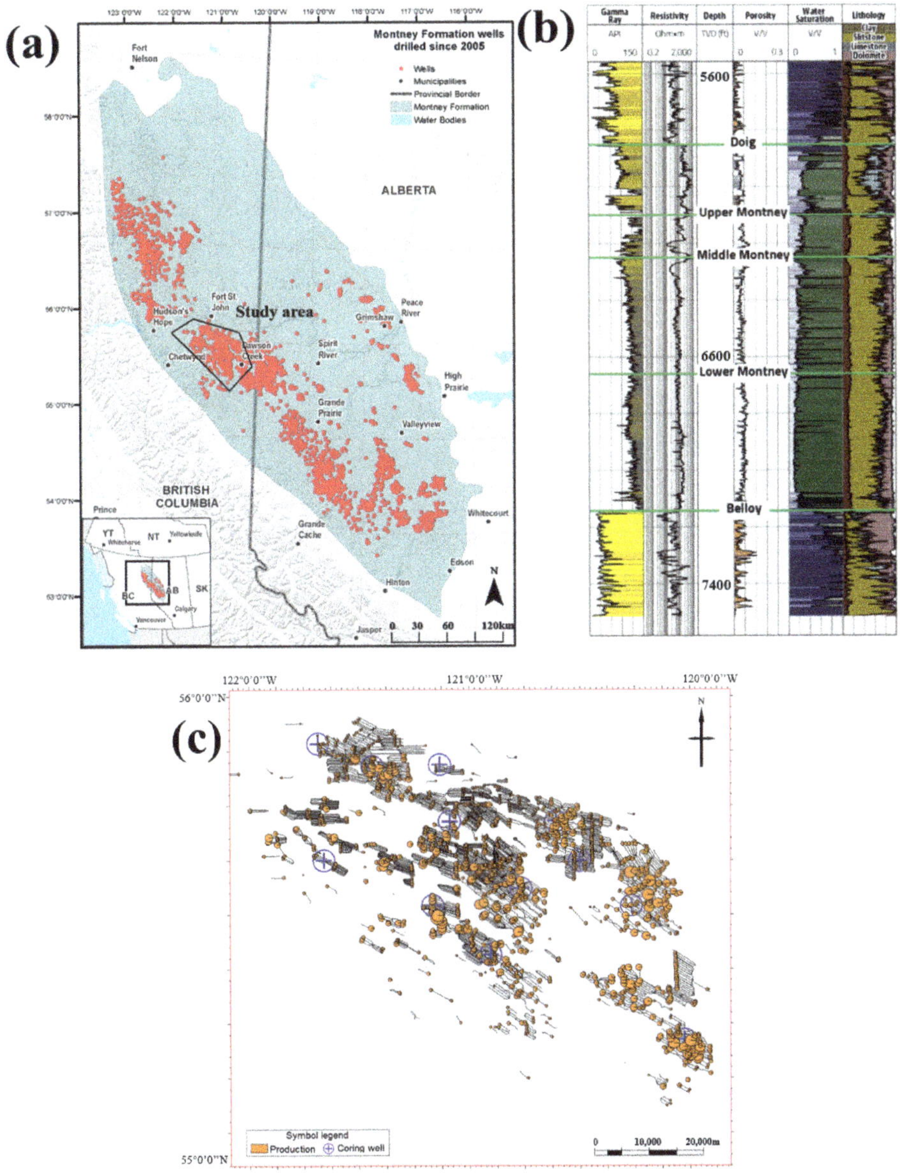

Figure 1. Geological information of the studied region. (**a**) Map view of the studied region, marked by the purple circle [23]; (**b**) stratigraphic and logging information of the Montney Formation [24]; (**c**) map view of studied horizontal wells. The orange circles denote the 12-month gas production, while the blue circles represent the location of coring wells. The black lines show the trajectory of 3146 fracturing horizontal wells.

The Montney gas field has burial depths of 1700–4000 m, a thickness of 300 m, a gas porosity of 1.0–6.0% and reserves of 80–700 trillion cubic feet (Figure 1b) [27,28]. Characterized by secondary pores and fracture-dominated seepage channels, the tight sandstone reservoir features high irreducible water saturation, elevated capillary pressure, dense rock matrices and complex pore–throat networks, resulting in poor fluid mobility and pronounced heterogeneity [29]. These attributes render conventional production-forecasting methods inadequate for tight gas well analysis.

This section summarizes the geological characteristics and production dynamics of the tight gas field in the Montney field, establishing a foundation for the subsequent research work. Data collected from the GeoScout database (https://www.geologic.com/products/geoscout/, accessed on 1 February 2025) include drilling, completion, fracturing and production records for 3146 wells, alongside core analysis data from 12 cored intervals (Figure 1c).

3. Methodology

This research aims at the tight sandstone oil and gas reservoirs in the Montney area. Based on previous research, it combines advanced theories and technologies like big data and artificial intelligence. The objective is to intelligently predict the production of tight gas wells in the researched area and construct a production capacity-prediction model for tight gas fracturing horizontal wells using a machine learning method (Figure 2). The data from the oil and gas field mainly includes geological and engineering data.

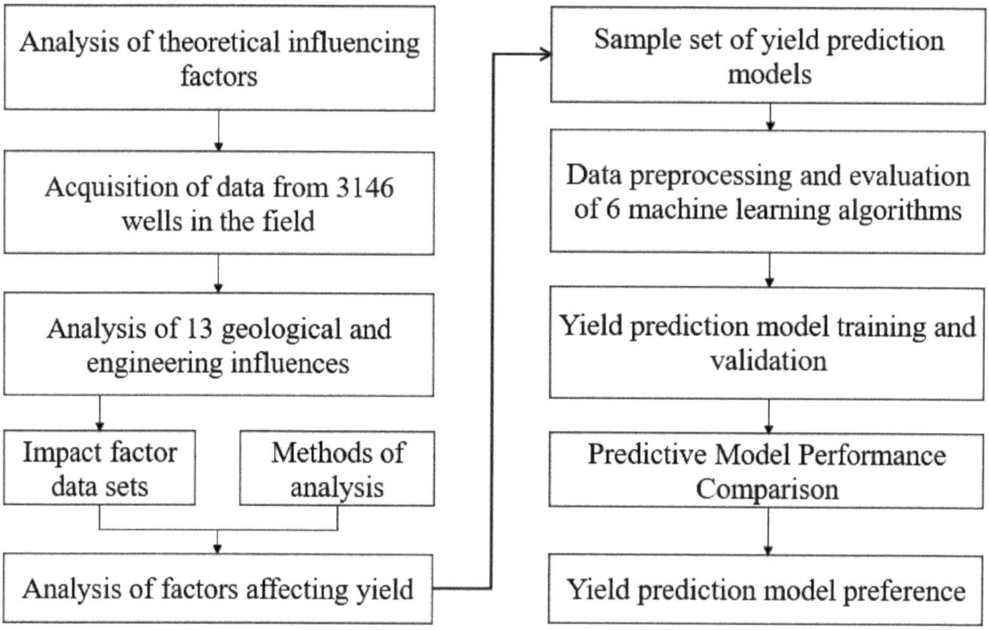

Figure 2. Flowchart of the integrated data-driven framework.

3.1. Parameter Characterization and Selection

Before constructing the production-prediction model for tight gas wells, it is essential to evaluate the feature dispersion of the dataset [30]. Feature selection requires systematic analysis of the relationship strength between parameters and production capacity, aiming to eliminate irrelevant variables while preserving geologically and statistically significant features. This process reduces the model complexity without sacrificing critical predictive factors [31]. The Pearson correlation coefficient is employed as a quantitative criterion to prioritize features strongly correlated with target variables and to exclude parameters exhibiting excessive mutual correlations, including multicollinearity.

3.1.1. Parameter Characterization

When constructing the production-prediction model, in order to incorporate the data sources that control the production rate, it is imperative to analyze the factors affecting the production rate of the production wells in the Montney work area. Through a comprehensive analysis of the correlation between various geological and engineering factors and the production rate, the geological factors are identified as reservoir physical parameters, specifically sandstone porosity and gas saturation; preservation conditions, as burial depth, pressure and thickness; and shale content, Poisson ratio, total organic carbon and young's modulus (Figure 3a) [32,33]. On the engineering side, the parameters include cumulative fluid injection, cumulative proppant injection, the number of stages and the horizontal length of the wells (Figure 3b) [34].

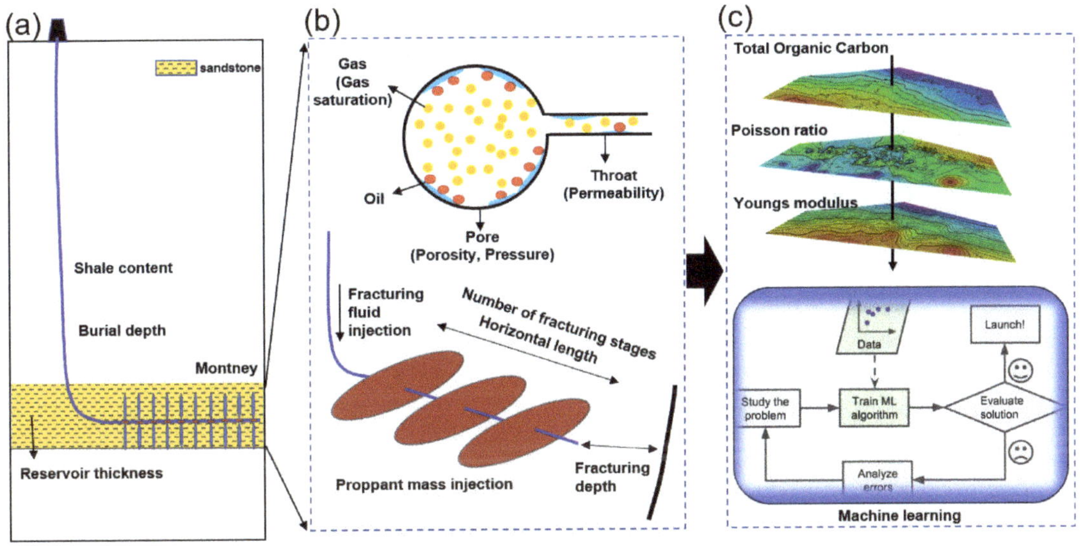

Figure 3. Related factors controlling the tight gas production and machine learning methods. (**a**) Fracturing horizontal well drilling the Montney Formation; (**b**) geological and operational parameters that contribute to gas productivity; (**c**) machine learning method that determines the primary controlling factors.

By ranking the importance of these features, the key factors controlling production were selected as the input features for machine learning, laying a solid foundation for the subsequent production-prediction modeling [35]. Previous studies have clearly demonstrated that geological and engineering factors affect the production of tight gas following the fracturing process [36]. Table 1 systematically lists the factors that impact tight gas production, along with the corresponding data sources for each factor. These parameters will serve as input parameters in the machine learning-based computing models (Figure 3c).

Table 1. Factors influencing tight gas production and data sources.

Typology	Influencing Factors	Data Sources	Representative Wells
Geological factors	Preservation conditions (burial depth, pressure and thickness)	Well-logging, well-completion and monitoring data	15-16-78-18
	Sandstone porosity and gas saturation Shale content, total organic carbon Poisson ratio, Young's modulus	Core analysis	6-26-78-18
Engineering factors	Cumulative fluid injection, cumulative proppant injection	Fracturing construction information	6-10-79-15
	Number of stages, horizontal length	Fracturing construction information	15-2-80-16

(1) Geological factors

In this paper, we primarily consider nine geological factors that affect oil production in sandstone reservoirs, which are formation pressure, sandstone thickness, burial depth, porosity, saturation, mud content, total organic carbon content, Poisson's ratio and Young's modulus. These geological factors were obtained through actual core measurements and well-log interpretation from the Montney sandstone field and relevant statistics were compiled.

The preservation conditions, serving as the external factors affecting the formation of tight gas, mainly include the regional tectonic background and the evolution of geological processes [37]. Previous literature predominantly focuses on the burial depth, pressure and sandstone thickness of tight gas reservoirs. The burial depth has an important impact on the economic value and benefits of tight gas reservoirs. Meanwhile, pressure significantly affects numerous characteristics of these reservoirs [38]. For example, temperature has an important effect on the adsorbed gas content with reservoirs [39]. The adsorbed gas content decreases as the temperature rises because the gas molecules move and the adsorbed gas content is lower. Similarly, higher pressure enhances gas content until reaching the desired point more quickly. In a saturation threshold, beyond which the adsorption rate diminishes, a similar vein grows with increasing pressure; however, this decreases beyond a point.

Porosity, permeability and saturation are three dominant characteristics of sandstone reservoirs. The micropores in sandstones contain a good quantity of crude oil and free gas; thus, the magnitude of the porosity generally dictates the free gas content [40,41]. The mineralogical composition of sandstones not only affects the petrophysical properties of shales but also influences the degree of fracture development, which in turn has an impact on the distribution of tight gas "deserts".

(2) Engineering factors

The fracturing engineering on tight gas horizontal wells directly affects both the production rate and its prediction. In this study, the predominant focus is on the engineering factors that impact gas production in sandstone reservoirs, including the length of the horizontal section, the number of fracturing sections, the volume of fracturing fluid injection and the volume of proppant injection [42,43]. The engineering factors of the Montney tight gas field are obtained from the actual well drilling data and the completed fracturing parameters [44].

Notably, as the fracturing fluid injection volume and the number of fracturing stages increase, more fractures can be produced within sandstone reservoirs. Meanwhile, an increase in the volume of proppant injection and the length of the horizontal section is typically associated with a larger reservoir volume, promoting production, thereby contributing to an increase in tight gas production. However, it is essential to optimize the design of each fracturing parameter by considering both geological factors and engineering factors rather than simply maximizing them.

3.1.2. Parameter Correlation Analysis

The gas production of sandstone reservoirs is affected by multiple factors, mainly geologic factors (e.g., reservoir quality) and engineering factors (e.g., permeability and stimulation efficiency). In this paper, the Pearson correlation coefficient is initially applied to describe the relationships between geologic factors, engineering factors and production.

(1) Correlation analysis expressions

The Pearson correlation coefficient, also known as the Pearson product–difference correlation coefficient, is commonly used to determine the correlation between two sets of continuous data that follow a bivariate normal distribution [45,46]. The Pearson correlation coefficient is symbolized as r, with its values ranging from -1 to 1. It is given by the following formula:

$$r = \frac{\sum_{i=1}^{n}(X_i - \overline{X})(Y_i - \overline{Y})}{\sqrt{\sum_{i=1}^{n}(X_i - \overline{X})^2}\sqrt{\sum_{i=1}^{n}(Y_i - \overline{Y})^2}} \quad (1)$$

where r is the correlation coefficient, n is the sample size, X and Y are random variables and $\overline{X}$ and $\overline{Y}$ are their respective means. The denominator is the product of the standard deviations of X and Y, while the numerator is the covariance. The covariance is used to characterize the correlation between two random variables X and Y. The variance is a special form of the covariance and when the two variables are identical, the covariance is equivalent to the variance.

(2) Classification of the degree of correlation

The Pearson correlation coefficient quantitatively describes the linear correlation between variables. A Pearson correlation coefficient $r > 0$ indicates a positive correlation between the two variables, $r < 0$ indicates a negative correlation and $r = 0$ indicates the absence of a linear correlation. Moreover, the larger the absolute value of r, the more significant the influence of the factor on gas production [47] (Table 2).

Table 2. Classification of degree of relevance [47].

Pearson Correlation Coefficient	Level of Relevance
$0.00 < \lvert r \rvert < 0.20$	Extremely weak correlation
$0.21 < \lvert r \rvert < 0.40$	Weak correlation
$0.41 < \lvert r \rvert < 0.60$	Moderately relevant
$0.61 < \lvert r \rvert < 0.80$	Strong correlation
$0.81 < \lvert r \rvert < 1.00$	Extremely Relevant

The Pearson correlation coefficient is used to analyze the geological and engineering parameters of tight gas, along with the production rate. This analysis can reveal the correlation between different geological and engineering parameters, as well as the correlation between each parameter and the production rate, thereby identifying the crucial parameters affecting the production rate of tight gas as the feature parameters in machine learning.

3.2. *Machine Learning Methods*

This work applies and compares six effective machine learning methods, namely Extreme Gradient Boosting Tree (XGBoost), Random Forest (RF), Gradient Boosting Decision Tree (GBDT), Artificial Neural Networks (ANN), Light Gradient Boosting Tree (LightGBM) and Extreme Randomized Tree (ET) [48,49] (Figure 4). The above-mentioned impacting parameters are adopted as the input variables and the output is natural gas production amounts over 6 months, 12 months and 18 months. The dataset is randomly divided into a training set and a test set. Thus, different computational models are assessed. The optimal prediction model is expected to exhibit the best performance in predicting natural gas production, possess the highest coefficient of determination and incur the smallest computational error [50–52].

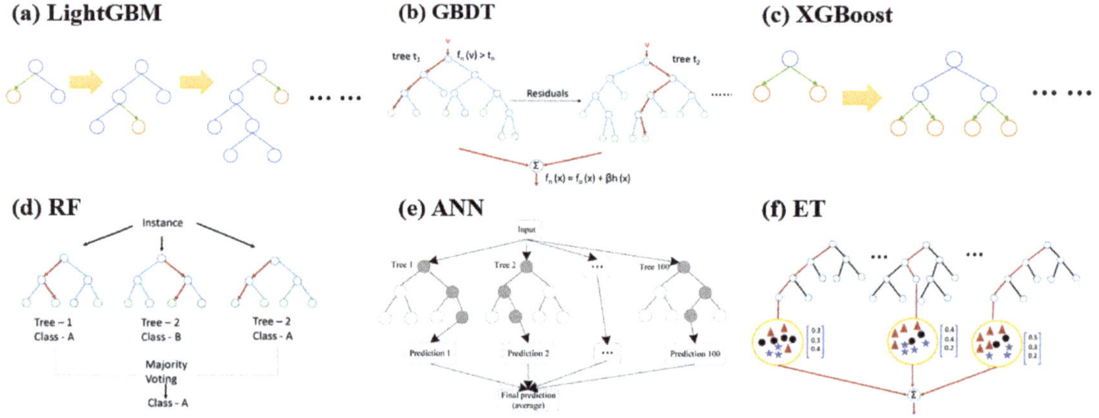

Figure 4. The frameworks of different machine learning algorithms. (**a**) LightGBM; (**b**) GBDT; (**c**) XGBoost; (**d**) RF; (**e**) ANN; (**f**) ET.

3.2.1. Data Preprocessing

(1) Dataset division

Considering the specificity of the oil and gas industry and the requirements for engineering inspection, the dataset is divided into two components with a ratio of 8:2 (training set:test set = 8:2). The training set is for model training and parameterization, establishing the relationship among geological parameters, fracturing construction parameters and the gas well production function [53]. The data within the test set are not involved in the model training process and its function is to evaluate the generalization ability of the model after the model has been built (the ability of the model to adapt to the unknown data). The entire process is calculated only once, effectively eliminating the interference of the human factor. And the prediction results of the test set can be used to verify the authenticity and applicability of the machine learning model [54,55].

(2) Data normalization

Data normalization is a commonly used data preprocessing technique. Oil and gas data, geological parameters, fracturing parameters and gas well production values have distinct scales and exhibit significant differences in their value intervals. This leads to a lower convergence rate or even prevents convergence when solving the machine learning model. The original data are processed using the data normalization method, aiming to eliminate the influence of the scale differences among various factors. Data normalization not only transforms the data of different variables into the same range but also ensures that the sample values of each parameter fall within the interval of 0 and 1. For a specific feature x in the sample set, the normalization formula is as follows:

$$x' = (x - X_{min})/(X_{max} - X_{min}) \tag{2}$$

where x' is the normalized data, x is the specific value in the data, X_{min} is the minimum value in the data and X_{max} is the maximum value in the data.

In this work, all machine learning models are trained and predicted with normalized data. Subsequently, all predicted data are reduced to the magnitude of the original data through inverse normalization.

3.2.2. Machine Learning Algorithms

(1) Extreme Gradient Boosting Tree (XGBoost)

XGBoost consists of multiple regression trees, specifically CART decision trees. Each decision tree predicts the residual, which is calculated as the difference between the true value and the sum of the predicted values of all preceding decision trees. The predicted values of all the decision trees are then aggregated to obtain the final result [19].

XGBoost is widely used in classification, regression and sorting problems. The regression trees generate continuous regression values and the predicted values of multiple decision trees are summed up for the final outcome. For classification problems, which are divided into binary classification and multi-classification, XGBoost is well suited. In binary classification, the sigmoid function is employed by XGBoost to map the cumulative value within the range of 0~1, where the value represents the probability of the binary classification result. In the multi-class classification, the softmax function is used to map the multi-class prediction value between 0 and 1, representing the probability that a sample belongs to a specific category. The sorting problem will not be discussed.

(2) Random Forest (RF)

Random Forest is an algorithm generated under the concept of ensemble learning. The core idea of ensemble learning is to provide a solution to the performance issues that a single model may encounter in certain aspects. It involves integrating a certain number of models to complement each other, thereby circumventing the limitations of the individual algorithms, especially reducing the likelihood of a single decision tree being one-sided and affecting the accuracy of judgment [47].

The principle of RF is to integrate the original data into the algorithm. The original data are divided into m training sets and for each training set, a corresponding decision tree model is established. These decision trees are independent of one another. During the splitting process of a decision tree, features are randomly searched at the nodes. Some features are randomly extracted from all available features to maximize the information gain and these extracted features are then used to search for the optimal solution for splitting. Essentially, this method samples all the samples and features, effectively avoiding overfitting. To ensure a sufficient model variance, a smaller set of samples selected for the model is preferable.

The algorithm combines the bagging ensemble learning theory with a random subspace algorithm. By introducing random attribute selection during the training of the decision tree as the base learner, it improves the generalization of the algorithm. RF can be applied to classification, regression and feature-selection problems.

(3) Gradient Boosting Decision Tree (GBDT)

GBDT (Gradient Boosting Decision Tree) is an iterative decision tree algorithm. All the trees in GBDT are regression trees and the core aim is to add up the results of all trees to obtain the final outcome [31]. Only the summation of the results of regression is meaningful, as the sum of the results of classification is meaningless. Although GBDT can be adapted to classification problems, it remains a regression tree.

Boosting is an ensemble method that combines weak classifiers to form a strong classifier. It is sequential, with several weak classifiers being trained one after another. The essence of GBDT is that each tree learns the residual of the sum of the conclusions of all previous trees. GBDT can be applied to both linear and nonlinear regression problems. It is also applicable to binary classification and multi-class classification, by setting a threshold where values greater than the threshold are classified as positive and others as negative.

(4) Artificial Neural Network (ANN)

Artificial Neural Network (ANN) is a human brain-like computational model constructed inspired by the neural networks in biology. It learns adaptively from a large number of samples through induction and constantly adjusts the weights of the connections between neurons. This process enables the neural network weights to converge within a stable range, ultimately allowing the model to acquire knowledge.

The basic information-processing unit of an ANN is the neuron. A neuron's structure can have multiple channels, each corresponding to its own connection weight, while it has only a single output. An artificial neural network model consists of five main components.

1. Input data: $x_1, x_2, x_3, \ldots, x_n$ are the n input data of the model, which can be formulated as $[x_1, x_2, x_3, \ldots, x_n, 1]^T$.
2. Connection weights: $w = [w_1, w_2, w_3, \ldots, w_n, b]$ is the connection weights of the model, which is the parameter for linear mapping, where b is the bias. The connection weights can reflect the connection strength between neurons. A positive weight indicates that the neuron is stimulated, while a negative weight means it is inhibited. During the model training, the connection weights are updated according to the loss function and learning rate until the loss function converges and the model achieves better performance.
3. Processing unit: This is used to calculate the weighted sum of each input signal.

$$z = \sum_{i=1}^{n} x_i \tag{3}$$

4. Activation function: The activation function plays the role of nonlinear mapping in neural networks, constraining the output value range to a reasonable interval. Sigmoid function, tanh function, ReLU function and Softmax function are several commonly used activation functions.
5. Output: This is the final result obtained after the input data undergoes linear and nonlinear mapping computations.

(5) Lightweight Gradient Boosting Tree (LightGBM)

The fundamental principle of LightGBM has a histogram-based decision tree algorithm and a leaf-wise algorithm with depth restriction [47].

The histogram-based decision tree algorithm includes Gradient-based One-Side Sampling (GOSS) and Exclusive Feature Bundling (EFB). GOSS can reduce a large number of data instances with relatively small gradients. As a result, when calculating the information gain, only the remaining data with high gradients are utilized. This approach significantly reduces time and space overheads compared to XGBoost, which traverses all feature values. Exclusive Feature Bundling (EFB) can combine numerous mutually exclusive features into a single feature. Compared with XGBoost, it saves plenty of time and space resources. EFB achieves the purpose of dimensionality reduction by binding multiple mutually exclusive features together.

Most of the GBDT tools use the inefficient leaf-wise Leaf Growth with Depth Restriction tree growth strategy. This strategy treats the leaves at the same level indiscriminately, leading to a significant amount of unnecessary overhead. Many leaves have low splitting gains and thus do not require searching and splitting. In contrast, LightGBM uses a leaf-wise algorithm with depth limitation. Additionally, LightGBM offers direct support for categorical features, enables efficient parallel processing and optimizes the cache hit rate.

(6) Extreme Random Tree (ET)

Extreme Random Trees (Extra Trees) algorithm, also known as Extra Trees or Extreme Random Trees, originates from the traditional Decision Trees with a simplified approach. In traditional decision tree algorithms, data objects are assigned to different sets based on their values across various features. The key to applying the algorithm is to select the optimal data features for decision-making and to determine their corresponding splitting points [47].

Like the classic extra tree growth model, the extra tree algorithm uses the classic top-down process to construct an untrained decision tree. However, it differs from other tree-based algorithms in two primary aspects. Firstly, the extra tree uses a completely randomized selection of node divisions for splitting the nodes. Secondly, it uses the entire set of training samples during the growth process.

The scoring mechanism in the node-division strategy of the extreme tree algorithm uses a special normalization of the information gain. Assuming a sample set D and a division node d, the scoring mechanism is formulated as follows.

$$Score_c(d, D) = \frac{2I_c^d(D)}{H_d(D) + H_C(D)} \tag{4}$$

where $H_C(D)$ represents the information entropy of the classification in the sample set D, $H_d(D)$ stands for the division entropy and $I_c^d(D)$ is the mutual information of the division result and the classification. Mutual information quantifies the amount of information that one random variable conveys about another.

3.2.3. Hyperparameter Tuning and Evaluation Criteria

The hyperparameter tuning method used in this study is mainly based on the balanced consideration of model complexity and computational resources. The steps of hyperparameter tuning are as follows.

(1) Set a particular hyperparameter as a variable each time and take multiple variable values; (2) perform a machine learning method regression using each variable value to build a regression model; (3) to calculate the error performance of this regression model, the root mean square error (MSE) is used in this study; (4) compare the values of MSE for different models to evaluate the performance of the model.

After the hyperparameter tuning, six production-prediction models for tight gas wells were established. The 10 influencing factors and the cumulative gas production over 6, 12 and 18 months, derived from the dataset of 3146 tight gas wells, were used as the input and output parameters for training and testing these six prediction models. The performance of the models was evaluated using three key indices to identify and optimize the best-performing model. These evaluation indices are the coefficient of determination (R^2), mean absolute percentage error (MAPE) and mean square error (MSE).

(1) Coefficient of determination (R^2)

The coefficient of determination (R^2) is an evaluation index that assesses the prediction results of machine learning models by calculating the weight of the regression sum of squares in the total variance. When the coefficient of determination evaluates the model, the closer its value is to 1, the better the explanation of the independent variable to the dependent variable in the machine learning model; in contrast, when its value is closer to 0, it means that the explanation is poor and the expression of the coefficient of determination is:

$$R^2 = 1 - \frac{\sum_{i=1}^{n}\left(y_i - \widehat{y_i}\right)^2}{\sum_{i=1}^{n}(y_i - \overline{y_i})^2} \tag{5}$$

(2) Mean Absolute Percentage Error (MAPE)

Mean Absolute Percentage Error (MAPE) is also a commonly used evaluation metric for machine learning models, which can be used to measure the deviation between the true value and the prediction; the smaller the value, the better, and its expression is:

$$MAPE = \frac{1}{n}\sum_{i=1}^{n}\left|\frac{y_i - \widehat{y_i}}{\widehat{y_i}}\right| \tag{6}$$

(3) Mean Square Error (MSE)

The mean square error is the average of the sum of the squares of the difference between the calculated predicted value and the true value. In a regression model, the mean square error can be used to assess how well the model fits the data and to find the combination of parameters that minimizes the mean square error, thereby obtaining the best-fitting straight line formula,

$$MSE = \frac{\sum_{i=1}^{n}\left(y_i - \widehat{y_i}\right)^2}{n} \tag{7}$$

4. Results and Discussion

4.1. Results of Parameter Characterization and Selection

4.1.1. Parameter-Characterization Results

Based on the horizontal well completion data, the Sequential Gaussian Simulation (SGSIM) method was employed to calculate the burial depth of the entire study area. The calculated burial depth was found to range from 1700.9 to 2874.9 m, with an average value of 2287.9 m. The depth of the reservoir was determined by using the SGSIM method. Based on the publicly available reservoir pressure monitoring data, the stratigraphic pressure gradient was calculated with the SGSIM method and ranged from 10.3 to 14.7 MPa/km, with an average value of 12.5 MPa/km. Stratigraphic analyses indicated that reservoir thicknesses were in the range from 119.6 to 263.6 m, with an average of 191.6 m.

Considering the relationship among the input factors, the outcomes of core analyses and the logging interpretations, the porosity and gas saturation were determined. The porosity ranged from 0.04 to 0.26 and the gas saturation from 15.9 to 99.3%, with average values of 0.15 and 57.6%, respectively.

The processed data indicated that the cumulative fluid-injection volume spanned from 17.3 to 43,307 m^3 and the proppant pumping volume from 25.6 to 14,133.2 t, with average values of 21,662.4 m^3 and 7079.4 t, respectively. The horizontal lengths ranged from 179 to 4636.5 m and the number of segments from 4 to 88, with average values of 2407.8 m and 44.5 segments, respectively. In addition, the 12-month gas-production ranged from 1.5 to 6765.2 million cubic feet (MMCF). Figure 5 illustrates the inferred distribution of the impact parameters and Table 3 lists the statistical details of the input and output parameters used in the machine learning process.

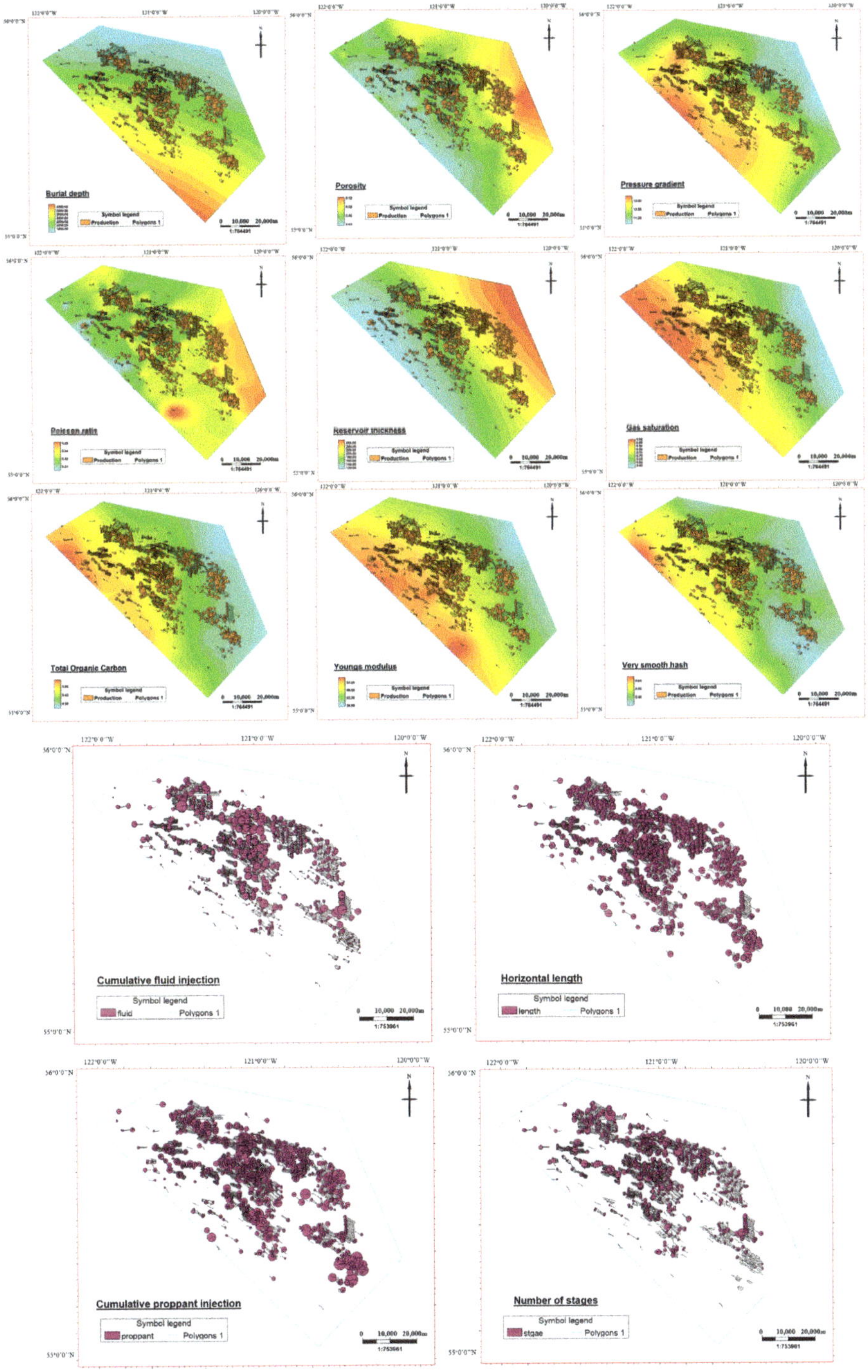

Figure 5. Map view of inferred distribution for influencing parameters. The base map represents the spatial features of related influencing parameters. The orange circles denote the 12-month gas production. The purple circle indicates the value of the parameter.

Table 3. Statistics of input and output parameters used for machine learning progress.

Type	Parameters	Unit	Minimum	Maximum	Average
Output variable	6-month gas production	MMCF	1.2	4847.1	629.3
	12-month gas production	MMCF	1.5	6765.2	1201.9
	18-month gas production	MMCF	2.4	9403.8	1609.2
Input geological parameters	Formation pressure gradient	MPa/km	10.3	14.7	12.5
	Reservoir thickness	m	119.6	263.6	191.6
	Burial depth	m	1700.9	2874.9	2287.9
	Porosity		0.04	0.26	0.15
	Gas saturation	%	15.9	95.3	57.6
	Shale content		0.41	0.66	0.54
	Total organic carbon		0.46	0.89	0.68
	Poisson ratio		0.21	0.25	0.23
	Youngs modulus	GPa	38.58	58.75	48.57
Input operational parameters	Horizontal length	m	179.0	4636.5	2407.8
	Number of stages		4	88	44.5
	Cumulative fluid injection	m^3	17.3	43,307.6	12,407
	Cumulative proppant injection	t	25.6	14,133.2	2367

4.1.2. Correlation Analysis Results

By analyzing the Pearson correlation coefficient, it is evident that among the geological factors, porosity, burial depth and formation pressure and among the engineering factors, fracturing fluid injection and horizontal section length exhibit a strong correlation with yield (Figure 6). Moreover, a strong correlation is observed between porosity and sandstone thickness, as well as between proppant injection volume and horizontal section length.

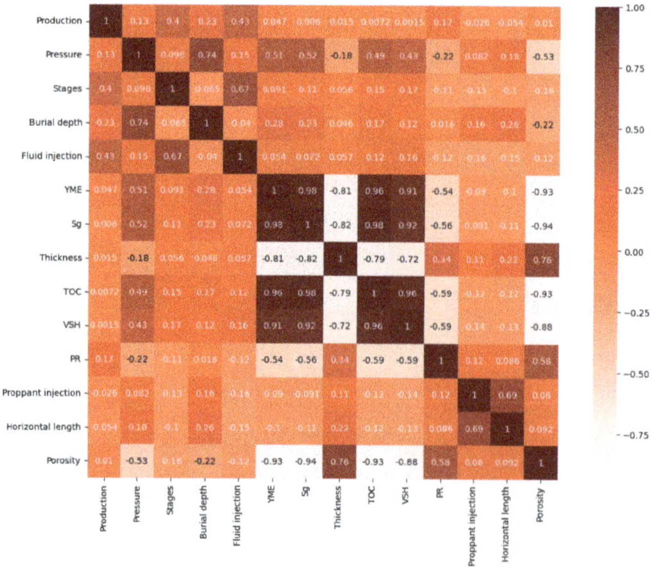

Figure 6. Pearson's correlation analysis of 13 geologic and engineering impact parameters and yield.

4.2. Machine Learning-Based Production Prediction

4.2.1. Feature Importance

The hyperparameter-tuning process is finished. For example, a Random Forest model is built using leaf sizes of 3, 6, 9 and 12, respectively (Figure 7). It shows the error performance after Random Forest regression using different leaf sizes. From the figure, it can be seen that the smaller the selected leaf size, the lower the model's error rate, and the model with a leaf size of 3 has the lowest final error, indicating that the model performs better. The leaf size of 3 was determined as the final parameter.

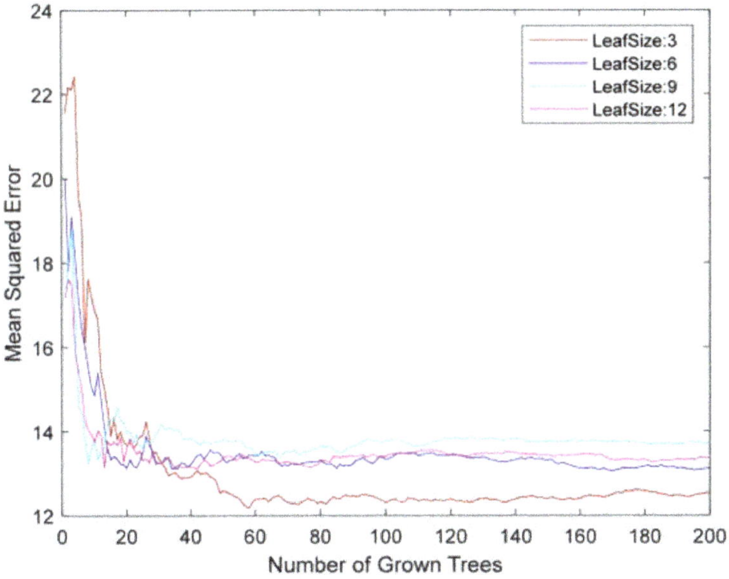

Figure 7. Training performance of Random Forest models with different leaf sizes.

The input variables consist of 13 relevant metrics from 3146 wells, whereas the output variables are the natural gas production of these wells. The machine learning procedure is performed by integrating the input and output variables. Figure 8 depicts the normalized frequencies of the 13 input variables, thereby highlighting the importance of the parameters in generating the predictive model. The findings indicate that fracturing fluid injection, burial depth and the number of fractured sections are the features that contribute most to tight gas production, with their normalized frequencies greater than 0.9. Young's modulus, formation pressure, saturation, sandstone thickness and total organic carbon content are the next most influential factors, with normalized frequency greater than 0.75 but less than 0.9. Other features, including mud content, Poisson's ratio, proppant injection volume, horizontal section length and porosity, had the least correlation, with normalized frequencies less than 0.5. These five parameters are considered to be of lesser importance because they may be directly proportional to certain more crucial values. The proportional relationship reduces their individual influence and importance in the analysis. For example, gas saturation, total organic carbon content and mud content are positively correlated with Young's modulus. Similarly, the cumulative proppant injection volume is directly proportional to the fracturing fluid-injection volume.

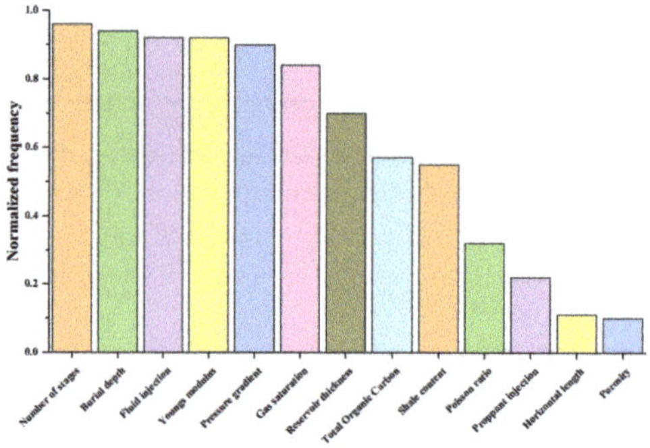

Figure 8. Feature importance influencing parameters. The normalized frequency of ten input parameters via machine learning. Feature importance in decreasing order is the number of stages, burial depth, cumulative fluid injection, Young's modulus, formation pressure, gas saturation, reservoir thickness, total organic carbon, shale content, Poisson ratio, cumulative proppant injection, horizontal length and porosity.

4.2.2. Comparison of Different Models

Figure 9 depicts the machine learning results obtained using the six methods, as a function of the number of parameters selected. The six algorithms consistently identified eight parameters that yielded the highest R^2. These parameters, including the number of fracturing sections, burial depth, fracturing fluid injection volume, Young's modulus, formation pressure and gas saturation, are the fundamental determinants for enhancing tight gas production.

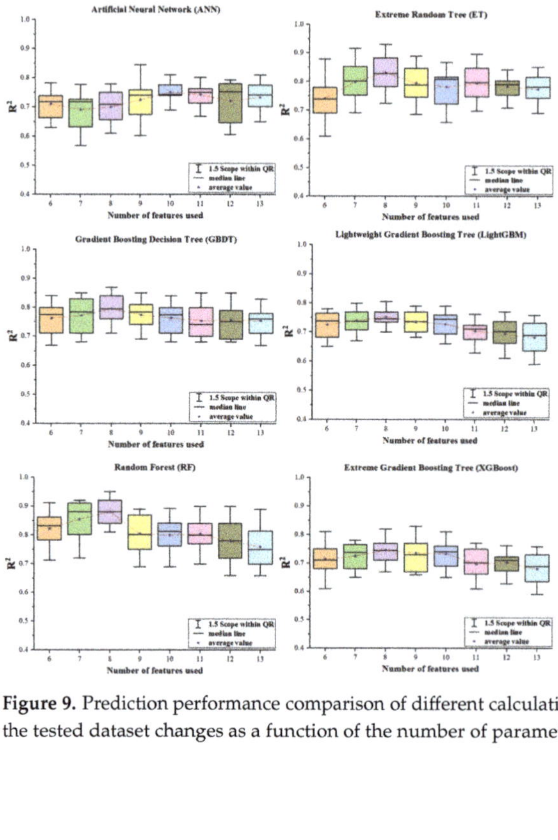

Figure 9. Prediction performance comparison of different calculating algorithms. The statistical R^2 of the tested dataset changes as a function of the number of parameters chosen.

The average R^2 of these eight parameters across the different methods are as follows (Figure 10): 0.688 for ANN, 0.853 for ET, 0.761 for GBDT, 0.823 for XGBoost, 0.783 for LightGBM and 0.886 for RF. Additionally, the average MSEs are 0.31 for ANN, 0.17 for ET, 0.21 for GBDT, 0.19 for XGBoost, 0.23 for LightGBM and 0.19 for LightGBM. LightGBM is 0.23 and RF is 0.12. Therefore, the RF stands out as the most optimal choice. With the selected eight parameters, it has the highest R^2 and the lowest MSE. Therefore, the RF was ultimately selected to construct the prediction model for tight gas production capacity.

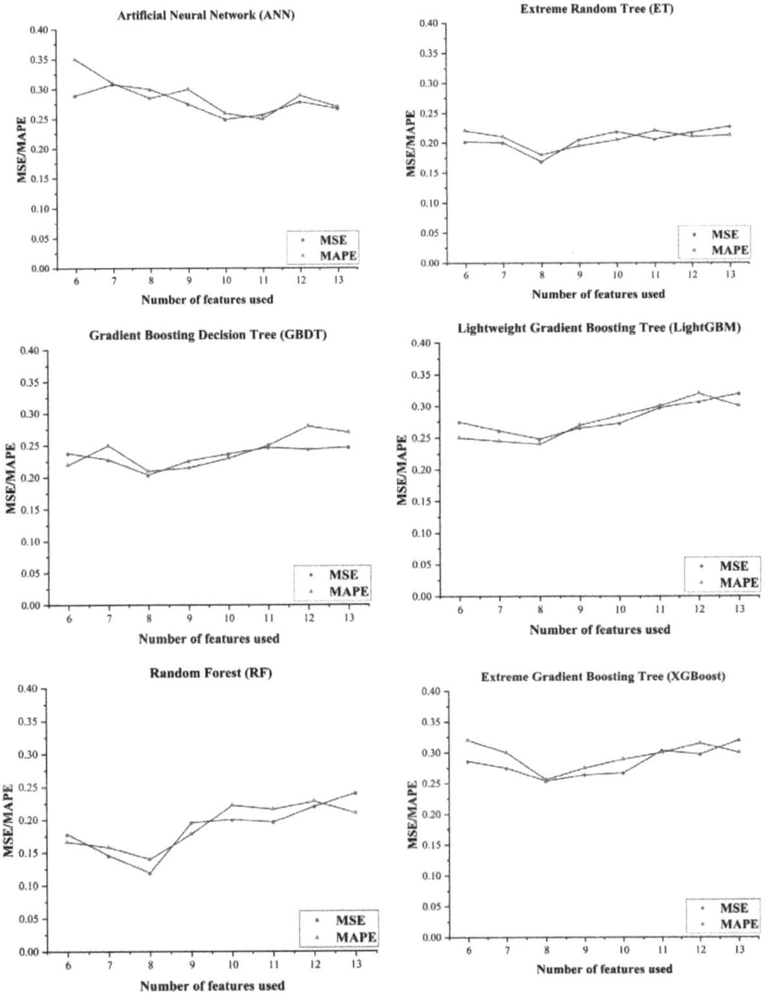

Figure 10. MSE/MAPE comparison of different machine learning algorithms. The statistical prediction performance of the tested dataset as a function of the number of parameters chosen. Among them, the Random Forest algorithm with eight parameters chosen has the lowest MSE/MAPE.

4.2.3. Prediction of Long-Term Production

The significance of the RF-based prediction model lies in its ability to predict natural gas-production rates, in terms of tight gas-production equivalents, over 12 months. Specifically, the RF-based prediction model predicts tight gas-production rates in the study area based on a prediction model of eight selected parameters (i.e., fracturing fluid injection, burial depth, number of fracturing sections, Young's modulus, formation pressure, saturation, sandstone thickness and total organic carbon content). The inferred gas-production capacity maps for the 6-month, 12-month and 18-month durations (Figure 11a–c) exhibit a high degree of correspondence with the observed spatial production capacity (Figure 11d). This correspondence serves to validate the effectiveness and reliability of the RF-based model. This predictive model can be utilized to guide the future positioning of horizontal wells in the Montney Formation. Additionally, this approach can also be applied to other scenarios of tight gas production.

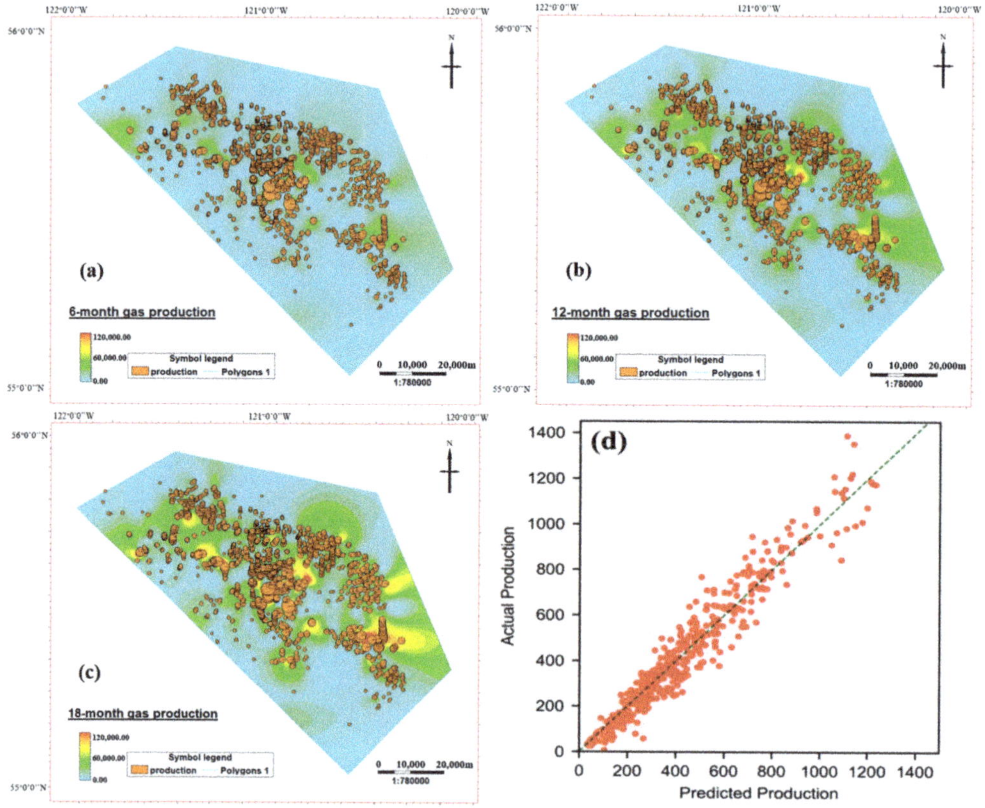

Figure 11. Map view of the Random Forest-based prediction model of gas production. (**a**–**c**) Map view of predicted 6-month, 12-month, 18-month gas production. (**d**) The 20% test set of the average prediction-production versus the actual production.

4.2.4. Optimization of Fracturing Parameters

The core aspect of this research is the optimization of fracturing parameters using the RF-based prediction model. Specifically, the fracturing fluid-injection volume and proppant mass are optimized to maximize gas productivity. Based on the relationship between fluid-injection volume and proppant mass in the studied region (Figure 12a), case studies of two test wells are conducted with the RF-based prediction model. Figure 12b,c present the details of fracturing parameter optimization for the two wells. The grey box marks the optimal range for engineering parameters corresponding to the maximum 12-month gas production. It is evident that the original fracturing parameters of Well W2 are similar to the optimal ones. The original fluid-injection volume was 4700 m^3 and the proppant mass was 800 t, compared to the optimal 5300 m^3 of fluid injection and 1000 t of proppant mass.

Therefore, the actual 12-month production of 1150 MMCF is close to that of 1270 MMCF. In contrast, Well W1 has relatively low production due to inappropriate fracturing parameters. If the fluid injection is increased by 97.5% and the proppant mass is increased by 243.8%, the optimal gas production of W1 would be 880 MMCF (Table 4). This work is highly significant as it determines the fracturing job scales under site-specific geological parameters of the studied region.

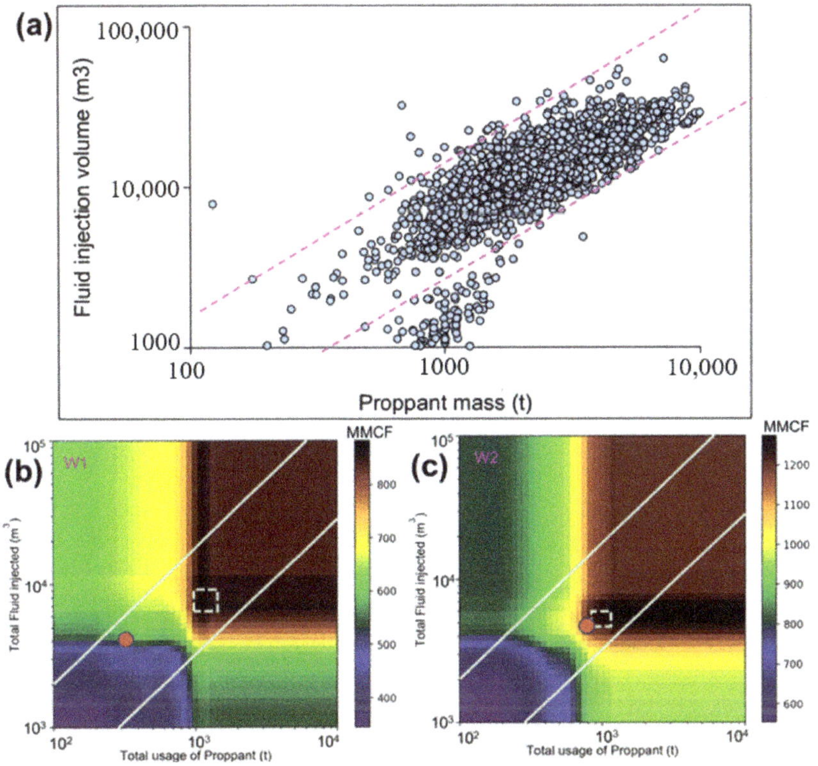

Figure 12. (**a**) Relationship between fluid-injection volume and proppant mass in the studied region. (**b**,**c**) Fracturing parameter optimization for two wells. The red circles represent 12-month gas production under original fracturing parameters. The white boxes denote the optimal engineering parameter range. The two grey lines illustrate the boundary lines shown in (**a**).

Table 4. Optimization of fracturing parameters for two horizontal wells.

Well	Original Parameters			Optimal Parameters			Parameter Comparison	
	12 Mo Prod (MMCF)	Fluid Volume (m^3)	Proppant Mass (t)	12 Mo Prod (MMCF)	Fluid Volume (m^3)	Proppant Mass (t)	Fluid Increment (%)	Proppant Increment (%)
W1	480	4000	320	890	7900	1100	97.5	243.8
W2	1150	4700	800	1270	5300	1000	12.8	25.0

4.2.5. Sensitivity Analysis of Fracturing Parameters

By applying the fracturing parameter-optimization model and the yield-prediction model above, a fracturing parameter sensitivity analysis can be realized. Take the above Well W2 as an example to illustrate the process of this sensitivity analysis. The set of scenario parameters with the highest predicted yield is the benchmark for the fracturing parameter sensitivity analysis. In the single-factor analysis, only one parameter is altered at a time. In Figure 13, when the cumulative injection volume of the fracturing fluid is 4700 m^3 and the cumulative pumping volume of proppant is 800 t, the maximum yield is attained when the number of designed fracturing sections is 18. Thus, under these fluid proppant conditions, the optimal number of fracturing sections is 18 and these scenario-specific fracturing parameters can be used as the basis for the sensitivity analysis. Furthermore, with the cumulative injection volume of fracturing fluid of 4700 m^3 and the cumulative pumped-in volume of proppant of 800 t, the highest yield can be obtained when the length of the designed horizontal section is 2500 m. Therefore, under these fluid proppant conditions, the optimal horizontal section length is 2500 m.

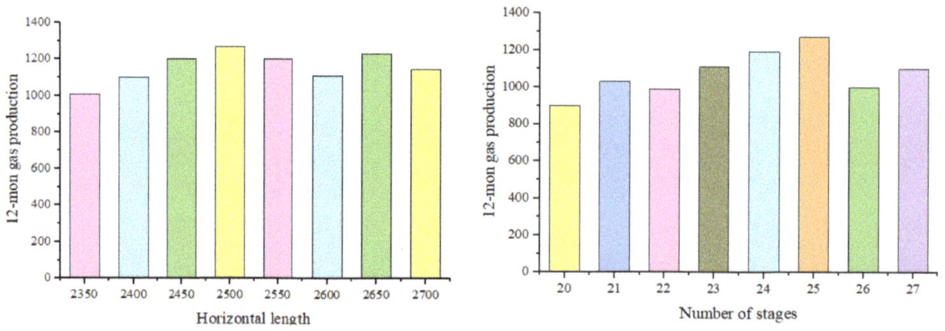

Figure 13. Sensitivity analysis of horizontal length and number of stages.

5. Conclusions

This paper proposes a novel data-driven approach for predicting tight gas production. It effectively integrates machine learning algorithms, feature selection, production prediction and fracturing parameter optimization. First, a database of geological and engineering parameters from 3146 horizontal wells in the Montney tight gas field was collected. Second, six machine learning methods were trained to predict production performance and quantify the contribution of geological and engineering factors to production outcomes. Operational plans were optimized to maximize productivity by adjusting the fracturing fluid-injection volume and proppant mass through the optimal calculating algorithm.

(1) The six machine learning algorithms selected eight parameters of the highest R^2 values. These parameters include fracturing fluid injection, burial depth, number of fractured sections, Young's modulus, formation pressure, saturation, sandstone thickness and total organic carbon (TOC) content and emerged as the key variables for tight gas production.

(2) After evaluating six machine learning algorithms, the Random Forest method was found to have the largest coefficient of determination, with an R^2 value of 0.886. A prediction model based on Random Forest was then developed to estimate tight gas productivity, which can be used to guide the well site selection for effective tight gas development.

(3) A case study of test wells demonstrated that the model can be used for fracturing parameter sensitivity analysis. By analyzing the impact of single- or multi-factor variations on production, it enables the optimal design of single or multiple parameters. Ultimately, increasing the fracturing fluid injection by 97.5% can nearly double the natural gas production. This work has provided accurate, evidence-based suggestions for optimizing the development plan.

Author Contributions: Conceptualization, F.Y. and G.H.; methodology, F.Y. and G.H.; software, Y.Z., X.Y. and Y.L.; validation, D.M., Y.R. and Y.Z.; formal analysis, D.M. Y.R. and P.B.; investigation, K.Z., Y.Z., D.W. and F.G.; data curation, C.G., K.Z. and Z.P.; writing—original draft preparation, F.Y.; writing—review and editing, G.H.; visualization, X.Y.; supervision, G.H.; funding acquisition, G.H. All authors have read and agreed to the published version of the manuscript.

Funding: This research was supported by the Science Foundation of China University of Petroleum, Beijing (No. 2462023BJRC001), NSFC "Intelligent thin-section identification method for oil and gas reservoir based on knowledge and data fusion" (42372175) and CNPC Technology Project "Research on Key Technologies of Artificial Intelligence for Oil and Gas Exploration and Development" (2023DJ84).

Data Availability Statement: The data presented in this study are available on request from the corresponding author.

Acknowledgments: We thank two anonymous reviewers and the editor for their instructive comments that considerably improved the manuscript's quality.

Conflicts of Interest: Authors Dewei Meng, Yili Ren and Fei Gu were employed by the company Research Institute of Petroleum Exploration & Development, PetroChina. The remaining authors declare that the research was conducted in the absence of any commercial or financial relationships that could be construed as a potential conflict of interest.

References

McGlade, C.; Speirs, J.; Sorrell, S. Unconventional Gas—A Review of Regional and Global Resource Estimates. *Energy* **2013**, *55*, 571–584. [CrossRef]

Wang, H.; Ma, F.; Tong, X.; Liu, Z.; Zhang, X.; Wu, Z.; Li, D.; Wang, B.; Xie, Y.; Yang, L. Assessment of Global Unconventional Oil and Gas Resources. *Pet. Explor. Dev.* **2016**, *43*, 925–940. [CrossRef]

Sun, L.; Zou, C.; Jia, A.; Wei, Y.; Zhu, R.; Wu, S.; Guo, Z. Development Characteristics and Orientation of Tight Oil and Gas in China. *Pet. Explor. Dev.* **2019**, *46*, 1073–1087. [CrossRef]

Di, C.; Wei, Y.; Wang, K.; Deng, P.; Chen, B.; Shen, L.; Chen, Z. The Impact of Pressurization-Induced Decrease of Capillary Pressure and Residual Saturation on Geological Carbon Dioxide Storage. *J. Clean. Prod.* **2025**, *486*, 144573. [CrossRef]

Hui, G.; Chen, S.; He, Y.; Wang, H.; Gu, F. Machine Learning-Based Production Forecast for Shale Gas in Unconventional Reservoirs via Integration of Geological and Operational Factors. *J. Nat. Gas Sci. Eng.* **2021**, *94*, 104045. [CrossRef]

Chen, J.; Wang, L.; Wang, C.; Yao, B.; Tian, Y.; Wu, Y.S. Automatic Fracture Optimization for Shale Gas Reservoirs Based on Gradient Descent Method and Reservoir Simulation. *Adv. Geo-Energy Res.* **2021**, *5*, 191–201. [CrossRef]

Hui, G.; Chen, Z.; Schultz, R.; Chen, S.; Song, Z.; Zhang, Z.; Song, Y.; Wang, H.; Wang, M.; Gu, F. Intricate Unconventional Fracture Networks Provide Fluid Diffusion Pathways to Reactivate Pre-Existing Faults in Unconventional Reservoirs. *Energy* **2023**, *282*, 128803. [CrossRef]

Weng, X.; Kresse, O.; Chuprakov, D.; Cohen, C.E.; Prioul, R.; Ganguly, U. Applying Complex Fracture Model and Integrated Workflow in Unconventional Reservoirs. *J. Pet. Sci. Eng.* **2014**, *124*, 468–483. [CrossRef]

Awoleke, O.; Lane, R. Analysis of Data from the Barnett Shale Using Conventional Statistical and Virtual Intelligence Techniques. *SPE Reserv. Eval. Eng.* **2011**, *14*, 544–556. [CrossRef]

Wang, Y.-F.; Xu, S.; Hao, F.; Liu, H.-M.; Hu, Q.-H.; Xi, K.-L.; Yang, D. Machine Learning-Based Grayscale Analyses for Lithofacies Identification of the Shahejie Formation, Bohai Bay Basin, China. *Pet. Sci.* **2025**, *22*, 42–54. [CrossRef]

Ozowe, W.; Daramola, G.O.; Ekemezie, I.O. Recent Advances and Challenges in Gas Injection Techniques for Enhanced Oil Recovery. *Magna Sci. Adv. Res. Rev.* **2023**, *9*, 168–178. [CrossRef]

Hui, G.; Chen, Z.; Wang, Y.; Zhang, D.; Gu, F. An Integrated Machine Learning-Based Approach to Identifying Controlling Factors of Unconventional Shale Productivity. *Energy* **2023**, *266*, 126512. [CrossRef]

Mohammed, A.I.; Bartlett, M.; Oyeneyin, B.; Kayvantash, K.; Njuguna, J. An Application of FEA and Machine Learning for the Prediction and Optimisation of Casing Buckling and Deformation Responses in Shale Gas Wells in an In-Situ Operation. *J. Nat. Gas Sci. Eng.* **2021**, *95*, 104221. [CrossRef]

Chen, Z.; Jiang, C. An Integrated Mass Balance Approach for Assessing Hydrocarbon Resources in a Liquid-Rich Shale Resource Play: An Example from Upper Devonian Duvernay Formation, Western Canada Sedimentary Basin. *J. Earth Sci.* **2020**, *31*, 1259–1272. [CrossRef]

15. Kalantari-Dahaghi, A.; Mohaghegh, S.; Esmaili, S. Coupling Numerical Simulation and Machine Learning to Model Shale Gas Production at Different Time Resolutions. *J. Nat. Gas Sci. Eng.* **2015**, *25*, 380–392. [CrossRef]
16. Tahmasebi, P.; Javadpour, F.; Sahimi, M. Data Mining and Machine Learning for Identifying Sweet Spots in Shale Reservoirs. *Expert Syst. Appl.* **2017**, *88*, 435–447. [CrossRef]
17. Meng, J.; Zhou, Y.; Ye, T.; Xiao, Y.; Lu, Y.; Zheng, A.W.; Liang, B. Hybrid Data-Driven Framework for Shale Gas Production Performance Analysis via Game Theory, Machine Learning, and Optimization Approaches. *Pet. Sci.* **2023**, *20*, 277–294. [CrossRef]
18. Saporetti, C.; Fonseca, D.; Oliveira, L.; Pereira, E.; Goliatt, L. Hybrid Machine Learning Models for Estimating Total Organic Carbon from Mineral Constituents in Core Samples of Shale Gas Fields. *Mar. Pet. Geol.* **2022**, *143*, 105783. [CrossRef]
19. Vikara, D.; Remson, D.; Khanna, V. Machine Learning-Informed Ensemble Framework for Evaluating Shale Gas Production Potential: Case Study in the Marcellus Shale. *J. Nat. Gas Sci. Eng.* **2020**, *84*, 103679. [CrossRef]
20. Mehana, M.; Guiltinan, E.; Vesselinov, V.; Middleton, R.; Hyman, J.D.; Kang, Q.; Viswanathan, H. Machine-Learning Predictions of the Shale Wells' Performance. *J. Nat. Gas Sci. Eng.* **2021**, *88*, 103819. [CrossRef]
21. Xiao, C.; Wang, G.; Zhang, Y.; Deng, Y. Machine-Learning-Based Well Production Prediction under Geological and Hydraulic Fracture Parameters Uncertainty for Unconventional Shale Gas Reservoirs. *J. Nat. Gas Sci. Eng.* **2022**, *106*, 104762. [CrossRef]
22. Yi, J.; Qi, Z.; Li, X.; Liu, H.; Zhou, W. Spatial Correlation-Based Machine Learning Framework for Evaluating Shale Gas Production Potential: A Case Study in Southern Sichuan Basin, China. *Appl. Energy* **2024**, *357*, 122483. [CrossRef]
23. Bachu, S.; Burwash, R.A. Regional-Scale Analysis of the Geothermal Regime in the Western Canada Sedimentary Basin. *Geothermics* **1991**, *20*, 387–407. [CrossRef]
24. González, P.; Furlong, C.; Gingras, M.; Playter, T.; Zonneveld, J. Depositional Framework and Trace Fossil Assemblages of the Lower Triassic Montney Formation, Northeastern British Columbia, Western Canada Sedimentary Basin. *Mar. Pet. Geol.* **2022**, *143*, 105822. [CrossRef]
25. Hui, G.; Chen, S.; Gu, F. Strike-Slip Fault Reactivation Triggered by Hydraulic-Natural Fracture Propagation during Fracturing Stimulations near Clark Lake, Alberta. *Energy Fuels* **2024**, *38*, 18547–18555. [CrossRef]
26. Egbobawaye, E. Sedimentology and Ichnology of Upper Montney Formation Tight Gas Reservoir, Northeastern British Columbia, Western Canada Sedimentary Basin. *IJG* **2016**, *07*, 1357–1411. [CrossRef]
27. Bao, P.; Hui, G.; Hu, Y.; Song, R.; Chen, Z.; Zhang, K.; Pi, Z.; Li, Y.; Ge, C.; Yao, F.; et al. Comprehensive Characterization of Hydraulic Fracture Propagations and Prevention of Pre-existing Fault Failure in Duvernay Shale Reservoirs. *Eng. Fail. Anal.* **2025**, *173*, 109461. [CrossRef]
28. Hui, G.; Yao, F.; Pi, Z.; Bao, P.; Wang, W.; Wang, M.; Wang, H.; Gu, F. Tight Gas Production Prediction in the Southern Montney Play Using Machine Learning Approaches. In Proceedings of the SPE Canadian Energy Technology Conference and Exhibition, Calgary, AB, Canada, 13 March 2024; p. D011S009R001.
29. Fang, M.; Shi, H.; Li, H.; Liu, T. Application of Machine Learning for Productivity Prediction in Tight Gas Reservoirs. *Energies* **2024**, *17*, 1916. [CrossRef]
30. Mao, S.; Chen, B.; Malki, M.; Chen, F.; Morales, M.; Ma, Z.; Mehana, M. Efficient Prediction of Hydrogen Storage Performance in Depleted Gas Reservoirs Using Machine Learning. *Appl. Energy* **2024**, *361*, 122914. [CrossRef]
31. LeCun, Y.; Bengio, Y.; Hinton, G. Deep Learning. *Nature* **2015**, *521*, 436–444. [CrossRef]
32. Hui, G.; Chen, Z.; Wang, H.; Song, Z.; Wang, S.; Zhang, H.; Zhang, D.M.; Gu, F. A Machine Learning-Based Study of Multifactor Susceptibility and Risk Control of Induced Seismicity in Unconventional Reservoirs. *Pet. Sci.* **2023**, *20*, 2232–2243. [CrossRef]
33. Deng, Y.; Wang, W.; Su, Y.; Sun, S.; Zhuang, X. An Unsupervised Machine Learning Based Double Sweet Spots Classification and Evaluation Method for Tight Reservoirs. *J. Energy Res. Technol.* **2023**, *145*, 072602. [CrossRef]
34. Pawley, S.; Schultz, R.; Playter, T.; Corlett, H.; Shipman, T.; Lyster, S.; Hauck, T. The Geological Susceptibility of Induced Earthquakes in the Duvernay Play. *Geophys. Res. Lett.* **2018**, *45*, 1786–1793. [CrossRef]
35. Brantson, E.T.; Ju, B.; Omisore, B.O.; Wu, D.; Selase, A.E.; Liu, N. Development of Machine Learning Predictive Models for History Matching Tight Gas Carbonate Reservoir Production Profiles. *J. Geophys. Eng.* **2018**, *15*, 2235–2251. [CrossRef]
36. Zou, C.; Yang, Z.; He, D.; Wei, Y.; Li, J.; Jia, A.; Chen, J.; Zhao, Q.; Li, Y.; Li, J.; et al. Theory, Technology and Prospects of Conventional and Unconventional Natural Gas. *Pet. Explor. Dev.* **2018**, *45*, 604–618. [CrossRef]
37. Song, N.; Li, S.; Zeng, B.; Duan, R.; Yang, Y. A Novel Grey Prediction Model with Four-Parameter and Its Application to Forecast Natural Gas Production in China. *Eng. Appl. Artif. Intell.* **2024**, *133*, 108431. [CrossRef]
38. Zhang, Z.; Tang, J.; Zhang, J.; Meng, S.; Li, J. Modeling of Scale-Dependent Perforation Geometrical Fracture Growth in Naturally Layered Media. *Eng. Geol.* **2024**, *336*, 107499. [CrossRef]
39. Song, R.; Liu, J.; Yang, C.; Sun, S. Study on the Multiphase Heat and Mass Transfer Mechanism in the Dissociation of Methane Hydrate in Reconstructed Real-Shape Porous Sediments. *Energy* **2022**, *254*, 124421. [CrossRef]
40. Su, X.; Zhou, D.; Wang, H.; Xu, J. Research on the Scaling Mechanism and Countermeasures of Tight Sandstone Gas Reservoirs Based on Machine Learning. *Processes* **2024**, *12*, 527. [CrossRef]

Hui, G.; Chen, Z.; Chen, S.; Gu, F. Hydraulic Fracturing-Induced Seismicity Characterization through Coupled Modeling of Stress and Fracture-Fault Systems. *Adv. Geo-Energy Res.* **2022**, *6*, 269–270. [CrossRef]

Cao, L.; Jiang, F.; Chen, Z.; Gao, Y.; Huo, L.; Chen, D. Data-Driven Interpretable Machine Learning for Prediction of Porosity and Permeability of Tight Sandstone Reservoir. *Adv. Geo-Energy Res.* **2025**, *16*, 21–35. [CrossRef]

Xie, C.; Du, S.; Wang, J.; Lao, J.; Song, H. Intelligent Modeling with Physics-Informed Machine Learning for Petroleum Engineering Problems. *Adv. Geo-Energy Res.* **2023**, *8*, 71–75. [CrossRef]

Omidkar, A.; Alagumalai, A.; Li, Z.; Song, H. Machine Learning Assisted Techno-Economic and Life Cycle Assessment of Organic Solid Waste Upgrading under Natural Gas. *Appl. Energy* **2024**, *355*, 122321. [CrossRef]

Wang, S.; Chen, S. Insights to Fracture Stimulation Design in Unconventional Reservoirs Based on Machine Learning Modeling. *J. Pet. Sci. Eng.* **2019**, *174*, 682–695. [CrossRef]

Naghizadeh, A.; Jafari, S.; Norouzi-Apourvari, S.; Schaffie, M.; Hemmati-Sarapardeh, A. Multi-Objective Optimization of Water-Alternating Flue Gas Process Using Machine Learning and Nature-Inspired Algorithms in a Real Geological Field. *Energy* **2024**, *293*, 130413. [CrossRef]

Liu, L.; Kang, W.; Wang, Y.; Zeng, L. Design of Tool Wear Monitoring System in Bone Material Drilling Process. *Coatings* **2024**, *14*, 812. [CrossRef]

Genuer, R.; Poggi, J.; Tuleau-Malot, C.; Villa-Vialaneix, N. Random Forests for Big Data. *Big Data Res.* **2017**, *9*, 28–46. [CrossRef]

Bakouregui, A.; Mohamed, H.; Yahia, A.; Benmokrane, B. Explainable Extreme Gradient Boosting Tree-Based Prediction of Load-Carrying Capacity of FRP-RC Columns. *Eng. Struct.* **2021**, *245*, 112836. [CrossRef]

Lawal, A.; Yang, Y.; He, H.; Baisa, N.L. Machine Learning in Oil and Gas Exploration: A Review. *IEEE Access* **2024**, *12*, 19035–19058. [CrossRef]

Tang, J.; Zhang, Z.; Xie, J.; Meng, S.; Xu, J.; Ehlig-Economides, C.; Liu, H. Re-Evaluation of CO2 Storage Capacity of Depleted Fractured-Vuggy Carbonate Reservoir. *Innov. Energy* **2024**, *1*, 100019-1–100019-11. [CrossRef]

Wang, Z.-Y.; Lu, S.-F.; Zhou, N.-W.; Liu, Y.-C.; Lin, L.-M.; Shang, Y.-X.; Wang, J.; Xiao, G.-S. Complementary Testing and Machine Learning Techniques for the Characterization and Prediction of Middle Permian Tight Gas Sandstone Reservoir Quality in the Northeastern Ordos Basin, China. *Pet. Sci.* **2024**, *21*, 2946–2968. [CrossRef]

Hu, X.; Meng, Q.; Guo, F.; Xie, J.; Hasi, E.; Wang, H.; Zhao, Y.; Wang, L.; Li, P.; Zhu, L.; et al. Deep Learning Algorithm-Enabled Sediment Characterization Techniques to Determination of Water Saturation for Tight Gas Carbonate Reservoirs in Bohai Bay Basin, China. *Sci. Rep.* **2024**, *14*, 12179. [CrossRef] [PubMed]

Liu, B.; Li, C. Mining and Analysis of Production Characteristics Data of Tight Gas Reservoirs. *Processes* **2023**, *11*, 3159. [CrossRef]

Zhao, X.; Chen, X.; Chen, W.; Liu, M.; Yao, Y.; Wang, H.; Zhang, H.; Yao, G. Quantitative Classification and Prediction of Diagenetic Facies in Tight Gas Sandstone Reservoirs via Unsupervised and Supervised Machine Learning Models: Ledong Area, Yinggehai Basin. *Nat. Resour. Res.* **2023**, *32*, 2685–2710. [CrossRef]

MDPI AG
Grosspeteranlage 5
4052 Basel
Switzerland
Tel.: +41 61 683 77 34

Processes Editorial Office
E-mail: processes@mdpi.com
www.mdpi.com/journal/processes

www.ingramcontent.com/pod-product-compliance
Lightning Source LLC
LaVergne TN
LVHW070501061126
851858LV00011B/276
* 9 7 8 3 7 2 5 8 8 2 2 9 8 *